D1064457

The Rat Nervous System

Volume 2

To the memory of Santiago Ramón y Cajal

The Rat Nervous System

Volume 2

Hindbrain and spinal cord

Edited by

GEORGE PAXINOS

University of New South Wales
Kensington, NSW, Australia

ACADEMIC PRESS
(Harcourt Brace Jovanovich, Publishers)

Sydney Orlando San Diego New York
London Toronto Montreal Tokyo

ACADEMIC PRESS AUSTRALIA
Centrecourt, 25–27 Paul Street North
North Ryde, N.S.W. 2113

United States Edition published by
ACADEMIC PRESS INC.
Orlando, Florida 32887

United Kingdom Edition published by
ACADEMIC PRESS, INC. (LONDON) LTD.
24/28 Oval Road, London NW1 7DX

Printed in Australia

National Library of Australia Cataloguing-in-Publication Data

---

The rat nervous system. Volume 2.
Hindbrain and spinal cord.

Includes bibliographies and index.
ISBN 0 12 547632 9.
ISBN 0 12 547634 5 (pbk).

1. Rats - Anatomy. 2. Brain - Anatomy. I. Paxinos, George, 1944- . II. Title. III. Title: Hindbrain and spinal cord.

599.32'33

Library of Congress Catalog Card Number: 84-73212

Contents

Contributors

The numbers in parentheses indicate the pages on which the authors' contributions begin.

Joseph A. Andrezik (1), Department of Anatomical Sciences, University of Oklahoma, Health Science Center, Box 26901, Biomedical Sciences Building Room 553, Oklahoma City, Oklahoma 73190, USA.

Alvin J. Beitz (1), Department of Anatomy, University of South Carolina Medical Science Building, Columbia, South Carolina 29208, USA.

Alan M. Brichta (293), Department of Zoology and Biomedical Sciences, Ohio University, Irvine Hall, Athens, Ohio 45701, USA.

Eva Bystrzycka (95), School of Anatomy, University of New South Wales, PO Box 1, Kensington, NSW 2033, Australia.

James H. Fallon (79), Department of Anatomy, College of Medicine, University of California, Irvine, California 92717, USA.

Brian A. Flumerfelt (221), Department of Anatomy, Health Sciences Centre, University of Western Ontario, London, Ontario N6A 5C1, Canada.

Giorgio Gabella (325), Department of Anatomy, University College London, Gower Street, London WC1E 6BT, United Kingdom.

N.M. Gerritts (251), Laboratory of Anatomy and Embryology, University of Leiden, Wassenaarseweg 62, 2333 AL Leiden, The Netherlands.

Gunnar Grant (293, 303), Department of Anatomy, Karolinska Institutet, PO Box 60400 S-104 01, Stockholm, Sweden.

A.W. Hrycyshyn (221), Department of Anatomy, Health Sciences Centre, University of Western Ontario, London, Ontario N6A 5C1, Canada.

Sandra E. Loughlin (79), Department of Pharmacology, College of Medicine, University of California, Irvine, California 92717, USA.

E. Marani (251), Laboratory of Anatomy and Embryology, University of Leiden, Wassenaarseweg 62, 2333 AL Leiden, The Netherlands.

George F. Martin (29), Department of Anatomy, The Ohio State University, 1645 Neil Avenue, Columbus, Ohio 43210, USA.

William R. Mehler (185), Department of Anatomy, School of Medicine, University of California, San Francisco, California 94143, USA.

Bruce S. Nail (95), School of Physiology and Pharmacology, University of New South Wales, PO Box 1, Kensington, NSW 2033, Australia.

Joseph A. Rubertone (185), Department of Anatomy, Hahnemann University School of Medicine, 230 North Broad Street, Philadelphia, Pennsylvania 19102, USA.

Istvan Törk (43), School of Anatomy, University of New South Wales, PO Box 1, Kensington, NSW 2033, Australia.

David J. Tracey (129, 311), School of Anatomy, University of New South Wales, PO Box 1, Kensington, NSW 2033, Australia.

Joseph B. Travers (111), The Milton S. Hershey Medical Center, The Pennsylvania State University, PO Box 850, Hershey, Pennsylvania 17033, USA.

Robert P. Vertes (29), Division of Basic Medical Sciences, Mercer University, School of Medicine, Macon, Georgia 31207, USA.

Jan Voogd (251), Laboratory of Anatomy and Embryology, University of Leiden, Wassenaarseweg 62, 2333 AL Leiden, The Netherlands.

Robert Waltzer (29), Department of Anatomy, The Ohio State University, 1645 Neil Avenue, Columbus, Ohio 43210, USA.

William R. Webster (153), Department of Psychology, Monash University, Clayton, Vic. 3168, Australia.

Preface

The great neuroanatomists of the nineteenth and early twentieth centuries masterfully outlined the basic organizational schemes of the mammalian brain. The revolutionary developments in connectivity and histochemical procedures in the past two decades make a comprehensive re-examination of the mammalian nervous system necessary and possible. In the present volumes the cytoarchitecture, chemoarchitecture and connectivity of the different regions of the nervous system of the rat are described by researchers who have an ardent affection for their area of specialization.

I am pleased to acknowledge the assistance received from the following colleagues: Charles Watson, for his insightful comments on the organization of the volumes and the selection of contributors; Bill Mehler, for providing a historical perspective to delineations and nomenclature and for sharing with a number of authors his excellent unpublished observations on the brain stem; Thomas Hökfelt, for his encouragement of this project while I was on sabbatical in his laboratory; and Kirsten Osen, Richard Bandler, Patrick Mantyh, John Parnavelas, Bill Staines, and Paul van Dogen for comments on a number of chapters. I thank Sharlane Velasco for her careful assistance with construction of figures. I would like to give a special thanks to Natalie Chabin for her dedication in ensuring comparable expression and nomenclature between chapters. My greatest debt is to my wife, Elly, for editorial assistance and for her total support for this project.

Contents of Volume 1

1

Reticular formation, central gray and related tegmental nuclei

JOSEPH A. ANDREZIK

University of Oklahoma
Oklahoma City, Oklahoma, USA

ALVIN J. BEITZ

University of Minnesota
St Paul, Minnesota, USA

I. THE RETICULAR FORMATION

Historically, the reticular formation was thought to be a diffuse group of neurons and various fiber systems occupying the central core of the brain stem. Meessen and Olszewski (1949) and Olszewski and Baxter (1954) were among the most recent investigators to describe discrete nuclei within the reticular formation. Most boundaries between these reticular nuclei are not distinct and most reticular nuclei are not homogeneous in cytoarchitecture. Olszewski and Baxter (1954) hypothesized that the differences in cytoarchitecture among the reticular nuclei reflected different functional roles. The reticular formation is indeed heterogeneous in function; distinct portions of it subserve various functions such as cardiac pressor and depressor roles, bladder and alimentary tract control, analgesia, aspects of respiration, and consciousness. The functional areas frequently fail to honor the cytoarchitectural borders of the reticular nuclei, resulting in functional maps which often do not coincide with anatomic structures. Functionally distinct groups of neurons may be comprised of portions of two or more reticular "nuclei" and neurons that lie next to each other in the same subdivision may show connectional as well as functional heterogeneity (Volume 2, Chapter 2).

In short, one reticular nucleus may have a number of functions, and, conversely, functional groups of neurons may also be dispersed among several

THE RAT NERVOUS SYSTEM
ISBN 0 12 547632 9

nuclei. A cytoarchitectural map of the reticular formation is necessary before one can suggest the roles reticular formation nuclei play in central nervous system function. Further study may prove that apparent discrepancies between functional and architectural maps are more imagined than real.

II. CYTOARCHITECTURE

A. Medullary reticular nuclei

1. *Ventral reticular nucleus of the medulla*

The ventral reticular nucleus of the medulla (MdV) is the rostral continuation of the deep layers of the cervical spinal cord gray matter. The nucleus begins at the pyramidal decussation and extends through levels of the caudal half of the inferior olivary nucleus (IO). Most neurons in the MdV are round or triangular in shape; the largest neurons are round. Clues to the three dimensional structure of neurons can often be obtained from observations of Nissl substance extensions into the proximal dendrites, although the method of choice for three dimensional analysis of neurons is Golgi impregnations.

The large round neurons, measuring 43 μm $\times$ 17 μm to 36 μm $\times$ 33 μm, generally have a central nucleus with large clumps of Nissl substance arranged concentrically around it. The Nissl substance appears heaviest along the periphery of the cell near the plasmalemma. In some large neurons the nucleus is located in an eccentric position, and in these cells the Nissl substance is aggregated into smaller parcels which are distributed evenly throughout the cytoplasm. It remains to be determined whether there are functional differences between these two classes of large neurons or whether these cell types simply represent opposite ends of a continuum.

Medium sized neurons are round or triangular shaped. The Nissl pattern in these cells is finer than that in most large neurons and it ranges from small clumps to evenly dispersed granules found throughout the cytoplasm. Some medium sized neurons have a number of neuroglial cells situated adjacent to them in satellite positions.

Small round neurons are scattered throughout the MdV. In preparations stained with cresyl violet, the neurons appear homogeneous. The cytoplasm of any one cell is uniform in color, with individual neurons exhibiting various hues of violet, or a nearly colorless cytoplasm.

There is no particular orientation of the neurons in the MdV and this characteristic, in conjunction with the size of the neurons, helps distinguish this nucleus from the dorsal reticular nucleus of the medulla.

2. *Dorsal reticular nucleus of the medulla*

The dorsal reticular nucleus of the medulla (MdD) begins at the level of the pyramidal decussation and, like the MdV, extends through caudal levels of the inferior olivary nucleus (IO). It is bordered dorsally by the cuneate nucleus (Cu), laterally by the caudal part of the spinal nucleus of the trigeminal nerve (Sp5C), and medially and ventrally by the ventral reticular nucleus of the medulla (MdV). The oral pole of the MdD blends imperceptibly with the parvocellular reticular nucleus (PCRt).

The MdD is composed mostly of small triangular shaped and round neurons. Occasionally medium sized neurons are found in the MdD, but their occurrence is much less frequent than in the MdV. Small neurons measure between 12 μm × 7 μm and 20 μm × 12 μm, contain a centrally located nucleus, and have little detectable Nissl substance in the cytoplasm. Medium sized neurons measure 33 μm × 24 μm or less (down to 20 μm in their longest dimension). They also have a central nucleus, and their Nissl substance, which is fine and granular, is spread throughout the cytoplasm.

The long axis of most neurons in the MdD is oriented from dorsomedial to ventrolateral. In cresyl violet stained sections the neurons in the MdD stain less intensely than those in the MdV.

3. *Paramedian reticular nucleus*

The paramedian reticular nucleus (PMn) commences at the junction of the caudal and middle thirds of the inferior olivary nucleus and extends through the rostral lateral reticular nucleus. In the cat (Brodal, 1953; Taber, 1961) and guinea pig (Petrovický, 1966), the PMn may be subdivided into three areas—a ventral group, a dorsal group, and an accessory group. Valverde (1961) comments that this nucleus (PMn) is difficult to see in the rat and that subdivisions described in other species are unwarranted. Mehler (1969) states that these subgroups are evident in the rat; we also recognize the three subgroups in Sprague-Dawley rats.

The ventral subgroup of the PMn in the rat contains neurons which are triangular or ovoid in shape. Most are small, measuring 13 μm × 7 μm to 17 μm × 15 μm. Some medium sized neurons are seen in the ventral PMn and in these neurons the Nissl substance is arranged in aggregate clumps which extend for some distance into the proximal dendrites.

The dorsal subgroup of the PMn contains neurons that are similar to those in the ventral PMn but with fewer medium sized neurons. The most striking feature distinguishing these two areas is that the packing density of neurons in the dorsal subdivision is less than in the ventral division. The dorsal subgroup also contains some neurons which are slightly smaller than those in the ventral subgroup, and these smaller neurons have more satellite glial cells than do other neurons in the PMn.

The accessory group is composed exclusively of small neurons which are dorsal and dorsomedial to the rostral portion of the dorsal subgroup. The number and occurrence of these cells is quite variable and the accessory group is difficult to demonstrate in some specimens.

4. *Interfascicular hypoglossal nucleus*

Throughout the course of the hypoglossal nerve root there are neurons which are entwined within the intramedullary fibers and the exiting rootlets. Clusters of these medium sized round or elongated neurons extend from the hypoglossal nucleus (the dorsal boundary of this group) to the ventral surface of the brain stem. Some authors describe these neurons as the interfascicular hypoglossal nucleus (IF 12) (Olszewski and Baxter, 1954; Taber, 1961) which is distinct from the PMn reticular nuclei of Brodal (1953, 1957).

It is difficult to separate the neurons of the interfascicular hypoglossal nucleus from those surrounding nuclei purely on cytoarchitectural grounds. Delineation of this cell group is more objectively made on the basis of fiber degeneration, retrograde axonal transport and cytochemical data.

5. *Gigantocellular reticular nucleus*

The gigantocellular reticular nucleus (Gi) (Figs 1 and 2) is readily distinguished from surrounding nuclei by the giant neurons from which it derives its name. These neurons measure 65 μm $\times$ 30 μm to 48 μm $\times$ 37 μm and contain large aggregates of Nissl substance which form concentric layers around a central nucleus. In fact, these giant neurons comprise only a small portion of the neurons in the Gi; most of the neurons in the Gi are medium sized or small! The gigantocellular nucleus begins at the midportion of the inferior olive and merges caudally with the MdV. Rostrally the Gi extends to the caudal pole of the superior olivary nucleus. The medial boundary of the Gi abuts the hypoglossal nerve and the PMn caudally, and the predorsal bundle fibers in more rostral sections. Dorsally, the Gi abuts the nuclei which comprise the dorsomedial surface of the medulla, the nucleus of Roller (Ro), the intercalated nucleus of the medulla (In) and the prepositus hypoglossal nucleus (PrH).

In the lateral portions of the Gi one sees few giant neurons. Here the cells are smaller than in other portions of the Gi and are more densely packed. On the lateral boundary (with PCRt) there are clusters of mainly medium sized round neurons with some clusters of small round neurons interspersed among them. Some of the clusters form the linear nucleus of the medulla (Li) (Fig. 1). Both the medium sized and small neurons stain intensely with cresyl violet or thionine. As mentioned, the lateral border of the Gi contacts the parvocellular reticular nucleus (PCRt). Ventrally, the Gi borders on three nuclei, the gigantocellular reticular nucleus, ventral part (GiV) (Fig. 1), the gigantocellular nucleus pars alpha (Giα) (Fig. 2), and the paragigantocellular reticular nucleus (PGi) (Figs 1 and 2).

6. *Gigantocellular reticular nucleus, ventral part*

The gigantocellular reticular nucleus, ventral part (GiV) is ventral to the Gi and

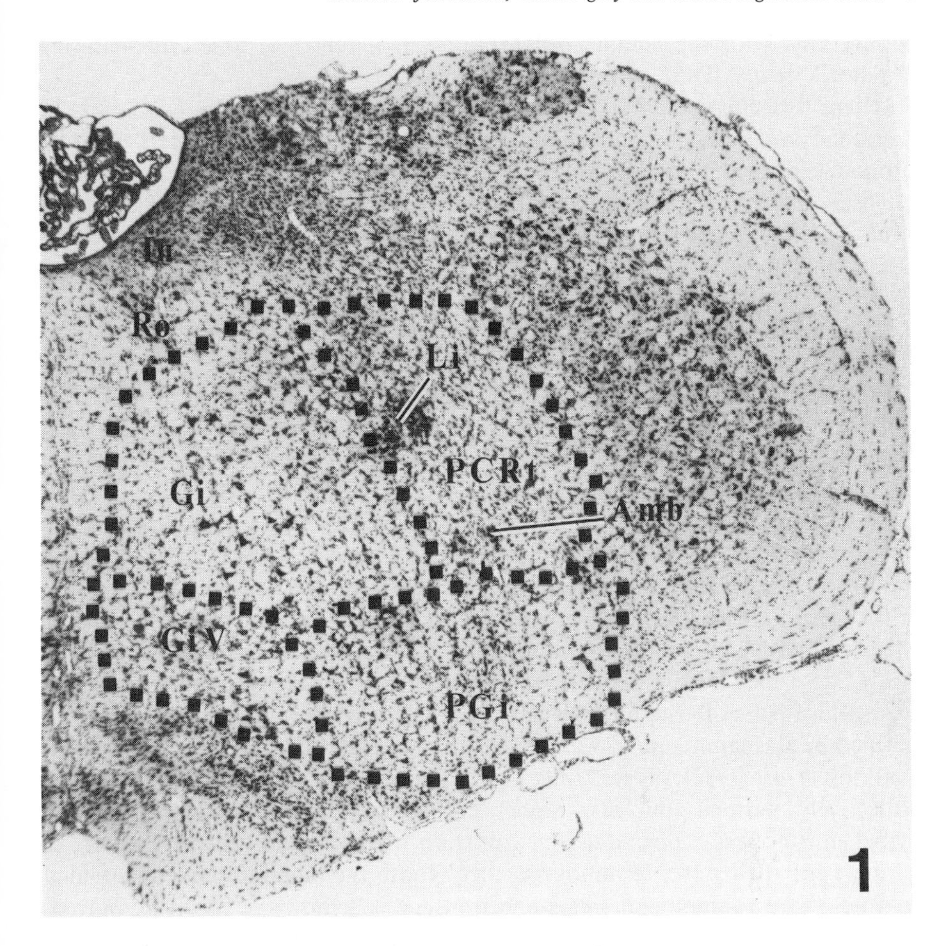

Fig. 1: Photomicrograph of a coronal section cut through the mid-portion of the medulla. The broken lines mark the boundaries of the reticular nuclei at this level. The gigantocellular nucleus, ventral part (GiV) lies just dorsal to the inferior olivary nucleus. Gigantocullular reticular nucleus (Gi) contains giant neurons and is found dorsal to GiV. The parvocellular reticular nucleus (PCRt) is lateral to Gi and is found in the lateral 1/3 of the reticular formation. The paragigantocellular nucleus (PGi) lies ventral to PCRt and extends to the ventral surface of the brainstem. The ambiguus nucleus (Amb) is found between the PGi and PCRt. In, intercalated nucleus; Li, linear nucleus of the medulla; Ro, nucleus of Roller.

dorsal to the rostral inferior olive (Fig. 1). The GiV is characterized by large neurons which have more delicate Nissl substance than neurons in the Gi. The Nissl bodies are dispersed around a central nucleus, and, in some cells which stain rather lightly, the Nissl substance is visible only in the peripheral cytoplasm. The concentration of neurons in the GiV is more dense than in the Gi and other neighboring reticular nuclei. The GiV corresponds to the caudal portion of the magnocellular tegmental field as described in the cat by Berman (1968). This subdivision is described here on the basis of cytoarchitecture, connections with the spinal cord (Volume 2, Chapter 2) and cytochemistry (Andrezik *et al.*, 1985).

There is a clustering of neurons on the lateral border of the GiV where it meets the paragigantocellular reticular nucleus (PGi), and these clusters are most pronounced on the dorsolateral margin of the GiV. Some of these clusters are intensely immunoreactive with antibodies to methionine enkephalin (Andrezik *et al.*, 1985). The GiV merges rostrally with the gigantocellular reticular nucleus, pars alpha (Giα).

7. Gigantocellular reticular nucleus, pars alpha

Rostral to the GiV lies the Giα (Fig. 2). It commences at the rostral tip of the inferior olivary nucleus and extends to the level of the superior olivary nucleus. The Giα is composed of medium sized and small neurons, many of which are elongated. The ellipticity ratio (ratio of somal length to width) exceeds 3:1 and in some neurons it exceeds 4:1. The long axis of these cells is oriented in the horizontal plane. Medially directed dendrites of some of these neurons cross the midline and penetrate the corresponding nucleus on the opposite side.

The Giα as defined here adheres to the criteria and terminology developed by Meessen and Olszewski (1949) in the rabbit. This area corresponds to the rostral Giα as described by Wünscher *et al.* (1965) in the rat, the rostral magnocellular tegmental field of Berman (1968) and the nucleus reticularis magnocellularis as defined by Basbaum *et al.* (1978) in the cat and Watkins *et al.* (1980) in the rat. Watkins *et al.* (1980) suggest that the Giα and the nucleus raphe magnus be collectively termed nucleus raphe alatus. Although they cite mutual cytochemistry and connections for a portion of the neurons in these nuclei as support for their novel terminology, they ignore the remaining neurons which may not share a commonality in connections, cytochemistry or function. Indeed, functional groups of neurons bridge cytoarchitectural boundaries, and thus we propose that the traditional boundaries of raphe and reticular nuclei be maintained.

Petrovický (1963a) considered Giα to be part of the paragigantocellular reticular nucleus (PGi) and drew the boundaries for the latter nucleus to the midline of the brain stem. Fifková and Maršala (1967) adopted Petrovický's (1963a, b; 1964) description of the reticular formation in their atlas of the rat brain stem and from that publication the terminology for reticular formation nuclei was adopted by a number of neurophysiologists and pharmacologists (see also Takagi, 1980). Multiple terminology for this nucleus has recently generated much confusion among neuroscientists and a common terminology should be adopted. The nomenclature of Giα would be preferable to that of PGi or reticularis magnocellularis because this area is quite different from the PGi (Andrezik *et al.*, 1981a) and the boundaries for the reticularis magnocellularis nucleus are not precisely described in any species.

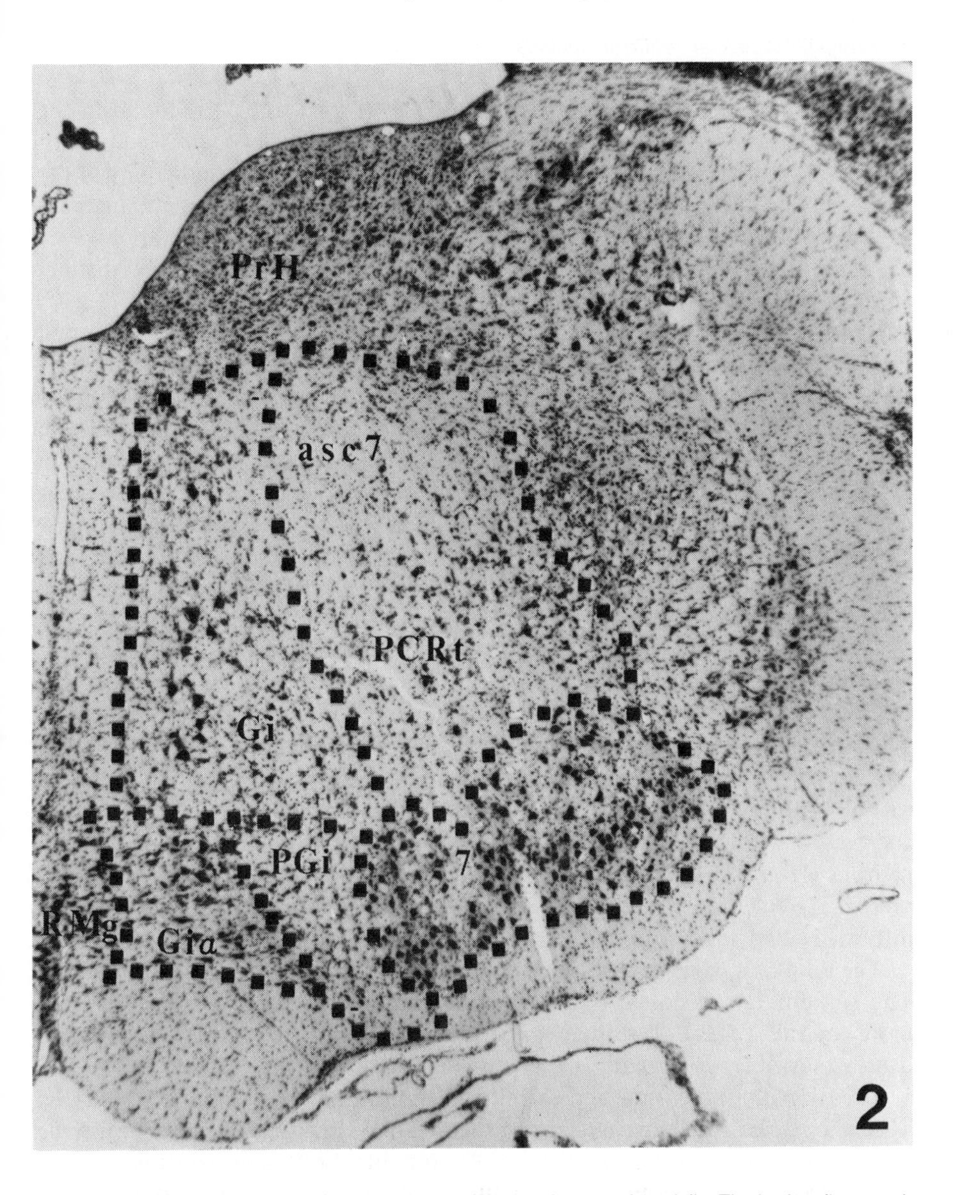

Fig. 2: Photomicrograph of a coronal section cut through the rostral medulla. The broken lines mark the boundaries of the reticular nuclei at this level. The gigantocellular nucleus, pars alpha (Giα) is located dorsal to the pyramid. The cytology of Giα and gigantocellular nucleus, ventral part (GiV) differ. At this level, the paragigantocellular nucleus (PGi) is found lateral to Giα and medial to the facial nucleus (7). The parvocellular reticular nucleus is dorsal to 7. 7n, the facial nerve; asc7, ascending root of the facial nerve; PCRt, parvocellular reticular nucleus; PnO, pontine reticular nucleus, oral part; PrH, prepositus hypoglossal nucleus; RMg, raphe magnus nucleus.

8. Paragigantocellular reticular nucleus

The paragigantocellular reticular nucleus (PGi) (Figs 1 and 2) is found in the ventrolateral quadrant of the medulla. The nucleus begins at the rostral pole of the lateral reticular nucleus (LRt) and extends rostrally to the caudal part of the trapezoid body. The PGi contains primarily small (20 μm × 8 μm) and medium sized neurons (21 μm to 35 μm) both of which stain lightly, giving the appearance of a paucity of cell bodies in the PGi. Several morphologically distinct populations of neurons are found in the PGi (Andrezik *et al.*, 1981a).

Some neurons in the caudomedial part of the PGi lie very close to the pial surface (5 μm) and their dendrites follow the periphery of the brain stem. This area corresponds to the pharmacologically sensitive "chemosensitive area" of the rostral medulla described in cats (Schlaefke, 1981), and is important in modulating respiration and the cardiovascular system. Its importance in autonomic regulation is evidenced by its numerous connections to other areas of the central nervous system known to regulate autonomic functions (Volume 2, Chapter 2). The PGi may also play a role in analgesia, presumably through the serotonergic and peptidergic systems of neurons and terminals prominent there.

9. Parvocellular reticular nucleus

The parvocellular reticular nucleus (PCRt) begins at the level of the interpositus portion of the spinal nucleus of the trigeminal nerve and extends through the rostral medulla to the level of the superior olivary nucleus. The nucleus is coextensive with the Gi (Figs 1 and 2). The PCRt is composed primarily of small neurons (11 μm × 7 μm to 18 μm × 8 μm) with some medium sized cells also present. The neurons stain lighter than cells in the Gi because of their sparsely dispersed Nissl substance.

The long axis of most neurons is oriented from dorsomedial to ventrolateral. The neurons in the dorsal PCRt are more densely packed than those located more centrally. The PCRt broadens at successively more rostral levels as the Gi decreases in relative breadth.

Occasionally there is an aggregate of intensely staining neurons in the PCRt which is part of the linear nucleus of the medulla (Li) (Fig. 1); the significance of these groups is not known. The PCRt contains the functional subgroups of salivatory nuclei at rostral medullary levels and some neurons project to cranial nerve motor nuclei which control oral and swallowing reflexes. Mehler (1969) has recently reviewed the connections of the PCRt and describes the PCRt as a vomiting center.

B. Pontine reticular nuclei

1. Pontine reticular nucleus, caudal part

The pontine reticular nucleus, caudal part (PnC) merges with the rostral Gi and extends from the level of the caudal pole of the trapezoid body to the level of the junction between the middle and the rostral thirds of the motor nucleus of the trigeminal nerve (Mo5). The PnC is composed primarily of small and medium sized neurons with a few outstandingly large cells interspersed among them. A distinctive feature of the PnC is the collection of giant neurons which populate its medial portion. As Valverde (1962) mentions, PnC giant neurons are among the largest cells of the reticular formation; in our material they measure from 65 μm–50 μm in their longest dimension.

In general, the Nissl pattern in PnC neurons is more delicate than that seen in the comparable giant and large neurons of the Gi. The Nissl substance is arranged in smaller units which are evenly distributed throughout the cytoplasm and it streams smoothly into the primary dendrites. The nuclei in PnC neurons are often eccentrically located, another feature which helps to distinguish the PnC from the Gi.

2. *Pontine reticular nucleus, ventral part*

The pontine reticular nucleus ventral part (PnV) is composed of groups of medium sized and small neurons which are arranged in clusters of four to six cells. This subdivision continues across the midline in some sections (Fig. 3) and occupies the entire ventromedial portion of the caudal pons. The overall appearance is that of a greater density of neurons in the PnV than in the PnC. In thionine preparations, the staining of PnV neurons is lighter than that of PnC cells. There are no giant neurons in the PnV. The long axis of PnV neurons is oriented in the horizontal plane and some of the cells look very similar to the neurons that populate the Giα. The distinction between the PnV and Giα is made largely on evidence of their different connections (see Volume 2, Chapter 2) and the PnV corresponds to a similar nuclear area described in the opossum by Oswaldo-Cruz and Rocha-Miranda (1969).

In summary, there exists in the rostral medulla and caudal pons a dorsal group of reticular nuclei formed by the Gi and PnC. A ventral tier of reticular nuclei is coextensive with the dorsal group and from caudal to rostral is composed of the GiV, Giα, and the PnV. The ventral tier is readily distinguished from the dorsal group by cytoarchitecture; parcelation into three nuclei is justified by cytoarchitectural and hodologic criteria (see Volume 2, Chapter 2).

3. *Pontine reticular nucleus, oral part*

The pontine reticular nucleus, oral part (PnO) extends from the junction of the rostral and the middle thirds of the Mo5 to the level of the trochlear nucleus. It is bordered by the raphe nuclei medially and the paralemniscal nuclei laterally (Fig. 4). The reticulotegmental nucleus of the pons (RtTg) and the pontine nuclei (Pn) border the PnO ventrally throughout most of its extent; there is no ventral part. The PnO contains primarily small neurons which stain very lightly with thionine, and a few medium sized neurons. In the ventrolateral part of the PnO there are some large neurons which measure between 40 μm and 36 μm in the

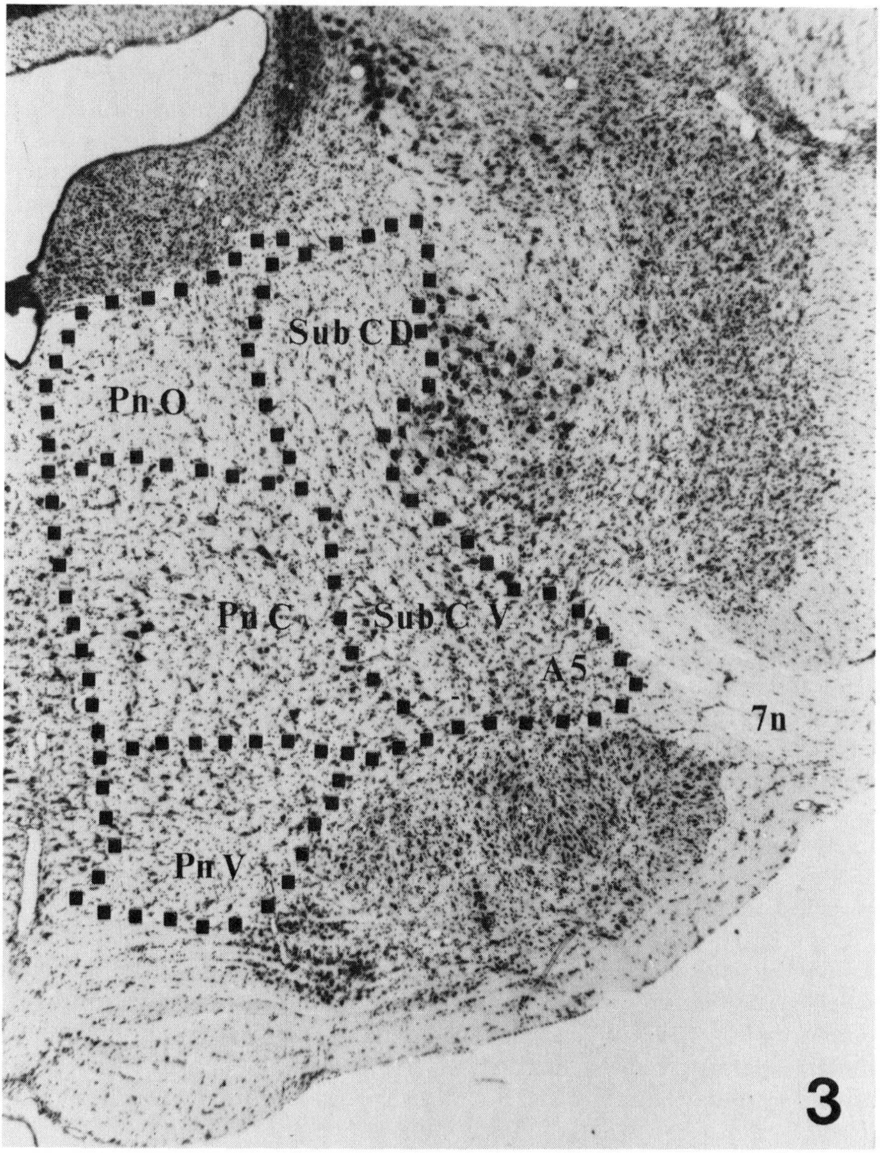

Fig. 3: Photomicrograph of the coronal section through the caudal pons. The pontine reticular nucleus, caudal part (PnC) is found in the dorsomedial reticular formation and is characterized by giant neurons. The pontine reticular nucleus, ventral part (PnV) is just ventral to PnC and it does not contain giant neurons; most neurons are oriented in the horizontal plane. Lateral to these nuclei are the subcoeruleus nuclei, dorsal part (SubCD) and ventral part (SubCV). A5, A5 noradrenergic cell group; 7n, root of the facial nerve; PnO, pontine reticular nucleus, oral part.

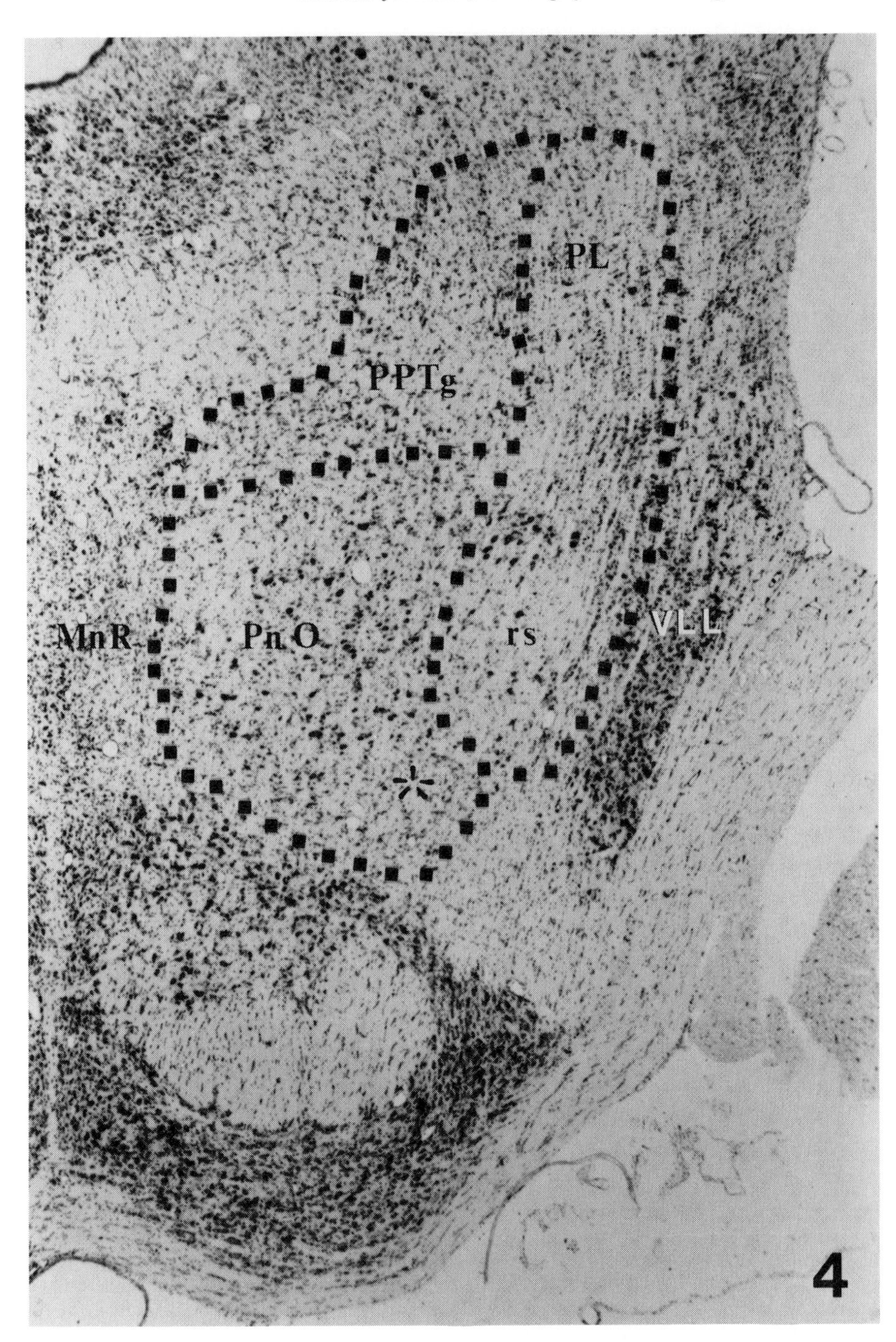
PL
PPTg
MnR
Pn O
rs
VLL
*
4

largest dimension. Valverde (1961) reports that a few giant neurons are present in the PnO in the rat; the same is true in the cat (Taber, 1961). The large neurons and some medium sized cells in the PnO show sizable clumps of Nissl substance around the periphery. These clumps have the appearance of large inlaid stones, a characteristic which helps to differentiate the PnO from the PnC.

We have based our definition of the pontine reticular nuclei (PnC, PnV, PnO) on cytoarchitecture. In general, our boundaries agree with those of Brodal (1957) and Taber (1961) in the cat, and Valverde (1962) and Wünscher *et al.* (1965) in the rat. Mehler (1969) extends the caudal boundary of the PnO further caudally to include the giant neurons in our medial PnC. Mehler bases these subdivisions on the pattern of innervation of these nuclei by fastigiobulbar and spinoreticular fibers. The most rostral part of the pontine tegmentum is generally labeled PnO in most atlases, but the connections of the greater part of the PnO at this level are presently unknown. The ventrolateral part of the PnO, marked by the asterisk in Fig. 4, is devoid of fastigiobulbar input whereas more caudal and medial parts of the pontine reticular formation are peppered with fastigial terminals (Achenbach and Goodman, 1968; Mehler, 1969). Parcelation of this area of the PnO awaits further experimentation.

4. *Ventral tegmental nucleus*

The ventral tegmental nucleus (VTg) is located in the dorsomedial portion of the rostral pontine reticular formation, just caudal to the decussation of the superior cerebellar peduncle. This nucleus surrounds the ventral and lateral borders of the medial longitudinal fasciculus (Hayakawa and Zyo, 1983; Petrovický, 1971). The VTg is composed primarily of round small neurons (14 μm $\times$ 12 μm) with a few medium sized neurons also present. The neurons have a greater packing density than those of the surrounding reticular nuclei. This characteristic, together with the differential staining properties of the neurons, allows the VTg to be readily distinguished. Most neurons have a large nucleus to cytoplasmic ratio and many exhibit large intensely staining clumps of Nissl substance around the periphery.

5. *Pedunculopontine tegmental nucleus*

The pedunculopontine tegmental nucleus (PPTg) extends from the level of the oral pole of the dorsal tegmental nucleus (DTg) and locus coeruleus to the caudal pole of the substantia nigra. This cell group is dorsal and lateral to the PnO. At middle levels of the PPTg the nucleus extends from the VTg medially to the PLO

◁ **Fig. 4:** Photomicrograph of a coronal section through the pontine-mesencephalic junction. The pontine reticular nucleus, oral part (PnO) ocupies most of the reticular formation. The pedunculopontine tegmental nucleus (PPTg) is dorsal and lateral to PnO. The paralemniscal nucleus (PL) is found lateral to PnO and PPTg and medial to the ventral nucleus of the lateral lemniscus (VLL). MnR, median raphe nucleus; rs, rubrospinal tract.

laterally. The PPTg is ventral and lateral to the superior cerebellar peduncle and the nucleus is found at all levels of the decussation. The PPTg is composed of small round and ovoid cells that stain moderately with thionine or cresyl violet. The long axis of most cells is in the horizontal plane. The PPTg contains mainly cholinergic neurons and is described in detail in Volume 1, Chapter 14.

6. Suprageniculate nucleus of the pons

The suprageniculate nucleus of the pons forms part of the periventricular gray matter which surrounds the ventricles from the aqueduct through to the fourth ventricle. The suprageniculate nucleus lies directly dorsal to the genu of the facial nerve. It is composed of small, intensely staining neurons which range in size from 6 μm $\times$ 6 μm to 10 μm $\times$ 5 μm and contain an evenly distributed Nissl substance. The medial part of the suprageniculate nucleus is more densely packed with neurons than is the lateral part. The suprageniculate neurons are round or ovoid, and when there is a polarity to the cell body, the long axis of the cell conforms to the adjacent margins of the facial nerve. The connections of the suprageniculate are not known.

7. Paralemniscal nuclei

The designation of the paralemniscal nuclei is based upon the description of a paralemniscal zone in the cat by Henkel and Edwards (1978). These authors describe a group of neurons medial to the medial lemniscus which ends rostrally at the retrorubral nucleus. This area is equivalent to the paralemniscal nucleus (PL) as described here. The PL extends caudally from the retrorubral nucleus to the level of the dorsal nucleus of the lateral lemniscus (DLL) and nucleus sagulum. It is composed mainly of small round neurons; clusters of darkly staining medium sized cells are also found here (Fig. 4), but they are in the minority. The neurons are arranged in cords which lie medial to the lateral lemniscus and in some cases are inserted between the fibers of the lateral lemniscus.

The paralemniscal nucleus, caudal part (PLC) extends through levels of the DLL to the oral pole of the motor nucleus of the trigeminal nerve. There are fewer neurons in the PLC compared with the PL and the orientation of the neurons in vertical cords is less obvious. The PLC is just medial to the caudal lateral lemniscus.

The parcelation into the PL and PLC is based on the different connections of these two areas. The PL projects to the facial nucleus and to other cranial nerve nuclei which innervate oral, labial, and auricular muscles, whereas the PLC projects primarily to other reticular nuclei and to the spinal cord (Volume 2, Chapter 2).

The paralemniscal nuclei as described here are different from the paralemniscal nuclei described by Olszewski and Baxter (1954) in the human and Taber (1961) in the cat. These authors describe a more superficial and dorsal cellular group which extends through the midbrain to end at thalamic levels. Wünscher *et al.* (1965) also describe the paralemniscal nucleus and it appears that the caudal portion of the nucleus that they describe is part of the PL as described here. The paralemniscal field of Berman (1968) may contain the PL and PLC, but it closely corresponds to our PnO.

C. Mesencephalic reticular nuclei

1. Cuneiform nucleus

The cuneiform nucleus (Cnf) begins at caudal levels of the inferior colliculus. It is bounded medially by the mesencephalic trigeminal nucleus and laterally by the dorsal nucleus of the lateral lemniscus. At more rostral levels, the lateral boundary of the CnF extends to the microcellular nucleus (Paxinos, 1983) which in turn abuts the parabigeminal nucleus.

The cuneiform nucleus is composed of small, lightly staining, round or ovoid cells. Within this nucleus dorsal and ventral subdivisions are identified, distinguishable by the relative sizes of their cells. The neurons of the dorsal subdivision are slightly larger than those of the ventral. At caudal levels, the dorsal subgroup consists of the dorsal surface layer of cells in the cuneiform nucleus, and more rostrally it expands to include a larger portion of the nucleus. The expansion of the dorsal subgroup is accompanied by a concomitant reduction in the size of the ventral subdivision.

The ventral subdivision corresponds to the cuneiform area as described by Gillilan (1943). The dorsal and ventral subdivisions detailed here are considered to be one by Taber (1961) and by Edwards (1975) in their respective descriptions of the cuneiform nucleus in the cat. Petrovický (1966) reports similar cuneiform boundaries to those of Taber (1961) and Edwards (1975) in guinea pigs.

2. Sagulum nucleus

Neurons of the sagulum nucleus occupy a pyramidal shaped area dorsolateral to the dorsal nucleus of the lateral lemniscus and ventrolateral to the inferior colliculus (Fig. 4). The sagulum nucleus extends to the periphery of the brain stem and abuts the pial membrane. In the anteroposterior direction the nucleus extends from the level of the ventral tegmental nucleus to the oral pole of the inferior colliculus. It is composed of a collection of small round or ovoid neurons which, due to their slight amount of cytoplasm, stain very lightly with thionine.

3. *Deep mesencephalic nucleus*

The deep mesencephalic nucleus is a heterogeneous population of neurons that occupies the lateral tegmentum of the rostral mesencephalon. The neurons are primarily small and round, some of which stain darkly with thionine. This nucleus is analogous to Berman's (1968) mesencephalic tegmental field in the cat; Petrovický (1964) named this area the reticular formation of the mesencephalon. The rostral connections of the deep mesencephalic nucleus course through the median forebrain bundle (Volume 2, Chapter 2). It may be possible to subdivide the deep mesencephalic nucleus based upon hodology but this awaits definition by further studies.

The reticular core does not extend into the area lying along the ventricular borders, but nuclei sharing some of the same functions as reticular neurons are easily recognized in the periventricular zone. Recently, there has been great interest in the central gray matter (periaqueductal gray) of the mesencephalon and its role in analgesia. The central gray and some related tegmental nuclei are considered with the reticular formation because of their analogous structure and connections. The connections and functions of reticular nuclei are discussed in another chapter (Volume 2, Chapter 2).

III. THE CENTRAL GRAY, DORSAL TEGMENTAL NUCLEUS AND DORSOLATERAL TEGMENTAL NUCLEUS

A. Cytoarchitecture

The mesencephalic central gray (CG) or periaqueductal gray in the rat is formed by a dense layer of relatively small neurons which surround the cerebral aqueduct. With the exception of the brief description by Gillilan (1943) and the more recent account by Mantyh (1982a), there is very little information available regarding the cytoarchitecture of this region in the rat. The following account, therefore, is based on the work of Mantyh (1982a). This description is supplemented with the results of recent work in our laboratory concerning the anatomic organization of the rodent central gray and with information reported by Tredici *et al.* (1983), Hamilton (1973a), and Liu and Hamilton (1980) regarding the anatomy of this region in the cat. The descriptions of the dorsal and dorsolateral tegmental nuclei are based on the reports by Hayakawa and Zyo (1983) and Tohyama *et al.* (1978).

1. *Nissl studies*

In Nissl preparations, the central gray is observed to begin caudally at the rostral level of the locus coeruleus and to extend cranially to the level of the posterior commissure. Mantyh (1982a) has recently shown that there is an increase in both the intensity of cellular staining and the apparent packing density from the central part outward to the periphery of the rodent central gray. The neurons comprising this region form a heterogeneous population of cells varying widely in their somatic shape and size. Mantyh further indicates that soma size varies between 10 μm and 35 μm with occasional large, 35μm–40 μm, oval cells. These results compare well with the 8 μm–30 μm size range reported by Hamilton (1973a) in the feline central gray. Embedded in the ventrocaudal portion of the central gray are several nuclear groups including the dorsal tegmental nucleus (DTg), the laterodorsal tegmental nucleus (LDTg) and the dorsal raphe nucleus. A discussion of the cytoarchitecture of the dorsal raphe nucleus can be found in Volume 2, Chapter 3.

The DTg is situated in the caudal central gray in a position dorsal to the medial longitudinal fasciculus, medial to the LDTg and lateral to the dorsal raphe nucleus. It is about 800μm long in the rat and is distinguished from the surrounding central gray matter by a ring of myelinated fibers (Hayakawa and Zyo, 1983). The LDTg is approximately 1000 μm long and is bordered at its caudal end by the locus coeruleus laterally and the DTg medially, while rostrally it lies between the dorsal raphe nucleus and the superior cerebellar peduncle. The neurons which comprise the LDTg are larger than those of the DTg and the surrounding central gray matter and are predominantly ovoid to fusiform in shape. See Volume 1, Chapter 14 for a detailed account of the dorsal tegmental nuclei on the basis of acetylcholinesterase distribution.

2. *Golgi and immunohistochemical studies of central gray neurons*

In Golgi impregnated material Mantyh (1982a) described four categories of neuronal cell types in the central gray (CG): (i) a fusiform neuron which is most prominent in the medial part of the CG; (ii) a multipolar neuron which is found in all areas; (iii) a stellate cell also found in all parts of the CG; and (iv) a pyramidal cell which is most prominent in the periphery of the CG. The fusiform neuron has a long elliptical soma with one to several dendritic processes emerging from each end. This cell type was originally described in the CG by Ramón y Cajal (1911) and appears to be equivalent to the spindle shaped type I neurons described by Liu and Hamilton (1980) and the type 4 neurons reported by Tredici *et al.* (1983). The multipolar neuron described by Mantyh (1982a) is comparable to the II and IIId neurons of Liu and Hamilton (1980) and includes the triangular shaped cells depicted by Ramón y Cajal (1911). This triangular cell type has also been observed in our laboratory (Prichard and Beitz, 1980) and typically gives rise to three or four dendrites. The stellate cells, which Mantyh (1982a) indicates have three to six processes radiating from an ovoid perikaryon, are similar to the type Ib cells of Liu and Hamilton (1980), the polygonal neurons described by Tredici *et al.* (1983), and are included in the multipolar neuronal category described by Prichard and Beitz (1980). The pyramidal cells, the final neuronal category, are also recognized in the feline CG (Tredici *et al.*, 1983) and are distinguished by the presence of seven to 10 dendrites arising from a pyramidal shaped cell body.

Regardless of the classification system used to distinguish CG neuronal types, it is important to appreciate that different neuronal varieties do exist in this periaqueductal region. The existence of these different cell classes raises obvious questions concerning the relationship of each cell type to the numerous neuropeptides and other putative neurotransmitters which have been identified in the CG (see below) and ultimately to the functional roles played by these neuronal varieties. In this regard, it is of interest that Tredici *et al.* (1983) have demonstrated that some neurons in all neuronal classes (except their type 4 cells) give off collaterals involved in intrinsic CG circuits. These collaterals would be able to influence the electrical activity of this region.

The central gray is an extremely interesting yet complex brain region in terms of neuropeptide distribution. It is one of the few extrahypothalamic sites that contains a majority of the two dozen or so recently discovered neuropeptides (see Snyder, 1980 for a review of the peptide neurotransmitter candidates). The gut-brain peptides, vasoactive intestinal polypeptide, cholecystokinin octapeptide, substance P, neurotensin, metenkephalin, leuenkephalin and pancreatic polypeptide have all been identified within the midbrain central gray (Hökfelt *et al.*, 1977; Jennes *et al.*, 1982; Loren *et al.*, 1979; Sims *et al.*, 1980; Vanderhaeghen *et al.*, 1980).

In addition to the above peptides, the hypothalamic releasing hormones LHRH and somatostatin (Finley *et al.*, 1981; Jennes and Stumpf, 1980) and the pituitary peptides ACTH, β-endorphin and β-MSH (Bloom *et al.*, 1978; Watson *et al.*, 1978) have been observed in the CG. Other peptides which are potential neurotransmitter candidates that have been found in the central gray include angiotensin II (Changaris *et al.*, 1978), bradykinin (Correa *et al.*, 1979), neurophysin (Swanson, 1977) and bombesin (Moody *et al.*, 1981). More recently, dynorphin (Watson *et al.*, 1982), motilin (Jacobowitz *et al.*, 1981), corticotrophin releasing factor (Cummings *et al.*, 1983), and the molluscan cardioexcitatory peptide FMRF-NH_2 (Weber *et al.*, 1981) have been detected in the CG.

Of these peptides the only ones that have been mapped in any detail in the CG are enkephalin, neurotensin, somatostatin, cholecystokinin, substance P and vasoactive intestinal polypeptide (Beitz, 1981; Beitz, 1982c; Beitz *et al.*, 1983a; Jennes *et al.*, 1982; Moss *et al.*, 1983; Moss and Basbaum, 1983). With regard to enkephalin localization, Moss and coworkers (1983) have described a ventral to dorsal shift in the enkephalin immunoreactivity from caudal to rostral levels of the feline CG. This ventral to dorsal shift also occurs in the rat CG (Beitz, 1981, 1982b), but enkephalinergic cells are not as prominent in the rostrodorsal portion of the rodent CG as in the cat. Neurotensin like and substance P like immunoreactive neurons have a similar pattern of distribution within the CG (Beitz, 1982b, c; Moss and Basbaum, 1983); that is, they are prominent in the ventral CG caudally and gradually shift dorsally at more rostral levels. Cholecystokinin immunoreactive neurons display a reverse pattern of distribution within the CG. The cells form a column which is prominent at rostroventral portions of the CG and extends in a dorsocaudal direction so that at inferior collicular levels these cells are less numerous and occupy a more dorsal position in the CG (Beitz *et al.*, 1983a). From the immunohistochemical analyses published to date, it is apparent that neurons containing specific neuropeptides within the CG do not occur in specific anatomic subdivisions, but rather form longitudinal neuronal columns through the CG that are unrelated to any of the proposed subdivisional schemes (see below).

Very few investigators have attempted to equate the neurons identified in their immunohistochemical material of the CG with the neuronal types described from Golgi preparations. Moss *et al.* (1983) have shown that enkephalin containing cells are of the bipolar or multipolar variety in the feline CG. Results from our laboratory have demonstrated that neurotensin like immunoreactivity can be identified in small bipolar neurons, large fusiform neurons and triangular shaped neurons, while somatostatin immunoreactivity is predominantly found in small and large multipolar neurons, as well as small bipolar cells (Beitz, 1982c; Beitz *et al.*, 1983a). Clearly, further studies are necessary to delineate the relationship of chemically specified neurons in the CG to the Golgi analyses of this region.

Regarding the localization of neurochemicals in the DTg and LDTg, many of the immunohistochemical mapping studies to date have ignored these nuclear groups, and so data concerning these two cell groups are limited (see Volume 1, Chapter 14). The data that are available, however, suggest that the DTg and LDTg are different with respect to their chemical composition. Substance P containing cell bodies (Ljungdahl *et al.*, 1978) and melanocyte stimulating hormone containing fibers (O'Donohue *et al.*, 1979), for instance, are present in the LDTg but are absent in the DTg. Conversely, cholecystokinin like immunoreactive neurons are present in the DTg, but not in the LDTg (Vanderhaeghen *et al.*, 1980). Finally, Steinbusch (1981) has shown that both the DTg and LDTg contain serotonin like immunoreactive axons, but the LDTg contains a much higher density of these fibers.

3. *Nuclear subdivisions*

The question of nuclear subdivisions within the central gray has been the basis for considerable controversy. As many as five subdivisions were recognized in the CG by Castaldi (1923) in his study of the mesencephalon of the guinea pig, while Mantyh (1982a) contends that no subnuclei exist in the CG of the rat, cat or monkey based on cytoarchitectural and hodologic studies. Between these two extremes are the phylogenetic studies of Crosby and Woodburne (1943) and the developmental studies of Altman and Bayer (1981) which suggest that the CG can be divided into dorsal, lateral and ventral portions. This scheme of three subdivisions receives additional support from the Nissl studies of the rodent CG by Gillilan (1943). Finally, the cytoarchitectural and connectivity studies of Hamilton (1973a, b) and Liu and Hamilton (1980) suggest that the feline CG should be divided into a rather acellular nucleus medialis, a cell rich nucleus lateralis and a dorsally located nucleus dorsalis. This model resembles the human CG subdivisions originally described by Olszewski and Baxter (1954).

The majority of cytoarchitectural and immunohistochemical studies to date, however, fail to provide convincing evidence for the existence of anatomic subdivisions within the CG. In addition, although some studies of the afferent and efferent connections of the CG suggest that differences do exist in the projections of this region, the majority of studies document quantitative rather than qualitative differences between proposed periaqueductal gray subdivisions. On the other hand, recent studies utilizing *in vitro* receptor binding techniques have demonstrated a differential localization of receptors in this region, which may well explain some of the functional differences observed among different portions of the CG. Wamsley *et al.* (1981) and Beitz *et al.* (1982) have shown that muscarinic cholinergic receptors are concentrated in the dorsolateral portion of the rodent CG. Similarly, opiate receptors are most dense in the dorsolateral portion of the CG, but are also somewhat dense ventrally (Beitz *et al.*, 1982; Herkenham and Pert, 1982). Neurotensin receptors also appear to be concentrated in the dorsolateral CG (Young and Kuhar, 1979) while GABA receptors are localized dorsomedially (Beitz *et al.*, 1982; Palacios *et al.*, 1981). Further study of receptor localization in the CG may lead to a pharmacologic parcelation of this region consistent with proposed anatomic subdivisions.

Tohyama and colleagues (1978) have divided the rodent LDTg into three regions based on cytoarchitectural and connectivity studies: (i) a parvocellular part situated peripherally and innervating mainly the anterior hypothalamus; (ii) a centrally located magnocellular portion which projects to the hypothalamus, cerebral cortex and the lateral posterior thalamic nucleus; and (iii) a peri-Barrington area situated laterally and projecting primarily to the lateral hypothalamus (see Chapter 14, Volume 1).

B. Connectivity

1. Afferent projections to the central gray

A detailed account of afferent input to the rodent CG can be found in a recent publication from our laboratory (Beitz, 1982a) as well as in the works of Morrell *et al.* (1981), Marchand and Hagino (1983) and Mantyh (1982b). The sources of afferent projections to the CG are extensive and allow this midbrain region to be influenced by motor, sensory and limbic structures. The most significant inputs are summarized in Fig. 5. Briefly, the greatest sources of descending input to the CG arise from the ventromedial hypothalamic and dorsal premammillary hypothalamic nuclei, the zona incerta, the anterior cingulate cortex, the lateral preoptic area, the lateral hypothalamic area, the parafascicular thalamic nucleus and the lateral habenular nucleus. Ascending input to the CG is quantitatively less than descending input and arises predominantly from the medullary and pontine reticular formation including certain raphe nuclei. Ascending input also arises from the dorsal horn of the spinal cord and the nucleus of the spinal tract of the trigeminal nerve. The CG receives significant projections from midbrain structures as well. The cuneiform nucleus, deep gray layer of the superior colliculus, substantia nigra and parabigeminal nucleus all contribute fibers to the CG. Finally, it is important to appreciate that the CG has a significant number of intrinsic connections. Connections between periaqueductal gray cells were suggested by retrograde transport studies (Beitz, 1982a) and have recently been demonstrated with Golgi and electron microscopic techniques (Tredici *et al.*, 1983).

2. Efferent projections

The most thorough study of efferent projections of the CG is the doctoral work of Ruda (1976) in the cat and this will be used as the basis for this account. The dorsal portion of the CG projects bilaterally and sends efferent axons to the dorsal and posterior hypothalamus, the midline nuclei and dorsomedial nucleus of the thalamus, the zona incerta, the cuneiform nucleus and the pontine reticular nuclei, oral and caudal parts. Recent studies have also shown a significant projection from this region to the raphe magnus (Beitz *et al.*, 1983c; Carlton *et al.*, 1983; Fardin *et al.*, 1984) in the rat. In addition the CG projects to the paragigantocellular reticular nucleus, the gigantocellular reticular nucleus, the gigantocellular reticular nucleus, pars α and the nucleus of the spinal tract of the trigeminal nerve (Beitz *et al.*, 1983b, c; Mantyh, 1982b). Moreover, we have demonstrated that the majority of the CG projections to these medullary nuclei, as well as to the raphe magnus, arise from separate neurons in the periaqueductal gray (Beitz *et al.*, 1983c). Finally, the CG sends some direct projections to the spinal cord (Watkins *et al.*, 1981).

Only recently have investigators begun to identify the neurotransmitters and neuropeptides associated with specific CG efferent pathways. Beitz (1982c) and Yezierski *et al.* (1982) have demonstrated a serotonergic projection from the CG to the raphe magnus. In addition, this periaqueductal gray–raphe magnus pathway contains somatostatin and neurotensin (Beitz *et al.*, 1983a). Skirboll and colleagues (1983) have shown that cholecystokinin and substance P

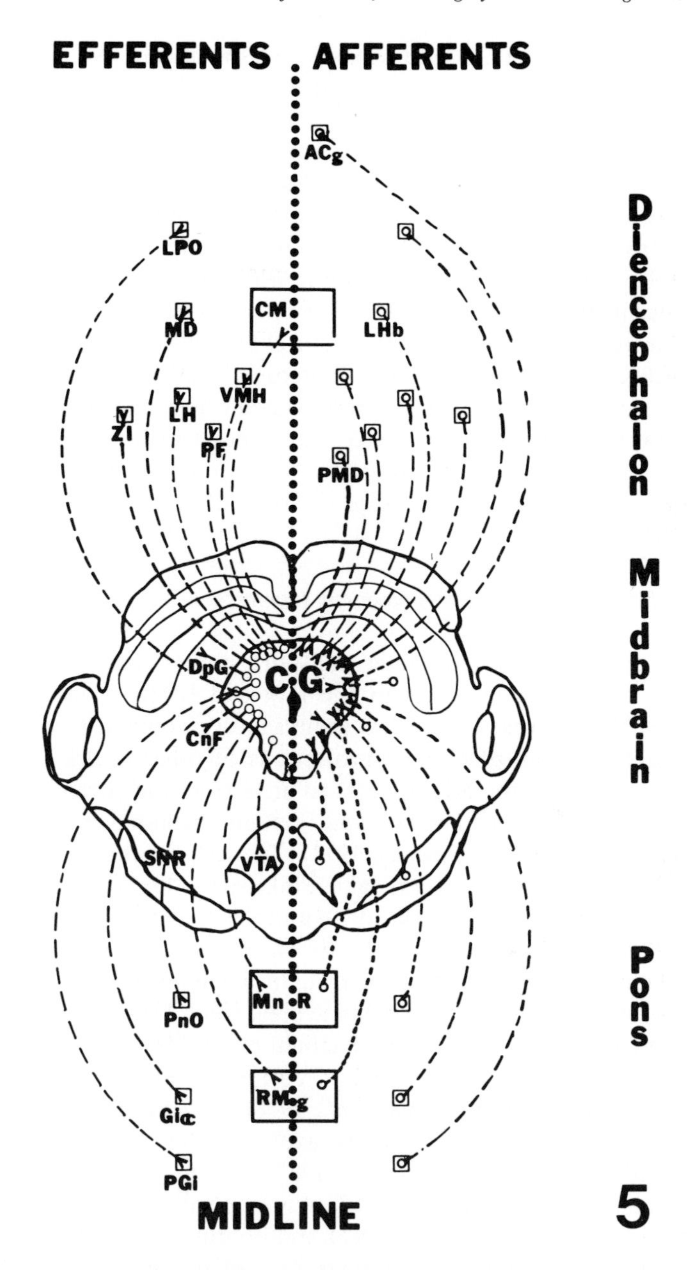

Fig. 5: A schematic diagram summarizing the major afferent and efferent connections of the central gray (CG). It can be seen from this diagram that the central gray has reciprocal connections with the majority of central nervous system structures with which it interacts. The abbreviations used are taken from Paxinos and Watson (1982).

containing neurons in the CG send axons to the spinal cord. No information is available concerning the specific neurochemicals associated with ascending CG axons.

A comparison of the source of the afferent connections and the targets of the efferent connections of the central gray shows that a majority of structures which supply axonal input to this complex midbrain region also receive efferent projections from it. For example, the CG projects to the raphe magnus nucleus, gigantocellular reticular nucleus, gigantocellular reticular nucleus, pars α paragigantocellular reticular nucleus and the nucleus of the spinal tract of the trigeminal nerve (Andrezik *et al.*, 1981a; Beitz *et al.*, 1983a, b, c; Ruda, 1976) and receives projections from all these structures (Beitz, 1982a). Reciprocal connections such as these have been demonstrated between the CG and all levels of the central nervous system. The CG thus appears to be influenced by and capable of influencing a wide spectrum of neuronal systems and its rich interconnections may explain the large diversity of functions that have been attributed to this region.

3. Connections of the dorsal and laterodorsal tegmental nuclei

We are not aware of any retrograde transport studies that have specifically analyzed the afferent connections of the DTg and LDTg. Thus, information regarding the input to these two nuclear groups must be obtained indirectly from autoradiographic studies of efferent projections of other nuclei. In brief, the major afferent input to the DTg appears to arise from the lateral mammillary, lateral habenular, interpeduncular and certain hypothalamic nuclei (see Hayakawa and Zyo, 1983; Petrovický, 1973). The LDTg receives afferents from the lateral hypothalamic area (Berk and Finkelstein, 1982), the ventromedial hypothalamic nucleus (Krieger *et al.*, 1979) and the A_1 catecholamine cell group (McKellar and Loewy, 1982).

The DTg sends efferent fibers to the median raphe nucleus, solitary nucleus, mammillary nuclei, posterior hypothalamic area, ventral tegmental area, medial septal nucleus and the nucleus of the diagonal band (Morest, 1961; Petrovický, 1973). The LDTg projects to the cerebral cortex, anterior hypothalamus and lateral hypothalamic area (Tohyama *et al.*, 1978). Recently the LDTg has been shown to send substance P containing axons to the medial frontal cortex (Paxinos *et al.*, 1978; Sakanaka *et al.*, 1983) and to the lateral septal area (Sakanaka *et al.*, 1981). Although the DTg may be involved in central autonomic regulation, as suggested by Morest (1961), more information must be obtained regarding both the connections and physiology of this nucleus and the LDTg before a clear understanding of their functions is possible.

C. Functions of the central gray

As alluded to above, the sources of afferent projections to this midbrain region are extensive and allow the central gray region to be influenced by motor, sensory and limbic structures. Many areas such as the hypothalamus and reticular formation which project to the central gray also receive reciprocal connections back from this midbrain area. This diverse interconnectivity probably underlies the numerous functional roles proposed for the central gray. This region has been implicated in central analgesic mechanisms (Fields and Basbaum, 1978), vocalization (Jurgens and Pratt, 1979), control of reproductive behavior (Sakuma and Pfaff, 1979), aggressive behavior (Mos *et al.*, 1982) and various autonomic functions (Skultety, 1959).

Although the CG appears to be involved in a diverse spectrum of functional roles, its involvement in analgesic mechanisms has received the most attention. Both electrical stimulation of, or morphine injection into, this midbrain region are capable of eliciting a potent analgesia (Fields and Basbaum, 1978; Mayer and Liebeskind, 1974). Moreover, it has been demonstrated that the dorsal and ventrolateral portions of the CG are the areas from which stimulation produced analgesia is best elicited (Cannon *et al.*, 1982; Mayer and Liesbeskind, 1974). Cannon *et al.* (1982) have shown that naloxone reliably elevates stimulation produced analgesia thresholds for ventral, but not dorsal, stimulation placements in the CG. This suggests that the dorsal portion mediates nonopioid mechanisms of analgesia while the ventral portion of the CG is involved in opioid mechanisms of analgesia. These findings are of interest in light of the results of *in vitro* receptor binding studies (see above) which show a high density of opiate binding sites ventrolaterally, but very few opiate binding sites dorsomedially (Beitz *et al.*, 1982). Moreover, Prieto and colleagues (1983) have recently shown that lesions of the raphe magnus nucleus disrupt stimulation produced analgesia from ventral but not dorsal portions of the CG. This finding suggests that the raphe magnus nucleus is a critical relay in the pain suppressive path from the ventral periaqueductal gray which mediates an opioid form of stimulation produced analgesia. Although the results of physiologic and pharmacologic experiments suggest a functional parcelation of the CG, further experimentation is required to ascertain if these functional areas correspond to anatomic subdivisions that have been proposed in the literature.

ACKNOWLEDGEMENT

The studies reported here from our laboratories were supported by NSF grant BNS-8311214 and NIH grants DE-06682 and NS-16868. The authors express appreciation to Drs G.F. Martin, W.R. Mehler and P. Mantyh for critical reading of the manuscript and to Dr George Paxinos for his constant aid and patience. The technical help of Ms M. Mullet, Ms S.G. Remington and Ms R. Shuey has been greatly appreciated and we thank Ms S. Allen for her secretarial help.

REFERENCES

Achenbach, K. E. and Goodman, D. C. (1968). Cerebellar projections to pons, medulla and spinal cord in the albino rat. *Brain Behav. Evol.* 1, 43–57.

Altman, J. and Bayer, S. A. (1981). Development of the brain stem in the rat. V. Thymidine-radiographic study of the time of origin of neurons in the midbrain tegmentum. *J. Comp. Neurol.* 198, 677–716.

Andrezik, J. A., Chan-Palay, V. and Palay, S. L. (1981a). The nucleus paragigantocellularis lateralis in rat. I. Conformation and Cytology. *Anat. Embryol.* 161, 355–371.

Andrezik, J. A., Chan-Palay, V. and Palay, S. L. (1981b). The nucleus paragigantocellularis lateralis in the rat. Demonstration of afferents by retrograde transport of horseradish peroxidase. *Anat. Embryol.* 161, 373–390.

Andrezik, J. A., Remington, S. G. and Ho, R. H. (1985). Localization of serotonin and peptide neurotransmitters in the ventral medullary reticular formation in the rat. *J. Comp. Neurol.* Submitted for publication.

Basbaum, A. I., Clanton, C. H. and Fields, H. L. (1978). Three bulbospinal pathways from the rostral medulla of the cat: An autoradiographic study of pain modulating systems. *J. Comp. Neurol.* 178, 209–224.

Beitz, A. J. (1981). An immunohistochemical study of the rodent periaqueductal gray. *Anat. Rec.* 199, 24A–25A.

Beitz, A. J. (1982a). The organization of afferent projections to the midbrain periaqueductal gray of the rat. *Neurosci.* 7, 35–159.

Beitz, A. J. (1982b). The nuclei of origin of brain stem enkephalin and substance-P projections to the rodent nucleus raphe magnus. *Neurosci.* 7, 2753–2768.

Beitz, A. J. (1982c). The sites of origin of brainstem neurotensin and serotonin projections to the rodent nucleus raphe magnus. *J. Neurosci.* 2, 829–842.

Beitz, A. J., Buggy, J., Terracio, L. and Wells, W. E. (1982). Autoradiographic localization of opiate, beta-adrenergic, cholinergic and GABA receptors in the midbrain periaqueductal gray. *Soc. Neurosci. Abstr.* 8, 265.

Beitz, A. J., Shepard, R. D. and Wells, W. E. (1983a). The Periaqueductal gray-raphe magnus projection contains somatostatin, neurotensin and serotonin but not cholecystokinin. *Brain Research* 261, 132–137.

Beitz, A. J., Wells, W. E. and Shepard, R. D. (1983b). The location of brainstem neurons which project bilaterally to the spinal trigeminal nuclei as demonstrated by the double fluorescent retrograde tracer technique. *Brain Research* 258, 305–312.

Beitz, A. J., Mullett, M. A. and Weiner, L. L. (1983c). The periaqueductal gray projections to the rat spinal trigeminal, raphe magnus, gigantocellular pars alpha and paragigantocellular nuclei arise from separate neurons. *Brain Research* 288, 307–314.

Berk, M. L. and Finkelstein, J. A. (1982). Efferent connections of the lateral hypothalamic area of the rat: An autoradiographic investigation. *Brain Res. Bull.* 8, 511–526.

Berman, A. L. (1968). *The brainstem of the cat: A cytoarchitectonic atlas with stereotaxic coordinates.* University of Wisconsin Press, Madison, Wis.

Bloom, F., Battenberg, E., Rossier, J., Ling, N. and Guillemin, R. (1978). Neurons containing β-endorphin in rat brain exist separately from those containing enkephalin: Immunocytochemical studies. *Proc. Nat. Acad. Sci. USA* 75, 159–1595.

Brodal, A. (1953). Reticulo-cerebellar connections in the cat. *J. Comp. Neurol.* 98, 113–153.

Brodal, A. (1957). *The reticular formation of the brain stem.* Oliver and Boyd, Edinburgh.

Cajal, S. Ramón y (1911). *Histologie du système nerveux de l'homme et des vertébrés.* Vol. 2, Maloine, Paris, 159–261.

Cannon, J. L., Prieto, G. J., Lee, A. and Liebeskind, J. C. (1982). Evidence for opioid and non-opioid forms of stimulation-produced analgesia in the rat. *Brain Res.* 243, 315–321.
Carlton, S. M., Leichnetz, G. R., Young, E. G. and Mayer, D. J. (1983). Supramedullary afferents of the nucleus raphe magnus in the rat: A study using the transcannula HRP gel and autoradiographic techniques. *J. Comp. Neurol.* 214, 43–58.
Castaldi, S. (1923). Studi sulca struttura e sullo sviluppo del mesencefalo. *Arch. Ital. Anat. Embriol.* 20, 23–226.
Changaris, D. C., Severs, W. B. and Keil, L. C. (1978). Localization of angiotensin in rat brain. *J. Histochem. Cytochem.* 26, 593–607.
Correa, F. M. A., Innis, R. B., Uhl, G. R. and Snyder, S. H. (1979). Bradykinin-like immunoreactive neuronal systems localized histochemically in rat brain. *Proc. Natl. Acad. Sci. USA* 76, 1489–1493.
Crosby, E. C. and Woodburne, R. T. (1943). General summary. *J. Comp. Neurol.* 78, 505–520.
Cummings, S., Elde, R., Ells, J. and Lindall, A. (1983). Corticotrophin releasing factor immunoreactivity is widely distributed within the central nervous system of the rat. *J. Neurosci.* 3, 1355–1368.
Edwards, S. B. (1975). Autoradiographic studies of the projections of the midbrain reticular formation: Descending projections of nucleus cuneiformis. *J. Comp. Neurol.* 161, 3e41–358.
Fardin, V., Oliveras, J. L. and Besson, J. M. (1984). Projections from the periaqueductal gray matter towards the B3 cellular area (nucleus raphe magnus and nucleus reticularis paragigantocellularis) as revealed by the retrograde transport of horseradish peroxidase in the rat. *J. Comp. Neurol.* (in press).
Fields, H. L. and Basbaum, A. L. (1978). Brainstem control of spinal pain transmission neurons. *Ann. Rev. Physiol.* 40, 193–221.
Fifková, E. and Maršala, J. (1967). Stereotaxic atlases for the cat, rabbit and rat. In J. Bures, M. Petran and J. Zacker (Eds). *Electrophysiological methods in biological research*, pp. 653–731. Academia, Prague.
Finley, J. C. W., Maderdrut, J. L., Roger, L. J. and Petrusz, P. (1981). The immunocytochemical localization of somatostatin-containing neurons in the rat central nervous system. *Neurosci.* 6, 2173–2192.
Gillilan, L. A. (1943). The nuclear pattern of the nontectal portions of the midbrain and isthmus in rodents. *J. Comp. Neurol.* 78, 213–252.
Hamilton, B. L. (1973a). Cytoarchitecture subdivisions of the periaqueductal gray matter in the cat. *J. Comp. Neurol.* 149, 1–28.
Hamilton, B. L. (1973b). Projections of the nuclei of the periaqueductal gray matter in the cat. *J. Comp. Neurol.* 152, 45–48.
Hayakawa, T. and Zyo, K. (1983). Comparative cytoarchitectonic study of Gudden's tegmental nuclei in some mammals. *J. Comp. Neurol.* 216, 233–244.
Henkel, C. K. and Edwards, S. B. (1978). The superior colliculus control of pinna movements in the cat: Possible anatomical connections. *J. Comp. Neurol.* 182, 763–776.
Herkenham, M. and Pert, C. B. (1982). Light microscopic localization of brain opiate receptors: A general autoradiographic method which preserves tissue quality. *J. Neurosci.* 2, 1129–1149.
Hökfelt, T., Ljundahl, A., Terenius, L., Elde, R. and Nilsson, G. (1977). Immunohistochemical analysis of peptide pathways possibly related to pain and analgesia: Enkephalin and substance P. *Proc. Natl. Acad. Sci. USA* 74, 3081–3085.
Jacobowitz, D. M., O'Donohue, T. L., Chey, W. Y. and Chang, T. (1981). Mapping of motilin-immunoreactive neurons of the rat brain. *Peptides* 2, 479–487.
Jennes, L. and Stumpf, W. E. (1980). LHRH-systems in the brain of the Golden Hamster. *Cell Tissue Res.* 209, 239–256.
Jennes, L., Stumpf, W. E. and Kalivas, P. W. (1982). Neurotensin: topographical distribution in the rat brain by immunohistochemistry. *J. Comp. Neurol.* 210, 211–224.
Jurgens, U. and Pratt, R. (1979). Role of the periaqueductal gray in vocal expression of emotion. *Brain Res.* 167, 367–378.
Krieger, M. S., Conrad, L. C. A. and Pfaff, D. W. (1979). An autoradiographic study of the efferent projections of the ventromedial nucleus of the hypothalamus. *J. Comp. Neurol.* 183, 785–816.

Ljungdahl, A., Hökfelt, T. and Nilsson, G. (1978). Distribution of substance P-like immunoreactivity in the central nervous system of the rat. I. Cell bodies and nerve terminals. *Neurosci.* 3, 861-943.

Loren, I., Alumts, J., Hakanson, R. and Sundler, F. (1979). Immunoreactive pancreatic polypeptide (PP) occurs in the central and peripheral nervous system: Preliminary immunocytochemical observations. *Cell Tissue Res.* 200, 179-186.

Liu, R. P. C. and Hamilton, B. L. (1980). Neurons of the periaqueductal grey matter as revealed by Golgi study. *J. Comp. Neurol.* 189, 403-418.

McKellar, S. and Loewy, A. D. (1982). Efferent projections of the A1 catecholamine cell group in the rat: An autoradiographic study. *Brain Res.* 241, 11-29.

Mantyh, P. W. (1982a). The midbrain periaqueductal gray in the rat, cat and monkey: A Nissl, Weil and Golgi analysis. *J. Comp. Neurol.* 204, 349-363.

Mantyh, P. W. (1982b). Forebrain projections to the periaqueductal gray in the monkey, with observations in the cat and rat. *J. Comp. Neurol.* 206, 146-158.

Mantyh, P. W. (1983). Connections of midbrain periaqueductal gray in the monkey II. Descending efferent projections. *J. Neurophysiol.* 49, 582-594.

Marchand, J. E. and Hagino, N. (1983). Afferents to the periaqueductal gray in the rat: A horseradish peroxidase study. *Neurosci.* 9, 95-106.

Mayer, D. J. and Leibeskind, J. C. (1974). Pain reduction by focal electrical stimulation of the brain: An anatomical and behavioural analysis. *Brain Res.* 68, 73-93.

Meessen, H. and Olszewski, J. (1949). *A cytoarchitectonic atlas of the rhombencephalon of the rabbit.* S. Karger, Basel

Mehler, W. R. (1969). Some neurological species differences—*a posteriori*. *Ann. N.Y. Acad. Sci.* 167, 424-468.

Moody, T. W., O'Donohue, T. L. and Jacobowitz, D. M. (1981). Biochemical localization and characterization of Bombesin-like peptides in discrete regions of rat brain. *Peptides* 2, 75-79.

Morest, D. K. (1961). Connexions of the dorsal tegmental nucleus in rat and rabbit. *J. Anat.* 95, 229-246.

Morrell, J. I., Greenberger, L. M. and Pfaff, D. W. (1981). Hypothalamic, other diencephalic and telencephalic neurons that project to the dorsal midbrain. *J. Comp. Neurol.* 201, 589-620.

Mos, J., Kruk, M. R., Van der Poel, A. M. and Meelis, W. (1982). Aggressive behavior induced by electrical stimulation in the midbrain central gray of male rat. *Aggress. Behav.* 8, 261-284.

Moss, M. S., Glazer, E. J. and Basbaum, A. I. (1983). The peptidergic organization of the cat periaqueductal gray. I. The distribution of immunoreactive enkephalin-containing neurons and terminals. *J. Neurosci.* 3, 603-616.

Moss, M. S. and Basbaum, A. I. (1983). The peptidergic organization of the cat periaqueductal gray. II. The distribution of immunoreactive substance P and vasoactive intestinal polypeptide. *J. Neurosci.* 3, 1437-1449.

O'Donohue, T. L., Miller, R. L. and Jacobowitz, D. M. (1979). Identification, characterization and stereotaxic mapping of intraneuronal-melanocyte stimulating hormone-like immunoreactive peptides in discrete regions of the rat brain. *Brain Res.* 176, 101-123.

Olszewski, J. and Baxter, D. (1954). *Cytoarchitecture of the human brain stem.* J. B. Lippincott Co., Philadelphia, Montreal.

Oswaldo-Cruz, E. and Rocha-Miranda, C. E. (1969). *The brain of the opossum (*Didelphis marsupialis*).* Instituto de Biofisica, Rio de Janiero.

Palacios, J. M., Wamsley, J. K. and Kuhar, M. J. (1981). High affinity GABA receptors autoradiographic localization. *Brain Res.* 222, 285-307.

Paxinos, G. (1983). Evidence for the existence of a parvocellular nucleus medial to the parabigeminal nucleus. *Neurosci. Lett. Suppl.* 14, 5276.

Paxinos, G. and Watson, C. (1982) *The rat brain in stereotaxic coordinates.* Academic Press, Sydney.

Paxinos, G., Emson, P. C. and Cuello, A. C. (1978). The substance P projections to the frontal cortex and substantia nigra. *Neurosci. Lett.* 7, 127-131.

Petrovický, P. (1963a). Formatio reticularis medulla oblongatae in rat. *Acta Univ. Carol. Med. (Praha)* 9, 733-740.

Petrovický, P. (1963b). Formatio reticularis pontis in rat. *Acta Univ. Carol. Med. (Praha)* 9, 741-749.

Petrovický, P. (1964). Formatio reticularis mesencephali in rat. *Acta Univ. Carol. Med. (Praha)* 10, 423-453.

Petrovický, P. (1966). A comparative study of the reticular formation of the guinea pig. *J. Comp. Neurol.* 128, 85–108.
Petrovický, P. (1971). Structure and incidence of Gudden's tegmental nuclei in some mammals. *Acta Anat.* 80, 273–286.
Petrovický, P. (1973). Note on the connections of Gudden's tegmental nuclei. I. Efferent ascending connections in the mamillary peduncle. *Acta Anat.* 86, 165–190.
Pritchard, S. M. and Beitz, A. J. (1980). A Golgi analysis of the rodent periaqueductal gray. *Soc. Neurosci. Abstr.* 6, 429.
Prieto, G. J., Cannon, J. T. and Leibeskind, J. C. (1983). Raphe magnus lesions disrupt stimulation-produced analgesia from ventral but not dorsal midbrain areas in the rat. *Brain Res.* 261, 53–57.
Ruda, M. T. (1976). Autoradiographic study of the efferent projections of the midbrain central gray in the cat. Doctoral Thesis, University of Pennsylvania.
Sakuma, Y. and Pfaff, D. W. (1979). Mesencephalic mechanisms for integration of female reproductuve behavior in the rat. *Am. J. Physiol.* 237, R285–R290.
Sakanaka, M., Shiosaka, S., Takatsuki, K., Inagaki, S., Takagi, H., Senba, E., Kawai, Y., Hara, Y., Iida, H., Minagawa, H., Matsuzaki, T. and Tohyama, M. (1981). Evidence for the existence of a substance P-containing pathway from the nucleus laterodorsalis tegmenti (Castaldi) to the lateral septal area of the rat. *Brain Res.* 230, 351–355.
Sakanaka, M., Shiosaka, S., Takatsuki, K. and Tohyama, M. (1983). Evidence for the existence of a substance P-containing pathway from the nucleus latero-dorsalis tegmenti (Castaldi) to the medial frontal cortex of the rat. *Brain Res.* 259, 123–126.
Schlaefke, M. E. (1981). Central chemosensitivity: a respiratory drive. *Rev. Physiol. Biochem. Pharmacol.* 90, 171–244.
Sims, K. B., Hoffman, D. L., Said, S. I. and Zimmerman, E. A. (1980). Vasoactive intestinal polypeptide in mouse and rat brain: An immunohistochemical study. *Brain Res.* 186, 165–182.
Skirboll, L., Hokfelt, T., Dockray, G., Rehfeld, J., Brownstein, M. and Cuello, A. C. (1983). Evidence for periaqueductal cholecystokinin-substance P neurons projecting to the spinal cord. *J. Neurosci.* 3, 1151–1157.
Skultety, F. M. (1959). Relation of periaqueductal gray matter to stomach and bladder motility. *Neurol.* 9, 190–197.
Snyder, S. H. (1980). Brain peptides as neurotransmitters. *Science* 209, 976–983.
Steinbusch, H. W. M. (1981). Distribution of serotonin-immunoreactivity in the central nervous system of the rat: Cell bodies and terminals. *Neurosci.* 6, 557–618.
Swanson, L. W. (1977). Immunohistochemical evidence for a neurophysin-containing autonomic pathway arising in the paraventricular nucleus of the hypothalamus. *Brain Res.* 128, 346–353.
Taber, R. (1961). The cytoarchitecture of the brain stem of the cat. I. Brain stem nuclei. *J. Comp. Neurol.* 116, 27–70.
Takagi, H. (1980). The nucleus reticularis paragigantocellularis as a site of analgesic action of morphine and enkephalin. *Trends in Pharmacol. Sciences*, 182–184.
Tohyama, M., Satoh, K., Sakumoto, T., Kimoto, Y., Takahashi, Y., Yamamato, K. and Itakura, T. (1978). Organization and projections of the neurons in the dorsal tegmental area of the rat. *J. für Hirnforsch.* 19, 165–176.
Tredici, G., Bianchi, R. and Gioia, M. (1983). Short intrinsic circuit in the periaqueductal gray matter of the cat. *Neurosci. Lett.* 39, 131–136.
Valverde, F. (1961). Reticular formation of the pons and medulla oblongata. A Golgi study. *J. Comp. Neurol.* 116, 71–99.
Valverde, F. (1962). Reticular formation of the albino rats' brain stem. Cytoarchitecture and corticofugal connections. *J. Comp. Neurol.* 119, 25–53.
Vanderhaeghen, J. J., Lotstra, F., DeMey, J. and Gilles, C. (1980). Immunohistochemical analysis of cholecystokinin and gastrin-like peptides in the brain and hypophysis of the rat. *Proc. Natl. Acad. Sci. USA* 77, 1190–1194.
Wamsley, J. K., Lewis, M. S., Young, W. S. and Kuhar, M. J. (1981). Autoradiographic localization of muscarinic cholinergic receptors in rat brain stem. *J. Neurosci.* 1, 176–191.
Watkins, L. R., Griffin, G., Leichnetz, G. R. and Mayer, D. J. (1980). The somatotopic organization of the nucleus raphe magnus and surrounding brain stem structures as revealed by HRP slow-release gels. *Brain Res.* 181, 1–15.

Watkins, L. R., Griffin, G., Leichnetz, G. R. and Mayer, D. J. (1981). Identification and somatotopic organization of nuclei projecting via the dorsolateral funiculus in rats: A retrograde tracing study using HRP slow-release gels. *Brain Res.* 223, 237–255.
Watson, S. J., Richard, C. W. and Brachas, J. D. (1978). Adrenocorticotropin in rat brain: Immunocytochemical localization in cells and axons. *Science* 200, 1180–1182.
Watson, S. J., Khachaturian, H., Akil, H., Coy, D. H. and Goldstein, A. (1982). Comparison of the distribution of dynorphin systems and enkephalin systems in brain. *Science* 218, 1134–1136.
Weber, E., Evans, C. J., Samuelsson, S. J. and Barchas, J. D. (1981). Novel peptide neuronal system in rat brain and pituitary. *Science* 213, 1247–1251.
Wünscher, W., Shober, W. and Werner, L. (1965). *Architectonisher Atlas vom Hirnstamm der Ratte.* S. Hirzel, Leipzig.
Yezierski, R. P., Bowker, R. M., Kevetter, G. A., Westlund, K. N., Coulter, J. D. and Willis, W. D. (1982) Serotonergic projections to the caudal brain stem: A double label study using horseradish peroxidase and serotonin immunocytochemistry. *Brain Res.* 239, 258–264.
Young, W. S. and Kuhar, M. J. (1979). Neurotensin receptors: Autoradiographic localization in rat CNS. *Eur. J. Pharmacol.* 59, 161–163.

2

Major projections of the reticular formation

GEORGE F. MARTIN
ROBERT WALTZER

The Ohio State University
Columbus, Ohio, USA

ROBERT P. VERTES

Mercer University
Macon, Georgia, USA

I. INTRODUCTION

In this chapter the projections of the reticular formation to the spinal cord, the cerebellum, the diencephalon and the basal forebrain, as well as to selected motor nuclei of cranial nerves, are examined. Data obtained from the rat will be used whenever possible, including unpublished observations by the senior author. When results obtained from experiments on other species are employed, that will be made clear.

II. SPINAL PROJECTIONS OF THE RETICULAR FORMATION

A. Projections of the medullary reticular formation

Medullary reticulospinal axons originate within the ventral and dorsal reticular nuclei of the medulla (MdV and MdD), the gigantocellular reticular nucleus (Gi), the interfascicular hypoglossal nucleus (IF 12), the gigantocellular reticular nuclei, ventral part and pars α (GiV and Giα; see Volume 2, Chapter 1), the paragigantocellular reticular nucleus (PGi), the paramedian reticular nucleus (PMn) and, to a lesser extent, the parvocellular reticular nucleus (PCRt) (Basbaum and Fields, 1979, 1982; Bowker *et al.*, 1981a, b; Huisman *et al.*, 1981;

THE RAT NERVOUS SYSTEM
ISBN 0 12 547632 9

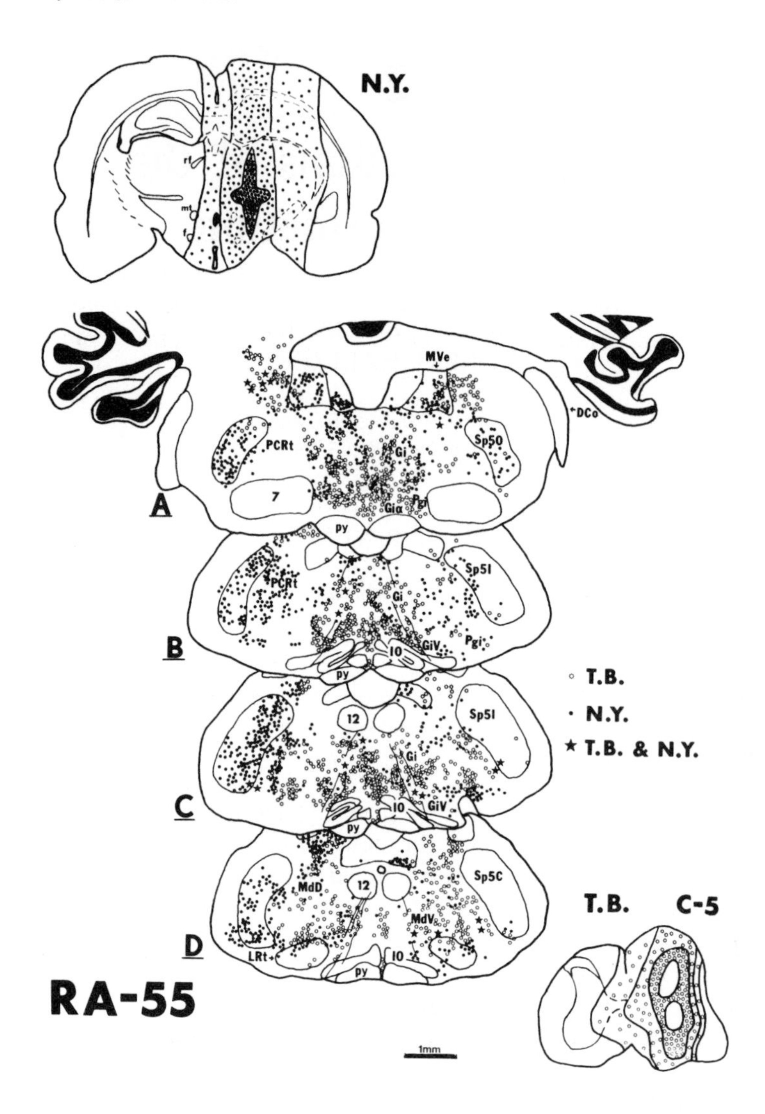

Fig. 1: Plot of neurons in the medulla (rostral A, to caudal, D) labeled by the spinal injection of true blue (TB) shown at the lower right and the diencephalic injection of nuclear yellow (NY) shown at the upper left. Neurons containing TB innervate the spinal cord, whereas neurons containing NY project to the diencephalon. Neurons containing both markers are presumed to innervate both the spinal cord and diencephalon via axonal collaterals. The symbols for neurons labeled by TB and NY alone, as well as by both markers, are indicated (revised with permission from Waltzer and Martin, 1984). 7, facial nucleus; 12, hypoglossal nucleus; DCo, dorsal cochlear nucleus; f, fornix; Gi, gigantocellular reticular nucleus; Giα and GiV, gigantocellular reticular nucleus α and ventral parts; IO, inferior olive; LRt, lateral reticular nucleus; MdD and MdV, reticular nucleus of the medulla, dorsal and ventral parts; MVe, medial vestibular nucleus; mt, mammillothalamic tract; PCRt, parvocellular reticular nucleus; PGi, paragigantocellular reticular nucleus; py, pyramidal tract; rf, fasciculus retroflexus; Sp5C, Sp5O, Sp5I, nucleus of the spinal tract of the trigeminal, caudal, oral and interpolar parts.

Leichnetz *et al.*, 1978; Martin *et al.*, 1985; Westlund *et al.*, 1982). Such neurons are indicated by a hollow star with eight points in Fig. 1. The spinal projections of the ventral (MdV) and dorsal (MdD) reticular nuclei have not been studied in the rat, but in the cat (Bowker and Coulter, 1978) and opossum (Martin *et al.*, 1981), axons from both nuclei project bilaterally to laminae 7, 8 and possibly 10. In both species, the retroambiguus nucleus, which is contained within the MdV, contributes to an additional, mainly crossed projection (Bowker and Coulter, 1978; Martin *et al.*, 1981). In the opossum and cat, the MdV and MdD, (as well as the retroambiguus nucleus) innervate mainly cervical and thoracic levels (Bowker and Coulter, 1978; Martin *et al.*, 1981). Autoradiographic results suggest the presence of bilateral projections from the reticular formation of the ventrolateral medulla to the phrenic nucleus and to thoracic motoneurons in the rat (Martin *et al.*, 1985). The latter projections may arise from the so called ventral respiratory area (for example the cat, Merrill, 1970).

The gigantocellular reticular nucleus (Gi) contributes axons to the ventral and lateral funiculi, bilaterally, with a decided ipsilateral predominance (Martin *et al.*, 1985; Zemlan and Pfaff, 1979; Zemlan *et al.*, 1982). It is well known that the gigantocellular axons distribute to medial parts of lamina 7, to lamina 8 and to lamina 10 throughout the length of the cord (Martin *et al.*, 1985; Zemlan *et al.*, 1982), but they also innervate lateral parts of laminae 3 to 7, the intermediolateral cell column and the sacral parasympathetic nucleus (Martin and Waltzer, 1984; Martin *et al.*, 1985). Evidence for lamina 9 projections also exists (Martin and Waltzer, 1984; Martin *et al.*, 1985). The GiV, as defined in Volume 2, Chapter 1, projects heavily to lamina 9 and to autonomic nuclei was well as to the other areas referred to above (Martin and Waltzer, 1984; Martin *et al.*, 1985). It has been shown electronmicroscopically that the region in question directly innervates motoneurons of the lumbar cord (Holstege, J. C. and Kuypers, 1982).

Serotonergic neurons are found within the Gi, GiV and PGi (Björklund and Skagerberg, 1982; Bowker *et al.*, 1981a, b, 1982; Steinbusch, 1981) and some of them apparently contribute to the heavy serotonergic innervation found within laminae 9 and 10 as well as autonomic nuclei (Bowker *et al.*, 1982; Steinbusch, 1981). In fact, Loewy and McKellar (1981) have presented evidence that the projections from such areas to the intermediolateral cell column are primarily serotonergic. Several peptides are also present within the Gi and raphe neurons (Ljungdahl *et al.*, 1978) and some of them coexist with serotonin (see review by Hökfelt *et al.*, 1982). It is also known that peptide neurons in such regions innervate the spinal cord (Bowker *et al.*, 1981b; Hökfelt *et al.*, 1979). One such peptide, substance P, is fairly abundant within the ventral horn, including lamina 9, and most of it disappears after spinal transection (see review by Hökfelt *et al.*, 1982) which suggests that it originates within the brain stem. The Gi is known to be involved in posture and balance (see review by Peterson, 1980 on the cat) and it has been suggested that its serotonergic projections modulate motor activity in highly motivated behaviors (Kuypers, 1982).

The ventral reticular nucleus (MdV) as well as the PGi and adjacent areas contain catecholamine neurons (Dahlström and Fuxe, 1964) and some of them innervate the spinal cord (Ross *et al.*, 1981). The rostral part of the region in question (part of the PGi and adjacent areas according to current nomenclature or the nucleus reticularis rostroventralis of Ross *et al.*, 1984) is close to the surface of the brain stem and has been implicated in cardiovascular control (see reviews by Loewy, 1982; Ross *et al.*, 1981). The spinal projections from this region are largely adrenergic (Ross *et al.*, 1981; Westlund *et al.*, 1982), arising from the C1 cell group described by Hökfelt *et al.* (1982), and targeted mainly on the intermediolateral cell column (Martin *et al.*, 1985; Ross *et al.*, 1984).

The Giα (see Volume 2, Chapter 1) and, possibly, the rostral part of PGi, have projections to lamina 1 and the outer part of lamina 2 (Martin and Waltzer, 1984; Martin *et al.*, 1985) as well as to some of the areas referred to above. The projections to laminae 1 and 2 course in the dorsal part of the lateral funiculus. The uniqueness of the region referred to herein as Giα has been emphasized by Watkins *et al.* (1980). Serotonergic (Bowker *et al.*, 1982; Steinbusch, 1981) as well as peptidergic neurons (Bowker *et al.*, 1981a, b; Ljungdahl *et al.*, 1978) are found in Giα (as well as within the adjacent PGi) and some of them innervate the spinal cord (Bowker *et al.*, 1981a, b). Johannessen *et al.* (1981) have emphasized that many of the Giα neurons which project through the dorsal part of the lateral funiculus are nonserotonergic. The literature on peptidergic projections from reticular and raphe nuclei to the spinal cord has been reviewed by Hökfelt *et al.* (1982). The Giα may play a role in centrally induced analgesia (see reviews by Basbaum and Fields, 1978 and Mayer and Price, 1976) including that involved in highly motivated behaviors (Kuypers, 1982).

B. Projections of the pontine reticular formation

Spinal axons arise within the caudal (PnC) and rostral (PnO) nuclei of the pontine reticular formation (Basbaum and Fields, 1979; Bowker *et al.*, 1982; Huisman *et al.*, 1981; Satoh, 1979; Westlund *et al.*, 1982; Zemlan *et al.*, 1979). Axons from dorsomedial portions of both nuclei project primarily through the sulcomarginal and ventral funiculi to lamina 8 and medial portions of layer 7 (autoradiographic data from our laboratory). This projection is mainly an ipsilateral one and extends the length of the spinal cord. The pontine reticular formation is known to effect somatic motor functions (see review by Peterson, 1980, on the cat). There is evidence in the opossum (Martin *et al.*, 1979, 1982) that the ventral most part of the PnC innervates the dorsal horn by way of axons in the dorsal part of the lateral funiculus. The comparable area in the rat may be the ventral pontine reticular nucleus (PnV) (see Volume 2, Chapter 1). The PnV contains serotonergic neurons and has been referred to by others as the lateral wing of the raphe magnus. The PnV is closely associated with auditory circuits and its spinal connections may be involved in the acoustic startle response (see Davis *et al.*, 1982, for a discussion relative to the cat).

The caudal paralemniscal nucleus (PLC; Volume 2, Chapter 1) projects through the lateral funiculi of the spinal cord (Basbaum *et al.*, 1978; Leichnetz *et al.*, 1978; Zemlan *et al.*, 1979). This projection is mainly crossed and innervates lateral portions of layers 3 to 7 (results from our laboratory). Based on its connections, the paralemniscal region may be involved in motor responses to auditory and/or visual stimuli (Davis *et al.*, 1982). Although the major projection of the pedunculopontine nucleus (PPTg) is to the forebrain and thalamic areas, it also innervates the spinal cord (Jackson and Crossman, 1983; Spann and Grofova, 1984).

Many neurons in the dorsolateral pons project spinalward (Basbaum and Fields, 1979; Saper and Loewy, 1980; Satoh, 1979; Zemlan *et al.*, 1979). Such neurons are located within the locus coeruleus (LC) (not classically a reticular nucleus), the nucleus subcoeruleus (SubC) and the Kölliker-Fuse nucleus (KF). Taken as a whole, the dorsolateral pons projects to intermediate areas of the spinal gray (layers 5 to 8), to layer 10, to layer 9 (sparsely), to autonomic cell groups and to layers 1 and 2 (results from our laboratory). Saper and Loewy (1980) have suggested that the projections to the intermediolateral cell column originate within the Kölliker-Fuse and the laterodorsal tegmental nucleus (LDTg). Many of the neurons within the LC and SubC which innervate the spinal cord are noradrenergic (Westlund *et al.*, 1981, 1982) and noradrenergic axons are present within all the spinal areas mentioned above (Westlund *et al.*, 1982). Catecholamine projections to layers 1 and 2 as well as 5 play a role in centrally induced analgesia (for example, Basbaum and Fields, 1978). It has been suggested that noradrenergic projections from the dorsolateral pons to the spinal cord modulate pain processing and motor activity in highly motivated behaviors (Kuypers, 1982). Holstege, G. and Kuypers (1982) have found evidence in the cat for projections from the reticular formation of the lateral pons to the lateral cervical nucleus. It is likely that such projections modify sensory processing in the spinocervical, cervicothalamic system.

Many neurons in ventrolateral parts of the pontine reticular formation are catecholaminergic (the A5 and A7 groups of Dahlström and Fuxe, 1964) and some of them innervate the spinal cord (Loewy *et al.*, 1979; Westlund *et al.*, 1982). Loewy *et al.* (1979) have provided evidence that the A5 group projects to the intermediolateral cell column. The organization of descending catecholamine systems in the rat has been reviewed by Björklund and Skagerberg (1982).

C. Projections of the mesencephalic reticular formation

The cuneiform nucleus (Cnf) (Satoh, 1979) and the deep mesencephalic nucleus (DpMe; Veazey and Severin, 1980; Waldron and Gwyn, 1969) project to the spinal cord. The terminal distribution of such projections has not been studied in detail in the rat, but in the opossum they primarily innervate lamina 7 (Martin *et al.*, 1979).

D. Do single reticular neurons innervate more than one level of the spinal cord?

Some of the studies referred to above have presented evidence that a number of reticular nuclei in the medulla and pons project the length of the spinal cord and Huisman *et al.* (1981) have shown that many neurons provide collaterals to more than one level.

III. CEREBELLAR PROJECTIONS OF THE RETICULAR FORMATION

The gigantocellular reticular nucleus (Gi), the gigantocellular reticular nucleus, ventral part (GiV), the interfascicular hypoglossal nucleus (IF12), the para-gigantocellular reticular nucleus (PGi), and the paramedian reticular nucleus project to the cerebellum (Martin and Waltzer, 1984). Reticulocerebellar neurons are located lateral and medial to the hypoglossal nerve and some of the latter are found in clusters. Some of the reticulocerebellar neurons located lateral to the rostral part of the inferior olive, as well as rostral to it, are serotonergic (Bishop and Ho, 1983). Reticulocerebellar neurons are also found in the rostral reticular nucleus of the pons (PnO) and some of them may be serotonergic (Bishop and Ho, 1983).

IV. DIENCEPHALIC PROJECTIONS OF THE RETICULAR FORMATION

A. Projections of the medullary reticular formation

The components of the medial forebrain bundle have been reviewed by Nieuwenhuys *et al.* (1982) and Veening *et al.* (1982). Neurons in the ventral (MdV) and dorsal (MdD) reticular nuclei, the gigantocellular (Gi, GiV and GIα) and adjacent paragigantocellular (PGi) reticular nuclei project rostrally and many of them contribute axons to the medial forebrain bundle (Loewy *et al.*, 1981; Takagi *et al.*, 1980; Vertes, 1982, 1984b). Such neurons are indicated by solid dots in Fig. 1. There is evidence that axons from one or more of these nuclei innervate the zona incerta, the lateral and dorsal hypothalamic nuclei, the median eminence, the paraventricular nucleus, the supraoptic nucleus, the medial preoptic area, the nucleus of the diagonal band, the accumbens nucleus, the lateral septal nuclei, the bed nucleus of the stria terminalis, the cingulate cortex, the pretectum, the laterodorsal nucleus of the thalamus, and the reuniens nucleus (Berk and Finkelstein, 1981; Loewy *et al.*, 1981; McKellar and Loewy, 1981; Sawchenko and Swanson, 1982; Vertes *et al.*, 1984). Some of the above projections are catecholaminergic (McKellar and Loewy, 1981; Sawchenko and Swanson, 1982), arising from neurons in the A1 group described by Dahlström and Fuxe (1964) and involved in neuroendocrine as well as autonomic functions (see review by Sawchenko and Swanson, 1982). The Gi contributes axons to a bundle in the midbrain which divides into dorsal and ventral branches (Vertes *et al.*, 1984) as described for the cat by Nauta and Kuypers (1958). The dorsal branch projects to the dorsal thalamus and the ventral one innervates the fields of Forel, the zona incerta, the hypothalamus and more rostral areas (Vertes *et al.*, 1984).

B. Projections of the pontine reticular formation

Neurons within the caudal (PnC) and rostral (PnO) reticular nuclei of the pons also contribute axons to the medial forebrain bundle (Kita and Oomura, 1982; Takagi *et al.*, 1980; Vertes, 1982, 1984a, b). Most of them are located lateral to the median raphe nucleus and within dorsal and lateral parts of the PnO (possibly including the A7 area of Dahlström and Fuxe, 1964, and the pedunculopontine tegmental nucleus).

The diencephalic projections of the PnC and PnO have not been reported in the rat. In the cat, however, they innervate certain intralaminar nuclei and the dorsomedial nucleus of the thalamus as well as various hypothalamic, subthalamic and basal forebrain areas (Graybiel, 1977; Langer and Kaneko, 1983; Robertson and Feiner, 1982). The region of the pontine reticular formation containing omnipause neurons (neurons involved in rapid eye movements) innervates different parts of the subthalamus than are innervated by adjacent regions (Langer and Kaneko, 1983). Results from our laboratory (not reported previously) suggest that comparable projections exist in the rat and that reticular axons of pontine origin reach the supramammillary nucleus, the intermediate and lateral mammillary nuclei, the lateral preoptic area, the septum, and the diagonal band region. The relative contribution of monoamine and nonmonoamine axons to such projections is not known. Projections to the dorsomedial nucleus of the thalamus from the pontine reticular formation may be involved in eye movements (Graybiel, 1977). The locus coeruleus (LC), the subcoeruleus nucleus (SubC), the parabrachial nuclei (VPB and DPB) and the nucleus Kölliker-Fuse (KF) also project through the medial forebrain bundle (Takagi *et al.*, 1980; Vertes, 1982), but the projections of these nuclei are beyond the scope of the present chapter (see Saper and Loewy, 1980, for a description).

The pedunculopontine nucleus (PPTg) projects to the globus pallidus, the entopeduncular nucleus, the subthalamic nucleus and the thalamus (Jackson and Crossman, 1983). Similar connections have been reported for other species (Ahlsen and Lo, 1982; de Vito *et al.*, 1980; Nomura *et al.*, 1980; Rodrigo-Angulo and Reinoso-Suarez, 1982).

C. Projections of the mesencephalic reticular formation

The mesencephalic reticular formation has extensive ascending projections. Axons from the deep mesencephalic nucleus (DpMe) have been reported to innervate the zona incerta, the subthalamic nucleus, the lateral thalamic nucleus and the dorsomedial nucleus of the thalamus (Veazey and Severin, 1980). In the cat (Edwards and de Olmos, 1976) the cuneiform nucleus (Cnf), as defined by Taber (1961), innervates the fields of Forel, the lateral, posterior and dorsal hypothalamic areas, the zona incerta, the preoptic area, the bed nucleus of the stria terminalis, the thalamic reticular nucleus, the ventral lateral geniculate nucleus, the intralaminar nuclei and a small portion of the dorsomedial nucleus of the thalamus. Ascending projections from the DpMe and/or Cnf may be involved in arousal and/or selective attention (Scheibel, A. B., 1980).

V. PROJECTIONS OF THE RETICULAR FORMATION TO SELECTED CRANIAL NERVE MOTOR NUCLEI

The ventral and dorsal reticular nuclei of the medulla (MdV and MdD), the lateral part of the gigantocellular (Gi) reticular nucleus (the medial portion of the parvocellular tegmental field of Holstege *et al.*, 1977), the region around the motor trigeminal nucleus, the nucleus of Kölliker-Fuse (KF) and the parabrachial nuclei (VPB, DPB) provide projections, in differing amounts, to the motor nuclei of the hypoglossal, vagal, facial and trigeminal nerves (Borke *et al.*, 1983; Travers and Norgren, 1983). A few neurons located lateral to the medullary raphe provides comparable projections (Travers and Norgren, 1983), some of which may be serotonergic. Neurons supplying catecholamine projections to the motor trigeminal nucleus have been reported within the lateral part of the oral pontine reticular nucleus and within the lateral lemniscus (Vernov and Sutin, 1983). The latter regions may be comparable to the paralemniscal nucleus, caudal part (PLC) (Volume 2, Chapter 1). Reticular projections to the cranial nerve nuclei referred to above are involved in mastication, lapping, swallowing and grooming (Travers and Norgren, 1983).

In the cat (Henkel and Edwards, 1978) and opossum (Panneton and Martin, 1979, 1983) an area which appears comparable to the oral paralemniscal nucleus (PLO; Volume 2, Chapter 1) provides contralateral projections to that portion of the facial nucleus which innervates caudal auricular muscles. It has been suggested that such connections are involved in pinna orientation to acoustic stimuli (Henkel and Edwards, 1978; Panneton and Martin, 1979).

Specific areas of the pontine reticular formation have been shown to innervate the abducens nuclei and/or closely adjacent regions in the cat (Graybiel, 1977). Connectional specificity seems to exist, that is, neurons at the junction of the PnO and PnC project to the ipsilateral abducens nucleus, whereas those in more caudal regions innervate the same nucleus contralaterally. These connections are probably involved in rapid eye movements (Graybiel, 1977).

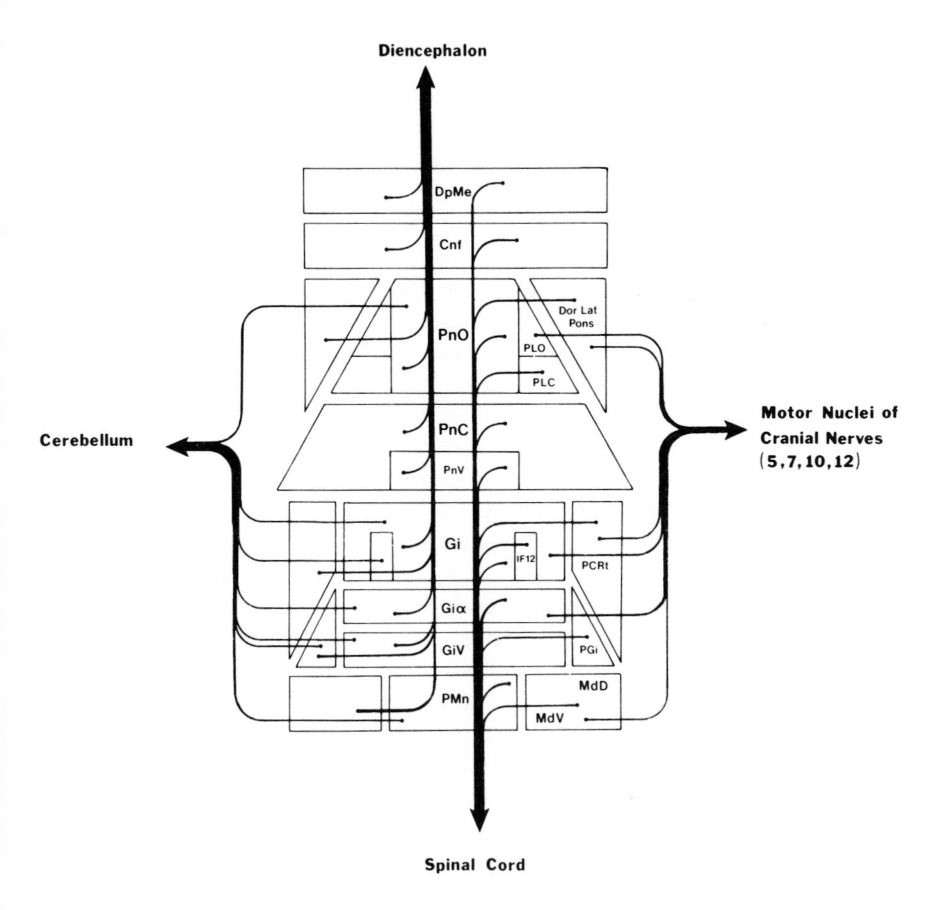

Fig. 2: Schematic illustration of projections of the reticular nuclei considered in this chapter. It should be noted that most nuclei project to several targets, but that connectional specificity also exists. Neurons within a single nucleus which project to different areas often have characteristic locations which are not illustrated. Reference to Fig. 1 shows that reticular neurons projecting to widely divergent targets may exist in close proximity and that some provide collateral innervation to more than one area. Cnf, cuneiform nucleus; DpMe, deep mesencephalic nucleus; Gi, gigantocellular reticular nucleus; Giα and GiV, gigantocellular reticular nucleus, α and ventral parts; IF12, interfascicular hypoglossal nucleus, MdD and MdV, reticular nucleus of the medulla, dorsal and ventral parts; PCRt, parvocellular reticular nucleus; PGi, paragigantocellular reticular nucleus; PMn, paramedial reticular nucleus; PnC, PnO and PnV, pontine reticular nucleus, caudal, oral and ventral parts; PLC and PLO, paralemniscal nucleus caudal and oral parts.

VI. DO RETICULAR NEURONS INNERVATE WIDELY DIVERGENT TARGETS VIA AXONAL COLLATERALS?

The gigantocellular reticular nucleus (Gi) projects to the spinal cord, the cerebellum and the diencephalon. In order to determine if single (Gi) neurons innervate more than one target, double labeling techniques have been employed

(Waltzer and Martin, 1982, 1983, 1984). Such experiments show that some of the Gi neurons which innervate the spinal cord also project to either the cerebellum or the diencephalon (presumably via axonal collaterals; see also Bentivoglio and Molinari, 1984), but most of them do not. The results from one case are plotted in Fig. 1.

VII. SUMMARY AND CONCLUSIONS

The reticular core of the rat's brain stem is cytoarchitecturally and chemoarchitecturally heterogeneous (Volume 2, Chapter 1). It is not surprising, therefore, that the projections of cytoarchitecturally distinct areas (nuclei) differ and that, to some extent, these differences reflect involvement in different behaviors (Peterson, 1980; Siegel, 1979). Individual reticular nuclei project to multiple targets, however, (Fig. 1) and as suggested by Brodal (1956) for the cat, overlap exists in the location of neurons within a single nucleus which innervate different targets (Fig. 1). As suggested by the Scheibels (Scheibel and Scheibel, 1958), some neurons provide collaterals to widely separate areas (Fig. 1) providing the basis for relatively global activation. It appears, however, that many reticular neurons do not provide such collaterals but have more restricted fields of innervation. An attempt to summarize the projections referred to herein is provided in Fig. 2.

Criteria for subdividing the reticular formation, or any other area of the neural axis, should include connectional as well as cytoarchitectural and chemoarchitectural considerations. As new data become available it may be possible to draw "boundaries" more clearly and to propose an even more meaningful system of nomenclature.

ACKNOWLEDGEMENTS

The authors are grateful to Ms Mary Ann Jarrell for the autoradiographic processing of rat brains and to Mrs Nina J. Anspaugh for typing the manuscript. The new data reported herein were obtained as a result of research supported by USPHS Grants BNS-8008675 and BNS-8309245 to Dr Martin.

REFERENCES

Ahlsen, G. and Lo, F.-S. (1982). Projections of brain stem neurons to the perigeniculate nucleus and the lateral geniculate nucleus in the cat. *Brain Res.* 238, 433–438.

Basbaum, A. I. and Fields, H. L. (1978). Endogenous pain control mechanisms: A review and hypothesis. *Ann. Neurol.* 4, 451–462.

Basbaum, A. I. and Fields, H. L. (1979). The origin of descending pathways in the dorsolateral funiculus of the spinal cord of the cat and rat: Further studies on the anatomy of pain modulation. *J. Comp. Neurol.* 187, 513–532.

Basbaum, A. I., Clanton, C. H. and Fields, H. L. (1978). Three bulbospinal pathways from the rostral medulla of the cat: An autoradiographic study of pain modulating systems. *J. Comp. Neurol.* 178, 209–224.

Bentivoglio, M. and Molinari, M. (1984). Fluorescent retrograde triple labeling of brain stem reticular neurons. *Neurosci. Lett.* 46, 121–126.
Berk, M. L. and Finkelstein, J. A. (1981). Afferent projections to the preoptic area and hypothalamic regions in the rat brain. *Neurosci.* 6, 1601–1624.
Bishop, G. A. and Ho, R. (1983). The origin of serotonergic afferents to the cerebellum of the rat. *Soc. Neurosci. Abst.* 9, 1092.
Björklund, A. and Skagerberg, G. (1982). Descending monoaminergic projections to the spinal cord. In B. Sjolund and A. Björklund (Eds). *Brainstem control of spinal mechanisms*, pp. 55–58. Elsevier, New York.
Borke, R. C., Nav, M. E. and Ringler, R. L. (1983). Brainstem afferents of hypoglossal neurons in the rat. *Brain Res.* 269, 47–55.
Bowker, R. M. and Coulter, J. D. (1978). Studies of descending projections from the caudal medulla in the cat. *Soc. Neurosci. Abst.* 4, 291.
Bowker, R. M., Westlund, K. N. and Coulter, J. D. (1981a). Origins of serotonergic projections to the spinal cord in rat: An immunocytochemical-retrograde transport study. *Brain Res.* 226, 187–199.
Bowker, R. M., Steinbusch, H. W. M. and Coulter, J. D. (1981b). Serotonergic and peptidergic projections to the spinal cord demonstrated by a combined HRP histochemical and immunocytochemical staining method. *Brain Res.* 211, 412–417.
Bowker, R. M., Westlund, K. N., Sullivan, M. C. and Coulter, J. D. (1982). Organization of descending serotonergic projections to the spinal cord. In H. G. J. M. Kuypers and G. F. Martin (Eds). *Progress in brain research: Descending pathways to the spinal cord*, pp. 239–265. Elsevier, New York.
Brodal, A. (1956). Anatomical aspects of the reticular formation of the pons and medulla oblongata. In J. Ariens Kappers (Ed.). *Prog. Neurobiol.* pp. 240–255. Elsevier.
Dahlström, A. and Fuxe, K. (1964). Evidence for the existence of monoamine-containing neurons in the central nervous system. I. Demonstration of monoamines in the cell bodies of brainstem neurons. *Acta Physiol. Scand.* 62, 5–55.
Davis, M., Gendelman, D. S., Tischler, M. D. and Gendelman, P. M. (1982). A primary acoustic startle circuit: Lesion and stimulation studies. *J. Neurosci.* 2, 791–805.
De Vito, J. L., Anderson, M. E. and Walsh, K. E. (1980). A horseradish peroxidase study of afferent connections of the globus pallilus in *Macaca mulatta*. *Exp. Brain Res.* 38, 65–73.
Edwards, S. B. and de Olmos, J. S. (1976). Autoradiographic studies of the projections of the midbrain reticular formation: Ascending projections of nucleus cuneiformis. *J. Comp. Neurol.* 165, 417–432.
Graybiel, A. M. (1977). Direct and indirect preoculomotor pathways of the brainstem: An autoradiographic study of the pontine reticular formation in the cat. *J. Comp. Neurol.* 175, 37–78.
Henkel, C. K. and Edwards, S. B. (1978). Superior colliculus control of pinna movement in cat ... possible anatomical connections. *J. Comp. Neurol.* 182, 763–776.
Hökfelt, T., Terenius, L., Kuypers, H. G. J. M. and Dann, O. (1979). Evidence for enkephalin immunoreactive neurons in the medulla oblongata projecting to the spinal cord. *Neurosci. Lett.* 14, 55–60.
Hökfelt, T., Skirboll, L., Dalsgaard, C., Johansson, O., Lundberg, J. M., Norell, G. and Jansco, G. (1982). Peptide neurons in the spinal cord with special reference to descending systems. In B. Sjolund and A. Björklund (Eds). *Brainstem control of spinal mechanisms*, pp. 89–117. Elsevier, New York.
Holstege, G. and Kuypers, H. G. J. M. (1982). The anatomy of brain stem pathways to the spinal cord in cat. A labeled amino acid tracing study. In H. G. J. M. Kuypers and G. F. Martin (Eds). *Progress in brain research: Descending pathways to the spinal cord*, pp. 145–175. Elsevier, New York.
Holstege, G., Kuypers, H. G. J. M. and Dekker, J. J. (1977). The organization of the bulbar fibre connections to the trigeminal, facial and hypoglossal motor nuclei. *Brain* 100, 265–286.
Holstege, J. C. and Kuypers, H. G. J. M. (1982). Brainstem projections to spinal motoneuronal cell groups in rat studied by means of electron microscopy autoradiography. In H. G. J. M. Kuypers and G. F. Martin (Eds). *Progress in brain research: Descending pathways to the spinal cord*, pp. 177–183. Elsevier, New York.

Huisman, A. M., Kuypers, H. G. J. M. and Verburgh, C. A. (1981). Quantitative differences in collateralization of the descending spinal pathways from red nucleus and other brainstem cell groups in rats as demonstrated with the multiple fluorescent retrograde tracer technique. *Brain Res.* 209, 271-286.

Jackson, A. and Crossman, A. R. (1983). Nucleus tegmenti pedunculopontinus: Efferent connections with special reference to the basal ganglia; studied in the rat by anterograde and retrograde transport of horseradish peroxidase. *Neurosci. Lett.* 10, 725-765.

Johannessen, J. N., Watkins, L. R. and Mayer, D. J. (1981). Non-serotonergic cells at the origin of the dorsolateral funiculus in the rat medulla. *Soc. Neurosci. Abstr.* 7, 533.

Kita, H. and Oomura, Y. (1982). An HRP study of the afferent connections to rat lateral hypothalamic region. *Brain Res. Bull.* 8, 63-71.

Kuypers, H. G. J. M. (1982). A new look at the organization of the motor system. In H. G. J. M. Kuypers and G. F. Martin (Eds). *Progress in brain research: Descending pathways to the spinal cord*, pp. 381-403. Elsevier, New York.

Langer, T. P. and Kaneko, C. R. S. (1983). Efferent projections of the cat oculomotor reticular omnipause neuron region: An autoradiographic study. *J. Comp. Neurol.* 217, 288-306.

Leichnetz, G. L., Watkins, L., Griffin, G., Murfin, R. and Mayer, D. J. (1978). The projection from nucleus raphe magnus and other brainstem nuclei to the spinal cord in the rat: A study using the HRP blue-reaction. *Neurosci. Lett.* 8, 119-124.

Ljungdahl, A., Hokfelt, T. and Nilsson, G. (1978). Distribution of substance P-like immunoreactivity in the rat. I. Cell bodies and nerve terminals. *Neurosci.* 3, 861-943.

Loewy, A. D. (1982). Descending pathways to the sympathetic preganglionic neurons. In H. G. J. M. Kuypers and G. F. Martin (Eds). *Progress in brain research: Descending pathways to the spinal cord*, pp. 267-277. Elsevier, New York.

Loewy, A. D. and McKellar, S. (1981). Serotonergic projections from the ventral medulla to the intermediolateral cell column in the rat. *Brain Res.* 211, 146-153.

Loewy, A. D., McKellar, S. and Saper, C. B. (1979). Direct projections from the A5 catecholamine cell group to the intermediolateral cell column. *Brain Res.* 174, 309-315.

Loewy, A. D., Wallach, J. H. and McKellar, S. (1981). Efferent connections of the ventral medulla oblongata. *Brain Res. Rev.* 3, 63-80.

McKellar, S. and Loewy, A. D. (1981). Organization of some brainstem afferents to the paraventricular nucleus of the hypothalamus in the rat. *Brain Res.* 217, 351-357.

McKellar, S. and Loewy, A. D. (1982). Efferent projections of the A1 catecholamine cell group in the rat: An autoradiographic study. *Brain Res.* 241, 11-29.

Martin, G. F., Humbertson, A. O., Laxson, C. L., Panneton, W. M. and Tschismadia, I. (1979). Spinal projections from the mesencephalic and pontine reticular formation in the North American opossum: A study using axonal transport techniques. *J. Comp. Neurol.* 187, 373-400.

Martin, G. F., Cabana, T., Humbertson, A. O., Laxson, L. C. and Panneton, W. M. (1981). Spinal projections from the medullary reticular formation of the North American opossum: Evidence for connectional heterogeneity. *J. Comp. Neurol.* 196, 663-682.

Martin, G. F., Cabana, T., DiTirro, F. J., Ho, R. H. and Humbertson, A. O. (1982). Raphe spinal projections in the North American opossum. Evidence for connectional heterogeneity. *J. Comp. Neurol.* 196, 67-84.

Martin, G. F. and Waltzer, R. (1984). Spinal projections of the gigantocellular reticular formation in the rat. Evidence for differential projections to laminae I, II and IX. *Soc. Neurosci. Abst.* 10, 29.

Martin, G. F., Vertes, R. P. and Waltzer, R. (1985). Spinal projections of the gigantocellular reticular formation in the rat. Evidence for projections from different areas to laminae I an II and Lamina IX. *Exp. Brain Res.* (in press).

Mayer, D. J. and Price, D. D. (1976). Central nervous system mechanisms of analgesia. *Pain* 2, 379-404.

Merrill, E. G. (1970). The lateral respiratory neurones of the medulla: their association with nucleus ambiguus, nucleus retroambiguus, the spinal accessory nucleus and the spinal cord. *Brain Res.* 24, 11-28.

Nauta, W. J. H. and Kuypers, H. G. J. M. (1958). Some ascending pathways in the brainstem reticular formation. In H. H. Jasper, L. D. Proctor, R. S. Knighton, W. S. Noshay and R. T. Costello (Eds). *Reticular formation of the brain*, pp. 13-30. (Henry Ford Hospital Symposium) Little Brown, Boston.

Nieuwenhuys, R., Geeraedts, L. M. G. and Veening, J. G. (1982). The medial forebrain bundle of the rat. *J. Comp. Neurol.* 206, 49–81.

Nomura, S., Mizuno, N. and Sugimoto, T. (1980). Direct projections from the pedunculo-pontine tegmental nucleus to the subthalamic nucleus in the cat. *Brain Res.* 196, 223–227.

Panneton, W. M. and Martin, G. F. (1979). Midbrain projections to the trigeminal, facial and hypoglossal nuclei in the opossum. A study using axonal transport techniques. *Brain Res.* 168, 493–511.

Panneton, W. M. and Martin, G. F. (1983). Brainstem projections to the facial nucleus of the opossum. A study using axonal transport techniques. *Brain Res.* 267, 19–83.

Peterson, B. (1980). Participation of pontomedullary reticular neurons in specific motor activity. Chairman's overview of Part III. In J. A. Hobson and M. A. B. Brazier (Eds). *The reticular formation revisited. Specifying function for a nonspecific system*, pp. 171–192. International Brain Research Organization (IBRO) monograph series 6, Raven Press, New York.

Robertson, R. T. and Feiner, A. R. (1982). Diencephalic projections from the pontine reticular formation: Autoradiographic studies in the cat. *Brain Res.* 239, 3–16.

Rodrigo-Angulo, M. L. and Reinoso-Suarez, F. (1982). Topographical organization of the brainstem afferents to the lateral posterior-pulvinar thalamic complex in the cat. *Neurosci.* 7, 1495–1508.

Ross, C. A., Ruggiero, D. A. and Reis, D. J. (1981). Projections to the spinal cord from neurons close to the ventral surface of the hindbrain in the cat. *Neurosci. Lett.* 21, 143–148.

Ross, R. A., Ruggiero, D. A., Joh, T. H., Park, D. H. and Reis, D. T. (1984). Rostral ventrolateral medulla: Selective projections to the thoracic autonomic cell column from the region containing Cl adrenaline neurons. *J. Comp. Neurol.* 228, 168–185.

Saper, C. B. and Loewy, A. D. (1980). Efferent connections of the parabrachial nucleus in the rat. *Brain Res.* 197, 291–317.

Satoh, K. (1979). The origin of reticulospinal fibers in the rat. A HRP study. *J. für Hirnforsch.* 20, 313–332.

Sawchenko, P. E. and Swanson, L. W. (1982). The organization of noradrenergic pathways from the brainstem to the paraventricular and supraoptic nuclei in the rat. *Brain Res. Rev.* 4, 275–325.

Scheibel, M. E. and Scheibel, A. B. (1958). Structural substrates for integrative patterns in the brainstem reticular core. In H. H. Jasper, L. D. Proctor, R. S. Knighton, W. S. Noshay and R. T. Costello (Eds). *Reticular formation of the brain*, pp. 31–55. (Henry Ford Hospital Symposium), Little Brown, Boston.

Scheibel, A. B. (1980). Anatomical and physiological substrates of arousal: A view from the bridge. Chairman's overview of Part III. In J. A. Hobson and M. A. B. Brazier (Eds). *The reticular formation revisited: Specifying function for a nonspecific system*, pp. 55–66. International Brain Research Organization (IBRO) Monograph Series 6, Raven Press, New York.

Siegel, J. M. (1979). Behavioral functions of the reticular formation. *Brain Res. Rev.* 1, 69–105.

Spann, B. and Grofova, I. (1984). Ascending and descending projections of the nucleus tegmenti pendunculopontinus in the rat. *Soc. Neurosci. Abst.* 10, 182.

Steinbusch, H. W. M. (1981). Distribution of serotonin-immunoreactivity in the central nervous system of the rat-cell bodies and terminals. *Neurosci.* 6, 557–618.

Taber, E. (1961). The cytoarchitecture of the brainstem of the cat. I. Brainstem nuclei of the cat. *J. Comp. Neurol.* 116, 27–70.

Takagi, H., Shiosaka, S., Tohyama, M., Senba, E. and Sakanaka, M. (1980). Ascending components of the medial forebrain bundle from the lower brainstem in the rat, with special reference to raphe and catecholamine cell groups. A study by the HRP method. *Brain Res.* 193, 315–337.

Travers, J. B. and Norgren, R. (1983). Afferent projections to the oral motor nuclei in the rat. *J. Comp. Neurol.* 220, 280–298.

Veazey, R. B. and Severin, C. M. (1980). Efferent projections of the deep mesencephalic nucleus (pars lateralis) in the rat. *J. Comp. Neurol.* 190, 231–244.

Veening, J. G., Swanson, L. W., Cowan, W. M., Nieuwenhuys, R. and Geeraedts, L. M. G. (1982). The medial forebrain bundle of the rat. II. An autoradiographic study of the topography of the major descending and ascending components. *J. Comp. Neurol.* 206, 81–108.

Vernov, J. J. and Sutin, J. (1983). Brainstem projections to the normal and noradrenergically hyperinnervated trigeminal motor nucleus. *J. Comp. Neurol.* 214, 198–208.

Vertes, R. P. (1982). At least 25 distinct brainstem nuclei project to the forebrain along the medial forebrain bundle (MFB). *Soc. Neurosci. Abstr.* 8, 642.

Vertes, R. P. (1983a). A lectin-HRP study of the origin of ascending fibers in the medial forebrain bundle of the rat. The upper brainstem. *Neurosci.* 11, 669–690. .

Vertes, R. P. (1983b). A lectin-HRP study of the origin of ascending fibers in the medial forebrain bundle of the rat. The lower brainstem. *Neurosci.* 11, 651–668..

Vertes, R. P., Waltzer, R. and Martin, G. F. (1984). An autoradiographic study of ascending nucleus gigantocellularis projections in the rat. *Soc. Neurosci. Abst.* 10, 901.

Waldron, H. A. and Gwyn, D. C. (1969). Descending nerve tracts in the spinal cord of the rat. I. Fibers from the midbrain. *J. Comp. Neurol.* 137, 143–154.

Waltzer, R. P. and Martin, G. F. (1982). A double-labelling study demonstrating that most cells in the nucleus reticularis gigantocellularis and adjacent raphe project to either the anterior lobe of the cerebellum or the spinal cord in the rat. *Neurosci. Abst.* 8, 874.

Waltzer, R. P. and Martin, G. F. (1983). A double labeling study demonstrating that most spinally projecting neurons of the nucleus reticularis gigantocellularis do not provide collateral innervation to the diencephalon in the rat. *Soc. Neurosci. Abst.* 9, 284.

Waltzer, R. P. and Martin, G. F. (1984). Collateralization of reticulospinal axons from the nucleus reticularis gigantocellularis to the cerebellum and diencephalon. A double labeling study in the rat. *Brain Res.* 293, 153–158.

Watkins, L. R., Griffin, G., Leichnetz, G. R. and Mayer, D. J. (1980). The somatotopic organization of the nucleus raphe magnus and surrounding brainstem structures as revealed by slow-release gels. *Brain Res.* 181, 1–15.

Westlund, K. N., Bowker, R. M., Ziegler, M. G. and Coulter, J. D. (1981). Origins of spinal noradrenergic pathways demonstrated by retrograde transport of antibody to dopamine-B-hydroxylase. *Neurosci. Lett.* 25, 243–249.

Westlund, K. N., Bowker, R. M., Ziegler, M. G. and Coulter, J. D. (1982). Descending noradrenergic projections and their spinal terminations. In H. G. J. M. Kuypers and G. F. Martin (Eds). *Progress in brain research: Descending pathways to the spinal cord*, pp. 219–238. Elsevier, New York.

Zemlam, F. P. and Pfaff, D. W. (1979). Topographical organization in medullary reticulospinal systems as demonstrated by the horseradish peroxidase technique. *Brain Res.* 174, 161–166.

Zemlan, F. P., Low, L., Morell, J. I. and Pfaff, D. W. (1979). Descending tracts of the lateral columns of the rat spinal cord: A study using the horseradish peroxidase and silver impregnation technique. *J. Anat.* 128, 489–512.

Zemlan, F. P., Behbehani, and Beckstead, R. M. (1982). Ascending and descending projections of N.R. Magnocellularis (NMC) and N.R. gigantocellularis (NGC): An autoradiographic and HRP study. *Soc. Neurosci. Abst.* 8, 92.

3

Raphe nuclei and serotonin containing systems

ISTVAN TÖRK

University of New South Wales
Kensington, NSW, Australia

I. INTRODUCTION

The midline groups of neurons associated with the midsagittal seam ('raphe') of the brain stem, extending from the level of the interpeduncular nucleus in the midbrain to the level of the pyramidal decussation in the medulla, form the system of the raphe nuclei. In ordinary Nissl stained preparations these nuclei appear as unpaired aggregations of neurons. Some of the raphe nuclei are clearly separated from the surrounding areas of the brain stem tegmentum, while others tend to blend with the neighboring areas of the brain stem reticular formation. A characteristic feature of the raphe nuclei is that a substantial part of their neuronal population is serotonergic, although the proportion of serotonin containing neurons in the different raphe nuclei is variable, and serotonin containing neurons are found outside these nuclei as well. There is now substantial evidence that several other putative neurotransmitters are present in the raphe neurons and in some cases they coexist in the same neuron with other transmitters. The terminal systems of the serotonin containing neurons in the brain are extensive: fibers containing serotonin are found virtually everywhere in the brain although the density of innervation varies from area to area.

There was little interest in the morphology and function of the raphe nuclei until Dahlström and Fuxe (1964) described in the brain stem of the rat the existence of an extensive system of monoaminergic neurons. Using the method of formaldehyde induced fluorescence these authors described a chain of catecholaminergic and serotonergic neurons in the brain stem. They realized that the serotonergic neurons only partially conformed with known clusters of cells in the reticular formation and raphe and called them groups B1–B9 in a caudorostral order.

THE RAT NERVOUS SYSTEM
ISBN 0 12 547632 9

The delineation of the raphe nuclei used in this chapter is based on the detailed descriptions of the raphe system by Brodal *et al.* (1960a, b), Taber *et al.* (1960), Taber (1961), Felten and Cummings (1979), Steinbusch and Nieuwenhuys (1983), Berman (1968), and our own serial sections of the rat brain which were immunostained for serotonin using a monoclonal antibody and the biotin–avidin technique, and counterstained with cresyl violet. The observations on other species are easily adaptable to the rat brain stem, since the raphe system shows little variation across species (Petrovicky, 1980). However, there is some inconsistency in the names given to the different nuclei of the raphe system which is side-stepped here by applying the terminology adopted in Paxinos and Watson (1982; henceforth *Atlas*), while giving the alternative terms for the nuclei.

In this chapter, the following raphe nuclei are discussed in a rostrocaudal order: the caudal linear nucleus; dorsal raphe nucleus (DR); median raphe nucleus (MnR); raphe pontis nucleus (RPn); raphe magnus nucleus (RMg); raphe obscurus nucleus (ROb); and raphe pallidus nucleus (RPa). In addition, attention is paid to those extraraphe areas that contain significant groups of serotonergic neurons. In the past, some authors dealing with the raphe system have included in their list of raphe nuclei two nuclei of the midbrain ventral tegmentum (Brodal *et al.*, 1960a, b; Petrovicky, 1980; Valverde, 1962). This practice is not followed here because these midline nuclei (interfascicular and rostral linear nuclei) are predominantly dopaminergic and are described in Volume 1, Chapter 9.

II. MORPHOLOGY OF THE RAPHE NUCLEI

A. Caudal linear nucleus (Central linear nucleus, N. linearis oralis, N. linearis intermedius)

The caudal linear nucleus (CLi) is the most rostral member of the raphe system. It extends from the dorsal aspect of the interpeduncular nucleus rostrally to the beginning of the dorsal raphe nucleus caudally. Anteriorly, the nucleus is bounded by the interfascicular nucleus of the midbrain ventromedial tegmentum. The caudal part of the caudal linear nucleus is dorsal to the decussation of the superior cerebellar peduncles (xscp). Anterolaterally, the nucleus is adjacent to the subnuclei of the midbrain ventromedial tegmentum; that is, the paired rostral linear nuclei, paranigral nuclei, and the parabrachialis pigmented nuclei. All these neighboring nuclei of the ventral tegmentum contain large numbers of dopaminergic cells, while in the caudal linear nucleus the proportion of dopaminergic cells is smaller (Swanson, 1982). The caudal linear nucleus contains small to medium sized cells scattered among the crossing fibers of the midbrain tegmentum. The dendrites of these cells are preferentially arranged in a dorsoventral direction. The caudal linear nucleus is the most rostral nucleus in the midline that contains a significant number of serotonergic cells (Steinbusch and Nieuwenhuys, 1983); the density of the 5-hydroxytryptamine (5HT) cells increases toward the caudal tip of the nucleus where it abuts onto the dorsal raphe nucleus. In the cat, Wiklund *et al.* (1981b) recorded over 2000 serotonergic neurons in this nucleus, about one tenth of the size of the 5HT cell population found in the dorsal raphe nucleus.

Due to the presence of many crossing fibers of the tegmental decussation, the caudal linear is very loosely organized. It is not known whether these crossing fibers make any contacts with the cells residing between them. Two dominant cell types can be found in the caudal linear nucleus: dopaminergic, and serotonergic. The dopaminergic cells belong to the caudal extension of the cell cluster A10 of Dahlström and Fuxe (1964) and could also be identified immunocytochemically with tyrosine hydroxylase antibodies (Swanson, 1982). The cluster of serotonergic cells is attached dorsally to the group B7 (the DR), and is clearly separated from the more ventrocaudal B8 group by the decussating fibers of the superior cerebellar peduncle. Steinbusch and Nieuwenhuys (1983) distinguish only one type of serotonergic neuron in the CLi: the small piriform cells. The dendrites of the 5HT neurons found in the dorsal part of this nucleus are oriented rostrocaudally while those of cells in the ventral part of the nucleus extend into the interpeduncular complex.

B. Dorsal raphe nucleus (Groups B7, B5 and B4)

The dorsal raphe nucleus (DR) has received much attention in the literature, possibly because it is the largest raphe nucleus and it contains the largest accumulation of serotonergic cell bodies in the brain. It is a long structure located in the caudal part of the central gray matter (CG) and the rostral part of periventricular gray of the 4th ventricle. The DR extends from the caudal pole of the oculomotor nuclear complex to the pons (*Atlas* Figs. 28 to 33). The great majority of the cells in the nucleus are located in the midbrain, dorsal to the medial longitudinal fasciculus (mlf), although some ventrally located midline cells are located between these two fiber bundles. The dorsal part of the nucleus extends laterally into the CG, making the delineation of the nucleus difficult. Steinbush *et al.* (1981) distinguish several groups of cells in the DR: four clusters in the mesencephalic part (ventromedial, dorsomedian, and two lateral groups), and one in the rhombencephalic part (caudal part). Cell sizes and shapes vary from small (14 μm mean soma diameter) and spherical, through to medium size (24 μm) fusiform cells, to large (35 μm) multipolar neurons. In Golgi impregnated material of the DR, Danner and Pfister (1980) and Diaz-Cintra *et al.* (1981) also identified three types of neurons according to the shape of the cell bodies and dendritic morphology. Type 1 neurons are polygonal in shape and possess spines on the cell body and on the proximal parts of the dendrites. The axons of these neurons are directed ventrally. Type 2 neurons have fusiform cell bodies with spines on the distal portions of the dendrites, and axons directed preferentially dorsally. Type 3 neurons are small, piriform cells with aspiny dendrites. These three types of neurons are also present in the other raphe nuclei (Hölzel and Pfister, 1981). The existence of somatic spines on some serotonergic raphe cells has been confirmed with intracellular horseradish peroxidase (HRP) labeling of identified neurons (Park *et al.*, 1982) and under the electron microscope (Descarries *et al.*, 1982). It has been agreed that the type 2 fusiform neurons are serotonergic and at least some cells of the other types may also contain this transmitter.

Serotonergic neurons are, of course, the most characteristic cell type in the DR (Pfister and Danner, 1980). Since the first observation by Dahlström and Fuxe (1964), who designated the cell group as B7 (rostral part) and B5–B4 (caudal part), the existence of serotonergic perikarya in this region has been amply confirmed using reuptake of ^{3}H-serotonin and autoradiography (Aghajanian and Bloom, 1967; Chan-Palay, 1977; Descarries *et al.*, 1982; Gamrani and Calas, 1980) and immunocytochemistry (Lidov *et al.*, 1980; Steinbusch, 1981; Steinbush and Nieuwenhuys, 1983; Steinbusch *et al.*, 1981; Takeuchi *et al.*, 1982a). Quantitative studies based on the transmitter specific labeling of the serotonergic neurons have clearly established that a large proportion of the neurons in the DR are not serotonergic. Wiklund *et al.* (1981b) reported that only approximately 40% of the neurons of the DR in the cat can be identified as serotonergic. A similar figure is true for the DR of the rat (Moore, 1981). In an autoradiographic study, Descarries *et al.* (1982) reported that out of 24 000 DR cells only about 11 500 (almost 48%) were found to be serotonergic. It is important to note that the serotonergic cell population in the DR constitutes about half of all serotonergic nerve cells of the brain (Descarries *et al.*, 1982; Wiklund *et al.*, 1981b).

The great sensitivity of the immunocytochemical technique allowed the study of the size and morphology of the 5HT positive cells. Serotonergic cells found in the dorsomedial part of the DR are medium sized and fusiform, while those found in the ventromedial part are predominantly small and round. However, in the lateral parts of the DR the 5HT neurons are of the large, multipolar type (Fig. 1A, B). These observations indicate that the presence of the transmitter is not restriced to any particular morphologic cell type in the raphe, although the most common serotonergic DR neuron seems to be the fusiform cell.

Fig. 1: A: Serotonin cell groups in the dorsal raphe nucleus (DR). The immunohistochemical staining reveals the four distinct cell groups found in the main body of the DR. The dorsomedian, ventromedial and lateral groups are shown separated from each other by dotted white lines. Magnification = × 64. B: Multipolar serotonergic neurons in the lateral cell group of the DR. The dendrites of the neurons extend dorsally and laterally into the CG well beyond the boundaries of the DR. Magnification = × 160. The tissue presented in these and subsequent figures was counterstained with cresyl violet. mlf, medial longitudinal fasciculus. ▷

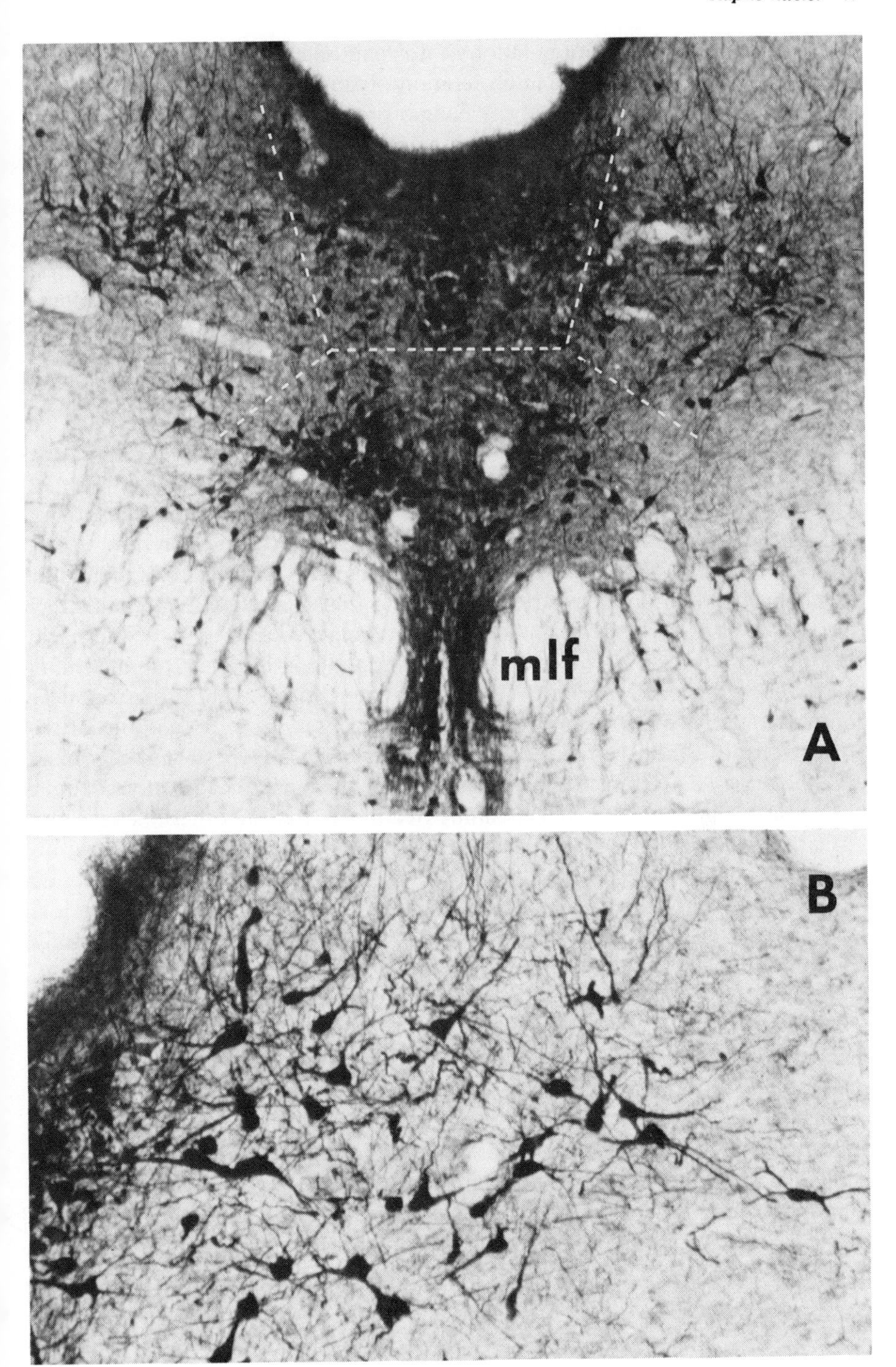
mlf
A
B

The presence of a large population of nonserotonergic neurons in the DR has important implications in the interpretation of functional studies involving DR lesions. Behavioral and physiologic changes produced by DR lesions cannot be interpreted as the result of the destruction of the serotonergic cells alone. Indeed, the list of other putative neurotransmitters identified in the DR is growing rapidly. With an antibody to dopamine-β-hydroxylase (DBH), noradrenergic neurons can be observed in the caudal tip of the lateral part of the DR (Grzanna and Molliver, 1980; Grzanna *et al.*, 1978; Steinbusch *et al.*, 1981). Most of these neurons are large and multipolar, and have their somata just at the lateral boundary of the DR, but they send their medial dendrites well into the domain of the DR cells. Contrary to the noradrenergic cells, the few dopaminergic cells that have been observed in the DR (Ochi and Shimizu, 1978) are found near the midline. Another transmitter whose presence in the DR has been predicted on the basis of physiologic observations is γ-aminobutyric acid (GABA). Using tritium labeled GABA, labeling could be demonstrated in a small number of neurons of the DR, especially those near the fourth ventricle (Belin *et al.*, 1979; Gamrani *et al.*, 1979). GABA labeling also appears in nerve fibers in the nucleus, although some of these fibers can be removed with 5,7-dihydroxytryptamine pretreatment (Gamrani *et al.*, 1979) indicating that some GABA accumulating fibers are also serotonergic. Antibodies developed against glutamic acid decarboxylase (GAD) have been used by Nanopoulos *et al.* (1982) and Belin *et al.* (1983) to demonstrate the GABAergic neurons in DR. These authors have confirmed earlier observations that GABA was present in many neurons of the DR and firmly established that any of the three basic cell types could be stained with the antiGAD serum. However, the GAD positive neurons are of two kinds: one sensitive to 5,7-dihydroxytryptamine treatment, and the other not sensitive.

During recent years substantial evidence emerged that many of the neurons in the DR contained peptides. Enkephalin like immunoreactive cells are found in the lateral parts of the nucleus (Glazer *et al.*, 1981; Hökfelt *et al.*, 1977; Moss *et al.*, 1983; Uhl *et al.*, 1979a), although many enkephalin immunoreactive cells are found outside the DR in the central gray matter (Moss *et al.*, 1983). Substance P has been demonstrated to coexist with 5HT in some DR raphe cells (Chan-Palay *et al.*, 1978; Hökfelt *et al.*, 1978). Neurotensin has also been demonstrated in the DR; a significant concentration of the neurotensin immunoreactive cells was found just posterior to the caudal end of the nucleus designated as nucleus recessus sulci medianus (Jennes *et al.*, 1982b; Minagawa *et al.*, 1983; Uhl *et al.*, 1979b), also called nucleus O (see Volume 1, Chapter 14). Also in the caudal part of the DR, intensively positive cholecystokinin octapeptide immunoreactive cells were observed (Innis *et al.*, 1979). The cholecystokinin positive neurons form a separate population to the 5HT containing neurons (Van der Kooy *et al.*, 1981).

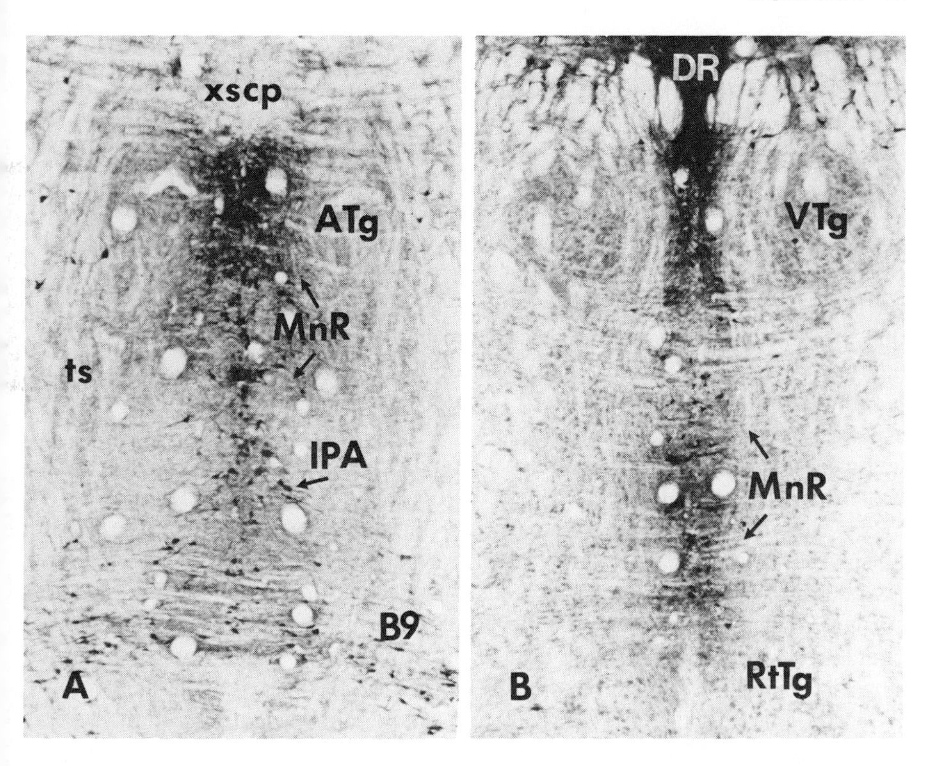

Fig. 2: Serotonin containing neurons and nerve fibers in the median raphe nucleus (MnR), seen in the center of the figures. A: The MnR on this section from its rostral pole appears continuous with the cell groups in the apical part of the interpeduncular nucleus (IPA) and B9. ts, tectospinal tract; xscp, decussation of the superior cerebellar peduncle. B: The caudal end of the MnR at the level of the ventral tegmental nucleus (VTg) and the reticular tegmental nucleus (RtTg). At this level and dorsally, the MnR is continuous with the dorsal raphe nucleus (DR). ts, tectospinal tract; xscp, decussation of the superior cerebellar peduncle. Magnification = × 75.

C. The median raphe nucleus (Central superior nucleus, medial raphe nucleus, group B8)

Unlike most other raphe nuclei, the median raphe nucleus (MnR) is easily defined. It is located in a central position in the cranial part of the pons, extending caudally from the decussation of the superior cerebellar peduncle to about the level of the ventral tegmental nucleus (VTg) (Fig. 2). For most of its rostrocaudal extent, the MnR is accompanied by the tectospinal tracts which run parallel to it on either side, but the nucleus itself is not contiguous with the tracts. The lateral boundaries of the MnR are enhanced by blood vessels which appear in a curved line just 200 μm to 300 μm from the midline (Fig. 2A, B).

The caudal part of the MnR is located between the two tegmental reticular nuclei (RtTg) (Fig. 2B), and is in contact with the rostral end of the raphe pontis nucleus. The morphology of the MnR has received much less attention than the DR, perhaps because only a minority of its cells is serotonergic (Moore, 1981; Wiklund *et al.*, 1981b). Steinbusch and Nieuwenhuys (1983) distinguished three cell types on the basis of size and shape: small ellipsoid, medium sized ellipsoid, and medium sized fusiform. The serotonin containing neurons are found directly on the midline, and belong mainly to the small ellipsoid and medium sized fusiform type. Recently, in Golgi studies, Hölzel and Pfister (1983) reported on the different types of neurons in this nucleus, and distinguished three types of neurons as in the DR. They found that the morphology of the cells in the MnR is rather similar to that of the cells in the DR. Somatic spines are also present, especially on the larger neurons.

Little is known about the internal organization of the MnR, although the presence of clusters of cells in it indicate the existence of subnuclei. In Nissl or 5HT stained preparations three distinct parts of the nucleus can be distinguished: oral, intermediate and caudal. The oral part is dorsal in position and contains medium sized to small cells oriented vertically or obliquely to the coronal plane (Köhler and Steinbusch, 1982). This part of the nucleus is also characterized by a paired bundle of serotonergic fibers which appear on either side of the midline, accompanying the cells. The intermediate part of the MnR is ventral and caudal to the oral part and extends further laterally than the oral group (Fig. 2A). Moreover, the cells in the intermediate part are horizontally oriented with dendrites extending for several hundred micrometers into the reticular formation.

The three cell types are not evenly distributed in the MnR, as the medium sized cells appear on and near the midline, while the majority of the small neurons extend from the midline towards the medial lemniscus where the B9 serotonergic cluster of Dahlström and Fuxe (1964) is found (Fig. 2B). The caudal part of the nucleus is narrow and is characterized by a lack of orientation of the dendritic fields of the neurons in it. In all three parts of the nucleus several neurons are in intimate contact with the blood vessels, and this is particularly well discernible in 5HT immunostained preparations since many of these cells are serotonergic.

Lateral to the dorsal part of the MnR, two pairs of well discernible magnocellular cell groups are found (Fig. 2A; ATg). Caudally, the well defined ventral tegmental nucleus of Gudden is seen (*Atlas* Fig. 31), but rostral to this there is another group of cells, less compact than the VTg and relatively poor in acetylcholinesterase (*Atlas* Plate 30, unmarked). This cluster has been designated by Petrovicky (1971) as the 'compact part' of the MnR, a view not shared by the writer since these lateral groups do not contain serotonergic cells. It is more appropriate to associate these lateral nuclei with the ventral tegmental nucleus since they both receive projections from the mammillary bodies. It has been suggested (Volume 1, Chapter 14) that the nucleus be called the anterior tegmental nucleus (ATg).

D. Raphe pontis nucleus (Group B4)

The small raphe pontis nucleus (RPn) is found in the rostral part of the pons, extending rostrally to the level of the tegmental reticular nucleus and caudally to the level of the raphe magnus nucleus. Its rostral part has a structure similar to that of the MnR, but the caudal part contains only a few scattered cells aligned parallel to the midline. Most of the nucleus is restricted to the dorsal half of the tegmentum. At the caudal and ventral end of the RPn there is a small cluster of larger neurons on either side of the midline, quite different from the rest of the cells found in the rostral part of the nucleus. These cells are not serotonergic, although serotonergic axons appear in between them in abundance. These cells have been erroneously enclosed within the boundaries of the RPn (*Atlas* Fig. 33). In transverse sections, this group of cells seems independent from the reticular tegmental nucleus (RtTg), but in sagittal sections a loose continuity with that structure can be observed, suggesting that the cell cluster is the dorsocaudal extension of the RtTg. It remains to be established whether the connections of this subnucleus are similar to those of the RtTg.

Steinbusch and Nieuwenhuys (1983) distinguished three types of cells in the RPn according to size (small, medium, and large). Most of these cells are fusiform or multipolar with only three to four primary dendrites. The number of serotonergic neurons in the nucleus is relatively small (10%–30%; Moore, 1981), but 5HT is still the major known transmitter in the neurons (Steinbusch and Nieuwenhuys, 1983; Wiklund *et al.*, 1981b).

E. Raphe magnus nucleus (Groups B2 and B3)

The raphe magnus nucleus (RMg) is a long and wide cell group in the rostral part of the medulla. In coronal sections, it is usually outlined as a triangular area just dorsal to the medial lemniscus (Fig. 3). Its most dorsal cells are in the midline about half the distance between the medial longitudinal fasciculus and the pyramids. Rostrocaudally, the RMg is coextensive with the facial nucleus; rostrally it reaches the nucleus of the trapezoid body, and caudally it is contiguous with the raphe obscurus nucleus. Laterally, the delineation of the nucleus is difficult since many of its cells seem to extend into a region dorsal to the pyramids; this is especially obvious in 5HT immunostained preparations. The chain of 5HT cells extends beyond the pyramids and continues into a group of smaller cells lateral to it. The cells above and lateral to the pyramids are considered part of the gigantocellular reticular nucleus pars α (Giα). The lateral

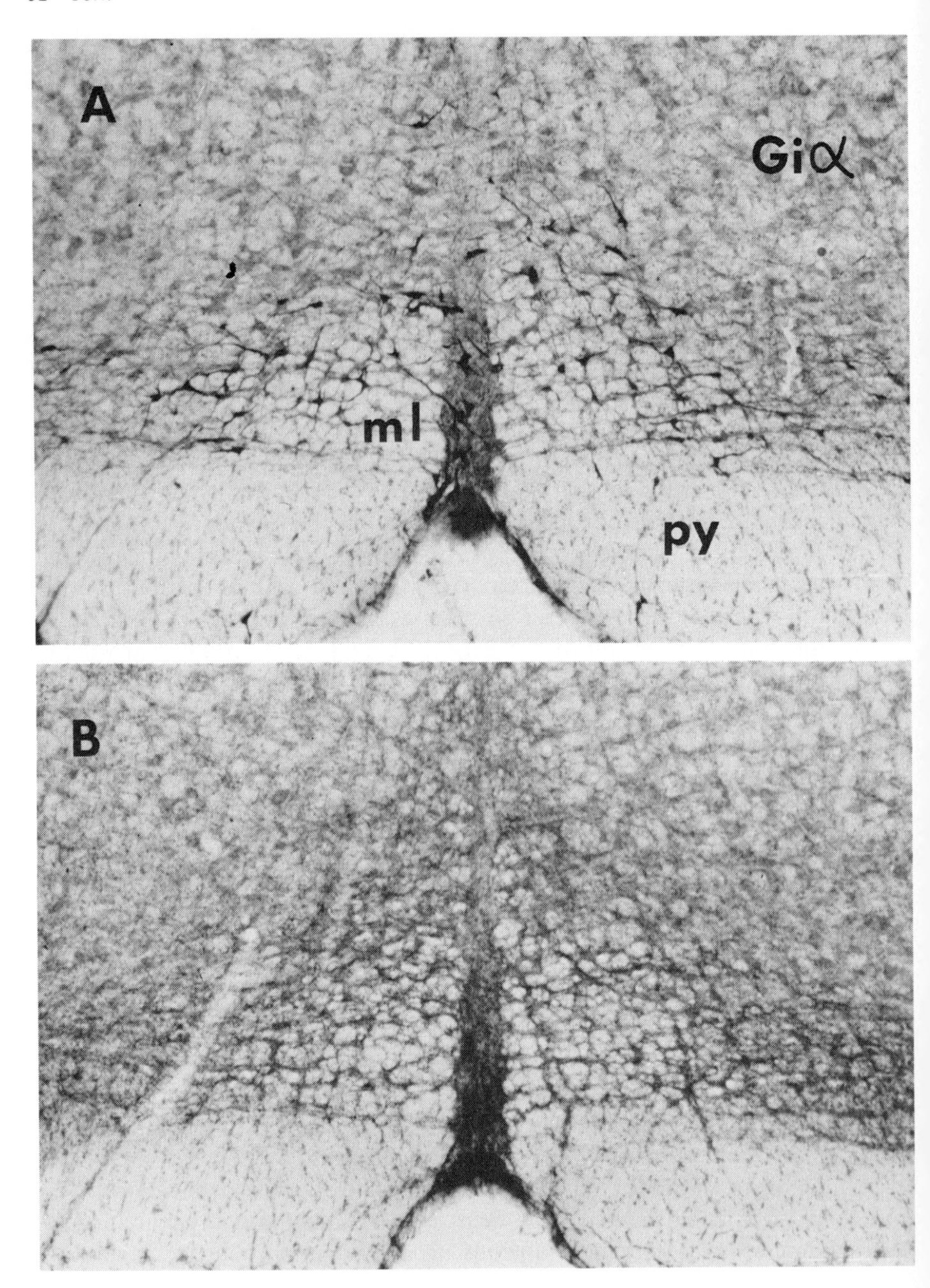

Fig. 3: The nucleus raphe magnus (RMg) demonstrated with two immunocytochemical reactions. In A, serotonin is demonstrated, and the 5-hydroxytryptamine positive cells are seen to occupy a wide triangular area, extending dorsal to the medial lemniscus (ml) and the pyramids (py) into the gigantocellular reticular nucleus pars α(Giα). In B (which is an adjacent section), the dense network of substance P axons surrounding the cells of the RMg can be seen. Magnification = × 60.

cluster of 5HT neurons were first seen by Dahlström and Fuxe (1964) and were designated as nucleus B3. Amongst the raphe nuclei, the RMg has received interest second only to the DR because of its strong projections to the spinal cord which, if stimulated, are thought to elicit analgesia (Basbaum and Fields, 1978, 1984).

The organization of the RMg is different from that of the raphe nuclei of the upper brain stem. In the RMg the cells are larger and are arranged in a rather loose network. The numerous serotonergic cells are oriented at right angles to the midline, extending their dendrites well into the reticular formation (Fig. 3A).

Several neuroactive compounds have been identified in the RMg. The presence of serotonin has been well known since the original work of Dahlström and Fuxe (1964) and has been confirmed in subsequent studies. Steinbusch and Nieuwenhuys (1983) localized the 5HT cells mainly in the ventrocaudal part of the nucleus and described them as a single class of large cells. The serotonergic cells have four to eight dendrites often placed at right angles to each other, with preferential horizontal and vertical orientations.

Other putative transmitters have also been observed in the RMg, and in some cases they coexist with 5HT. Thus the coexistence of serotonin with either substance P or thyrotropin releasing hormone (TRH) was reported by Johansson *et al.* (1981). The quantitative studies of these authors revealed that slightly more than 50% of the immunoreactive neurons in the RMg were serotonergic, while approximately a quarter of RMg neurons are substance P and TRH positive. Rather similar ratios were found for the other two medullary raphe nuclei, ROb and RPa. Apart from substance P and TRH, there is evidence for the existence of neurons containing other peptides in the RMg. Thus, enkephalin (Finley *et al.*, 1981a; Hökfelt *et al.*, 1979; Khachaturian *et al.*, 1983), somatostatin (Finley *et al.*, 1981b), and recently cholecystokinin (Kubota *et al.*, 1983; Mantyh and Hunt, 1984) have been identified immunohistochemically in RMg neurons. The demonstration of the various peptides in the RMg neurons is possible only with colchicine pretreatment, which results in an accumulation of cell products in the perikarya of neurons.

Substance P containing axon terminals surrounding the cell bodies and dendrites of neurons of the RMg have recently been identified. The density of the terminals is best appreciated in preparations taken from animals not treated with colchicine (Fig. 3B). The direct axosomatic contact of substance P and enkephalin axon terminals on the neurons of the RMg and other medullary 5HT neurons has been confirmed by Hancock (1984).

F. Raphe obscurus nucleus (Group B2)

The raphe obscurus nucleus (ROb) succeeds the RMg caudally, but its organization is different from that of the RMg. The cells in the ROb are arranged in two paramedian lines and do not extend laterally into the gigantocellular reticular nucleus. Ventrolaterally, the ROb is flanked by the inferior olive. Dorsal to the inferior olive, in the reticular formation adjacent to ROb, no 5HT cells are found. Caudally, the nucleus extends as far as the medulla itself: its most caudal part is displaced by the bulky pyramidal decussation. Dorsally, the ROb is attenuated and virtually disappears a few hundred micrometers from the floor of the fourth ventricle or the central canal. Ventrally, it is accompanied by the RPa, although the two nuclei are separated from each other for most of their rostrocaudal extent by the inferior olivary complex. The majority of the cells in the ROb are serotonergic and are of medium (18 μm–27 μm) to large (28 μm–36 μm) size (Steinbusch and Nieuwenhuys, 1983). The cells are arranged in two distinct paramedian rows with dendrites extending along the midline, into the neighboring reticular formation, and contralaterally (Fig. 4). Many dendrites appear to form dendritic bundles in the midline (Cummings and Felten, 1979).

Serotonin is not the only putative neurotransmitter found in the neurons of the ROb. Substance P and thyrotropin releasing hormone (TRH)-like immunoreactive neurons have been reported by Johansson *et al.* (1981) and cholecystokinin immunoreactive neurons were reported by Kubota *et al.* (1983) and Mantyh and Hunt (1984). Some of these neuroactive compounds were found to coexist with 5HT and with each other, although the number of SP, TRH and cholecystokinin immunoreactive cells is less than the 5HT positive cells, indicating that many 5HT neurons contain this transmitter only. The peptides have not been demonstrated in the cell bodies of the ROb without colchicine pretreatment. Substance P immunoreactivity in the ROb of untreated animals can be visualized only in nerve terminals, many of which surround the somata and principal dendrites of 5HT containing neurons.

G. Raphe pallidus nucleus (Group B1, postpyramidal nucleus of the raphe)

The raphe pallidus nucleus (RPa) is a dense, rod like nucleus found near the ventral surface of the medulla interposed between the two pyramids. Its rostrocaudal extent is approximately the same as that of the ROb, although its rostral pole reaches the caudal levels of the RMg. For most of its length, the RPa is separated from the ROb by fibers of the medial lemniscus and the olivocerebellar tract as well as by the inferior olivary nucleus. Some dorsoventrally oriented dendrites of 5HT neurons in the ROb extend along the midline to the RPa. In the cat, Taber (1961) distinguished a ventral and dorsal part of the RPa; the ventral part corresponds to the description of the entire nucleus given above, and the dorsal part to a group of cells adjoining the caudal end of the RMg. In the rat, the cells located more dorsally are better considered part of the caudal RMg and rostral ROb because they maintain continuity with those groups of cells. Some cells of the RPa extend their dendrites towards the surface, and a number of somata can be observed on the surface of the pyramids.

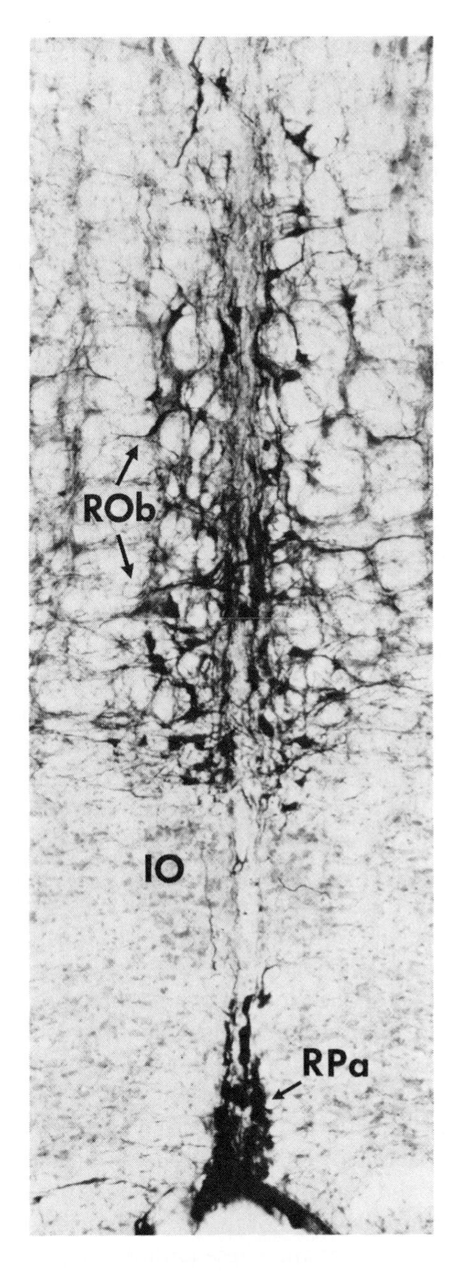

Fig. 4: The nucleus raphe obscurus (ROb) and pallidus (RPa), demonstrated with serotonin immunocytochemistry. ROb has large, multipolar neurons with dendrites extending into the midline region and the adjacent reticular formation, while the RPa contains small, densely packed cells. The two nuclei are separated from each other by the interior olive (IO). Magnification = × 145.

The RPa itself is very compact, with cell density similar to that found in the RD. It contains small (less than 18 μm), medium, and large (more than 36 μm) neurons. The serotonergic cells belong mainly to the medium sized and large neurons (Bowker *et al.*, 1982; Steinbusch and Nieuwenhuys, 1983), but other neurotransmitters have also been identified in this nucleus. Thus, TRH, substance P and enkephalin containing neurons have been reported by Hökfelt *et al.* (1977, 1979), Johansson *et al.* (1981) and Bowker *et al.* (1983). As in the neurons of the RMg and the ROb, peptides coexist with serotonin in some RPa cells.

III. SEROTONERGIC NEURONS OUTSIDE THE RAPHE SYSTEM

Although most serotonergic neurons are found within the raphe nuclei, one can find significant numbers of serotonergic neurons outside the raphe system of the brain stem and in the hypothalamus. Most of these brain stem cells were first observed by Dahlström and Fuxe (1964), but some smaller groups received attention later. The nonraphe 5HT neurons are described here in a craniocaudal order.

A. Serotonergic neurons in the hypothalamus

The existence of hypothalamic cell bodies that can accumulate intraventricularly injected 5HT was first described in Fuxe and Ungerstedt (1968). Their observations were confirmed and extended by Chan-Palay (1977) and Beaudet and Descarries (1979) using tritium labeled serotonin and autoradiography. The serotonin accumulating neurons found near the third ventricle in the dorsomedial hypothalamic nucleus were described by Frankfurt *et al.* (1981). According to this study, the 5HT immunoreactive cells of the hypothalamus are different from the raphe 5HT cells in several respects. First, they are a population of small neurons (average diameter, 9 μm); second, these cells have only been demonstrated after the application of monoamine oxidase inhibitors and tryptophan and they cannot be demonstrated in colchicine treated animals (Steinbusch and Nieuwenhuys, 1981); finally, they constitute a rather small group of cells, approximately 380 cells on each side. The connectivity of the hypothalamic 5HT neurons is not known and they may well be small local neurons of the hypothalamus, similar to the tuberoinfundibular dopaminergic neurons. Indeed, Kent and Sladek (1978) reported on the existence of serotonin histofluorescence in some perikarya of the arcuate nucleus following administration of pargyline and l-tryptophan. These authors suggest that the existence of these hypothalamic 5HT neurons explains why the serotonin content of the hypothalamus is only reduced, but not abolished, after surgical interruption of the brain stem afferents to it (Palkovits *et al.*, 1977).

B. The B9 Group and the serotonergic neurons of the mesencephalic reticular formation

Group B9 is a large cluster of serotonergic neurons found in the ventrolateral tegmentum of the midbrain, associated with the medial lemniscus. The serotonergic cells are medium sized and rather uniform. The rostrocaudal extent of the cluster is about 2.5 mm, but most cells are grouped in the rostral half of group B9, 2.2 mm–1.2 mm anterior to the interaural line. In coronal sections of the midbrain, the cluster appears crescent shaped, with a pointed end laterally, and a concave outline facing dorsally (Fig. 5B). The main characteristic of the B9 group is its close relationship to the fibers of the dorsal part of the medial lemniscus but several cells are scattered into neighboring ventral tegmental area and retrorubral fields.

C. Serotonergic neurons in the interpeduncular complex

It was only recently that the presence of serotonergic neurons in the interpeduncular complex (IP) was noticed. Steinbusch (1981) was the first to report that in the dorsal and lateral parts of the IP serotonergic neurons can be found. These neurons were further characterized by Shinghaniyom *et al.* (1982) with the use of a histofluorescence technique. According to these authors, the 5HT containing neurons in the IP have an average diameter of about 13 μm and are found in the dorsal magnocellular part of the complex, corresponding to the apical nucleus of the interpeduncular complex (IPA) (Fig. 5A). The serotonergic cell cluster appearing in the apical part of the interpeduncular nucleus is continuous caudally with the MnR or the B8 group of Dahlström and Fuxe (1964).

D. Serotonergic neurons of the pontine reticular formation

Many authors have reported the occurrence of serotonergic neurons in several areas of the reticular formation and periventricular gray matter of the pons. Numerous serotonergic neurons are associated with the raphe system, lateral to the RMg (see RMg), while other cells are found mixed with the neurons of the locus coeruleus (LC) and in the vicinity of the dorsal tegmental nucleus (Steinbusch, 1981).

E. Serotonergic neurons of the medullary reticular formation

Most serotonergic neuronal somata are in areas of the reticular formation adjacent to the raphe system. A large group of horizontally arranged serotonergic neurons are found in the Giα just dorsal and dorsolateral to the

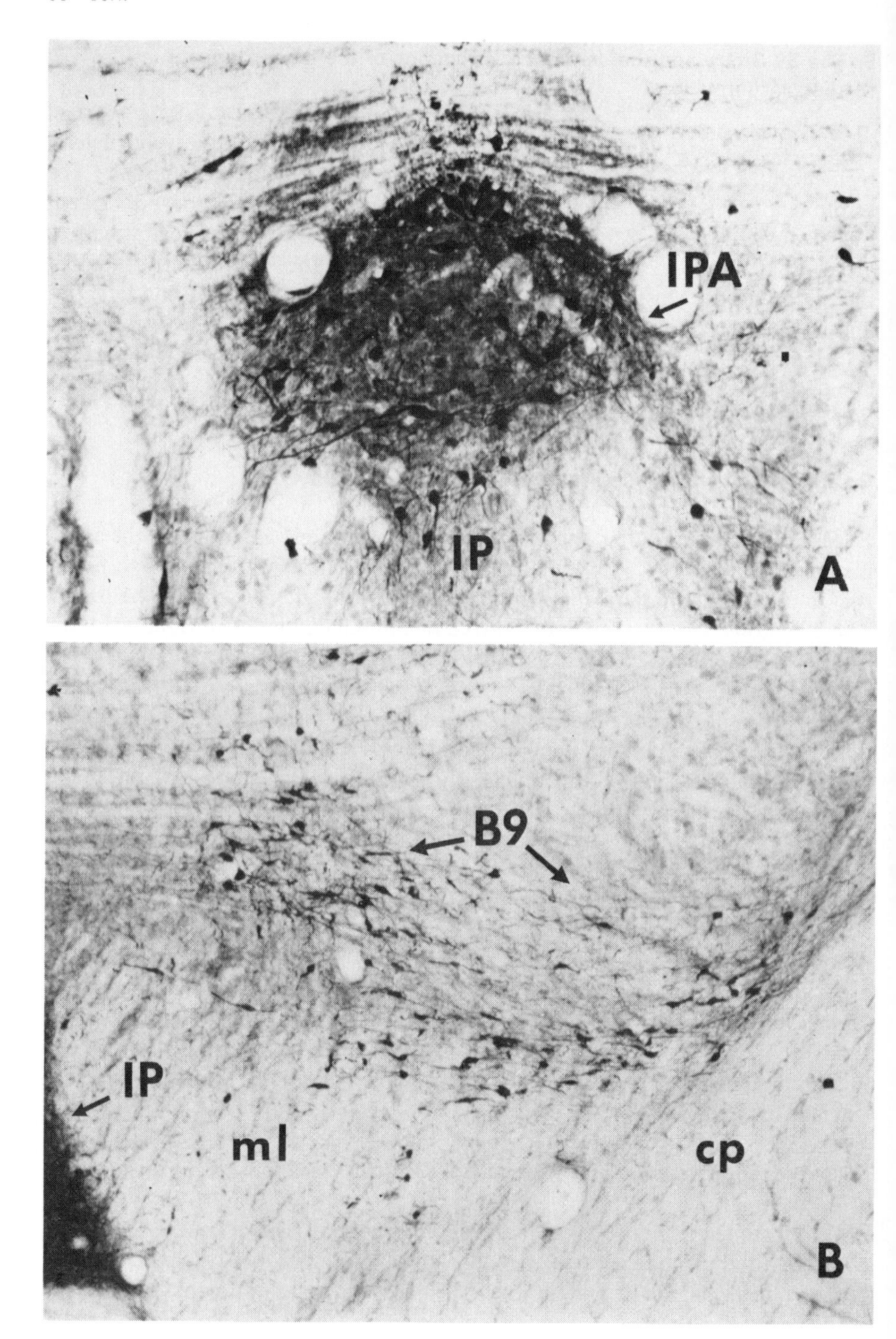
IPA
IP
A
B9
IP
ml
cp
B

pyramids, along the RMg. These large, triangular cells, just like the cells of the RMg itself, belong to group B3 of Dahlström and Fuxe (1964), but more caudally there is evidence for group B1 to extend laterally as well. There is a conspicuous accumulation of serotonin containing neurons just at the surface of the medulla lateral to the pyramids. These cells have characteristic long dendrites that run parallel to the surface and shorter ones that run at right angles to the surface. This spectacular accumulation of 5HT neurons has been described using autoradiography (Chan-Palay, 1977), histofluorescence, and immunocytochemistry (Loewy and McKellar, 1981; Steinbusch, 1981).

F. Serotonergic neurons in the area postrema

The area postrema, a circumventricular organ, is found at the caudal end of the fourth ventricle. The presence in the area postrema of small serotonin containing somata along with neurons containing catecholamines was first reported by Fuxe and Owman (1965). Further studies confirmed the presence of serotonin containing neurons in the area postrema, but revealed little about their structure or connections. On the basis of immunocytochemical investigations it can be concluded that there is a single population of small (6 μm to 8μm) serotonergic cells scattered in all parts of the area postrema (Steinbusch, 1981).

IV. CONNECTIONS OF THE RAPHE NUCLEI

The afferent and efferent connections of the raphe system are complex since the relatively long chain of nuclei receives diverse afferent inputs from many regions of the brain and the projection systems of the nuclei are extensive. Judged from many biochemical and immunohistochemical studies, virtually all regions of the brain receive serotonergic axons, although the density of the serotonergic innervation in the different regions of the brain is very variable. The use of most modern neuroanatomic techniques was needed to unravel the complex hodology of the serotonergic pathways. Thus, the description presented here is a synthesis of data obtained from investigations using autoradiography of ^{3}H-5HT, axonal transport of HRP, labeled compounds such as ^{3}H-leucine, proline and serotonin, and immunohistochemical demonstration of serotonin.

Although the serotonergic projections are the best known efferent pathways emerging from the raphe system, there is substantial evidence that projections emerge from the nonserotonergic cells of the raphe as well. Only in some cases do we have data on the putative transmitters of these systems.

Fig. 5: Two non-raphe serotonergic cell groups. A: The small 5-hydroxytryptamine cells of the interpeduncular complex are concentrated in the apical subnucleus, IPA. Magnification = 190×. B: Group B9 of Dahlström and Fuxe (1964) is a crescent shaped cluster of medium sized multipolar neurons just dorsal to the medial lemniscus (ml) and dorsomedial to the cerebral peduncle (cp). IP: interpeduncular nucleus. Magnification = × 65.

Since virtually all areas of the brain receive raphe projections, the efferent projections of the raphe nuclei can be conveniently divided into three parts: the ascending projections, the brain stem, and descending projections.

A. Ascending projections of the dorsal raphe and median raphe nuclei

The ascending projections of the raphe nuclei travel via two main routes which meet each other at the level of the hypothalamus in the medial forebrain bundle (mfb) (Takagi *et al.*, 1980). The dorsal fiber system is found in the central gray matter embedded in a dense network of fine serotonergic fibers. Axons emerging from this system supply the superior and inferior colliculi, and the epithalamic region (especially the subcommissural organ) before turning ventrally to join the second group of ascending fibers which travel in the mfb (Azmitia and Segal, 1978; Bobillier *et al.*, 1975, 1979; Moore *et al.*, 1978; Nieuwenhuys *et al.*, 1982; Parent *et al.*, 1981). The main target areas of the ascending projections are in the olfactory bulb, hypothalamus, septal area, thalamus, caudoputamen, hippocampal region, and cerebral cortex. The detailed morphology of the raphe projections to these areas has been studied by autoradiography, immunocytochemistry, and retrograde tracing.

The existence of major ascending projections from the dorsal raphe nucleus was first suggested by Brodal *et al.* (1960a) on the basis of fiber degeneration studies following lesions in the DR. Later, histofluorescence studies by Dahlström and Fuxe (1964) and Ungerstedt (1971) also described a major ascending indoleaminergic tract emerging from the DR and other raphe nuclei. However, a more exact description of the ascending pathways followed the introduction of the axonal tracing techniques, transmitter specific autoradiography, and immunocytochemistry. Using ^{3}H-labeled amino acids as tracers, Conrad *et al.* (1974), Bobillier *et al.* (1975, 1979), and Taber-Pierce *et al.* (1976) demonstrated the route and termination of the DR projections in the forebrain. Extensive terminal fields were seen in the midbrain ventral tegmental area, central gray matter, lateral hypothalamus, habenular region, lateral geniculate nucleus, and more rostrally, the septal area and basal forebrain. Labeling was also observed in the caudate nucleus, amygdala, cerebral cortex, and hippocampal region. In fact, all the major fiber bundles in the forebrain were found to contain axons of raphe origin. Thus, labeling could be found in the fasciculus retroflexus, mammillothalamic tract, fornix, stria medullaris, stria terminalis, and cingulum. All these fiber systems receive axons of raphe origin through the mfb. Also using the autoradiographic technique, Azmitia and Segal (1978) established that some brain areas receive raphe afferents via several routes. The hippocampus receives afferents from both the DR and the MnR, and the smaller DR projections reach the dentate gyrus through the entorhinal cortex and perforant path. Projections from the MnR to the hippocampal region are more numerous than from the DR; these MnR projections have been shown to reach the hippocampal region via two routes: the cingulum bundle, and the fornix (Azmitia and Segal, 1978). Additional projections reach the hippocampal region through the region of the amygdala, which in turn receives its raphe projection through the ansa peduncularis–ventral amygdaloid bundle (Moore, 1981; Moore and Halaris, 1975).

1. Raphe projections to the hypothalamus and epithalamus

The presence of medium concentrations of serotonin and of serotonergic axons in the hypothalamus has been shown by several authors (Fuxe, 1965; Parent *et al.*, 1981; Saavedra *et al.*, 1974; Steinbusch and Nieuwenhuys, 1981; Steinbusch *et al.*, 1982; Ungerstedt, 1971). The densest regions of 5HT innervation are the ventromedial part of the suprachiasmatic nucleus and the lateral part of the medial preoptic nucleus (Simerly *et al.*, 1984; Steinbusch, 1981) and the dorsal and ventral premammillary nuclei; the density of 5HT fibers in the rest of the hypothalamus, with the exception of the periventricular region, is relatively low. There is a remarkable paucity of serotonergic fibers in the magnocellular part of the paraventricular nucleus and supraoptic nuclei. There is a medium dense 5HT fiber plexus in the parvocellular part of the paraventricular nucleus and other regions of the hypothalamus (Ajika and Ochi, 1978; Sawchenko *et al.*, 1983). The origin of serotonergic fibers innervating the paraventricular nucleus is in the DR, MnR and the B9 group (Van de Kar and Lorens, 1979), however, these and the raphe obscurus and raphe magnus nuclei also send nonserotonergic projections to the hypothalamus (Sawchenko *et al.*, 1983).

In the infundibular region, the density of 5HT fibers is relatively low; the rostral part of the median eminence contains more 5HT axons than the caudal part. In the pituitary, a concentration of 5HT axons is seen in the pars nervosa and pars intermedia (Léránth *et al.*, 1983; Mezey *et al.*, 1984; Steinbusch and Nieuwenhuys, 1983).

In the anterior wall of the third ventricle, just ventral to the lamina terminalis, a dense serotonergic innervation of the vascular organ is found (Bosler, 1978; Moore, 1977). The subcommissural organ, which lies at the rostral end of the cerebral aqueduct in the epithalamic region, is densely innervated by serotonergic fibers originating mainly from the DR and MnR. These axons form synaptic connections with the ependymal cells of the subcommissural organ presumably influencing secretory activity (Bouchoud and Arluison, 1977; Calas *et al.*, 1978; Leger *et al.*, 1983; Møllgard and Wiklund, 1979).

2. Raphe projections to the thalamus

The thalamic projections of the raphe system are differentially organized. Using histochemical techniques which demonstrate serotonin fibers, a dramatic map of the thalamic nuclei can be obtained. Recently, Cropper *et al.* (1984) reported on the thalamic distribution of 5HT fibers and extended earlier studies on the serotonergic innervation of the thalamus (Anden *et al.*, 1966; Chan-Palay, 1977; Fuxe, 1965; Parent *et al.*, 1981). They established that the most densely innervated regions in the dorsal thalamus were the intralaminar nuclei, the periventricular nuclei and the anteroventral nucleus, while in the ventral thalamus the most prominently labeled nuclei were the reuniens nucleus and the rhomboid nucleus. In the metathalamus, a remarkably dense network of 5HT fibers can be seen in the ventral lateral geniculate nucleus (Parent *et al.*, 1981), especially in its lateral part and in the intrageniculate leaflet (Mantyh and Kemp, 1983).

It is important to emphasize that the demonstration of 5HT fiber systems in the thalamus does not reveal all raphe projections; neither does it reveal the precise origin of the projections. Anterograde and retrograde tracing studies demonstrate that the raphe projections to the thalamus are differential. Thus, the dorsal raphe nucleus projects strongly to the pretectal area, the intralaminar nuclei next to the fasciculus retroflexus (especially the parafascicular nucleus), the rhomboid nucleus, the mediodorsal thalamic nucleus and the subthalamic nucleus. There is also a significant projection to the lateral geniculate nucleus (Lüth *et al.*, 1977; Mackay-Sim *et al.*, 1983; Pasquier and Villar, 1982) and this projection has been shown to inhibit the transmission of visual impulses to the cortex (Yoshida *et al.*, 1984).

The median raphe nucleus does not send significant projections to the intralaminar nuclei. Rather, the strongest MnR input reaches the ventrobasal complex, the medial and lateral geniculate nuclei, the reticular nucleus, and the medial thalamic nuclei, particularly the mediodorsal nucleus; the subthalamic nucleus receives an input similar to the one from the DR (Azmitia and Segal, 1978; Conrad *et al.*, 1974; Moore *et al.*, 1978; Peschanski and Besson, 1971).

There is a less substantial, but still significant, diencephalic projection from the RMg (Peschanski and Besson, 1971; Takagi *et al.*, 1981a). This projection is strongest to the anterior intralaminar regions, gelatinosus nucleus, reuniens nucleus and pretectal area. The thalamic relay nuclei do not receive any RMg projections.

3. *Projections of the raphe nuclei to the basal telencephalon*

The 5HT projections to the caudoputamen and septal area arise from medial and rostral regions of the DR and travel through the dorsal ascending pathway in the dorsolateral part of the mfb. The caudoputamen receives a dense, mainly ipsilateral projection from the DR, but only a sparse one from the MnR (Jacobs *et al.*, 1978; Miller *et al.*, 1975). About one third of the serotonergic neurons of the DR project to the caudoputamen; in addition there are nonserotonergic raphe neurons (8%) which project to the caudoputamen (Steinbusch *et al.*, 1980, 1981). There is a distinct laterality on the DR projections to the caudoputamen:

the raphe cells project ipsilaterally or contralaterally, but not bilaterally (Loughlin and Fallon, 1982; Van der Kooy and Hattori, 1980a). However, some neurons projecting to the caudoputamen send axon collaterals to the medial striatum and globus pallidus (Van der Kooy and Kuypers, 1979) and substantia nigra (Van der Kooy and Hattori, 1980b).

The septal area also receives a dense serotonergic input. Both the medial and lateral septal nuclei are innervated, but the ventral lateral septal nucleus (LSV) receives the densest supply. Further, in the dorsomedial and dorsal lateral septal nucleus the serotonergic varicosities tend to surround some nerve cells forming "baskets" around them (Köhler *et al.*, 1982; Gall and Moore, 1984; Parent *et al.*, 1981). In the medial septum, serotonin containing fibers are in contact with neurons containing gonadotropin releasing hormone (Jennes *et al.*, 1982a). Many 5HT fibers coursing through the septum subsequently enter the fornix and the stria terminalis. The majority of the ascending serotonergic and raphe axons enter the cingulum bundle passing rostral and dorsal to the corpus callosum; this fiber system supplies raphe afferents to the dorsal and posterior neocortex and hippocampus (Azmitia and Segal, 1978; Bobillier *et al.*, 1979; Moore *et al.*, 1978; Ungerstedt, 1971). Other fibers, or possibly axon collaterals move laterally through the ventral septum and diagonal band to the olfactory tubercle and olfactory bulb.

The major part of the serotonergic projection to the septum arises from the dorsal and median raphe nuclei (Swanson and Cowan, 1979). However, using a combination of 5HT immunocytochemistry and fluorescent retrograde tracing, Köhler *et al.* (1982) demonstrated that there were additional 5HT afferents from the reticular tegmental nucleus, RPn, and RMg. In each raphe nucleus there were also nonserotonergic neurons projecting to the septum, but the transmitter of these cells is not yet known (see also Halaris *et al.*, 1976).

The 5HT innervation in the olfactory tubercle is particularly rich around the islands of Calleja (Parent *et al.*, 1981). The serotonergic innervation of the olfactory bulb is also very rich, and can be demonstrated with anterograde tracing techniques (Moore *et al.*, 1978), autoradiography of labeled 5HT (Halász *et al.*, 1978; Parent *et al.*, 1981), and immunohistochemistry (Steinbusch, 1981).

4. Serotonergic innervation of the cerebral cortex

The existence of a serotonergic innervation of the cortex was first demonstrated by histofluorescence (Dahlström and Fuxe, 1964; Ungerstedt, 1971), but the full extent of this innervation was recognized only later with the introduction of the retrograde tracing techniques. It is now established that there is a distinct projection from the dorsal and median raphe nuclei to the cortex (Arikuni and Ban, 1978; Bentivoglio *et al.*, 1978; Divac and Kosmal, 1978; Gerfen and Clavier, 1979; Kellar *et al.*, 1977; Pasquier and Reinoso-Suarez, 1977, 1978; Segal, 1977; Törk *et al.*, 1979). The histofluorescence and immunocytochemical investigations clearly established that the serotonergic projections from the dorsal and median raphe nuclei reach the cortex via several routes. Initially, all fibers run in the ventral serotonergic bundle which is part of the mfb, but then a group of them moves laterally from the lateral hypothalamus to the amygdala–entorhinal cortical area. Branches from this lateral bundle ascend in the external capsule to the lateral cortex. Axons that continue in the mfb enter the medial and posterior cortex through the supracallosal cingulum bundle; if they continue rostrally they reach the frontal cortex (Lidov and Molliver, 1982; Tohyama *et al.*, 1980; Ungerstedt, 1971).

Detailed characterization of the serotonin innervation of the rat cerebral cortex (Lidov *et al.*, 1980; Lidov and Molliver, 1982) has revealed that the density of 5HT axons is highest in the superficial layers of the cortex, although serotonin containing axons are found through the whole thickness of the cortex (see also Lamour *et al.*, 1983). Most of the 5HT axonal varicosities in the superficial layers are not involved in any synaptic contact (Beaudet and Descarries, 1976; Descarries *et al.*, 1975). In the lateral neocortex, the orientation of the 5HT fibers varies with the layers. In layer 1, most fibers run parallel to the pial surface while in layers 2 and 3, the axons run in parallel bundles normal to the surface. The fibers in layers 4 and 5 are extremely thin and highly tortuous while in layer 6, they, again, are arranged parallel to the pial surface. In the cingulate gyrus there is a higher density of 5HT axons in the anterior part, while in the posterior cingulate cortex (retrosplenial cortex) alternate layers of the cortex (1, 3 and 6) have a high density of serotonergic axons. The regional variation of the serotonergic innervation of the cortex is further supported by biochemical (Reader, 1980) and immunocytochemical studies (in the cat, observations by Mulligan and Törk, 1983; in the monkey, Takeuchi and Sano, 1983, 1984). In the monkey visual cortex, a complementary distribution of serotonergic and noradrenergic axons exists: for example, layer 4 of this cortex received extremely dense 5HT supply and is characterized by the relative paucity of noradrenergic axons (Morrison *et al.*, 1982). In many regions of the cat cortex, serotonin immunoreactive fibers densely innervate cortical neurons to the extent that the morphology of the postsynaptic cell is clearly discernible (Törk and Mulligan, 1984).

The serotonergic innervation of the hippocampal region is highly organized (Lidov *et al.*, 1980; Saavedra *et al.*, 1974). The entorhinal cortex contains the densest network of serotonergic axons among all cortical areas: these axons belong to serotonergic neurons located in the DR and MnR (Köhler *et al.*, 1980; Köhler, 1981). In the hippocampus, the serotonergic fibers are differentially distributed in the four cytoarchitectural areas and in the dentate gyrus. The layer of granule cells is almost completely devoid of 5HT innervation. A very dense band is seen in the stratum lacunosum moleculare of areas CA1 and CA2. A band of fibers is also present in the stratum oriens, in the region of the basal dendrites of the pyramidal cells. Area CA3 and the subicular region receive a similar but noticeably less dense serotonergic supply than areas CA1 and CA2 (Lidov *et al.*, 1980). It is generally agreed that it is the median and dorsal raphe nuclei that project to the hippocampus (Wyss *et al.*, 1979) and that the serotonergic fibers reach the area via three routes: (i) the cingulum bundle; (ii) the fornix–fimbria (mostly from the MnR and from the ventral part of the DR); and (iii) the lateral hypothalamus and perforant path (from the DR) (Azmitia, 1981; Azmitia and Segal, 1978; Köhler, 1981; Moore and Halaris, 1975; Pasquier and Reinoso-Suarez, 1977). It has been shown that some of the nonserotonergic neurons of the raphe also project to the hippocampal region. Both serotonergic and nonserotonergic projections may be crossed, but are not bilateral, and at least 10% of the raphe cells project to both the hippocampus and entorhinal area (Köhler and Steinbusch, 1982).

B. Brain stem and cerebellar projections

Histochemical methods specific to serotonin (histofluorescence, autoradiography, and immunohistochemistry) have revealed that there is a highly patterned 5HT innervation of brain stem structures. Such studies, unfortunately, do not indicate the specific nucleus of origin in the 5HT system. Thus, although we have good maps of serotonin distribution in the brain stem, the exact location of cells of origin is not always known. Studies using anterograde and retrograde tracing techniques have revealed some of the direct connections of the raphe nuclei with structures in the brain stem, but it is not known whether these connections are collaterals of ascending or descending connections of 5HT neurons of the raphe or are independent from them.

In the midbrain, the following regions demonstrate a dense network of serotonergic axons: the two most superficial layers of the superior colliculus, the CG, ventral tegmental area, interpeduncular complex and the substantia nigra (SN) (Baack *et al.*, 1983; Steinbusch, 1981; observations by Törk). Indeed, all these areas have been shown to receive projections from the raphe nuclei, especially from the DR and the MnR (Beitz, 1982a; Bobillier *et al.*, 1975, 1979; Bunney and Aghajanian, 1976; Conrad *et al.*, 1974; Dray *et al.*, 1976; Fibiger and Miller, 1977; Marchand *et al.*, 1980; Moore *et al.*, 1978; Phillipson, 1979; Simon *et al.*, 1979; Törk *et al.*, 1984; Van der Kooy and Hattori, 1980b). The axons projecting to the SN are, in part, collaterals of projections to the caudate-putamen (Van der Kooy and Hattori, 1980b; Van der Kooy and Kuypers, 1979). In the ventral tegmental area the serotonin containing fibers demonstrate a differential distribution. The 5HT fibers are most dense in the parabrachial pigmented and interfascicular nuclei, but they are also present in the rostral and caudal linear nuclei (observations by Halliday and Törk).

In the hindbrain, three regions contain dense innervation by serotonergic fibers: (i) the CG, especially the circumference of the dorsal tegmental nucleus and the locus coeruleus; (ii) the reticular formation, especially in the retrorubal field, parabrachial nuclei, ventral part of the parvocellular reticular nucleus of the medulla in the vicinity of the nucleus ambiguus, and solitary complex; and (iii) some of the cranial nerve nuclei.

The raphe projections to the locus coeruleus are well documented (Bobillier *et al.*, 1979; Conrad *et al.*, 1974; Leger and Descarries, 1978; Leger *et al.*, 1980; Pasquier *et al.*, 1977). After HRP injections into the locus coeruleus, retrogradely labeled cells were found in all nuclei of the raphe, although the majority of the cells were in the DR, the MnR and the RMg (Cedarbaum and Aghajanian, 1978; Morgane and Jacobs, 1979; Sakai *et al.*, 1977b). Indeed, the intensity of serotonergic innervation to this region of the brain stem exceeds that of most other densely innervated nuclei of the brain. Another region where high density of 5HT containing fibers is present is the nucleus of the solitary tract (Fuxe, 1965; Maley and Elde, 1982; Pickel *et al.*, 1984; Steinbusch, 1981). However, it appears that at least some of the 5HT axons in this region may be of peripheral origin as they may be the axons of serotonergic ganglion cells in the nodose ganglion of the vagus nerve (Gaudin-Chazal *et al.*, 1982).

The branchiomotor nuclei, especially the facial nucleus and the motor trigeminal nucleus, are so richly supplied with serotonergic afferents that distinct perisomatic baskets of the 5HT terminals can be recognized (Fuxe, 1965). Histochemical and electron microscopic examinations have demonstrated that the serotonergic axon terminals contact the somata and proximal dendrites of the motoneurons (Aghajanian and McCall, 1980; Schaffar *et al.*, 1984). However, there is some doubt with regard to the raphe origin of these serotonin containing axons since HRP injections into the motor trigeminal and facial nuclei fail to reveal retrogradely labeled neurons in any of the raphe nuclei (Travers and Norgren, 1983).

The nucleus of the spinal tract of the trigeminal nerve also contains many serotonergic axons, especially in its caudal part. The 5HT axons are mostly in the superficial layers of the nucleus where they form conventional axosomatic and axodendritic synapses. The origin of the serotonergic innervation is primarily the medullary raphe system and the gigantocellular nucleus pars α (Beitz, 1982c).

Another structure in the medulla that contains many 5HT fibers is the inferior olive. The density of the 5HT fibers is regionally organized; it is highest in the lateral part of the dorsal accessory olivary nucleus and the medial accessory olivary nucleus, and lowest in the principal olive (Bishop and Ho, 1984; King *et al.*, 1984; Takeuchi *et al.*, 1982a; Wiklund *et al.*, 1981a). As in the cortex, the majority of the axonal varicosities are without synaptic specializations; that is, they are nonjunctional (Wiklund *et al.*, 1981a).

The serotonin containing axons in the cerebellum reach the deep nuclei as well as the cortex. The highest density of 5HT axons in the cortex is in the granule layer. In the molecular layer the axons are arranged parallel to the pial surface, in a way similar to the parallel fibers. In some cases, serotonergic axons follow dendrites of Purkinje cells, suggesting synaptic contacts. Electron microscopic studies on the 5HT axons in the cerebellum revealed a relative paucity of synaptic contacts, similar to the findings in the cerebral cortex, hypothalamus and inferior olive (Chan-Palay, 1975; Hökfelt and Fuxe, 1969; Takeuchi *et al.*, 1982b).

C. Spinal projections

In recent years, the spinal projections of the raphe nuclei received much attention because of their possible involvement in the endogenous pain modulating system of the brain (Basbaum and Fields, 1978, 1984). However, the projection from the raphe magnus nucleus and adjacent medullary reticular formation to the dorsal horn of the spinal cord is only one of the descending systems from the raphe. The spinal cord receives a highly organized serotonergic input. Although serotonergic axons are present throughout the gray matter, the highest density of terminals is found in lamina 1 of Rexed in the dorsal horn, the intermediolateral column, lamina 10 surrounding the central canal, and the ventral horn (Bowker *et al.*, 1982; Hoffert *et al.*, 1983; Maxwell *et al.*, 1983; Steinbusch, 1981). From retrograde tracing studies it is now clear that with the exception of the median raphe, all raphe nuclei project to the spinal cord, but most of the descending projections arise from the groups B1 to B3 of Dahlström and Fuxe (1964); these groups correspond to the raphe magnus, raphe obscurus and raphe pallidus nuclei, and the pars α region of the paragigantocellular reticular nucleus (Bowker *et al.*, 1981a, b, 1982; Loewy, 1981, 1982; Loewy and McKellar, 1981; Satoh, 1979). The midbrain serotonergic groups B7 and B9 project only to the cervical cord, and the descending projections from the B9 cluster are much more significant.

The descending raphe–spinal fibers travel along two main routes. The projections from the RMg to the dorsal horn (and which are presumed to be involved in the endogenous pain control mechanisms) travel in the dorsolateral fasciculus. The projections from the raphe obscurus and pallidus, and to some extent from the raphe magnus as well, to the ventral and intermediolateral horns travel in the ventral funiculus (Basbaum *et al.*, 1978; Bobillier *et al.*, 1979; Leichnetz *et al.*, 1978; Martin *et al.*, 1979; Watkins *et al.*, 1980). Most RMg neurons projecting to the spinal cord dorsal horn also project to the caudal part of the nucleus of the spinal tract of the trigeminal nerve (Lovick and Robinson, 1983).

The combination of retrograde tracing with immunocytochemistry provided new evidence regarding the neurotransmitters found in the raphe neurons that project to the spinal cord. Thus, evidence is available that the raphe–spinal pathways contain, apart from serotonin, several other putative neurotransmitters, especially peptides. Projections containing substance P, thyrotropin releasing hormone, enkephalins and cholecystokinin have been positively identified (Bowker *et al.*, 1983; Helke *et al.*, 1982; Hökfelt *et al.*, 1979; Johansson *et al.*, 1981; Mantyh and Hunt, 1984).

D. Afferent connections of the raphe nuclei

The dorsal raphe nucleus (DR) receives descending and ascending projections. A major descending projection to the DR is from the habenular region (Aghajanian and Wang, 1977; Herkenham and Nauta, 1979; Neckers *et al.*, 1979; Pasquier *et al.*, 1976). Other descending afferents come from the prefrontal and anterior cingulate cortex, hypothalamus (Beckstead, 1979; Sakai *et al.*, 1977a; Wyss and Sripanidkulchai, 1984) and retina (Foote *et al.*, 1978). There are many brain stem afferents to the DR. These predominantly arise from the locus coeruleus, laterodorsal tegmental nucleus, parabrachial nuclei, pontine central gray, substantia nigra, medullary reticular formation and other raphe nuclei (Aghajanian and Wang, 1977; Sakai *et al.*, 1977a; Saper and Loewy, 1980). The projections to the median raphe nucleus are similar to those terminating in the dorsal raphe nucleus, although some additional projections exist. Thus, there is a strong projection from the interpeduncular nucleus, the dorsal tegmental region, the pontine central gray and the medullary catecholamine cell groups (Maciewicz *et al.*, 1981a, b; Massari *et al.*, 1979; Shibata and Suzuki, 1984).

The afferent input to the medullary raphe nuclei has been studied in detail, not only with morphologic but also with physiologic methods. In general, a significant part of the projections to raphe magnus, obscurus and pallidus arises from the reticular formation and the midbrain central gray (Gallager and Pert, 1978; Shah and Dostrovsky, 1980). The projections arising from the central gray come from all areas around the aqueduct, but from cells different from those innervating the neighboring reticular formation (Beitz *et al.*, 1983). Additional sources of midbrain afferents to the raphe magnus are the B9 group, DR and MnR (Beitz, 1982b; Carlton *et al.*, 1983; Fardin *et al.*, 1984; Yezierski *et al.*, 1982). The cells of the RMg receive a dense substance P (Fig. 3B) and neurotensin innervation. The origin of the neurotensin afferents is the central gray, solitary nucleus, parabrachial nuclei and cuneiform nucleus (Beitz, 1982b). In the ventral portion of the medullary raphe there is a dense noradrenergic plexus present; the source of these fibers is found in the nearby ventrolateral medulla, and in the catecholamine neuron groups A1 and A3 (Takagi *et al.*, 1981b).

ACKNOWLEDGEMENT

The original work presented here was supported by a grant from the National Health and Medical Research Council. The author is grateful to Mirella Fabbri for skillful technical assistance, and to Natalie Chabin and Sharlane Velasco for editorial help.

REFERENCES

Aghajanian, G. K. and Bloom, F. E. (1967). Localization of tritiated serotonin in rat brain by electron microscope radioautography. *J. Pharmacol. Exp. Ther.* 156, 23–30.

Aghajanian, G. K. and McCall, R. B. (1980). Serotonergic synaptic input to facial mononeurons: Localization by electron microscopic autoradiography. *Neurosci.* 5, 2155–2162.

Aghajanian, G. K. and Wang, R. Y. (1977). Habenular and other midbrain raphe afferents demonstrated by a retrograde tracing technique. *Brain Res.* 122, 229–242.

Ajika, K. and Ochi, J. (1978). Serotonergic projections to the suprachiasmatic nucleus and the median eminence of the rat: Identification by fluorescence and electron microscopy. *J. Anat.* 127, 563–576.

Anden, N. E., Dahlström, A., Fuxe, K., Larsson, K., Olson, L. and Understedt, U. (1966). Ascending monoamine neurons to the telencephalon and diencephalon. *Acta Physiol. Scand.* 67, 313–326.

Arikuni, T. and Ban, T. (1978). Subcortical afferents to the prefrontal cortex in rabbits. *Exp. Brain Res.* 32, 69–75.

Azmitia, E. C. (1981). Bilateral serotonergic projections to the dorsal hippocampus of the rat: Simultaneous localization of ^{3}H-5HT and HRP after retrograde transport. *J. Comp. Neurol.* 203, 737–743.

Azmitia, E. C. and Segal, M. (1978). An autoradiographic analysis of the differential ascending projections of the dorsal and median raphe nuclei in the rat. *J. Comp. Neurol.* 179, 641–668.

Baack, J. C., Dey, R. D. and Waterhouse, B. D. (1983). Organization of dorsal raphe projection neurons and serotonergic innervation of the rat tectum. *Soc. Neurosci. Abstr.* 9, 1149.

Basbaum, A. I. and Fields, H. L. (1978). Endogenous pain control mechanisms: Review and hypothesis. *Ann. Neurol.* 4, 451–462.

Basbaum, A. I. and Fields, H. L. (1984). Endogenous pain control systems: Brainstem spinal pathways and endorphin circuitry. *Ann. Rev. Neurosci.* 7, 309–338.

Basbaum, A. I., Clanton, C. H. and Fields, H. L. (1978). Three bulbospinal pathways from the rostral medulla of the cat: An autoradiographic study of pain modulating systems. *J. Comp. Neurol.* 178, 209–224.

Beaudet, A. and Descarries, L. (1976). Quantitative data on serotonin nerve terminals in adult rat neocortex. *Brain Res.* 111, 301–309.

Beaudet, A. and Descarries, L. (1979). Radiographic characterization of a serotonin-accumulating nerve cell group in adult rat hypothalamus. *Brain Res.* 160, 231–243.

Beckstead, R. M. (1979). An autoradiographic examination of corticocortical and subcortical projections of the mediodorsal-projection (prefrontal) cortex in the rat. *J. Comp. Neurol.* 184, 43–62.

Beitz, A. J. (1982a). The organization of afferent connections to the midbrain periaqueductal gray of the rat. *Neurosci.* 7, 133–159.

Beitz, A. J. (1982b). The sites or origin of brain stem neurotensin and serotonin projections to the rodent nucleus raphe magnus. *J. Neurosci.* 2, 829–842.

Beitz, A. J. (1982c). The nuclei of origin of brainstem serotonergic projections to the rodent spinal trigeminal nucleus. *Neurosci. Lett.* 32, 223–228.

Beitz, A. J., Mullett, M. A. and Weiner, L. L. (1983). The periaqueductal gray projections to the rat spinal trigeminal, raphe magnus, gigantocellular pars alpha and paragigantocellular nuclei arise from separate neurons. *Brain Res.* 288, 307–314.

Belin, M.-F., Aguera, M., Tappaz, M., McRae-Degueurce, A., Bobilliier P. and Pujol, J.-F. (1979). GABA-accumulating neurons in the nucleus raphe dorsalis and periaqueductal gray in the rat: A biochemical and radioautographic study. *Brain Res.* 170, 279–297.

Belin, M. F., Nanopoulos, D., Didier, M., Aguera, M., Steinbusch, H., Verhofstad, A., Maitre, M. and Pujol, J. -F. (1983). Immunohistochemical evidence for the presence of gamma-aminobutyric acid and serotonin in one nerve cell. A study on the raphe nuclei of the rat using antibodies to glutamate decarboxylase and serotonin. *Brain Res.* 275, 329–339.

Bentivoglio, M., Macchi, G., Rossini, P. and Tempesta, E. (1978). Brain stem neurons projecting to neocortex: A HRP study in the cat. *Exp. Brain Res.* 13, 489–498.

Berman, A. L. (1968). *The brain stem of the cat: A cytoarchitectonic atlas in stereotaxic coordinates.* University of Wisconsin Press, Madison, Wis.
Bishop, G. A. and Ho, R. H. (1984). Substance P and serotonin immunoreactivity in the rat inferior olive. *Brain Res. Bull.* 12, 105–114.
Bobillier, P., Petitjean, F., Salvert, D., Ligier, M. and Seguin, S. (1975). Differential projections of the nucleus raphe dorsalis and raphe centralis as revealed by autoradiography. *Brain Res.* 85, 205–210.
Bobillier, P., Seguin, S., Degueurce, A., Lewis, D. B. and Pufol, J. -F. (1979). The efferent connections of the nucleus raphe centralis superior in the rat as revealed by radioautography. *Brain Res.* 166, 1–8.
Bosler, O. (1978). Radioautographic identification of serotonin axon terminals in the rat organon vasculosum laminae terminalis. *Brain Res.* 150, 177–181.
Bouchoud, C. and Arluison, M. (1977). Serotoninergic innervation of ependymal cells in the rat subcommissural organ. A fluorescence, electron microscopic and radioautographic study. *Biol. Cell.* 30, 61–64.
Bowker, R. M., Steinbusch, H. W. M. and Coulter, J. D. (1981a). Serotonergic and peptidergic projections to the spinal cord demonstrated by a combined retrograde HRP histochemical and immunocytochemical staining method. *Brain Res.* 211, 412–417.
Bowker, R. M., Westlund, K. N. and Coulter, J. D. (1981b). Serotonergic projections to the spinal cord from the midbrain in the rat: An immunocytochemical and retrograde transport study. *Neurosci. Lett.* 24, 221–226.
Bowker, R. M., Westlund, K. N., Sullivan, M. C. and Coulter, J. D. (1982). Organization of descending serotonergic projections to the spinal cord. *Progr. Brain Res.* 57, 239–265.
Bowker, R. M., Westlund, K. N., Sullivan, M. C., Wilber, J. F. and Coulter, J. D. (1983). Descending serotonergic, peptidergic and cholinergic pathways from the raphe nuclei: A multiple transmitter complex. *Brain Res.* 288, 33–48.
Brodal, A., Taber, E. and Walberg, F. (1960a). The raphe nuclei of the brain stem in the cat. II. Efferent connections. *J. Comp. Neurol.* 114, 239–259.
Brodal, A., Taber, E. and Walberg, F. (1960b). The raphe nuclei of the brain stem in the cat. III. Afferent connections. *J. Comp. Neurol.* 114, 261–281.
Bunney, B. S. and Aghajanian, G. K. (1976). The precise localization of nigral afferents in the rat as determined by a retrograde tracing technique. *Brain Res.* 117, 423–435.
Calas, A., Bosler, O., Arluison, M. and Bouchaud, C. (1978). Serotonin as a neurohormone in circumventricular organs and supraependymal fibers. In D. E. Scott, G. P. Kozlowski and A. Weindl (Eds). *Brain-Endocrine Interaction III. Neural Hormones and Reproduction*, pp. 238–250. Karger, Bassel.
Carlton, S. M., Leichnetz, G. R., Young, E. C. and Mayer, D. J. (1983). Supramedullary afferents of the nucleus raphe magnus in the rat: A study using the transcannula HRP gel and autoradiographic techniques. *J. Comp. Neurol.* 214, 43–58.
Cedarbaum, J. M. and Aghajanian, G. K. (1978). Afferent projections to the rat locus coeruleus as determined by a retrograde tracing technique. *J. Comp. Neurol.* 178, 1–16.
Chan-Palay, V. (1975). Fine structure of labeled axons in the cerebellar cortex and nuclei of rodents and primates after intraventricular infusions with tritiated serotonin. *Anat. Embryol.* 148, 235–265.
Chan-Palay, V. (1977). Indoleamine neurons and their processes in the normal rat brain and in chronic diet-induced thiamine deficiency demonstrated by uptake of ^{3}H-serotonin. *J. Comp. Neurol.* 176, 467–494.
Chan-Palay, V., Jonsson, G. and Palay, S. L. (1978). Serotonin and substance P coexist in neurons of the rat's central nervous system. *Proc. Natl. Acad. Sci. USA* 75, 1582–1586.
Conrad, L. C. A., Leonard, C. M. and Pfaff, D. W. (1974). Connections of the median and dorsal raphe nuclei in the rat: An autoradiographic and degeneration study. *J. Comp. Neurol.* 156, 179–206.
Cropper, E. C., Eisenman, J. S. and Azmitia, E. C. (1984). An immunocytochemical study of the serotonergic innervation of the thalamus of the rat. *J. Comp. Neurol.* 224, 38–50.
Cummings, J. P. and Felten, D. L. (1979). A raphe dendrite bundle in the rabbit medulla. *J. Comp. Neurol.* 183, 1–24.

Dahlström, A. and Fuxe, K. (1964). Evidence for the existence of monoamine-containing neurons in the central nervous system. I. Demonstration of monoamines in cell bodies of brain stem neurons. *Acta Physiol. Scand. Suppl.* 232, 1-55.
Danner, H. and Pfister, C. (1980). Untersuchungen zur Zytoarchitektonik des Nucleus raphe dorsalis der Ratte. *J. Hirnforsch.* 21, 655-664.
Descarries, L., Beaudet, A. and Watkins, K. C. (1975). Serotonin nerve terminals in adult rat neocortex. *Brain Res.* 100, 536-588.
Descarries, L., Watkins, K. C., Garcia, S. and Beaudet, A. (1982). The serotonin neurons in nucleus raphe dorsalis of adult rat: A light and electron microscope radioautographic study. *J. Comp. Neurol.* 207, 239-254.
Diaz-Cintra, S., Cintra, L., Kemper, T., Resnick, O. and Morgane, P. J. (1981). Nucleus raphe dorsalis: A morphometric Golgi study in rats of three age groups. *Brain Res.* 207, 1-16.
Divac, I. and Kosmal, A. (1978). Subcortical projections to the prefrontal cortex in the rat as revealed by the horseradish peroxidase technique. *Neurosci.* 3, 785-796.
Dray, A., Gonye, T. J., Oakley, N. R. and Tanner, T. (1976). Evidence for the existence of a raphe projection to the substantia nigra in rat. *Brain Res.* 113, 45-57.
Fardin, V., Oliveras, J.-L. and Besson, J. M. (1984). Projections from the periaqueductal gray matter to the B_3 cellular area (nucleus raphe magnus and nucleus reticularis paragigantocellularis) as revealed by the retrograde transport of horseradish peroxidase in the rat. *J. Comp. Neurol.* 223, 483-500.
Felten, D. L. and Cummings, J. P. (1979). The raphe nuclei of the rabbit brain stem. *J. Comp. Neurol.* 187, 199-244.
Fibiger, H. C. and Miller, J. J. (1977). An anatomical and electrophysiological investigation of the serotonergic projection from the dorsal raphe nucleus to the substantia nigra in the rat. *Neurosci.* 2, 975-987.
Finley, J. C., Maderdrut, J. L. and Petrusz, P. (1981a). The immunocytochemical localization of enkephalin in the central nervous system of the rat. *J. Comp Neurol.* 198, 541-565.
Finley, J. C., Maderdrut, J. L., Roger, L. J. and Petrusz, P. (1981b). The immunocytochemical localization of somatostatin-containing neurons in the rat central nervous system. *Neurosci.* 6, 2173-2192.
Foote, W. E., Taber-Pierce, E. and Edwards, L. (1978). Evidence for a retinal projection to the midbrain raphe of the cat. *Brain Res.* 156, 135-140.
Frankfurt, M., Lauder, J. M. and Azmitia, E. C. (1981). The immunocytochemical localization of serotonergic neurons in the rat hypothalamus. *Neurosci. Lett.* 24, 227-232.
Fuxe, K. (1965). Evidence for the existence of monoamine neurons in the central nervous system—IV. Distribution of monoamine nerve terminals in the central nervous system. *Acta Physiol. Scand. Suppl.* 247, 37-85.
Fuxe, K. and Owman, Ch. (1965). Cellular localization of monoamines in the area postrema of certain mammals. *J. Comp. Neurol.* 125, 337-353.
Fuxe, K. and Ungerstedt, U. (1968). Histochemical studies on the distribution of catecholamine and 5-hydroxytryptamine after intraventricular injections. *Histochemie* 13, 16-28.
Gall, C. and Moore, R. Y. (1984). Distribution of enkephalin, substance P, tyrosine hydroxylase and 5-dydroxytryptamine immunoreactivity in the septal region of the rat. *J. Comp. Neurol.* 225, 212-227.
Gallagher, D. W. and Pert, A. (1978). Afferents to brain stem nuclei (brain stem raphe, nucleus reticularis pontis caudalis and nucleus gigantocellularis) in the rat as demonstrated by microiontophoretically applied horseradish peroxidase. *Brain Res.* 144, 257-275.
Gamrani, H. and Calas, A. (1980). Cytochemical, stereological and radioautographic studies of rat raphe neurons. *Mikroskopie* 36, 1-11.
Gamrani, H., Calas, A., Belin, M. F., Agnara, M. and Pujol, J.-F. (1979). High resolution radioautographic identification of (^{3}H)GABA labeled neurons in the rat nucleus raphe dorsalis. *Neurosci. Lett.* 15, 43-48.
Gaudin-Chazal, G., Seyfritz, N., Araneda, S., Vigier, D. and Puizillout, J. -J. (1982). Selective retrograde transport of ^{3}H-serotonin in vagal afferents. *Brain Res. Bull.* 8, 503-509.
Gerfen, C. R. and Clavier, R. M. (1979). Neural inputs to the prefrontal agranular insular cortex in the rat: Horseradish peroxidase study. *Brain Res. Bull.* 4, 347-353.

Glazer, E. J., Steinbusch, H., Verhofstad, A. and Basbaum, A. I. (1981). Serotonergic neurons of the cat nucleus raphe dorsalis and paragigantocellularis contain enkephalin. *J. Physiol. (Paris)* 77, 241–245.

Grzanna, R. and Molliver, M. E. (1980). Cytoarchitecture and dendritic morphology of central noradrenergic neurons. In J. A. Hobson and M. A. B. Brazier (Eds). *The reticular formation revisited*, pp. 83–97. Raven Press, New York.

Grzanna, R., Molliver, M. E. and Coyle, J. T. (1978). Visualization of central noradrenergic neurons in thick sections by the unlabeled antibody method: A transmitter-specific Golgi image. *Proc. Natl. Acad. Sci.* 75, 2502–2506.

Halaris, A. E., Jones, B. E. and Moore, R. Y. (1976). Axonal transport in serotonin neurons of the midbrain raphe. *Brain Res.* 107, 554–574.

Halasz, N., Ljungdahl, A. and Hökfelt, T. (1978). Transmitter histochemistry of the rat olfactory bulb—II. Fluorescence histochemical, autoradiographic and electron microscopic localization of monoamines. *Brain Res.* 154, 253–271.

Hancock, M. B. (1984). Visualization of peptide-immunoreactive processes on serotonin-immunoreactive cells using two-color immunoperoxidase staining. *J. Histochem. Cytochem.* 32, 311–314.

Helke, C. J., Neil, J. J., Massari, V. J. and Loewy, A. D. (1982). Substance P neurons project from the ventral medulla to the intermediolateral cell column and ventral horn in the rat. *Brain Res.* 243, 147–152.

Herkenham, M. and Nauta, W. J. H. (1979). Efferent connections of the habenular nuclei in the rat. *J. Comp. Neurol.* 187, 19–48.

Hoffert, M. J., Miletic, V., Ruda, M. A. and Dubner, R. (1983). Immunocytochemical identification of serotonin axonal contacts on characterized neurons in laminae I and II of the cat dorsal horn. *Brain Res.* 267, 361–364.

Hökfelt, T. and Fuxe, K. (1969). Cerebellar monamine nerve terminals, a new type of afferent fibers in the cortex cerebelli. *Exp. Brain Res.* 9, 63–72.

Hökfelt, T., Elde, R., Johansson, O., Terenius, L. and Stein, L. (1977). The distribution of enkephalin-immunoreactive cell bodies in the central nervous system. *Neurosci. Lett.* 5, 25–31.

Hökfelt, T., Ljungdahl, A., Steinbusch, H., Verhofstad, A., Nilsson, G., Brodin, E., Pernow, B. and Goldstein, M. (1978). Immunohistochemical evidence of substance P-like immunoreactivity in some 5-hydroxytryptamine-containing neurons in the rat central nervous system. *Neurosci.* 3, 517–538.

Hökfelt, T., Terenius, L., Kuypers, H. G. J. M. and Dann, O. (1979). Evidence for enkephalin immunoreactive neurons in the medulla oblongata projecting to spinal cord. *Neurosci. Lett.* 14, 55–60.

Hölzel, B. and Pfister, C. (1981). Untersuchungen zur Topographie and Zytoarchitektonik der Raphe-Kerne der Ratte. *J. Hirnforsch.* 22, 697–708.

Hölzel, B. and Pfister, C. (1983). Zur Neurotypisierung des Nucleus centralis superior (Nucleus raphe medianus) der Ratte. *J. Hirnforsch.* 24, 593–598.

Innis, R. B., Correa, F. M. A., Uhl, G. R., Schneider, B. and Snyder, S. H. (1979). Cholecystokinin octapeptide-like immunoreactivity: Histochemical localization in rat brain. *Proc. Nat. Acad. Sci. USA* 76, 521–525.

Jacobs, B. L., Foote, S. L. and Bloom, F. E. (1978). Differential projections of neurons within the dorsal raphe nucleus of the rat: A horseradish peroxidase (HRP) study. *Brain Res.* 147, 149–153.

Jennes, L., Beckman, W. C., Stumpf, W. E. and Grzanna, R. (1982a). Anatomical relationships of serotoninergic and noradrenalinergic projections in the GnRH system in septum and hypothalamus. *Exp. Brain Res.* 46, 331–338.

Jennes, L., Stumpf, W. E. and Kalivas, P. W. (1982b). Neurotensin: Topographic distribution in rat brain by immunohistochemistry. *J. Comp. Neurol.* 210, 211–224.

Johansson, O., Hökfelt, T., Pernow, B., Jeffcoate, S. L., White, N., Steinbusch, H. W. M., Verhofstad, A. A. J., Emson, P. C. and Spindel, E. (1981). Immunohistochemical support for three putative neurotransmitters in one neuron: Coexistence of 5-hydroxytryptamine, substance P and thyrotropin releasing hormone-like immunoreactivity in medullary neurons projecting to the spinal cord. *Neurosci.* 6, 1857–1881.

Kellar, K. J., Brown, P. A., Madrid, J., Bernstein, M., Vernikos-Danellis, J. and Mehler, W. R. (1977). Origins of serotonin innervation of forebrain structures. *Exp. Neurol.* 56, 52–62.

Kent, D. L. and Sladek, J. R. (1978). Histochemical, pharmacological and microspectrofluorometric analysis of new sites of serotonin localization in the rat hypothalamus. *J. Comp. Neurol.* 180, 221-236.

King, J. S., Ho, R. H. and Burry, R. W. (1984). The distribution and synaptic organization of serotoninergic elements in the inferior olivary complex of the opossum. *J. Comp. Neurol.* 227, 357-368.

Khachaturian, H., Lewis, M. E. and Watson, S. J. (1983). Enkephalin systems in the diencephalon and brainstem of the rat. *J. Comp. Neurol.* 220, 318-320.

Köhler, C. (1981). On the serotonergic innervation of the hippocampal region. An analysis employing immunohistochemistry and retrograde fluorescent tracing in the rat brain. In S. L. Palay and V. Chan-Palay (Eds). *Cytochemical methods in neuroanatomy.* Alan R. Liss, New York.

Köhler, C. and Steinbusch, H. W. M. (1982). Identification of serotonin and non-serotonin-containing neurons of the midbrain raphe projecting to the entorhinal area and the hippocampal formation. A combined immunohistochemical and fluorescent retrograde tracing study in the rat brain. *Neurosci.* 7, 951-975.

Köhler, C., Chan-Palay, V., Haglund, L. and Steinbusch, H. (1980). Immunohistochemical localization of serotonin nerve terminals in the lateral entorhinal cortex of the rat. Demonstration of two separate patterns of innervation from the midbrain raphe. *Anat. Embryol.* 160, 121-129.

Köhler, C., Chan-Palay, V. and Steinbusch, H. (1981). The distribution and orientation of serotonin fibers in the entorhinal and other retrohippocampal areas. An immunohistochemical analysis with anti-serotonin antibodies in the rat brain. *Anat. Embryol.* 161, 237-264.

Köhler, C., Chan-Palay, V. and Steinbusch, H. (1982). The distribution and origin of serotonin-containing fibers in the septal area: A combined immunohistochemical and fluorescent retrograde tracing study in the rat. *J. Comp. Neurol.* 209, 91-111.

Kubota, Y., Inagaki, S., Shiosaka, S., Cho, S. J., Tateishi, K., Hashimura, E., Hamaoka, T. and Tohyama, M. (1983). The distribution of cholecystokinin octapeptide-like structures in the lower brain stem of the rat: An immunohistochemical analysis. *Neurosci.* 9, 587-604.

Lamour, Y., Rivot, J. P., Pointis, D. and Ory-Lavollee, L. (1983). Laminar distribution of serotonergic innervation in rat somatosensory cortex, as determined by in vivo electrochemical detection. *Brain Res.* 259, 163-166.

Leger, L. and Descarries, L. (1978). Serotonergic nerve terminals in the locus coeruleus of adult rat. A radioautographic study. *Brain Res.* 145, 1-13.

Leger, L., McRae-Degueurce, A. and Pujol, J. -F. (1980). Origine de l'innervation sérotonergique du locus coeruleus chez le rat. *C.R. Acad. Sci. Paris* 290, 807-810.

Leger, L., Degueurce, A., Lundberg, J. J., Pujol, J. F. and Møllgard, K. (1983). Origin and influence of serotoninergic innervation of the subcommissural organ in the rat. *Neurosci.* 10, 411-423.

Leichnetz, G. R., Watkins, L., Griffin, G., Murfin, R. and Mayer, D. J. (1978). The projections from nucleus raphe magnus and other brainstem nuclei to the spinal cord in the rat: A study using the HRP-blue reaction. *Neurosci. Lett.* 8, 119-124.

Léránth, Cs., Palkovits, M. and Krieger, D. T. (1983). Serotonin immunoreactive nerve fibres and terminals in the rat pituitary—light and electron microscopic studies. *Neurosci.* 9, 289-296.

Lidov, H. G. W. and Molliver, M. E. (1982). Immunohistochemical study of the development of serotonergic neurons in the rat CNS. *Brain Res. Bull.* 9, 559-604.

Lidov, H. G. W., Grzanna, R. and Molliver, M. E. (1980). The serotonin innervation of the cerebral cortex in the rat—an immunohistochemical analysis. *Neurosci.* 5, 207-227.

Loewy, A. D. (1981). Raphe pallidus and raphe obscurus projections to the intermediolateral cell column in the rat. *Brain Res.* 222, 129-133.

Loewy, A. D. (1982). Descending pathways to the sympathetic preganglionic neurons. *Prog. Brain Res.* 57, 267-277.

Loewy, A. D. and McKellar, S. (1981). Serotonergic projections from the ventral medulla to the intermediolateral column in the rat. *Brain Res.* 211, 146-152.

Loughlin, S. E. and Fallon, J. H. (1982). Mesostriatal projections from ventral tegmentum and dorsal raphe: cells project ipsilaterally or contralaterally but not bilaterally. *Neurosci. Lett.* 32, 11-16.

Lovick, T. A. and Robinson, J. P. (1983). Bulbar raphe neurons with projections to the trigeminal nucleus caudalis and the lumbar cord in the rat: A fluorescence double-labelling study. *Exp. Brain Res.* 50, 299-308.

Lüth, von H. -J., Seidel, J. and Schober, W. (1977). Morphologisch-histochemische Characterisierung der Verbindungen des Corpus geniculatum laterale pars dorsalis (Cgld), mit dem Hirnstamm bei Albinoratten. *Acta Histochem.* 60, 91–102.
Maciewicz, R., Foote, W. E. and Bry, J. (1981a). Excitatory projection from the interpeduncular nucleus to central superior raphe neurons. *Brain Res.* 225, 179–183.
Maciewicz, R., Taber-Pierce, E., Ronner, S. and Foote, W. E. (1981b). Afferents to the central superior raphe nucleus in the cat. *Brain Res.* 216, 414–421.
Mackay-Sim, A., Sefton, A. J. and Martin, P. R. (1983). Subcortical projections to lateral geniculate and thalamic reticular nuclei in the hooded rat. *J. Comp. Neurol.* 213, 24–35.
Maley, B. and Elde, R. (1982). Immunohistochemical localization of putative neurotransmitters within the feline nucleus tractus solitarii. *Neurosci.* 7, 2469–2490.
Mantyh, P. W. and Hunt, S. P. (1984). Evidence for cholecystokinin-like immunoreactive neurons in the rat medulla oblongata which project to the spinal cord. *Brain Res.* 291, 49–54.
Mantyh, P. W. and Kemp, J. A. (1983). The distribution of putative neurotransmitters in the lateral geniculate nucleus of the rat. *Brain Res.* 288, 344–348.
Marchand, E. R., Riley, J. N., Moore, R. Y. (1980). Interpeduncular nucleus afferents in the rat. *Brain Res.* 193, 339–352.
Martin, G. F., Humbertson, Jr., A. O., Laxson, L. C., Linauts, M., Panneton, M. and Tschimadia, I. (1979). Spinal projections from the mesencephalic and pontine reticular formation in the North American opossum: A study using axonal transport techniques. *J. Comp. Neurol.* 187, 373–400.
Massari, V. J., Tizabi, Y., Jacobowitz, D. M. (1979). Potential noradrenergic regulation of serotonergic neurons in the median raphe nucleus. *Exp. Brain Res.* 162, 45–54.
Maxwell, D. J., Léránth, Cs. and Verhofstad, A. A. J. (1983). Fine structure of serotonin-containing axons in the marginal zone of the rat spinal cord. *Brain Res.* 266, 252–259.
Mezey, E., Léránth, Cs., Brownstein, M. J., Friedman, E., Krieger, D. T. and Palkovits, M. (1984). On the origin of the serotonergic input to the intermediate lobe of the rat pituitary. *Brain Res.* 294, 231–238.
Miller, J. J., Richardson, T. L., Fibiger, H. C. and McLennan, H. (1975). Anatomical and electrophysiological identification of a projection from the mesencephalic raphe to the caudate-putamen in the rat. *Brain Res.* 97, 133–138.
Minagawa, H., Shiosaka, S., Inagaki, S., Sakanaka, M., Takatsuki, K., Ishimoto, I., Senba, E., Kawait, Y., Hara, Y., Matsuzaki, T. and Tohyama, M. (1983). Ontogeny of neurotensin-containing neuron system of the rat: Immunohistochemical analysis—II. Lower brain stem. *Neurosci.* 8, 467–486.
Møllgard, K. and Wiklund, L. (1979). Serotoninergic synapses on ependymal and hypendymal cells of the rat subcommissural organ. *J. Neurocytol.* 8, 445–467.
Moore, R. Y. (1977). Organon vasculosum laminae terminalis: Innervation by serotonergic neurons of the midbrain raphe. *Neurosci. Lett.* 5, 297–302.
Moore, R. Y. (1981). The anatomy of central serotonin neuron systems in the rat brain. In B. L. Jacobs and A. Gelperin (Eds). *Serotonin neurotransmission and behavior*, pp. 35–71. MIT Press, Cambridge, Mass.
Moore, R. Y. and Halaris, A. E. (1975). Hippocampal innervation by serotonin neurons of the midbrain raphe in the rat. *J. Comp. Neurol.* 164, 171–184.
Moore, R. Y., Halaris, A. E. and Jones, B. E. (1978). Serotonin neurons of the midbrain raphe: Ascending projections. *J. Comp. Neurol.* 180, 417–438.
Morgane, P. J. and Jacobs, M. S. (1979). Raphe projections to the locus coeruleus in the rat. *Brain Res. Bull.* 4, 519–534.
Morrison, J. H., Foote, S. L., Molliver, M. E., Bloom, F. E. and Lidov, H. G. W. (1982). Noradrenergic and serotonergic fibers innervate complementary layers in monkey primary visual cortex: An immunohistochemical study. *Proc. Nat. Acad. Sci. USA* 79, 2401–2405.
Moss, M. S., Glazer, E. J. and Basbaum, A. I. (1983). The peptidergic organization of the cat periaqueductal gray. I. The distribution of immunoreactive enkephalin-containing neurons and terminals. *J. Neurosci.* 3, 603–616.
Mulligan, K. A. and Törk, I. (1983). Immunocytochemical demonstration of serotonergic innervation of the cat neocortex. *Proc. Int. Union Physiol. Sci.* 15, 445.

Nanopoulos, D., Belin, M. F., Maitre, M., Vincendon, G. and Pujol, J.-F. (1982). Immunocytochemical evidence for the existence of GABAergic neurons in the nucleus raphe dorsalis. Possible existence of neurons containing serotonin and GABA. *Brain Res.* 232, 375–389.

Neckers, L. M., Schwartz, J. P., Wyatt, R. J. and Speciale, S. G. (1979). Substance P afferents from habenula innervate and dorsal raphe nucleus. *Exp. Brain Res.* 37, 619–623.

Nieuwenhuys, R., Geeraedts, L. M. G. and Veening, J. G. (1982). The medial forebrain bundle of the rat. I. General introduction. *J. Comp. Neurol.* 206, 49–81.

Ochi, J. and Shimizu, K. (1978). Occurrence of dopamine-containing neurons in the midbrain raphe nuclei of the rat. *Neurosci. Lett.* 8, 317–320.

Palkovits, M., Browstein, M. and Saavedra, J. M. (1974). Serotonin content of the brain stem nuclei in the rat. *Brain Res.* 80, 237–249.

Palkovits, M., Saavedra, J. M., Jacobowitz, D. M., Kizer, J. S., Zaborszky, L. and Brownstein, M. J. (1977). Serotonergic innervation of the forebrain: effect of lesions on serotonin and tryptophan hydroxylase levels. *Brain Res.* 130, 121–134.

Parent, A., Descarries, L. and Beaudet, A. (1981). Organization of ascending serotonin systems in the adult rat brain. A radioautographic study after intraventicular administration of (^{3}H)5-hydroxytryptamine. *Neurosci.* 6, 115–138.

Park, M. R., Imai, H. and Kitai, S. T. (1982). Morphology and intracellular responses of an identified dorsal raphe projection neuron. *Brain Res.* 240, 321–326.

Pasquier, D. A. and Reinoso-Suarez, F. (1977). Differential efferent connections of the brain stem to the hippocampus in the cat. *Brain Res.* 120, 540–548.

Pasquier, D. A. and Reinoso-Suarez, F. (1978). The topographic organization of hypothalamic and brain stem projections to the hippocampus. *Brain Res. Bull.* 3, 373–389.

Pasquier, D. A. and Villar, M. J. (1982). Specific serotonergic projections to the lateral geniculate body from the lateral cell groups of the dorsal raphe nucleus. *Brain Res.* 249, 142–146.

Pasquier, D. A., Anderson, C., Forbes, W. B. and Morgane, P. J. (1976). Horseradish peroxidase tracing of the lateral habenular-midbrain raphe nuclei connections in the rat. *Brain Res. Bull.* 1, 443–451.

Pasquier, D. A., Kemper, T. L., Forber, W. N. and Morgane, P. J. (1977). Dorsal raphe, substantia nigra and locus coeruleus: Interconnections with each other and the neostriatum. *Brain Res. Bull.* 2, 323–339.

Paxinos, G. and Watson, C. R. R. (1982). *The rat brain in stereotaxic coordinates.* Academic Press, Sydney.

Peschanski, M. and Besson, J. -M. (1984). Diencephalic connections of the raphe nuclei of the rat brainstem: An anatomical study with reference to the somatosensory system. *J. Comp. Neurol.* 224, 509–534.

Petrovicky, P. (1971). Structure and incidence of Gudden's tegmental nuclei in some mammals. *Acta Anat.* 80, 273–286.

Petrovicky, P. (1980). Reticular formation and its raphe system. I. Cytoarchitectonics with comparative aspects. *Acta Univ. Carol. Med. (Praha)* 99, 1–117.

Pfister, C. and Danner, H. (1980). Fluoreszenzhistochemische und neuro-histologische Untersuchungen am Nucleus raphe dorsalis der Ratte. *Acta Histochem.* 66, 253–261.

Phillipson, O. T. (1979). Afferent projections to the ventral tegmental area of Tsai and interfascicular nucleus: A horseradish peroxidase study in the rat. *J. Comp. Neurol.* 187, 117–144.

Pickel, V. M., Joh, T. H., Chan, J. and Beaudet, A. (1984). Serotonergic terminals: Ultrastructure and synaptic interaction with catecholamine-containing neurons in the medial nuclei of the solitary tracts. *J. Comp. Neurol.* 225, 291–301.

Reader, T. A. (1980). Serotonin distribution in rat cerebral cortex; radioenzymatic assays with thin-layer chromatography. *Brain Res. Bull.* 5, 609–613.

Saavedra, J. M., Brownstein, M. and Palkovits, M. (1974). Serotonin distribution in the limbic system of the rat. *Brain Res.* 79, 437–441.

Sakai, K., Salvert, D., Touret, M. and Jouvet, M. (1977a). Afferent connections of the nucleus raphe dorsalis in the cat as visualized by the horseradish peroxidase technique. *Brain Res.* 137, 11–35.

Sakai, K., Touret, M., Salvert, D., Leger, L. and Jouvet, M. (1977b). Afferent projections to the cat locus coeruleus as visualized by the horseradish peroxidase technique. *Brain Res.* 119, 21–41.

Saper, C. B. and Loewy, A. D. (1980). Efferent connections of the parabrachial nucleus in the rat. *Brain Res.* 197, 291–317.
Satoh, K. (1979). The origin of reticulospinal fibers in the rat: A HRP study. *J. Hirnforsch.* 20, 313–322.
Sawchenko, P. E, Swanson, L. W., Steinbusch, H. W. M. and Verhofstad, A. A. J. (1983). The distribution and cells of origin of serotonergic inputs to the paraventricular and supraoptic nuclei of the rat. *Brain Res.* 277, 355–360.
Schaffar, N., Jean, A. and Calas, A. (1984). Radioautographic study of serotonergic axon terminals in the rat trigeminal motor nucleus. *Neurosci. Lett.* 44, 31–36.
Segal, M. (1977). Afferents to the entorhinal cortex of the rat studied by the method of retrograde transport of horseradish peroxidase. *Exp. Neurol.* 57, 750–765.
Shah, Y. and Dostrovsky, J. P. (1980). Electrophysiological evidence for a projection of the periaqueductal grey matter to nucleus raphe magnus in cat and rat. *Brain Res.* 193, 534–538.
Shibata, H. and Suzuki, T. (1984). Efferent projections of the interpeduncular complex in the rat, with special reference to its subnuclei: A retrograde horseradish peroxidase study. *Brain Res.* 296, 345–349.
Shinghaniyom, W., Wreford, N. G. M. and Güldner, F.-H. (1982). Distribution of 5-hydroxytryptamine-containing neuronal perikarya in the rat interpeduncular nucleus. *Neurosci. Lett.* 30, 51–55.
Simerly, R. B., Swanson, L. W. and Gorski, R. A. (1984). Demonstration of a sexual dimorphism in the distribution of serotonin-immunoreactive fibers in the medial preoptic nucleus of the rat. *J. Comp. Neur.* 225, 151–166.
Simon, H., Le Moal, M. and Calas, A. (1979). Efferents and afferents of the ventral tegmental-A10 region studied after local injection of (^{3}H)-leucine and horseradish peroxidase. *Brain Res.* 178, 17–40.
Steinbusch, H. W. M. (1981). Distribution of serotonin-immunoreactivity in the central nervous system of the rat. Cell bodies and terminals. *Neurosci.* 4, 557–618.
Steinbusch, H. W. M. and Nieuwenhuys, R. (1981). Localization of serotonin-like immunoreactivity in the central nervous system and pituitary of the rat, with special reference to the innervation of the hypothalamus. In B. Haber, S. Gabay, M. R. Issidorides and S. G. A. Alivisatos (Eds). *Serotonin: Current aspects of neurochemistry and function*, pp. 7–36. *Adv. Exp. Med. Biol.* 133. New York: Plenum Press.
Steinbusch, H. W. M. and Nieuwenhuys, R. (1983). The raphe nuclei of the rat brainstem: A cytoarchitectonic and immunohistochemical study. In P. C. Emson (Ed.). *Chemical Neuroanatomy*, pp. 131–207. Raven Press, New York.
Steinbusch, H. W. M., Van der Kooy, D., Verhofstad, A. A. J. and Pellegrino, A. (1980). Serotonergic and non-serotonergic projections from the nucleus raphe dorsalis to the caudate-putamen complex in the rat, studied by combined immunofluorescence and fluorescent retrograde axonal labeling technique. *Neurosci. Lett.* 19, 137–142.
Steinbusch, H. W. M., Nieuwenhuys, R., Verhoftad, A. A. J. and Van der Kooy, D. (1981). The nucleus raphe dorsalis of the rat and its projection upon the caudo-putamen. A combined cytoarchitectonic, immunocytochemical and retrograde transport study. *J. Physiol. (Paris)* 77, 157–174.
Steinbusch, H. W. M., Verhofstad, A. A. J., Joosten, H. W. J. and Goldstein, M. (1982). Serotonin-immunoreactive cell bodies in the nucleus dorsomedialis hypothalami, in the substantia nigra and in the area tegmentalis ventralis of Tsai: Observations after pharmacological manipulations in the rat. In S. Palay and V. Chan-Palay (Eds). *Cytochemical methods in neuroanatomy*, pp. 407–421. Alan R. Liss, New York.
Swanson, L. W. (1982). The projections of the ventral tegmental area and adjacent regions: A combined fluorescent retrograde tracer and immunofluorescence study in the rat. *Brain Res. Bull.* 9, 321–353.
Swanson, L. W. and Cowan, W. M. (1979). The connections of the septal region in the rat. *J. Comp. Neurol.* 186, 621–656.
Taber, E. (1961). The cytoarchitecture of the brain stem of the cat. I. Brain stem nuclei of cat. *J. Comp. Neurol.* 116, 27–69.

Taber, E., Brodal, A. and Walberg, F. (1960). The raphe nuclei of the brain stem in the cat. I. Normal topography and general discussion. *J. Comp. Neurol.* 114, 161–187.

Taber-Pierce, E., Foote, W. E. and Hobson, J. A. (1976). The efferent connections of the nucleus raphe dorsalis, *Brain Res.* 107, 137–144.

Takagi, H., Shiosaka, S., Tohyama, M., Senba, E. and Sakanaka, M. (1980). Ascending components of the medial forebrain bundle from the lower brain stem in the rat, with special reference to raphe and catecholamine cell groups. *Brain Res.* 193, 315–337.

Takagi, H., Senba, E., Shiosaka, S., Sakanaka, M., Inagaki, S., Takatsuki, K. and Tohyama, M. (1981a). Ascending and cerebellar non serotonergic projections from the nucleus raphe magnus of the rat. *Brain Res.* 206, 161–165.

Takagi, H., Yamamoto, K., Shiosaka, S., Senba, E., Takatsuki, K., Inagaki, S., Sakanaka, M. and Tohyama, M. (1981b). Morphological study of noradrenaline innervation in the caudal raphe nuclei with special reference to fine structure. *J. Comp. Neurol.* 203, 15–22.

Takeuchi, Y. and Sano, Y. (1983). Immunohistochemical demonstration of serotonin nerve fibers in the neocortex of the monkey (*Macaca fuscata*). *Anat. Embryol.* 166, 155–168.

Takeuchi, Y. and Sano, Y. (1984). Serotonin nerve fibres in the primary visual cortex of the monkey: Quantitative and immunoelectronmicroscopical analysis. *Anat. Embryol.* 169, 1–8.

Takeuchi, Y., Kumura, H. and Sano, Y. (1982a). Immunocytochemical demonstration of the distribution of serotonin neurons in the brainstem of the rat and cat. *Cell Tiss. Res.* 224, 247–267.

Takeuchi, Y., Kimura, H. and Sano, Y. (1982b). Immunocytochemical demonstration of serotonin-containing nerve fibres in the cerebellum. *Cell Tiss. Res.* 226, 1–12.

Tohyama, M., Shiosaka, S., Sakanaka, M., Takagi, H., Senba, E., Satoh, Y., Takahashi, Y., Sakumoto, T. and Shimizu, N. (1980). Detailed pathways of the raphe dorsalis neuron to the cerebral cortex with use of horseradish peroxidase-3,3′,5,5′ tetramethylbenzidine reaction as a tool for the fiber tracing technique. *Brain Res.* 181, 433–439.

Törk, I. and Mulligan, K. A. (1984). Dense serotonergic innervation of select cortical neurons in cat neocortex. *Soc. Neurosci. Abst.* 10, 63.

Törk, I., Leventhal, A. G. and Stone, J. (1979). Brainstem projections to visual cortical areas 17, 18 and 19 in the cat, demonstrated by horseradish peroxidase. *Neurosci. Lett.* 11, 247–252.

Törk, I., Halliday, G. M., Scheibner, T. and Turner, S. (1984). The organization and connections of the mesencephalic ventromedial tegmentum (VMT) in the cat. In R. Bandler (Ed.). *Modulation of sensorimotor activity during alterations in behavioral states*, pp. 39–73. Alan R. Liss, New York.

Travers, J. B. and Norgren, R. (1983). Afferent projections to the oral motor nuclei in the rat. *J. Comp. Neurol.* 220, 280–298.

Uhl, G. R., Goodman, R. R., Kuhar, M. J., Childers, S. R. and Snyder, S. H. (1979a). Immunohistochemical mapping of enkephalin containing cell bodies, fibers and nerve terminals in the brain stem of the rat. *Brain Res.* 166, 75–94.

Uhl, G. R., Goodman, R. R. and Snyder, S. H. (1979b). Neurotensin-containing cell bodies, fibers and nerve terminals in the brain stem of the rat: Immunohistochemical mapping. *Brain Res.* 167, 77–91.

Ungerstedt, U. (1971). Stereotaxic mapping of the monoamine pathways in the rat brain. *Acta Physiol. Scand. Suppl.* 367, 1–48.

Valverde, F. (1962). Reticular formation of the albino rat's brain stem. Cytoarchitecture and corticofugal connections. *J. Comp Neurol.* 119, 25–53.

Van de Kar, L. D. and Lorens, S. A. (1979). Differential serotonergic innervation of individual hypothalamic nuclei and other forebrain regions by the dorsal and medial midbrain raphe nuclei. *Brain Res.* 162, 45–54.

Van der Kooy, D. and Hattori, T. (1980a). Bilaterally situated dorsal raphe cell bodies have only unilateral forebrain projections in rat. *Brain Res.* 192, 550–554.

Van der Kooy, D. and Hattori, T. (1980b). Dorsal raphe cells with collateral projections to the caudate-putamen and substantia nigra: A fluorescent retrograde double labeling study in the rat. *Brain Res.* 186, 1–7.

Van der Kooy, D. and Kuypers, H. G. J. M. (1979). Fluorescent retrograde double labeling: Axonal branching in the ascending raphe and nigral projections. *Science* 204, 873–875.

Van der Kooy, D., Hunt, S. P., Steinbusch, W. M. and Verhofstad, A. J. (1981). Separate populations of cholecystokinin and 5-hydroxytryptamine-containing neuronal cells in the rat dorsal raphe and their contribution to the ascending raphe projections. *Neurosci. Lett.* 26, 25–30.
Watkins, L. R., Griffin, G., Leichnetz, G. R. and Mayer, D. J. (1980). The somatotopic organization of the nucleus raphe magnus and surrounding brain stem structures as revealed by HRP slow release gels. *Brain Res.* 181, 1–15.
Wiklund, L., Descarries, L. and Mollgard, K. (1981a). Serotoninergic axon terminals in the rat dorsal accessory olive: normal ultrastructure and light microscopic demonstration of regeneration after 5,6 dihydroxytryptamine lesioning. *J. Neurocytol.* 10, 1009–1027.
Wiklund, L., Leger, L. and Persson, M. (1981b). Monoamine cell distribution in the cat brain stem. A fluorescence histochemical study with quantification of indoleaminergic and locus coeruleus cell groups. *J. Comp. Neurol.* 203, 613–647.
Wyss, J. M. and Sripanidkulchai, K. (1984). The topography of the mesencephalic and pontine projections from the cingulate cortex of the rat. *Brain Res.* 293, 1–15.
Wyss, J. M., Swanson, L. W. and Cowan, W. M. (1979). A study of subortical afferents to the hippocampal formation in the rat. *Neurosci.* 4, 463–476.
Yezierski, R. P., Bowker, R. M., Kevetter, G. A., Westlund, K. N., Coulter, J. D. and Willis, W. D. (1982). Serotonergic projections to the caudal brain stem: A double label study using horseradish peroxidase and serotonin immunocytochemistry. *Brain Res.* 239, 258–264.
Yoshida, M., Sasa, M. and Takaori, S. (1984). Serotonin-mediated inhibition from dorsal raphe nucleus of neurons in dorsal lateral geniculate and thalamic reticular nuclei. *Brain Res.* 290, 95–105.

4

Locus coeruleus

SANDRA E. LOUGHLIN
JAMES H. FALLON

University of California
Irvine, California, USA

I. INTRODUCTION

The small, nondescript cluster of cells in the pontine tegmentum known as the nucleus locus coeruleus (LC) has been the subject of intense research for the past 20 years. This attention began with the development of a technique for visualizing cellular monoamines (Carlsson *et al.*, 1962; Falck, 1962) and the subsequent demonstration that the cells of LC exhibited monoamine fluorescence (Dahlström and Fuxe, 1964; Ungerstedt, 1971). The LC is unique among monoamine containing nuclei in that it is relatively compact and homogeneous and gives rise to a massive norepinephrine (NE) containing efferent projection throughout the neuraxis. These characteristics have led many investigators to assume a simplicity of organization which recent work does not support. The LC consists of subdivisions with distinct cell morphologies, and afferent and efferent connections. The organization of the LC terminal arborizations is also complex, and the neurochemical content of the LC cells has been recently shown to be more varied than was previously known.

The LC is situated in the pontine tegmentum at the ventrolateral border of the central gray rostrally, and the lateral border of the central gray caudally (*Atlas* Figs 32 to 35).* It contains approximately 1650 cells in the rat, including the densely packed central LC or A6 cell group (according to the nomenclature of Dahlström and Fuxe, 1964), the posterior A4 cells, and the rostroventral subcoeruleus or A7. Scattered noradrenergic cells also extend from the anterior pole of the LC along the central gray border to the level of the inferior colliculus. The superior cerebellar peduncle lies at the dorsal border of the LC. The mesencephalic nucleus of 5 is at the lateral border of the rostral and middle LC, and the parabrachial and Kölliker-Fuse nuclei are at the lateral border of the

*The *Atlas* referred to here is Paxinos and Watson (1982).

THE RAT NERVOUS SYSTEM
ISBN 0 12 547632 9

caudal LC. Three dimensional computer graphic reconstructions of the distribution of cells in the rat LC have demonstrated that individual nuclei from different animals are remarkably similar both visually and statistically (Foote *et al.*, 1980b; Loughlin *et al.*, in press, a).

Similar cell groups containing NE have been described in all species of mammals examined, and this nucleus is thought to be homologous with NE cells in the pons of other vertebrates. Marked similarities exist in the appearance, afferents and efferents of the most studied species, the rat, the cat, and the monkey (Felten *et al.*, 1974; German and Bowden, 1975; Hubbard and DiCarlo, 1973). The present review describes the LC of the albino rat unless otherwise noted. In addition to their well-known NE content, recent findings indicate that some LC cells in the rat contain the peptides vasopressin, neurophysin (Caffe and van Leeuwen, 1983), neurotensin (Uhl *et al.*, 1979), and corticotropin releasing factor (Swanson *et al.*, 1983).

II. NEURONAL TYPES AND CYTOARCHITECTURE

The LC consists of two major segments, the dorsal or compact LC, and the ventral division (Swanson, 1976). The dorsal segment, which measures just less than 1 mm in anterior to posterior extent in paraffin sections, contains approximately 1400 cells. Cells are densely packed and oriented obliquely from dorsolateral to ventromedial as well as anterior to posterior. The ventral division measures less than 500 μm in length and consists of approximately 250 larger cells. These cells are not clearly oriented and cell density is lower than in the dorsal segment.

Several morphologic cell types can be defined in the LC (Shimizu *et al.*, 1978; Swanson, 1976). The major type of cells impregnated in Golgi-Cox material are large multipolar cells. Three or four long, rather thin, dendrites leave the cell body and branch once or twice, extending well beyond the boundaries of the nucleus. These are studded with infrequent, regularly spaced spines. Axons give off collaterals within the LC. A smaller, fusiform cell type, which has been called bipolar, actually bears several dendrites arising from each end of the cell in Golgi material. Small round neurons, with several radiating processes confined within the boundaries of the nucleus, have also been described.

Immunohistochemical staining for the NE synthetic enzyme dopamine beta hydroxylase (DBH) (Grzanna and Molliver, 1980a, b) has allowed further characterization of the LC morphology. In the A4 region extending dorso-posteriorly from the LC in the superior cerebellar peduncle, longitudinally elongated cells have dendritic branches extending between ependymal cells to the ventricular surface. The cells of the ventral region are large and multipolar, as are the subcoeruleus cells extending ventrally and rostrally from the LC, but the largest cells are those extending anteriorly along the lateral edge of the central gray. Multipolar and fusiform neurons in both dorsal and ventral segments of the LC exhibit DBH reactivity. The distribution of DBH neurons is generally coextensive with the distribution of Nissl stained neurons. It thus appears that the majority of cells in the LC contains norepinephrine. The small round cells, however, have not been conclusively shown to contain norepinephrine (Loughlin *et al.*, in press, b).

The ultrastructure of the LC has also been the subject of several reports (Groves and Wilson, 1980a, b; Shimuzu *et al.*, 1979). Cell bodies are predominantly medium sized, exhibiting a large nucleus with a prominent nucleolus. The cytoplasm contains a highly developed Golgi apparatus around the nucleus and prominent Nissl bodies are clustered at the poles of the soma. Somatic and dendritic spine like appendages are frequent, often clustered and connected by puncta adherentia. The LC cells exhibit mainly longitudinally directed dendritic branches. These are predominantly oriented rostrocaudally within the brain stem, creating disc shaped dendritic fields.

Several types of presynaptic profiles synapse onto the LC cells. These are mainly concentrated on small dendrites and many are located on the spines. The comprehensive study of Groves and Wilson (1980b) reported relatively few synapses onto the LC somata, proximal dendritic trunks, or axons within the LC. They observed predominantly axodendritic synapses in the LC. When synaptic profiles are examined in serial sections, all profiles containing vesicles show membrane specializations. This is in contrast to past reports (Leger and Descarries, 1978) which sampled random sections through the LC and reported a paucity of monoaminergic synapses.

When animals are pretreated with 5-hydroxydopamine and tissue is stained with osmium tetroxide, monoaminergic synaptic vesicles exhibit an electron dense core and membranes, and membrane specializations associated with synapses, are visible (Koda *et al.*, 1978). In such tissue, dendrodendritic synapses between the LC cells can be demonstrated. Monoaminergic axons are also observed to contact the LC cells. Though these may be recurrent collaterals, they may also be axons of other monoamine cells in the brain stem.

There are also close contacts between the LC cells and blood vessels. In the monkey (Felten and Crutcher, 1979), neurons and their processes directly abut the capillary basement membrane with no intervening glial sheath. No vesicles are observed in regions of close apposition, suggesting that these are not presynaptic elements. It has been demonstrated, however, that the density of capillaries in the LC is no higher than in the surrounding tissue and is much lower than many brain regions in the rat (McGinty *et al.*, 1983). In addition, the LC dendrites appear to perforate the ependymal wall to contact the ventricular surface (Grzanna and Molliver, 1980b).

To summarize, the LC can be divided into two major (dorsal and ventral) regions or five subregions (dorsal, ventral, anterior, A4, and subcoeruleus).

Several cell types can be defined within the nucleus on the basis of morphology, location, and orientation. Whether these cell types or regions correspond to any functionally distinct populations is not known. It is also not known whether they can be segregated on the basis of their afferents. A few reports have commented on whether they can be defined with respect to their efferents; these will be reviewed later.

III. AFFERENT CONNECTIONS

The afferent projections to the LC are quite diverse. Many anterograde and retrograde transport studies have confirmed and extended the results of the comprehensive study of Cedarbaum and Aghajanian (1978). Immunohistochemical techniques have also contributed to our knowledge of the transmitters which are localized to terminals within the boundaries of the LC. The first section below summarizes findings from anterograde and retrograde methods. The second section describes those afferents which have been chemically defined. Both errors of omission and false positives are possibilities in examining the afferents to such a small and undifferentiated nucleus. Retrograde methods assume that the injection did not extend outside the boundaries of the LC, but did label all cells projecting to the nucleus. Similar problems occur with anterograde methods. Injections in other brain regions may extend outside a given area and the label may not be transported to all terminal fields. Ideally, both methods should be applied to confirm each afferent.

There is a direct projection from the insular cortex to the LC. The central nucleus of the amygdala and the bed nucleus of the stria terminalis (Swanson and Saper, 1978) also innervate this structure. There is a relatively extensive projection from the hypothalamus; in particular, the lateral preoptic area (Clavier, 1979; Saper *et al.*, 1979), the ventromedial nucleus (Saper *et al.*, 1979), the posterior lateral hypothalamic area (Clavier, 1979; Saper *et al.*, 1979), the dorsomedial nucleus (Swanson and Saper, 1978) and the paraventricular nucleus (Swanson, 1977) project to the LC. Thalamic projections from ventrolateral and parafascicular nuclei have been described (Clavier, 1979). Several reticular formation structures also contribute to the afferent projections to the LC. These include the cuneiform nucleus, the lateral reticular nucleus, and the central gray (Saper *et al.*, 1979). There is good evidence for serotonergic projections to the LC from the raphe nuclei, especially from the dorsal raphe (Groves and Wilson, 1980b; Sakai *et al.*, 1977). Some of these may contain substance P as well (Pickel *et al.*, 1979). There is also evidence for a projection from the VTA to the LC (Simon *et al.*, 1979). Spinal cord projections appear to originate from the marginal zone of the dorsal horn. The solitary tract, the vestibular nucleus, and the deep cerebellar nuclei project to the LC (Clavier, 1979).

Several opioid peptides have been identified in terminals in or near the LC. β-endorphin fibers, presumably originating in the arcuate nucleus, terminate in the anterior LC (Bloom *et al.*, 1978). Metenkephalin and leuenkephalin terminals innervate the LC (Pickel *et al.*, 1979). Their origin is unknown. Dynorphin fibers are also present (Watson *et al.*, 1982).

Acetylcholine and its degradative enzyme acetylcholinesterase have been demonstrated to be present in the LC in high concentrations. In fact, this enzyme is localized to the LC cell bodies (Albanese and Butcher, 1979). The synthetic enzyme choline acetyltransferase is also present in the nucleus, but concentrations are quite low (Cheney *et al.*, 1978). These findings imply that the LC cells are not cholinergic but cholinoceptive.

The dopamine synthetic enzyme tyrosine hydroxylase is present in terminals in the LC (Pickel *et al.*, 1977), but since this is also a synthetic enzyme for NE it is difficult to demonstrate that DA terminals are present. Dopamine-β-hydroxylase (DBH), a synthetic enzyme unique to NE and epinephrine (E) cells, is also present in terminals (Grzanna and Molliver, 1980a). These could be recurrent collaterals of the LC cells, projections from the contralateral LC, or terminals from other NE or epinephrine cell groups such as A1 (C1) or A2 (C2). Phenylethanolamine *n*-methyl transferase, the final synthetic enzyme for epinephrine, has been reported to be contained in terminals within the LC (Hökfelt *et al.*, 1974). These are probably afferents from the brain stem epinephrine groups C1 or C2.

Thyrotropin releasing hormone containing terminals are found in the LC but these are quite scarce (Hökfelt *et al.*, 1978). They probably originate from hypothalamic nuclei containing this substance. Vasopressin and oxytocin are found separately in different cells in the hypothalamus (Choy and Watkins, 1977) and terminals containing each are found in the LC (Swanson, 1977). Angiotensin II fibers have also been observed in the LC (Fuxe *et al.*, 1976).

There is, thus, an extremely diverse input to the LC. In light of the fact that the nucleus has been shown to respond to sensory stimuli of all modalities with a relatively short latency (Foote *et al.*, 1980a), however, it is of interest that sensory input does not predominate. The input from the vestibular nucleus might be thought of as sensory, and that from the marginal zone of the spinal cord and the central gray might suggest a role in pain perception or modulation. The majority of information reaching the LC, however, might be thought of as either "reticular" (see above) or as "visceral" or "limbic" (hypothalamus, amygdala, parafascicular thalamus). There are also projections from areas considered to be motor areas (medial cerebellar nucleus, ventrolateral thalamic nucleus). In addition, dendrites of LC cells are found in close contact with capillaries and even extend into the ventricular ependyma. Such interactions might provide an opportunity for hormonal or chemotaxic inputs. It has been reported, for instance, the LC cells concentrate ^{3}H-estradiol when this substance is injected into the blood (Heritage *et al.*, 1977). The LC is thus in a position to receive extremely diverse information about the needs, state, and environment of the animal.

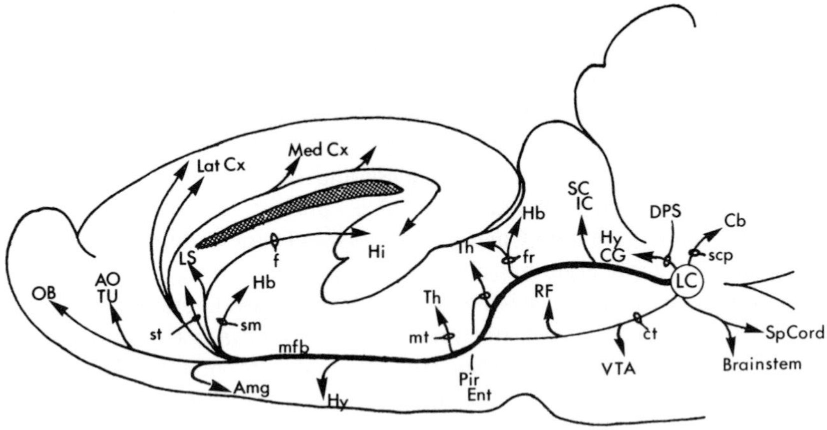

Fig. 1: Schematic representation of the major projections of the locus coeruleus (LC) collapsed on a sagittal section. Amg, amygdala; AO, anterior olfactory nucleus; Cb, cerebellum; CG central gray; ct, central tegmental tract; cx, cortex; DPS, dorsal periventricular system (see text); Ent, entorhinal cortex; f, fornix; fr, fasciculus retroflexus; Hb, habenula; Hi, hippocampus; Hy, hypothalamus; IC, inferior colliculus; LS, lateral septum; mfb, median forebrain bundle; mt, mammillothalamic tract; OB, olfactory bulb; Pir, piriform cortex; RF, reticular formation; SC, superior colliculus; scp, superior cerebellar peduncle; sm, stria medullaris; st, stria terminalis; Th, thalamus; Tu, olfactory tubercle; VTA, ventral tegmental area.

IV. EFFERENT CONNECTIONS

Catecholamine fluorescence, in combination with various lesions and pharmacologic manipulations, first revealed the extensive projections of the LC (Lindvall and Björklund, 1974; Loizou, 1969; Ungerstedt, 1971). These findings were confirmed and extended by anterograde (Jones *et al.*, 1977; Pickel *et al.*, 1974) and retrograde transport methods (see below). The difficulties with such methods have been noted above. Several studies have also utilized biochemical measurements of NE in combination with lesions to demonstrate LC projections to particular regions of the brain (Kobayashi *et al.*, 1974). Since it is difficult to differentiate dopamine from NE by the catecholamine fluorescence method, the development of a method for the immunocytochemical localization of the NE synthetic enzyme DBH has added much information on the organization of LC terminal fields (Jacobowitz, 1978; Swanson and Hartman, 1975).

A. Five major efferent bundles

Five major efferent bundles form an apparently diffuse, monosynaptic projection throughout the neuraxis (Moore and Bloom, 1979). In fact, this projection is not diffuse; it is highly selective. Certain structures are avoided entirely and only portions of other brain regions are innervated. The five major pathways will first be summarized, followed by a discussion of particular terminal fields of interest (Fig. 1).

1. *Major descending component*

The major descending component follows the central tegmental tract. Some evidence suggests that fibers leave this bundle to innervate some brain stem nuclei such as the inferior olive (Swanson and Hartman, 1975), and the medullary reticular formation (Westlund and Coulter, 1980), but exactly which NE terminal fields are contributed by other NE cell groups is not yet clear. This fiber bundle descends in the anterior funiculus and ventrolateral funiculus of the spinal cord and distributes terminals to all segments.

2. *Innervation of the cerebellar cortex*

A small bundle of fibers ascends laterally from the LC in two fascicles within the superior cerebellar peduncle to innervate the cerebellar cortex.

3. *Dorsal periventricular system*

There is some evidence for a "dorsal periventricular system" (Lindvall and Björklund, 1974). Axons from subcoeruleus cells, and probably from the LC proper, join and ascend along the dorsal longitudinal fasciculus to distribute to the central gray. These probably extend anteriorly to the periventricular nuclei of the hypothalamus and may contribute to the innervation of more rostral hypothalamic nuclei.

4. *Other nuclei*

Another group of axons leaves the LC to turn ventrally and join the ascending central tegmental tract. Fibers leave this bundle to innervate the ventral tegmental area (VTA) (Phillipson, 1979; Simon *et al.*, 1979) and probably the substantia nigra (Fig. 1). The bundle continues rostrally to join the medial forebrain bundle (mfb).

In addition, LC fibers give rise to terminals in other pontine and midbrain nuclei. There is good evidence for LC innervation of the cochlear nucleus (Kromer and Moore, 1976); the principal sensory trigeminal nucleus, pontine nuclei and central gray (Levitt and Moore, 1979). There is a moderately dense terminal field in the inferior and superior colliculi (Jacobowitz, 1978) and there may be LC terminals in the dorsal raphe (Barban and Aghajanian, 1981; Saavedra *et al.*, 1976; Sakai *et al.*, 1977; Westlund and Coulter, 1980).

5. *Dorsal bundle*

The major projection of the LC is known as the "dorsal bundle". This large group of axons ascends to traverse the midbrain tegmentum ventrolateral to the periaqueductal gray and then turns ventrally at the level of the fasciculus retroflexus to join the mfb.

Fibers leave the dorsal bundle and ascend along the fasciculus retroflexus to innervate the habenula. Some leave this small bundle to contribute to LC innervation of the thalamus (Lindvall *et al.*, 1974). This appears to be moderately dense in the majority of the dorsal thalamus, especially the anterior thalamus, and the lateral (Kromer and Moore, 1980) and medial geniculate nuclei. The anterior thalamus projections also ascend along the mammillothalamic tract.

Anterior to the point at which the dorsal bundle joins the mfb many fibers exit along the ventral amygdalofugal pathway. These innervate the amygdala, especially the central nucleus as well as the basal and lateral nuclei. Fibers also supply the entorhinal and piriform cortices by this route (Fallon and Moore, 1978; Fallon *et al.*, 1978) and this bundle has been shown to contribute to several portions of the complex innervation of the hippocampus (Loy *et al.*, 1980). As the mfb ascends, fibers may leave to terminate in several hypothalamic nuclei.

At the level of the caudal septum, the mfb divides to form five major branches. Fibers turn caudally along the stria medullaris, probably continuing as far as the habenula. Some LC fibers course along the stria terminalis to innervate its bed nucleus and the amygdala. Fibers also continue medially along the diagonal band to the septum, especially the medial nucleus, and continue on into the fornix. These fibers contribute to the innervation of the hippocampus. Some axons traverse the diagonal band and Zuckerkandl's bundle to turn caudally around the corpus callosum and run along the cingulum (Lindvall and Stenevi, 1978). This bundle gives off fibers to medial cortex, but probably not to dorsal neocortex (Morrison *et al.*, 1978). It continues through the subiculum to Ammon's horn and contributes to the innervation of the dentate gyrus. The fifth branch continues in the mfb and gives off fibers to the olfactory tubercle and anterior olfactory nucleus, ascends in the external capsule to the neocortex and continues anteriorly as far as the olfactory bulb (Fallon and Moore, 1978). Here it provides terminals to only the granular and external plexiform layers (Halasz *et al.*, 1978). The extensive innervation of the neocortex derives almost exclusively from those fibers which ascend in the frontal pole. These direct themselves longitudinally in layer 6 throughout the neocortex (Morrison *et al.*, 1981).

While horseradish peroxidase (HRP) transport from the caudate-putamen to the LC has been reported (Mason and Fibiger, 1979; Loizou, 1969), it appears that the majority of the striatum is not innervated by the LC or any NE group.

3. Distribution of locus coeruleus terminals

The distribution of the LC terminals in several major brain structures requires further discussion. The cells of origin of projections to these structures have been shown to be clustered within the LC (Mason and Fibiger, 1979; Loughlin *et al.*; in press, a) and to arise from distinct morphologic subpopulations Loughlin *et al.*, in press, a) and to arise from distinct morphologic subpopulations (Loughlin *et al.*, in press, b).

1. *Hippocampus*

The LC innervation of the hippocampus is similar throughout all layers and regions of this structure (Loy *et al.*, 1980). The densest innervation is to the hilus of the dentate gyrus; it is also relatively dense in the stratum lucidum of field CA3 of Ammon's horn and the subiculum. The density of NE is slightly greater in septal regions than in temporal regions. There is a minor contralateral projection, amounting to approximately 10% of the ipsilateral. Terminals of the LC cells appear to innervate the granule cells of the dentate gyrus and the pyramidal cells of the hippocampus proper (Loy *et al.*, 1980; Koda *et al.*, 1978). They end on both the dendrites and somata of these cells. While there is a minor disagreement among authors, LC innervation of the hippocampus appears to arise from the dorsal two thirds of the compact LC and from the A4 region Haring and Davis, 1983; Loughlin *et al.*, in press, b; Mason and Fibiger, 1979). Fusiform cells and medium sized multipolar cells contribute to the projection.

2. *Neocortex*

The LC innervation of the cortex is complex, but a basic pattern can be described (Levitt and Moore, 1978; Morrison *et al.*, 1978). Layer 1 contains a dense grid like plexus of fibers. Layers 2 and 3 contain sparse, radially oriented fibers. Layers 4 and 5 have moderately dense, short, oblique fibers, and layer 6 exhibits long fibers which are oriented from anterior to posterior within the cortex. DBH positive fibers are rarely observed in white matter. This pattern is uniform throughout the cortex, with only minor variations between areas, most notably in medial cortex. The pattern corresponds closely to the termination of thalamocortical afferents, even in the somatosensory barrel fields (Lidov *et al.*, 1978).

Axons from the LC enter the neocortex at the frontal pole and run tangentially rostral to caudal in layer 6 (Morrison *et al.*, 1981). Medial and lateral areas are innervated by different axons (Morrison *et al.*, 1979), arising from different LC cells (Loughlin *et al.*, 1982). A crude topography may also exist in the LC such that more posterior cells within the dorsal compact LC innervate more posterior cortical regions (Waterhouse *et al.*, 1983). Although only 5% of cortical innervation arises from the contralateral LC, one LC cell may give rise to bilateral projections to the cortex (Ader *et al.*, 1980).

3. Hypothalamus

Norepinephrine afferents to this area arise from the dorsal periventricular system as well as the mfb. Those arising from the dorsal periventricular system probably innervate midline structures and fibers from the mfb probably contribute to NE innervation of the lateral nuclei (Palkovits and Jacobowitz, 1974). Contributions from the other NE (non-LC) cell groups are prominent in this area, making it difficult to assess which axons arise from the LC itself. To summarize the results of various studies, the LC projects to the paraventricular, periventricular, and supraoptic nuclei. There is some evidence for a contribution to the innervation of the dorsomedial nuclei and the median eminence (Iijima and Ogawa, 1980; Moore and Bloom, 1979; Palkovits *et al.*, 1977). The LC probably does not contribute extensively to the innervation of other areas of the hypothalamus. This projection arises from medium and large multipolar cells throughout the compact dorsal LC and the anterior pole. Small, round cells in the LC also project to the hypothalamus (Loughlin *et al.*, in press, b).

4. Cerebellum

A sparse plexus of catecholamine fluorescent fibers has been reported throughout the cerebellar hemispheres (Hökfelt and Fuxe, 1969). Fluorescent fibers within the cerebellum are located mainly within the molecular layer and appear to make contacts with Purkinje cell dendrites, especially on the proximal dendrite (Bloom *et al.*, 1974). Injections of HRP into the cerebellum label cells throughout the LC, including the extreme ventral tip. Large multipolar cells are extremely densely labeled (Mason and Fibiger, 1979).

5. Spinal cord

The LC innervates the ventral columns, the intermediate gray, and the ventral half of the dorsal column of the spinal cord while the subcoeruleus is the source of NE in the intermediolateral cell column (IML) (Commissiong *et al.*, 1978; Loewy *et al.*, 1979a, b; Nygren and Olson, 1977; Satoh *et al.*, 1977). Shared projections are issued to the ventral columns, the central gray and the dorsal columns (Westlund and Coulter, 1980). The LC and subcoruleus projections to the spinal cord arises mainly from large multipolar cells in the ventral segment of the LC, as well as from subcoeruleus cells (Ader *et al.*, 1980; Kuypers and Maisky, 1975; Loughlin *et al.*, in press, a; Satoh *et al.*, 1978; Westlund *et al.*, 1983). Retrograde labeling from spinal cord is mainly ipsilateral (90%).

V. CONCLUSION

To conclude, the massive innervation of diverse brain structures by this nucleus is quite striking. Whether those subpopulations of LC cells which contain certain peptides in addition to NE sustain different efferents remains to be determined. While many terminal fields arise from cells in restricted regions with the LC, significant evidence has accumulated that at least some LC cells sustain highly divergent collaterals. Individual cells simultaneously innervate, for example, cerebellum and spinal cord, frontal cortex and cerebellum, and hippocampus and spinal cord (Nagai *et al.*, 1981; Steindler and Deniau, 1980). These projections appear to arise, however, from the densely packed core of the LC and it should not be assumed that all LC cells innervate all terminal fields. Instead, this unique nucleus can be conceived of as clusters of cells with overlapping fields of influence throughout the neuraxis.

ACKNOWLEDGEMENTS

The authors gratefully acknowledge the expert secretarial assistance of Ms Natalie Sepion. Supported by NIH grant #NS15321 and a grant from the United Parkinson Foundation.

REFERENCES

Ader, J. -P., Room, R., Postema, F. and Korf, J. (1980). Bilaterally diverging axon collaterals and contralateral projections from rat coeruleus locus and norepinephrine metabolism. *J. Neur. Trans.* 49, 207–218.

Albanese, A. and Butcher, L. (1979). Locus coeruleus somata contain both acetylcholinesterase and norepinehrine: Direct histochemical demonstration on the same tissue section. *Neurosci. Lett.* 14, 101–104.

Barban, J. and Aghajanian, G. (1981). Noradrenergic innervation of serotonergic neurons in the dorsal raphe: Demonstration by electron microscopic autoradiography. *Brain Res.* 204, 1–11.

Bloom, F. E., Krebs, H., Nicholson, J. and Pickel, V. (1974). The noradrenergic innervation of cerebellar purkinje cells: Localization, function, synaptogenesis, and axonal sprouting of locus coeruleus. In K. Fuxe, L. Olson and Y. Zotterman (Eds). *Dynamics of degeneration and growth in neurons*, pp. 413–423. Pergamon Press, New York.

Bloom, F. E., Battenberg, E., Rossier, J., Ling, N. and Guillemiri, R. (1978). Neurons containing B-endorphin in rat brain exist separately from those containing enkephalin: Immunocytochemical studies. *Proc. Natl. Acad. Sci.* 75, 3, 1591–1595.

Caffe, A. R. and van Leeuwen, F. W. (1983). Vasopressin-immunoreactive cells in dorsomedial hypothalamic region, medial amygdaloid nucleus and locus coeruleus of the rat. *Cell Tissue Res.* 233, 23–33.

Carlsson, A., Falck, B. and Hillard, N. A. (1962). Cellular localization of brain monoamines. *Acts Physiol. Scand.* 56, 1–26.

Cedarbaum, J. M. and Aghajanian, C. K. (1978). Afferent projections to the rat locus coeruleus as determined by a retrograde tracing technique. *J. Comp. Neurol.* 187, 1–16.

Cheney, D. L., LeFevre, H. F. and Racagni, G. (1978). Choline acetyltransferase activity and mass fragmentographic measurement of acetylcholine in specific nuclei and tracts of rat brain. *Neuropharm.* 14, 801–809.

Choy, V. and Watkins, W. (1977). Immunocytochemical study of the hypothalamoneurohypophysial system. *Cell Tiss. Res.* 180, 467–490.

Clavier, R. (1979). Afferent projections to the self-stimulation regions of the dorsal pons, including the LC, in the rat as demonstrated by the HRP technique. *Brain Res. Bull.* 4, 497–504.
Commissiong, J. W., Hellstrom, S. O. and Neff, N. H. (1978). A new projection from locus coeruleus to the spinal ventral columns: histochemical and biochemical evidence. *Brain Res.* 148, 207–213.
Dahlstrom, A. and Fuxe, K. (1964). Evidence for the existence of monoamines in the cell bodies of brain stem neurons. *Acta Physiol. Scand.* 62, 5232.
Falck, B. (1962). Observations on the possibilities of the cellular localization of monoamines by a fluorescence method. *Acta Physiol. Scand.* 56, 1–25.
Fallon, J. H. and Moore, R. Y. (1978). Catecholamine innervation of the basal forebrain. III. Olfactory bulb, anterior olfactory nuclei, olfactory tubercle and piriform cortex. *J. Comp. Neurol.* 180, 533–544.
Fallon, J. H., Koziell, D. and Moore, R. Y. (1978). Catecholamine innervation of the basal forebrain. II. Amygdala, suprahinal cortex and entohinal cortex. *J. Comp. Neurol.* 180, 509–531.
Felten, D. L. and Crutcher, K. A. (1979). Neuronal-vascular relationships in the raphe nuclei, locus coeruleus and substantia nigra in primates. *Am. J.Anat.* 155, 467–482.
Felten, D. L., Laties, A. M. and Carpenter, M. B. (1974). Monoamine-containing cell bodies in the squirrel monkey brain. *Am. J. Anat.* 139, 153–166.
Foote, S. E., Loughlin, S. E., Cohen, P. S., Bloom, F. E. and Livingston, R. B. (1980a). Accurate three-dimensional reconstruction of neuronal distributions in brain: Reconstruction of the rat nucleus locus coeruleus. *J. Neurosci. Meth.* 3, 159–153.
Foote, S. L., Aston-Jones, G. and Bloom, F. E. (1980b). Impulse activity of locus coeruleus neurons in awake rats and monkeys is a function of sensory stimulation and arousal. *Proc. Natl. Acad. Sci. USA* 77, 5, 3033–3037.
Fuxe, K., Ganten, D., Hökfelt, T. and Bolme, P. (1976). Immunohistochemical evidence for the existence of angiotensin II—containing nerve terminals in the brain and spinal cord of the rat. *Neurosci. Lett.* 2, 229–234.
German, D. and Bowden, D. (1975). Locus coeruleus in Rhesus monkey (*Macaca mulatta*): A combined histochemical, fluorescence, Nissl and silver study. *J. Comp. Neurol.* 161, 19–30.
Groves, P. M. and Wilson, C. J. (1980a). Fine structure of locus coeruleus. *J. Comp. Neurol* 193, 841–852.
Groves, P. M. and Wilson, C. J. (1980b). Monoaminergic presynaptic axons and dendrites in rat locus coeruleus seen in reconstructions of serial sections. *J. Comp. Neurol.* 193, 853–862.
Grzanna, R. and Molliver, M. E. (1980a). The locus coeruleus in the rat: An immunohistochemical delineation. *Neurosci.* 5, 21–40.
Grzanna, R. and Molliver, M. E. (1980b). Cytoarchitecture and dendritic morphology of central noradrenergic neurons. *In* J. A. Hobson and M. A. B. Brazier (eds), *The reticular formation revisited.* Raven Press, New York.
Halasz, N., Ljundahl, A. and Hökfelt, T. (1978). Transmitter histochemistry of the rat olfactory bulb. II. Fluorescence histochemical, autoradiographic and electron microscopic localization of monoamines. *Brain Res.* 154, 253–271.
Haring, J. H. and Davis, J. N. (1983). Topography of locus coeruleus neurons projecting to the area dentata. *Exp. Neurol.* 79, 785–860.
Heritage, A. S., Grant, L. D. and Stumpf, W. E. (1977). 3H-estradiol in catecholamine neurons of rat brainstem: Combined localization by autoradiography and formaldehyde-induced fluorescence. *J. Comp. Neurol.* 176, 607–630.
Hokfelt, T. and Fuxe, K. (1969). Cerebellar monoamine nerve terminals, a new type of afferent fibers to the cortex cerebelli. *Exp. Brain Res.* 9, 63–72.
Hökfelt, T., Fuxe, K., Goldstein, M. and Johansson, O. (1974). Immunohistochemical evidence for the existence of adrenaline neurons in the rat brain. *Brain Res.* 66, 235–251.
Hökfelt, T., Elde, R., Fuxe, K., Johansson, O., Ljundahl, A., Goldstein, M., Luft, R., Efendic, E., Nilsson, G., Terenius, L., Gauten, D., Jeffcoate, S. L., Rehfeld, J., Said, S., Perez de la Mora, M., Possani, L., Tapia, R., Teran, L. and Palacios, R. (1978). Aminergic and peptidergic pathways in the nervous system with special reference to the hypothalamus. *In* S. Reichlin, R. Baldessarini and J. Martin (eds), *The hypothalamus*, pp. 69–135. Raven Press, New York.

Hubbard, J. E. and DiCarlo, V. (1973). Fluorescence histochemistry of monoamine-containing cell bodies in the brainstem of the squirrel monkey (*Saimiri sciureus*) I. The locus coeruleus. *J. Comp. Neurol.* 147, 553–566.

Iijima, K. and Ogawa, T. (1980). An HRP study on cell types and their regional topography within the locus coeruleus innervating the supraoptic nucleus of the rat. *Acta. Histochem.* 67, 127–138.

Jacobowitz, D. M. (1978). Monoaminergic pathways in the central nervous system. *In* M. A. Lipson, K. F. Killam and A. DeMascio (eds), *Psychopharmacology: A generation of progress*, pp. 39–66. Raven Press, New York.

Jones, B. E., Halaris, A. E., McInany, M. and Moore, R. Y. (1977). Ascending projections of the locus coeruleus in the rat. I. Axonal transport in central noradrenaline neurons. *Brain Res.* 127, 1–21.

Kobayashi, R. M., Palkovits, M., Kopin, I. J. and Jacobowitz, D. M. (1974). Biochemical mapping of noradrenergic nerves arising from the rat locus coeruleus. *Brain Res.* 77, 269–279.

Koda, L. T., Schulman, J. A. and Bloom, F. E. (1978). Ultrastructural identfiction of noradrenergic terminals in the rat hippocampus: Unilateral destruction of the locus coeruleus with 6-hydroxydopamine. *Brain Res.* 145, 190–195.

Kromer, L. and Moore, R. (1976). Cochlear nucleus innervation by central norepinephrine neurons in the rat. *Brain Res.* 118, 531–537.

Kromer, L. F. and Moore, R. Y. (1980). A study of the organization of the locus coeruleus projections to the lateral geniculate nuclei in the albino rat. *Neurosci.* 5, 255–271.

Kuypers, H. G. J. M. and Maisky, V. A. (1975). Retrograde axonal transport of horseradish peroxidase from spinal cord brainstem cell groups in the cat. *Neurosci. Lett.* 1, 9–14.

Leger, L. and Descarries, L. (1978). Serotonin nerve terminals in the locus coeruleus of adult rat: A radioautographic study. *Brian Res.* 145, 1–13.

Levitt, P. and Moore, R. Y. (1978). Noradrenaline neuron innervation of the neocortex in the rat. *Brain Res.* 139, 219–231.

Levitt, P. and Moore, R. V. (1979). Origin and organization of brainstem catecholamine innervation in the rat. *J. Comp. Neurol.* 186, 4, 505–528.

Lidov, H. G. W., Rice, F. L. and Molliver, M. E. (1978). The organization of the catecholamine innervation of somatosenstory cortex: The barrel field of the mouse. *Brain Res.* 153, 577–584.

Lindvall, O. and Björklund, A. (1974). The organization of the ascending catecholamine neuron systems in the rat brain. *Acta. Physiol. Scand.* 412, 3–48.

Lindvall, O., Björklund, A., Nobin, A. and Stenevi, U. (1974). The adrenergic innervation of the rat thalamus as revealed by the glyoxylic acid fluorescence method. *J. Comp. Neurol.* 154, 317–348.

Lindvall, O. and Stenevi, U. (1978). Dopamine and noradrenaline neurons projecting to the septal area in the rat. *Cell Tiss. Res.* 190, 383–407.

Loewy, A. D., Saper, C. B. and Baker, R. P. (1979a). Descending projections from the pontine micturition center. *Brain Res.* 172, 533–538.

Loewy, A. D., McKellar, S. and Saper, C. B. (1979b). Direct projections from the A5 catecholamine cell group to the intermediolateral cell column. *Brain Res.* 174, 309–314.

Loizou, L. (1969). Projections of the nucleus locus coeruleus in the albino rat. *Brain Res.* 15, 563–566.

Loughlin, S. E., Foote, S. L. and Fallon, J. H. (1982). Locus coeruleus projections to cortex: Topography, morphology and collateralization. *Brain Res. Bull.* 9, 287–294.

Loughlin, S. E., Foote, S. L. and Bloom, F. E. (in press, a). Efferent projections of nucleus locus coeruleus: Topographic organization of cells of origin demonstrated by three-dimensional reconstruction. *Neurosci.*

Loughlin, S. E., Foote, S. L. and Grzanna, R. (in press, b). Efferent projections of nucleus locus coeruleus: Morphologic subpopulations have different efferent targets. *Neurosci.*

Loy, R., Koziell, D., Lindsey, J. and Moore, R. Y. (1980). Noradrenergic innervation of the adult rat hippocampal formation. *J. Comp. Neurol.* 189, 699–710.

McGinty, J. F., Koda, L. Y. and Bloom, F. E. (1983). A combined vascular-catecholamine fluorescence method reveals the relative vascularity of rat locus coeruleus and the paraventricular and supraoptic nuclei of the hypothalamus. *Neurosci. Lett.* 36, 117–123.

Mason, S. T. and Fibiger, H. C. (1979). Regional topography within noradrenergic locus coeruleus as revealed by retrograde transport of horseradish peroxidase. *J. Comp. Neurol.* 187, 703–724.

Moore, R. Y. and Bloom, F. E. (1979). Central catecholamine neuron systems: Anatomy and physiology of the norepinephrine and epinephrine systems. *Ann. Rev. Neurosci.* 2, 113–168.

Morrison, J. H., Grzanna, R., Molliver, M. E. and Coyle, J. T. (1978). The distribution and orientation of noradrenergic fibers in neocortex of the rat: An immunofluorescence study. *J. Comp. Neurol.* 181, 17–40.

Morrison, J., Molliver, M. and Grzanna, R. (1979). Noradrenergic innervation of cerebral cortex: Widespread effects of local cortical lesions. *Science*, 202, 313–316.

Morrison, J. H., Molliver, M. E., Grzanna, R. and Coyle, J. T. (1981). The intracortical trajectory of the coeruleo-cortical projection in the rat. A tangentially organized cortical afferent. *Neurosci.* 6, 139–158.

Nagai, T., Satoh, K., Imamoto, K. and Maeda, T. (1981). Divergent projections of catecholamine neurons of the locus coeruleus as revealed by fluorscent retrograde double-labeling technique. *Neurosci. Lett.* 23, 117–123.

Nygren, L. and Olson, L. (1977). A new major projection from locus coeruleus: The main source of noradrenergic nerve terminals in the ventral and dorsal columns of the spinal cord. *Brain Res.* 132, 85–93.

Palkovits, M. and Jacobowitz, D. (1974). Topographic atlas of catecholamine and acetylcholinesterase containing neurons in the rat brain. *J. Comp. Neurol.* 157, 29–42.

Palkovits, M., Leranth, C., Zaborszky, L. and Brownstein, M. (1977). Electron microscopic evidence of direct neuronal connections from the lower brainstem to the median eminence. *Brain Res.* 136, 339–334.

Phillipson, O. T. (1979). Afferent projections to the ventral tegmental area of Tsai and interfascicular nucleus: A horseradish peroxidase study in the rat. *J. Comp. Neurol.* 187, 117–144.

Pickel, V. M., Segal, M. and Bloom, F. E. (1974). An autoradiographic study of the efferent pathways of the nucleus locus coeruleus. *J. Comp. Neurol.* 155, 15–42.

Pickel, V. M., Joh, T. H. and Reis, D. J. (1977). A serotonergic innervation of noradrenergic neurons in nucleus locus coeruleus: Demonstration by immunocytochemical localization of the transmitter specific enzymes tyrosine and tryptophan hydroxylase. *Brain Res.* 131, 197–214.

Pickel, V. M., Joh, T. H., Reis, D. J., Leeman, S. E. and Miller, R. J. (1979). Electron microscopic localization of substance P and enkephalin in axon terminals related to dendrites of catecholaminergic neurons. *Brain Res.* 160, 387–400.

Saavedra, J. M., Grobecker, H. and Zivin, J. (1976). Catecholamines in the raphe nuclei of the rat. *Brain Res.* 114, 339–345.

Sakai, K., Salvert, D., Touret, M. and Jouvet, M. (1977). Afferent connections of the nucleus raphe dorsalis in the cat as visualizd by the horseradish peroxidase technique. *Brain Res.* 137, 21–41.

Saper, C. B., Swanson, L. W. and Cowan, W. M. (1979). An autoradiographic study of the efferent connections of the lateral hypothalamic area in the rat. *J. Comp. Neurol.* 183, 689–706.

Satoh, K., Tohyama, M., Yamamoto, K., Sakumoto, T. and Shimizu, N. (1977). Noradrenaline innervation of the spinal cord studied by the horseradish peroxidase method combined with monoamine oxidase staining, *Exp. Brain Res.* 30, 175–186.

Satoh, K., Tohyama, M., Sakumotor, T., Yamamotoh, K. and Shimizu, N. (1978). Descending projection of the nucleus tegmentalis laterodorsalis to the spinal cord: Studies by the horseradish peroxidase method following 6-hydroxy-dopa administration. *Neurosci. Lett.* 8, 9–15.

Shimizu, N., Ohnisihi, S., Satoh, K. and Tohyama, M. (1978). Cellular organization of locus coeruleus in the rat as studied by the Golgi method. *Arch. Histol. Jap.* 41, 103–112.

Shimizu, N., Katoh, Y., Hida, T. and Satch, K. (1979). The fine structural organization of the locus coeruleus in the rat, with reference to noradrenaline content. *Exp. Brain Res.* 37, 139–148.

Simon, H., Le Moal, M. Stinus, L. and Calas, A. (1979). Anatomical relationships between the ventral mesencephalic tegmentum-A10 region and the locus coeruleus as demonstrated by anterograde and retrograde tracing techniques. *J. Neurol. TRans.* 44, 77–86.

Steindler, D. A. and Deniau, J. M. (1980). Anatomical evidence for collateral branching of substantia nigra neurons: A combined horseradish peroxidase and wheat germ agglutinin axonal transport. *Brain Res.* 196, 228–236.

Swanson, L. W. (1976). The locus coeruleus: A cytoarchitectonic, Golgi and immunohistochemical study in the albino rat. *Brain Res.* 110, 39–56.
Swanson, L. W. (1977). Immunohistochemical evidence for a neurophysin-containing autonomic pathway arising in the paraventricular nucleus of the hypothalamus. *Brain Res.* 128, 346–353.
Swanson, L. W. and Hartman, B. K. (1975). The central adrenergic system: An immunofluorescence study of the location of cell bodies and their efferent connections in the rat utilizing dopamine-B-hydroxylase as a marker. *J. Comp. Neurol.* 163, 467–505.
Swanson, L. W. and Saper, C. B. (1978). Direct neural inputs to locus coeruleus from basal forebrain. *Soc. Neurosci. Abst.* 4.
Swanson, L. W., Sanchenko, P. E., Rivier, J. and Vale, W. W. (1983). Organization of ovine corticotropin releasing factor immunoreactive cells and fibers in the rat brain: An immunohistochemical study. *Neuroendocrinology* 36, 165–186.
Uhl, G. R., Goodman, R. R. and Snyder, S. (1979). Neurotensin-containing cell bodies, fibers and nerve terminals in the brainstem of the rat: Immunohistochemical mapping. *Brain Res.* 167, 77–91.
Ungerstedt, U. (1971). Stereotaxic mapping of the monoamine pathways in the rat brain. *Acta. Physiol. Scand.* 367, 1–48.
Waterhouse, B. D., Lin, C. -S., Burne, R. A. and Woodward, D. J. (1983). The distribution of neocortical projection neurons in the locus coeruleus. *J. Comp. Neurol.* 217, 418–431.
Watson, S. J., Khachaturian, H., Akil, H., Coy, D. H. and Goldstein, A. (1982). Comparison of the distribution of dynorphin systems and enkephalin systems in brain. *Science* 218, 1134–1136.
Westlund, K. N. and Coulter, J. D. (1980). Descending projections of the locus coeruleus and subcoeruleus/medial parabrachial nuclei in monkey: Axonal transport studies and dopamine-B-hydroxylase immunocytochemistry. *Brain Res. Rev.* 2, 235–264.
Westlund, K. N., Bowker, R. M., Ziegler, M. G. and Coulter, J. D. (1983). Noradrenergic projections to the spinal cord of the rat. *Brain Res.* 263, 15–31.

5

Brain stem nuclei associated with respiratory, cardiovascular and other autonomic functions

EVA K. BYSTRZYCKA
BRUCE S. NAIL

University of New South Wales
Kensington, NSW, Australia

I. INTRODUCTION

Brain stem neurons associated with autonomic, cardiovascular and respiratory functions are difficult to describe because often they do not aggregate in histologically well defined nuclei. The introduction of more specific techniques, for example, the transport of the label horseradish peroxidase (HRP) and immunocytochemistry and autoradiography, has revealed important functional relations between aggregates of brain stem neurons, but has also uncovered further complexities. The location of the lacrimal and salivatory neurons, for example (for which there was scarcely any agreement as to nuclear definition among earlier workers), has been clearly established by use of modern histochemical techniques, and functional connections of these cell groups are now better understood. However, it has become apparent that these scattered cells do not constitute "nuclei" in the conventional sense.

The uncertainty concerning the exact location of some of the brain stem nuclei may be related to the migration of the cells during development. In mammals, the primitive dorsal visceral efferent column of the medulla separates into a dorsal and ventral portion. The general efferent neurons remain in the dorsal motor nucleus of vagus (10), while the special visceral efferent neurons supplying branchial musculature migrate to form the ambiguus (Amb) nucleus (Kappers *et al.*, 1960). This ventrolateral migration of part of the original dorsal efferent cell column proceeds to a variable extent in mammalian species so that few cells assume an intermediate position between 10 and the Amb, that is, in

THE RAT NERVOUS SYSTEM
ISBN 0 12 547632 9

the parvocellular reticular nucleus (PCRt), whereas others reach a position ventrolateral to the branchial motoneurons, so forming a periambigual group of neurons with cardiorespiratory and other autonomic functions.

II. NUCLEUS OF THE SOLITARY TRACT

The rostral limit of the nucleus of the solitary tract (Sol) is at the level of the caudal pole of the facial motor nucleus (7), and here the nucleus is located dorsomedial to the spinal trigeminal nucleus; its caudal limit lies at the level of the lower end of the pyramidal decussation. The right and left Sol fuse caudally to form the commissural nucleus of Cajal, which is continuous with the nucleus of the dorsal commissure of the spinal cord. The medial and dorsal borders of the Sol are easy to determine, but its ventrolateral border merges with the cells of the reticular formation and is ill defined. The nucleus can be divided into a larger medial (SolM), and a smaller lateral (SolL), portion (Torvik, 1956). The border between these divisions is medial to the solitary tract. The SolM is situated dorsolateral to 10 and at a short distance rostral to the area postrema it comes close (100 μm–150 μm) to the surface of the fourth ventricle. At the level of the obex, the right and left part of the nucleus are close to each other. Further distally (0.8 mm–0.9 mm) the cells merge in the midline, forming the commissural nucleus, a caudal extension of the SolM. The lateral division of the Sol is larger and better outlined in its rostral extent; caudally, it ends at the level of the area postrema.

The classical division of the nucleus of the solitary tract (Sol) is into a medial and a lateral subnucleus; the morphology of these two parts differs considerably. The SolM consists of densely packed, small, pale cells whose somata are round, oval, triangular or fusiform whereas the SolL is composed of less densely packed, larger multipolar cells which have a darker appearance in Nissl stained preparations (Torvik, 1956). Light and electron microscopic studies (Palkovits and Zaborszky, 1977) have shown that large numbers of SolM cells are multipolar with two or three long dendrites which remain within the nucleus while the axons (which originate from the soma) can be traced to the nucleus ambiguus and reticular formation. These multipolar cells have axosomatic and axodendritic synaptic terminals onto which the primary axons of the 9th and 10th nerves terminate. Fusiform and small round cells with short axons and (usually) two primary dendrites are also seen. Their axons appear to be confined to the nucleus where they make axodendritic and axoaxonic synaptic contact. More recently the Sol has been further subdivided using Nissl stained material and transganglionic transport of HRP (Kalia and Sullivan, 1982). Dorsomedial to the solitary tract, pale, medium sized oval cells form a medial subnucleus which extends for over 2 mm in a rostrocaudal direction. Lateral to this subnucleus, a compact cluster of large cells forms an intermediate subnucleus which is most prominent at a level rostral to the obex. A ventral subnucleus is situated ventral to the solitary tract and is composed of medium sized multipolar cells. A ventrolateral subnucleus is located about 150 μm ventral to the solitary tract and is composed of large multipolar neurons. On the lateral border of the solitary tract are concentrations of medium sized oval cells which form an interstitial subnucleus while on the dorsal border, clusters of oval and small cells form respectively, the dorsolateral and the dorsal subnuclei. Transganglionic transport of HRP injected into a nodose ganglion revealed a strong projection of vagal afferents to the ipsilateral and contralateral medial and commissural subnuclei, a more moderate projection to the ipsilateral intermediate, ventral and ventrolateral subnuclei, and a relatively weak projection to the interstitial and dorsolateral subnuclei.

Recent neuroanatomic and neuropharmacologic studies have demonstrated a substantial number of different neurotransmitters within the Sol. The neurons of the A_2 catecholamine group (Dahlström and Fuxe, 1964) are primarily distributed in both the Sol and 10. Norepinephrine (NE) and dopamine (DA) cells are present throughout the rostrocaudal extent of the Sol; the highest density being observed in the caudal part of the SolM, and in the commissural nucleus. There are also substantial numbers of catecholamine nerve terminals which form axodendritic synapses within the Sol, particularly in its medial and ventrolateral divisions. Some of these axons arise from the sensory neurons of the nodose ganglion (Chiba and Kato, 1978; Palkovits and Jacobowitz, 1974; Rea *et al.*, 1982; Ritchie *et al.*, 1982; Snyder *et al.*, 1978; Sumal *et al.*, 1983).

Serotonergic terminals in the Sol arise from the cells of the medullary raphe nuclei (Dahlström and Fuxe, 1964; Rea *et al.*, 1982). Other transmitters found to be present in the Sol include: glutamate and γ-aminobutyric acid (GABA) (Perrone, 1981; Siemers *et al.*, 1982; Talman *et al.*, 1980); substance P, particularly in the medial and ventrolateral subnuclei and in the commissural nucleus (Haeusler and Osterwalder, 1980; Kalia *et al.*, 1984; Palkovits, 1980; (see also Volume 1 Chapters 12 and 14, for distribution in the nucleus of the solitary tract); vasopressin (Loewy, 1981); neurotensin, enkephalin, somatostatin, neurophysin and vasoactive intestinal polypeptide (Beitz, 1982; Kalia *et al.*, 1984; Palkovits *et al.*, 1982; Simantov *et al.*, 1977); and cholecystokinin, probably of vagal origin (Kiss *et al.*, 1982).

In addition, autoradiographic studies have demonstrated opiate receptors in the Sol and commissural nuclei (Atweh and Kuhar, 1977; Oley *et al.*, 1982).

A. Connections of the nucleus of the solitary tract

1. Afferents

The afferent connections of the nucleus of the solitary tract (Sol), studied with the Nauta method for terminal degeneration (Torvik, 1956), or with more recent anterograde transport techniques (Bereiter *et al.*, 1980; Contreras *et al.*, 1982; Leslie *et al.*, 1982), have been shown to be arranged in a topographic order. Afferents from the 5th, 7th, 9th and 10th nerves form the solitary tract and terminate successively rostrocaudally, that is, sensory axons of the facial nerve terminate in the rostral part of the SolL prior to the termination of the trigeminal and glossopharyngeal afferents; vagal, and some glossopharyngeal afferents terminate in the SolM at the levels rostral to obex, and in the SolL at more caudal levels. The contralateral SolM, commissural nucleus and the area postrema receive substantial number of vagal afferent terminals (Kalia and Sullivan, 1982). Supramedullary afferent projections to the Sol arise from contralateral cerebral cortex, the lateral hypothalamic area and the pontine locus coeruleus (LC) (Bereiter *et al.*, 1980; Torvik, 1956; Vigier and Portalier, 1979).

2. Efferents

Projections of the nucleus of the solitary tract (Sol) have been extensively studied using autoradiographic, fluorescent and HRP techniques (Cedarbaum and Aghajanian, 1978; Loewy, 1981; Norgren, 1978; Ross *et al.*, 1979). It has been shown that some of the axons of the Sol ramify extensively within the nucleus itself, while others project to the different parts of the central nervous system. Axons originating in the rostral part of the Sol project to the facial, trigeminal and hypoglossal nuclei. Those originating in the caudal part project to 10 and the Amb. Projections to the contralateral ventral horn of the cervical, thoracic and lumbar spinal cord have been demonstrated using the HRP technique. There is a substantial projection from the Sol to the ventral surface of the medulla (a region implicated in cardiovascular and respiratory control), to the LC, and the parabrachial (VPB, DPB) and Kölliker-Fuse (KF) nuclei of the pons.

Forebrain structures which receive projections from the Sol include the periventricular nucleus of the thalamus, the bed nucleus of the stria terminalis, the central nucleus of the amygdala and the paraventricular, dorsomedial and arcuate nuclei of the hypothalamus. Most of these ascending solitary projections are ipsilateral (Ricardo and Koh, 1978).

III. DORSAL MOTOR NUCLEUS OF VAGUS

The dorsal motor nucleus of vagus (10) forms the ventral boundary of the SolM and it is separated from the hypoglossal nucleus (12) by a small, low cell density area. The nucleus is characteristically spindle shaped with large numbers of cells present in the middle of its rostrocaudal extent. Recent HRP studies (application of the enzyme to the cervical vagus nerve) have shown that 10 extends rostrally to the level of the facial motor nucleus; caudally, labeled cells have been found in the commissural nucleus, and beyond in the dorsal central gray matter of the first to sixth cervical segments (Contreras *et al.*, 1980; Dennison *et al.*, 1981; Kalia and Sullivan, 1982; Leslie *et al.*, 1982; Ritchie *et al.*, 1982).

Histologic techniques and retrograde HRP labeling of the vagus nerve have revealed a heterogeneous mixture of cell types and sizes in this nucleus. In both the rostral and caudal parts of the nucleus, multipolar cells are more numerous. Bipolar neurons predominate in the central part. The transverse diameter of cells in 10 varies between 12 μm and 30 μm (Lewis *et al.*, 1970; Ritchie *et al.*, 1982).

The use of choline acetyltransferase (the synthetizing enzyme for acetylcholine [ACh]) allowed precise identification of cholinergic neurons and their axons within 10 (Houser *et al.*, 1983). On the other hand the use of a combined HRP and AChE technique has allowed identification of noncholinergic neurons within the nucleus (Hoover and Barron, 1982). This is in agreement with immunocytochemical studies which showed that the rostral part of 10, and the dorsal part of the commissural nucleus, contain substantial number of NE and DA cells (Palkovits and Zaborszky, 1977; Rea *et al.*, 1982; Ritchie *et al.*, 1982; Swanson and Hartman, 1975). Neurons of 10 and the commissural nucleus receive dense innervation from higher centers of the brain. Neurotransmitters such as neurotensin (Higgins *et al.*, 1982), serotonin (Dahlström and Fuxe, 1964), vasopressin (Loewy, 1981), substance P, somatostatin, enkephalin and neurophysin have been identified in 10 (Kalia *et al.*, 1984; Sawchenko and Swanson, 1982). A number of opiate receptors have been found in 10, and in the commissural nucleus (Atweh and Kuhar, 1977).

A. Connections of the dorsal motor nucleus of vagus

1. Afferents

Electrophysiologic and degeneration studies (Nosaka *et al.*, 1978; Palkovits and Zaborszky, 1977) have shown that the dorsal motor nucleus of vagus (10) and the commissural nucleus receive myelinated and nonmyelinated vagal afferents. These observations have been confirmed and extended by autoradiographic and HRP studies which showed that vagal afferents terminate in the ipsilateral and contralateral 10, the commissural nucleus and the dorsal gray matter of the cervical spinal cord (Contreras *et al.*, 1982; Kalia and Sullivan, 1982; Leslie *et al.*, 1982; Rogers *et al.*, 1980). Other projections include those from the parvocellular and gigantocellular reticular nuclei, the principal sensory trigeminal nucleus and the A_5 noradrenergic cell group (Loewy *et al.*, 1979). Hypothalamic projections arise from the magnocellular part of the paraventricular nucleus (Rogers *et al.*, 1980; Saper *et al.*, 1976; Swanson and Hartman, 1975).

2. Efferents

Axons of the preganglionic cells of the dorsal motor nucleus of vagus travel in the 10th cranial nerve and terminate in the parasympathetic ganglia of the thoracic and abdominal viscera. The precise location of 10 cell bodies has been studied by HRP application to the vagus nerve, or to the nodose ganglion (Contreras *et al.*, 1980; Kalia and Sullivan, 1982; Karim and Leong, 1980; Leslie *et al.*, 1982; Nosaka *et al.*, 1979; Stuesse, 1982). HRP was confined to the ipsilateral and contralateral 10. The largest number of labeled cells was located in the rostral and medial part of the nucleus. Application of HRP to the cardiac branches of the vagus nerve, or injection of the enzyme into the myocardium, produces relatively sparse labeling in the middle to rostral part of 10, whereas injection of HRP into the subdiaphragmatic vagus, stomach or pancreas resulted in heavy bilateral labeling throughout the rostrocaudal extent of the nucleus (Coil and Norgren, 1979; Dennison *et al.*, 1981; Leslie *et al.*, 1982; Weaver, 1980). These results indicate that in the rat 10 contributes substantially to the innervation of the abdominal viscera.

IV. AMBIGUUS NUCLEUS AND THE PERIAMBIGUAL AREA

The ambiguus nucleus (Amb) appears in Nissl stained sections as an aggregation of large multipolar neurons lying ventral to the PCRt and dorsal to the lateral reticular nucleus. Although classically described as being composed of large motoneurons which innervate the laryngeal and pharyngeal muscles, recent HRP studies have shown that some parasympathetic motoneurons which innervate thoracic and abdominal viscera, and other neurons which supply spinal respiratory motoneurons, are located in or at the ventrolateral order of the Amb, just dorsal to the lateral reticular nucleus. Accordingly, the earlier finding of Bell (1960) that the Amb contained visceral as well as branchial efferent neurons, is confirmed and the region ventrolateral to the Amb is now referred to as the periambigual area (Bradley and Lucy, 1983; Coil and Norgren, 1979; Kalia and Sullivan, 1982; Nosaka *et al.*, 1979; Stuesse, 1982). The nucleus extends from the caudal pole of the facial nucleus to the pyramidal decussation. As for other species, the most rostral part of the Amb is referred to as the "retrofacial" nucleus, and its caudal part as the "retroambiguus" nucleus (Kalia and Sullivan, 1982; Taber, 1961; Takayama *et al.*, 1982).

A. Neurons associated with laryngeal muscles

HRP labeled neurons visualized after injection of the enzyme into different laryngeal muscles or into superior or recurrent laryngeal nerve (Hinrichsen and Ryan, 1981) form separate ipsilateral pools. Two types of neurons can be seen inside these pools: large multipolar cells with three to six primary dendrites, located in the ventral part of the Amb; and smaller neurons with fewer dendrites located in the dorsal part of the nucleus.

B. Neurons associated with preganglionic parasympathetic efferents

These neurons have been visualized after HRP application to the vagus nerve, nodose ganglion or thoracic and abdominal viscera (Coil and Norgren, 1979; Leslie *et al.*, 1982; Nosaka *et al.*, 1979; Ritchie *et al.*, 1982; Takayama *et al.*, 1982). Apart from dense labeling of 10, cells positively labeled by HRP were located in the "retrofacial" nucleus, the Amb, the periambigual area and the PCRt. The cells of the Amb and periambigual area are small to medium (18 μm–30 μm) multipolar with few primary dendrites, and their axons extend dorsally. The PCRt cells, more numerous in the caudal medulla, are larger (25 μm–36 μm), and are bipolar or multipolar.

Cholinergic neurons in the Amb have been identified by both histochemical and immunocytochemical methods (Houser *et al.*, 1983; Hoover and Barron, 1982; Palkovits and Jacobowitz, 1974; Paxinos and Watson, 1982). There is also evidence that adrenergic dopamine-β-hydroxylase (DBH) positive cells are present in the area between the Amb and the A_1 cell group (presumbly in the periambigual area) and that some of them are vagal efferents (Ritchie *et al.*, 1982). Other transmitters localized within the Amb include 5-hydroxytryptamine (5HT) and substance P. Opiate receptors have also been reported (Atweh and Kuhar, 1977; Sawchenko and Swanson, 1982).

C. Connections of the ambiguus nucleus

1. *Afferents*

An afferent input from the Sol to the Amb has been well established. Following small lesions in the SolM, terminal nerve degeneration was observed around motoneurons of the Amb (Palkovits and Zaborsky, 1977). Audioradiographic and HRP studies have shown that the rostral part of the Amb and periambigual area receive projections from the medial and, to a lesser extent, from the ventrolateral portion of the Sol (Norgren, 1978; Stuesse and Powell, 1982). Other projections arise from the VPB, DPB and Kölliker-Fuse nuclei of the pons (Saper and Loewy, 1980).

2. *Efferents*

a. Laryngeal motoneurons

Cells which innervate the cricothyroid muscles are located in the rostral part of the Amb whereas those innervating the posterior and lateral cricoarytenoid and thyroarytenoid muscles lie in the medial and caudal parts of the Amb. The representation is entirely ipsilateral (Hinrichsen and Ryan, 1981).

b. Preganglionic parasympathetic neurons

Large numbers of preganglionic cells have been localized in the Amb and periambigual area, in a column which extends from the caudal pole of the facial nucleus to the pyramidal decussation (Contreras *et al.*, 1980; Karim and Leong, 1980; Ritchie *et al.*, 1982). By applying HRP onto cardiac vagal branches or into the myocardium, Nosaka *et al.* (1979) and Stuesse (1982) produced sparse labeling in 10 and very heavy labeling in the Amb and periambigual area. In such experiments, cells positively labeled by HRP extended from the level of the "retrofacial" nucleus to the caudal medulla where they were also present in the PCRt. Exposure of hepatic vagal branches, the wall of the stomach, or the pancreas to HRP showed a relatively small fraction of labeled neurons to be located in the rostral part of the Amb and periambigual area, as compared to the dense labeling of 10 (Coil and Norgren, 1979; Rogers and Hermann, 1983; Takayama *et al.*, 1982; Weaver, 1980).

V. LACRIMAL AND SALIVATORY PREGANGLIONIC NEURONS

Early anatomic and physiologic studies which attempted to locate the lacrimal and salivatory "nuclei" produced conflicting results, due to the limitations of the available techniques. More accurate determination of the cells involved in lacrimation and salivation has been possible using axonal transport of HRP (Contreras *et al.*, 1980; Eisenman and Azmitia, 1982; Hiura, 1977; Hosoya *et al.*, 1983; Mitchell and Templeton, 1981; Nicholson and Severin, 1981). Cells so determined are scattered throughout the rostral medulla and caudal pons and do not form discrete "nuclei".

A. Lacrimal "nucleus"

Following exposure of the greater petrosal nerve to HRP, labeled cells were distributed in the ipsilateral parvocellular reticular nucleus (PCRt), dorsolateral to the rostral third of the facial nucleus, and medial to the spinal nucleus of the trigeminal nerve. Smaller numbers of cells positively labeled by HRP were also present at the rostral and caudal ends of the facial nucleus, and at the rostral end of the Amb. In addition, a separate small population of labeled neurons could be seen in the most dorsal part of the PCRt. The labeled cells were of medium size (10 μm–25 μm), round and fusiform with four to six primary dendrites oriented dorsoventrally.

B. Salivatory "nuclei"

The superior salivatory neurons have been visualized after application of HRP onto the chorda tympani nerve and were located in the PCRt, medial to the spinal nucleus of the trigeminal nerve and lateral to the ascending root of the facial nerve, on the ipsilateral side. The cells form a column which extends between the caudal end of the facial nucleus and the genu and root of the facial nerve. These efferent neurons appear to be located between the ventral and dorsal group of lacrimal cells. The inferior salivatory neurons have been identified by the retrograde transport of HRP applied to the otic ganglion, or to the tympanic branch of the 9th nerve. The location of labeled cells was similar to that observed after exposure of the chorda tympani to HRP except that a higher concentration of cells is evident at the caudal end of the facial nucleus. There are no distinguishable differences in the morphology of the salivatory and lacrimal preganglionic neurons.

VI. PARABRACHIAL AND KÖLLIKER-FUSE NUCLEI OF THE PONS

These nuclei may be subdivided into three parts: the ventral (or medial) parabranchial nucleus (VPB) located ventromedial to the superior cerebellar peduncle (scp) and ventrolateral to the LC; the dorsal (or lateral) parabranchial nucleus (DPB) situated dorsolateral to the scp; and the Kölliker–Fuse nucleus (KF) located ventrolateral to the scp.

The ventral parabrachial nucleus neurons can be divided into two categories: medium size multipolar; and small, spindle shaped cells. They form a distinctive group which can be easily distinguished from the neighboring cells of the reticular formation. The DPB cells are round, oval or spindle shaped and they tend to lie in clusters of homogeneous cell type. The average diameter of cells in both parabrachial nuclei is 16 μm. The Kölliker-Fuse neurons are larger (12 μm–20 μm), multipolar and less densely packed (Saper and Loewy, 1980; Westlund *et al.*, 1981).

The neurotransmitters identified in the VPB and DPB include neurotensin, enkephalin and substance P (Kawai *et al.*, 1982; Uhl *et al.*, 1977).

B. Connections of the nuclei

1. Afferents

Autoradiographic and electrophysiologic studies have shown that the parabrachial nuclei receive a substantial afferent input from the Sol. Axons of cells located in the rostral part of the Sol terminate on both the VPB and DPB neurons whereas those from the caudal part of the Sol and AP terminate in the rostral part of the DPB (Koh and Ricardo, 1978; Norgren, 1978; Ogawa and Kaisaku, 1982; Shapiro and Miselis, 1982). There is a bilateral projection to the DPB and Kölliker-Fuse nuclei from the Amb and the periambigual area (McKellar and Loewy, 1982). Another source of afferent fibers to the VPB and DPB is from the paraventricular nucleus of the hypothalamus and the amygdala (Berk and Finkelstein, 1982; Conrad and Pfaff, 1976; Krettek and Price, 1978).

2. *Efferents*

Parabrachial nuclei have extensive connections with the higher centers of the brain, the brain stem and the spinal cord. Rostral projections have been described from both the VPB and DPB to the midbrain raphe nuclei, the thalamus, the hypothalamus, the amygdala, and the frontal cortical areas. The Kölliker-Fuse nucleus projects only to the hypothalamus and amygdala (Divac *et al.*, 1978; Koh and Ricardo, 1978; Norgren, 1976; Voshart and Van der Kooy, 1981). Caudal projections of the PB, DPB and Kölliker-Fuse nuclei include those to the ventrolateral medullary reticular formation (that is, the Amb and periambigual area), sparse projections to the ventral part of the Sol, and extensive, mainly ipsilateral, projections to all levels of the spinal cord (Saper and Loewy, 1980; Westlund *et al.*, 1982).

VII. CENTRAL ORGANIZATION AND FUNCTIONAL ASPECTS OF THE NUCLEI

A. Respiratory functions

Areas associated with respiratory functions have not been defined in the rat with the same accuracy as in the cat, rabbit or sheep (for review, see Cohen, 1979). Electrophysiologic and histochemical techniques have located phasic respiratory neurons in the ventrolateral medulla of the rat (that is, 1.5 mm–2 mm lateral from the midline), the Amb, the lateral reticular and the neighboring reticular formation, at a distance of 1.5 mm rostral and 1 mm caudal to the obex (Bradley and Lucy, 1979, 1983; Howard and Tabatabai, 1975). No such cells were observed in the Sol, or in other areas of the brain stem known to contain a high density of respiratory related neurons in other species. The afferent and efferent connections of the respiratory cells so identified (with exception of respiratory phased laryngeal motoneurons) have not yet been demonstrated. A large proportion of the neurons in the rat medulla which possess a respiratory rhythm have been shown to be cholinoceptive (Bradley and Lucy, 1983).

Chemosensitive structures in the ventral, superficial layer of the rostral medulla, implicated in both cardiovascular (Loewy, 1981) and respiratory (Fukuda and Loeschcke, 1979; Fukuda *et al.*, 1980; Hori *et al.*, 1970; Laha *et al.*, 1977; Loeschcke, 1982) functions have attracted more attention in this species. Two areas have been described in the rat in relation to ventilatory responses. An increase in ventilation is observed during electric or chemical stimulation of an area located at the lateral border of the pyramidal tract (0.5 mm–1 mm lateral to the midline and rostral to the roots of the 12th nerve). This area corresponds to the most ventral part of the paragigantocellular nucleus. A decrease in ventilation occurs when an area ventral to the lateral reticular nucleus is stimulated. The neurons located in these two areas are sensitive to hydrogen ions and acetylcholine (ACh). The excitation of these neurons by hydrogen depends largely upon intact cholinergic transmission in the ventral superficial medullary layer. Application of ACh to the rostral part of the ventral medullary surface leads to an increase in ventilation.

B. Cardiovascular functions

The caudal part of the nucleus of the solitary tract, together with the commissural nucleus, forms the primary medullary center for the multisynaptic cardiovascular reflex arc. Lesions of these areas cause acute hypertension in rats, while electric stimulation results in hypotension and bradycardia (for review see Palkovits, 1980; Palkovits and Zaborszky, 1977; Spyer, 1981). Primary neurons of the baroreceptor reflex arc have perikarya in the nodose ganglia and connect the peripheral baroreceptor sites with the Sol via fibers of the glossopharyngeal and vagus nerves. These afferent fibers synapse on second order neurons located in or around the caudal and commissural part of the SolM. Secondary neurons in turn terminate in various medullary nuclei, for example, 10, the Amb and periambigual area, the lateral reticular, gigantocellular and paragigantocellular reticular nuclei and, either directly or by multisynaptic pathways, also reach higher centers involved in modulation of the baroreceptor reflex. It has been shown that the Sol projects to at least four sites which have direct connections with the interomediolateral column of the spinal cord. These sites are the paraventricular nucleus of the hypothalamus, the Kölliker–Fuse nucleus, the A_5 cell group and those areas of the ventral medulla involved in chemosensitive functions. These four sites modulate sympathetic and vagal parasympathetic cardiovascular activities (Loewy, 1981).

C. Other autonomic functions

The Sol, a site of termination of primary afferent fibers from visceral receptors, plays an essential role in the central neural integration of autonomic function. The rostral part of the Sol receives gustatory afferents and relays to the pontine parabrachial nuclei from whence the thalamocortical system and the hypothalamolimbic system may be affected. The caudal (visceral) part of the Sol sends projections either directly to the hypothalamus and limbic system, or via the parabrachial nuclei (Contreras *et al.*, 1982; Koh and Ricardo, 1978; Norgren, 1978; Ricardo and Koh, 1978; Saper and Loewy, 1980). Forebrain structures, including the hypothalamus, the central nucleus of the amygdala, the bed nucleus of the stria terminalis and the medial preoptic area receive direct inputs from the Sol and have been implicated in visceral effector mechanisms relating to alterations in blood pressure and heart rate, modification of respiratory functions and alterations of motility and secretory activity of the stomach. There have been suggestions that the solitaro-forebrain anatomic pathways can influence the release of hormones by the hypothalamus, and in addition may influence integrated behavior patterns such as the defense reactions, feeding and drinking, and the sleep-wakefulness cycle (Ricardo and Koh, 1978). The parabrachial nuclei appear to be involved in parallel processing of gustatory and interoceptive afferents, and project to virtually all hypothalamic nuclei except the mammillary nucleus. In turn, the parabrachial nuclei receive inputs from autonomic hypothalamic neurons and the central nucleus of the amygdala, regions which also have direct projections to the SolM. These interconnections between the solitary, parabrachial, hypothalamic and limbic structures underline the highly organized and integrated nature of the substrate which serves autonomic control.

Histochemical studies of the location of preganglionic neurons in the rat brain stem, and of their associated efferent pathways, provided important data on the organization of motor nuclei involved in autonomic function. It is now clear that 10, the Amb and the periambigual area, supply both thoracic and abdominal viscera. A more substantial innervation to abdominal, compared with thoracic, viscera is provided by 10. In addition to representing the site of pharyngeal and laryngeal motoneurons, the Amb participates in the innervation of both visceral compartments, particularly with reference to the parasympathetic innervation of the heart. These ventrolateral preganglionic motoneurons are to some extent intermingled with branchial motoneurons, but the majority of cells are outside the Amb in the periambigual area; a few are in the PCRt. The site of these preganglionic neurons is similar to those described in other species such as the cat and the rabbit, and reflects the pattern of origin and mode of migration of the cells from the dorsal surface of the medulla to the ventrolateral reticular formation (Kappers *et al.*, 1960).

A similar pattern can be observed in the position of the motoneurons supplying the lacrimal and the salivary glands. These cells, located predominantly in the PCRt, form a dispersed longitudinal column oriented radially with respect to the fourth ventricle, a finding which may explain why there has been little agreement over the years concerning the precise localization of this nuclear mass.

ACKNOWLEDGEMENTS

The authors are grateful to Ms Martina Wong for help in the literature search, and to Ms Lorraine Brooks for typing the manuscript.

REFERENCES

Atweh, S. F. and Kuhar, M. J. (1977). Autoradiographic localization of opiate receptors in rat brain. I. Spinal cord and lower medulla. *Brain Res.* 124, 53–67.

Beitz, A. J. (1982). The sites of origin of brainstem neurotensin and serotonin projections to the rodent nucleus raphe magnus. *J. Neurosci.* 2, 829–842.
Bell, F. R. (1960). The localization within the dorsal motor nucleus of the vagus of the efferent fibers to the ruminant stomach. *J. Anat* 94, 410–417.
Bereiter, D., Berthoud, H. R. and Jeanrenaud, B. (1980). Hypothalamic input to brainstem neurons responsive to oropharyngeal stimulation. *Exp. Brain Res.* 39, 33–39.
Berk, M. L. and Finkelstein, J. A. (1982). Efferent connections of the lateral hypothalamic area of the rat: An autoradiographic investigation. *Brain Res. Bull.* 5, 511–526.
Bradley, P. B. and Lucy, A. P. (1979). An iontophoretic study of the cholinoceptive properties of respiratory neurons in the rat medulla. *Brit. J. Pharmacol.* 66, 111–112.
Bradley, P. B. and Lucy, A. P. (1983). Cholinoceptive properties of respiratory neurons in the rat medulla. *Neuropharmacol.* 22, 853–858.
Cedarbaum, J. M. and Aghajanian, G. K. (1978). Afferent projections to the rat locus coeruleus as demonstrated by a retrograde tracing technique. *J. Comp. Neurol.* 178, 1–16.
Chiba, T. and Kato, M. (1978). Synaptic structures and quantification of catecholaminergic axons in the nucleus tractus solitarius of the rat: Possible modulatory roles of catecholamines in baroreceptor reflex. *Brain Res.* 151, 323–338.
Cohen, M. I. (1979). Neurogenesis of respiratory rhythm in the mammal. *Physiol. Rev.* 59, 1105–1173.
Coil, J. D. and Norgren, R. (1979). Cells of origin of motor axons in the subdiaphragmatic vagus of the rat. *J. Auton. Nerv. Sys.* 1, 203–210.
Conrad, L. C. A. and Pfaff, D. W. (1976). Efferents from medial based forebrain and hypothalamus in the rat. II. An autoradiographic study of the anterior hypothalamus. *J. Comp. Neurol.* 169, 221–262.
Contreras, R. J., Gomez, M. and Norgren, R. (1980). Central origins of cranial parasympathetic neurons in the rat. *J. Comp. Neurol.* 190, 373–394.
Contreras, R. J., Beckstead, R. M. and Norgren, R. (1982). The central projections of the trigeminal, facial, glossopharyngeal and vagus nerves: an autoradiographic study in the rat. *J. Auton. Nerv. Sys.* 6, 303–322.
Dahlström, A. and Fuxe, K. (1964). Evidence for the existence of monoamine-containing neurons in the central nervous system. I. Demonstration of monoamines in the cell bodies of the brainstem neurons. *Acta Physiol. Scand.* 62, 1–55.
Dennison, S. J., O'Connor, B. L., Aprison, M. H. and Merritt, V. E. (1981). Viscerotopic localization of preganglionic parasympathetic cell bodies of origin of the anterior and posterior subdiaphragmatic vagus nerves. *J. Comp. Neurol.* 197, 259–269.
Divac, I., Kosmal, A., Bjorklund, A. and Lindvall, O. (1978). Subcortical projections to the prefrontal cortex in the rat as revealed by the horseradish peroxidase technique. *Neurosci.* 3, 785–796.
Eisenman, J. S. and Azmitia, E. C. (1982). Physiological stimulation enhances HRP marking of salivary neurons in the rat. *Brain Res. Bull.* 8, 73–78.
Fukuda, Y. and Loeschcke, H. (1979). A cholinergic mechanism involved in the neuronal excitation by H^+ in the respiratory chemosensitive structures of the ventral medulla oblongata of rats in vitro. *Pflugers Arch.* 370, 125–135.
Fukuda, Y., See, W. R. and Honda, Y. (1980). H^+-sensitivity and pattern of discharge of neurons in the chemosensitive areas of the ventral medulla oblongata of rats in vitro. *Pflugers Arch.* 388, 53–61.
Haeusler, G. and Osterwalder, R. (1980). Evidence suggesting a transmitter or neuromodulatory role for substance P at the first synapse of the baroreceptor reflex. *Naunyn Schmiedebergs Arch. Pharmacol.* 314, 111–121.
Higgins, G. A., Hoffman, G. E., Wray, S. and Schwaber, J. S. (1982). Immunochemical analysis of the central connections of the vagus nerve in the rat: Neurotensin-immunoreactivity within the nucleus tractus solitarius and vagal motor nuclei. *Soc. Neurosci. Abst.* 8, 426.
Hinrichsen, C. F. L. and Ryan, T. (1981). Localization of laryngeal motoneurons in the rat: morphologic evidence for dual innervation. *Exp. Neurol.* 74, 341–355.
Hiura, T. (1977). Salivatory neurons innervate the submandibular and sublingual glands in the rat: Horseradish peroxidase study. *Brain Res.* 137, 145–149.

Hoover, D. B. and Barron, S. E. (1982). Localization of acetylcholinesterase content of vagal efferent neurons. *Brain Res. Bull.* 8, 279–284.
Hori, T., Roth, G. I. and Yamamoto, W. S. (1970). Respiratory sensitivity of rat brainstem surface to chemical stimuli. *J. Appl. Physiol.* 28, 721–724.
Hosoya, Y., Matsushita, M. and Sugiara, Y. (1983). A direct hypothalamic projection to the superior salivatory nucleus neurons in the rat. A study using anterograde autoradiographic and retrograde HRP methods. *Brain Res.* 266, 329–333.
Houser, C. R., Crawford, G. D., Barber, R. P., Salvaterra, M. and Vaughn, J. E. (1983). Organization and morphological characteristics of cholinergic neurons: An immunocytochemical study with a monoclonal antibody to cholineacetyltransferase. *Brain Res.* 266, 97–119.
Howard, B. R. and Tabatabai, M. (1975). Localization of the medullary respiratory neurons in rats by microelectrode recording. *J. Appl. Physiol.* 39, 812–817.
Kalia, M. and Sullivan, M. J. (1982). Brain stem projections of sensory and motor components of the vagus nerve in the rat. *J. Comp. Neurol.* 211, 248–264.
Kalia, M., Fuxe, K., Hökfelt, T., Johansson, O., Lang, R. E., Ganten, D., Cuello, C. and Terenius, L. (1984). Distribution of neuropeptide immunoreactive nerve terminals within the subnuclei of the nucleus of the tractus solitarius of the rat. *J. Comp. Neurol.* 222, 409–444.
Kappers, C. U. A., Huber, G. C. and Crosby, E. C. (1960). *The comparative anatomy of the nervous system of vertebrates, including man*, Vol. I. Hafner, New York.
Karim, M. A. and Leong, S. K. (1980). Neurons of origin of cervical vagus nerves in the rat and monkey. *Brain Res.* 186, 208–210.
Kawai, Y., Inagaki, S., Shiosaka, S., Senba, E., Takatsuki, K., Sakanaka, M., Umegki, K. and Tohyama, M. (1982). Multiple innervation by substance P-containing fibers in the parabrachial area of the rat. *Neurosci. Lett.* 35, 271–317.
Kiss, J. Z., Williams, T. H., Beinfeld, M. C. and Palkovits, M. (1982). Localization of cholecystokinin (CCK) in nucleus tractus solitarius (NTS) of the rat and its possible vagal origin. *Soc. Neurosci. Abst.* 8, 809.
Koh, E. T. and Ricardo, J. A. (1978). Afferents and efferents of the parabrachial region in the rat: Evidence for parallel ascending gustatory versus viscereceptive systems arising from the nucleus of the solitary tract. *Anat. Rec.* 190, 449.
Krtettek, J. E. and Price, J. L. (1978). Amygdaloid projections to subcortical structures within the basal forebrain and brainstem in the rat and cat. *J. Comp. Neurol.* 178, 225–254.
Laha, P. K., Nayar, Y., China, G. S. and Singh, B. (1977). Carbon dioxide sensitivity of the central chemosensitive mechanisms: An exploration by direct stimulation in rats. *Pflugers Arch.* 367, 241–247.
Leslie, R. A., Gwyn, D. G. and Hopkins, D. A. (1982). The central distribution of the vagus nerve and gastric afferent and efferent projections in the rat. *Brain Res. Bull.* 8, 37–43.
Lewis, P. R., Scott, J. A. and Navaratnam, V. (1970). Localization in the dorsal motor nucleus of the vagus in the rat. *J. Anat.* 107, 197–208.
Loeschcke, H. H. (1982). Central chemosensitivity and the reaction theory. *J. Physiol. (London)* 322, 1–24.
Loewy, A. D. (1981). Descending pathways to sympathetic and parasympathetic preganglionic neurons. *J. Auton. Nerv. Sys.* 3, 265–275.
Loewy, A. D., McKellar, S. and Saper, C. B. (1979). Direct projections from the A5 catecholamine cell group to the interomediolateral cell column. *Brain. Res.* 174, 309–314.
McKellar, S. and Loewy, A. D. (1982). Projections from the ventrolateral medulla to the dorsolateral pons and periaqueductal grey in the rat. *Soc. Neurosci. Abst.* 8, 269.
Mitchell, J. and Templeton, D. (1981). The origin of the preganglionic parasympathetic fibers to the mandibular and sublingual salivary glands in the rat: A horseradish peroxidase study. *J. Anat.* 132, 513–518.
Nicholson, J. E. and Severin, C. M. (1981). The superior and inferior salivatory nuclei in the rat. *Neurosci. Lett.* 21, 149–154.
Norgren, R. (1976). Taste pathways to hypothalamus and amygdala. *J. Comp. Neurol.* 166, 17–30.
Norgren, R. (1978). Projections from the nucleus of the solitary tract in the rat. *Neurosci.* 3, 207–218.
Nosaka, S., Kamaike, T. and Yasunaga, K. (1978). Central vagal organization in rats: An electrophysiological study. *Exp. Neurol.* 60, 405–419.

Nosaka, S., Yamamoto, T. and Yasunga, K. (1979). Localization of vagal cardioinhibitory preganglionic neurons within rat brainstem. *J. Comp. Neurol.* 186, 79–92.

Ogawa, H. and Kaisaku, H. (1982). Physiological characteristics of the solitarioparabranchial relay neurons with tongue efferent inputs in rats. *Exp. Brain Res.* 48, 362–368.

Oley, N., Cordova, C., Kelly, M. L. and Bronzino, J. D. (1982). Morphine administration to the region of the solitary tract nucleus produces analgesia in rats. *Brain Res.* 236, 511–515.

Palkovits, M. (1980). The anatomy of central cardiovascular neurons. *In* K. Fuxe, M. Goldstein, B. Hokfelt and T. Hökfelt (eds), *Central adrenaline neurons: Basic aspects of their role in cardiovascular functions*, pp. 3–17. Wenner-Gren Center, Int. Symp. Ser., Pergamon Press, Oxford.

Palkovits, M. and Jacobowitz, D. M. (1974). Topographic atlas of catecholamines and acetylcholinesterase containing neurons in the rat brain. II Hindbrain (mesencephalon, rhombencephalon). *J. Comp. Neurol.* 157, 29–42.

Palkovits, M. and Zaborszky, L. (1977). Neuroanatomy of central cardiovascular control. Nucleus tractus solitarii: Afferent and efferent neuronal connections in relation to the baroreceptor reflex arc. *Prog. Brain Res.* 47, 9–34.

Palkovits, M., Leranth, C., Eiden, L. E., Rotsztejn, W. and Williams, T. H. (1982). Intrinsic vasoactive interstinal polypeptide (VIP)-containing neurons in the baroreceptor nucleus of the solitary tract in rat. *Brain Res.* 244, 351–355.

Paxinos, G. and Watson, C. (1982). *The rat brain in stereotaxic coordinates.* Academic Press, Sydney.

Perrone, M. H. (1981). Biochemical evidence that L-glutamate is a neurotransmitter of primary vagal afferent nerve fibers. *Brain Res.* 230, 283–293.

Rea, M. A., Aprison, M. H. and Felton, D. L. (1982). Catecholamines and serotonin in the caudal medulla of the rat: Combined neurochemical-histofluorescence study. *Brain Res. Bull.* 9, 227–236.

Ricardo, J. A. and Koh, E. T. (1978). Anatomical evidence of direct projections from the nucleus of the solitary tract to the hypothalamus, amygdala and other forebrain structures in the rat. *Brain Res.* 153, 1–26.

Ritchie, T. C., Westlund, K. N., Bowker, R. M., Coulter, J. D. and Leonard, R. B. (1982). The relationship of the medulary catecholamine containing neurones to the vagal motor nuclei. *Neurosci.* 7, 1471–1482.

Rogers, R. C. and Hermann, G. E. (1983). Central connections of the hepatic branch of the vagus nerve: A horseradish peroxidase histochemical study. *J. Auton. Ner. Sys.* 7, 165–174.

Rogers, R. C., Kita, H., Butcher, L. L. and Novin, D. (1980). Afferent projections to the dorsal motor nucleus of the vagus. *Brain Res. Bull.* 5, 363–373.

Ross, C., Ruggiero, D. A. and Reis, D. J. (1979). Descending connections from the brain to the spinal cord in the rat, as studied with horseradish peroxidase. *Soc. Neurosci. Abst.* 5, 49.

Saper, C. B. and Loewy, A. D. (1980). Efferent connections of the parabrachial nucleus in the rat. *Brain Res.* 197, 291–317.

Saper, C. B., Loewy, A. D., Swanson, L. W. and Cowan, W. M. (1976). Direct hypothalamo-autonomic connections. *Brain Res.* 117, 305–312.

Sawchenko, P. E. and Swanson, L. W. (1982). Anatomical relationship between vagal preganglionic neurons and aminergic and peptidergic neural system in the brainstem of the rat. *Soc. Neurosci. Abst.* 8, 119.

Shapiro, R. E. and Miselis, R. R. (1982). An efferent projection from the area postrema and the cudal medial nucleus of the solitary tract to the parabrachial nucleus in rat. *Soc. Neurosci. Abst.* 8, 269.

Siemers, E. R., Rea, M. A., Felton, D. L. and Aprison, M. H. (1982). Distribution and uptake of glycine, glutamate and gamma-aminobutyric acid in the vagal nuclei and eight other regions of the rat medulla oblongata. *Neurochem. Res.* 455–468.

Simantov, R., Kuhar, M. J., Uhl, G. R. and Snyder, S. H. (1977). Opioid peptide enkephalin: Immunohistochemical mapping in rat central nervous system. *Proc. Nat. Acad. Sci.* 74, 2167–2171.

Snyder, D. W., Nathan, M. A. and Reis, D. L. (1978). Chronic lability of the catecholamine innervation of the nucleus tractus solitarii in the rat. *Circ. Res.* 43, 662–671.

Spyer, K. M. (1981). Neural organization and control of the baroreceptor reflex. *Rev. Physiol. Biochem. Pharmacol.* 88, 23–124.

Stuesse, S. L. (1982). Origins of cardiac vagal preganglionic fibers: A retrograde transport study. *Brain. Res.* 236, 15–25.

Stuesse, S. L. and Powell, K. S. (1982). Brainstem areas which project to cardiac regions of the nucleus ambiguus in the rat. *Soc. Neurosci. Abst.* 8, 76.

Sumal, K., Blessing, W. W., John, T. H., Reis, D. J. and Pickel, V. M. (1983). Synaptic interaction of vagal afferents and catecholaminergic neurons in the rat nucleus tractus solitarius. *Brain Res.* 277, 31–40.

Swanson, L. W. and Hartman, B. K. (1975). The central adrenergic system. An immunofluorescence study of the location of cell bodies and their efferent connections in the rat utilizing dopamine-beta-hydroxylase as a marker. *J. Comp. Neurol.* 163, 467–505.

Taber, E. (1961). The cytoarchitecture of the brain stem of the cat. I. Brain stem nuclei of the cat. *J. Comp. Neurol.* 116, 27–68.

Takayama, K., Ishikawa, N. and Miura, M. (1982). Sites of origin and termination of gastric vagus preganglionic neurons: An HRP study in the rat. *J. Auton. Nerv. Sys.* 6, 211–223.

Talman, W. T., Perrone, M. H. and Reis, D. J. (1980). Evidence for L-glutamate as the neurotransmitter of baroreceptor afferent nerve fibers. *Science* 209, 813–815.

Torvik, A. (1956). Afferent connections to the sensory trigeminal nuclei, the nucleus of the solitary tract and adjacent structures. *J. Comp. Neurol.* 106, 51–141.

Uhl, G. R., Kuhar, M. J. and Snyder, S. H. (1977). Immunohistochemical localization of neurotensin and enkephalin in rat CNS. *Soc. Neurosci. Abst.* 3, 1332.

Vigier, D. and Portalier, P. (1979). Afferent projections of the area postrema demonstrated by autoradiography. *Arch. Ital. Biol.* 117, 308–324.

Voshart, K. and Van der Kooy, D. (1981). The organization of the efferent projections of the parabrachial nucleus to the forebrain in the rat: A retrograde fluorescent double-labeling study. *Brain Res.* 212, 271–286.

Weaver, F. C. (1980). Localization of parasympathetic preganglionic cell bodies innervating the pancreas within the vagal nucleus ambiguus of the rat brainstem: Evidence on dual innervation based on the retrograde axonal transport of horseradish peroxidase. *J. Auton. Nerv. Sys.* 2, 61–69.

Westlund, K. N., Bowker, R. M., Ziegler, G. and Dan, J. (1981). Origin of spinal noradrenergic pathways demonstrated by retrograde transport of antibody to dopamine-β-hydroxylase. *Neurosci. Lett.* 25, 243–249.

Westlund, K. N., Bowker, R. M., Ziegler, M. G. and Coulter, J. D. (1982). Descending noradrenergic projections and their spinal termination. *In* H. G. J. M. Kuypers and G. F. Martin (eds), *Anatomy of descending pathways to the spinal cord*, pp. 219–238. *Prog. Brain Res.* Elsevier, Amsterdam.

6

Organization and projections of the orofacial motor nuclei

JOSEPH B. TRAVERS

Pennsylvania State University
Hershey, Pennsylvania, USA

I. INTRODUCTION

This chapter will describe the intrinsic organization of the trigeminal (Mo5), facial (7) and hypoglossal (12) motor nuclei and identify the source of central nervous system afferent projections to these nuclei. The myotopic organization of these orofacial motor nuclei in the rat is common to mammals in general. Motor pools within the nuclei are topographically related, reflecting peripheral musculature. Functionally synergistic or spatially adjacent muscles have adjacent central representation, sometimes extending between motor nuclei; for example, trigeminal motoneurons innervating the anterior digastric muscle are nearly spatially continuous with facial motoneurons innervating the functionally related posterior digastric. Similarly, hypoglossal motoneurons innervating the geniohyoid are nearly continuous with ventral horn motoneurons innervating the peripherally adjacent thyrohyoid muscle.

Certain central afferent projections are common to the orofacial motor nuclei. Projections from the sensory trigeminal nuclei and reticular formation provide a common substrate for coordinating synergistic activities among the orofacial musculature. Nevertheless, specificity of central projections is also apparent. This is particularly clear in the facial nucleus which is more functionally heterogeneous than either the Mo5 or 12. Well defined afferent projections differentiate between subdivisions of the facial nucleus that innervate muscles of the orofacial region and those that innervate facial muscles of the ear and eye.

Many of the afferent projections to each of the orofacial motor nuclei can also be differentiated pharmacologically. Specific reticular formation projections have been identified as adrenergic, cholinergic, enkephalinergic or

THE RAT NERVOUS SYSTEM
ISBN 0 12 547632 9

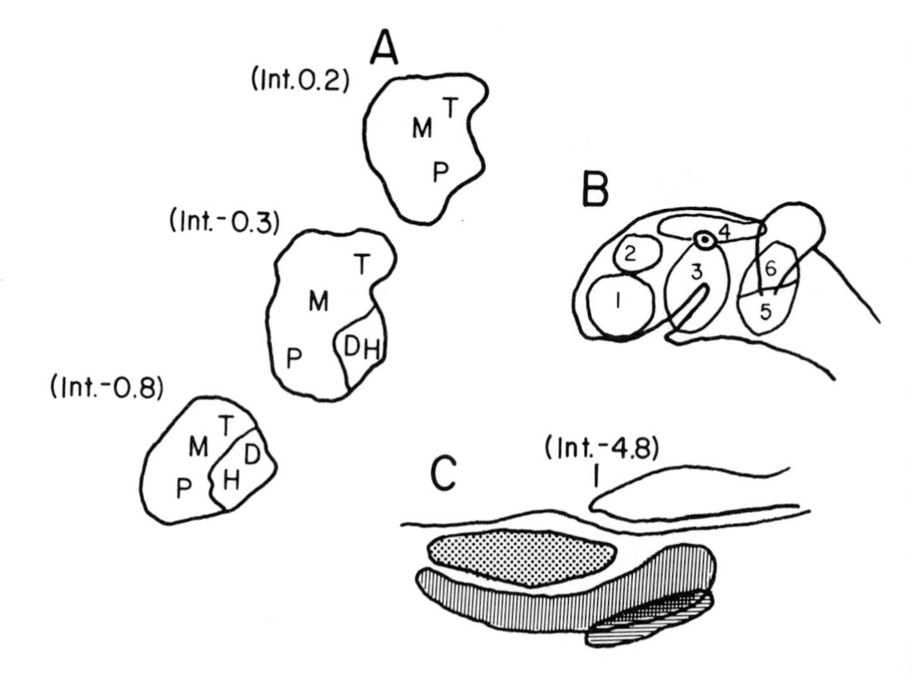

Fig. 1: A: Location of motoneurons innervating muscles of mastication at three levels of the motor trigeminal nucleus. Interaural levels correspond to *Atlas* (Paxinos and Watson, 1982): temporalis (T); masseter (M); pterygoids (P); anterior digastric (D); mylohyoid (H) (from Sasamoto, 1979). Lateral is left for both A and B. B: Topographic representation of subdivisions of facial nucleus: 1– lateral, 2– dorsolateral, 3 intermediate, 4– dorsal intermediate, 5– ventral medial, 6– dorsal medial. Refer to Table 1 for additional detail (from Watson *et al.*, 1982). C: Parasagittal view of the three columns of cells of hypoglossal nucleus. Stippled area innervates tongue retractor muscles, vertical and horizontal lines depict areas innervating tongue protruders. Rostral to left (from Krammer *et al.*, 1979).

serotonergic. While the functional significance of these various pathways is frequently unknown, the presence of complex orofacial behaviors in the decerebrate rat suggests a high level of integration in the caudal brain stem.

II. MOTOR TRIGEMINAL NUCLEUS

A. Cytoarchitecture and functional organization

The motor trigeminal nucleus (Mo5) of the rat is conventionally divided into a large dorsolateral division which extends for the rostrocaudal length of the nucleus and a smaller ventromedial division (in the caudal half only) (Limwongse and DeSantis, 1977; Mizuno *et al.*, 1975; Sasamoto, 1979). These divisions correspond to the dorsolateral and ventromedial (Taber, 1961) or dorsal and ventral divisions (Berman, 1968) in the cat and the γ and β nuclei in the rabbit (Meessen and Olszewski, 1949). Fig. 1(A) depicts the myotopic organization of the Mo5 at three different levels. The jaw closing muscles (the masseter (M), temporalis (T), and medial (external) pterygoid (P) muscles) are innervated from the dorsolateral division. The jaw opening muscles (the anterior digastric [D] and mylohyoid [M]) are innervated from the ventromedial division. A third jaw opening muscle, the lateral (internal) pterygoid (Nakamura, 1980), is grouped with jaw closing motoneurons in the ventral aspect of the dorsolateral division (Mizuno *et al.*, 1975; Sasamoto, 1979). Within the dorsolateral divison of Mo5, temporalis motoneurons are located in the dorsomedial corner, and masseter motoneurons somewhat more ventral and lateral (Mizuno *et al.*, 1975; Sasamoto, 1979). In the guinea pig, medial pterygoid motoneurons are located ventral to lateral pterygoid motoneurons in the dorsolateral division of Mo5 (Uemura-Sumi *et al.*, 1982), and in the rabbit dorsolateral division, medial pterygoid motoneurons are grouped lateral to those innervating the lateral pterygoid muscle (Matsuda *et al.*, 1978). In the rat, the central location of medial and lateral pterygoid motoneurons has not been determined individually, but these motoneurons are grouped ventral to masseter motoneurons within the dorsolateral division (Mizuno *et al.*, 1975; Sasamoto, 1979).

Within the ventromedial division of the motor trigeminal nucleus, anterior digastric motoneurons are dorsomedial to mylohyoid motoneurons (Mizuno *et al.*, 1975; Sasamoto, 1979). Though direct evidence in the rat is lacking (Mizuno *et al.*, 1975; Sasamoto, 1979), in cats, the tensor veli palatini is also innervated by motoneurons in the ventromedial division (Keller *et al.*, 1983). The tensor tympani is innervated by trigeminal motoneurons located ventral to Mo5 and medial to existing trigeminal rootlets (Spangler *et al.*, 1982). This distribution of cells probably corresponds to the accessory trigeminal nucleus (A5) identified in the *Atlas* (Fig. 34).*

Each motor trigeminal nucleus contains approximately 2500 neurons (Limwongse and DeSantis [1977] estimate from 2265 to 2979 neurons). It appears that most of these cells innervate the jaw closing muscles as injections of horseradish peroxidase (HRP) into the anterior digastric muscle (Kemplay and Cavanagh, 1983), or immersion of the mandibular nerve (which innervates both the anterior digastric and mylohyoid muscles) in cobalt (Szekely and Matesz, 1982) labeled only 160–250 cells. There is some indication, however, that these latter studies may have underestimated the total innervation of jaw opening muscles because application of HRP directly to the mylohyoid nerve labeled a larger more rostral distribution of cells in the motor trigeminal nucleus than observed with the cobalt technique (Jacquin *et al.*, 1983). There is also evidence for a small crossed projection (approximately 30 cells) of the anterior digastric muscle (Kemplay and Cavanagh, 1983).

---

*Reference to the *Atlas* is to Paxinos and Watson (1982). Reference to figures here is to figures in the *Atlas*.

Neurons in the dorsolateral and ventromedial divisions of the motor trigeminal nucleus can be distinguished according to certain cytoarchitectural criteria. The dorsolateral division contains both small (8 μm–14 μm) and large (28 μm–42 μm) diameter multipolar cells; this correlates with a bimodal distribution of nerve fiber diameters (Limwongse and DeSantis, 1977). The demonstration of muscle spindles in the jaw closing muscles of rats (Karlsen, 1965) is consistent with the presence of gamma efferents in the dorsolateral Mo5. Two classes of multipolar cells in the ventromedial division are also evident, though the large cells (24 μm–34 μm diameter) are not as large as their dorsolateral counterparts and the small cells (16 μm–20 μm diameter) are not as small. An absence of very small neurons in the ventromedial division is consistent with the lack of muscle spindles in jaw opening muscles (Limwongse and DeSantis, 1977).

Intracranial axonal trajectory further differentiates between dorsolateral and ventromedial cells. Axons from the dorsolateral division exit the brain directly, forming a few large roots that pass just ventral to the principal trigeminal sensory nucleus (Pr5). Axons from the ventromedial division first course dorsally to form a genu rostral to the more prominent seventh nerve (7n) genu before turning ventrolaterally as a separate root, parallel to those of the dorsolateral division (Jacquin *et al.*, 1983; Szekely and Matesz, 1982).

B. Afferent projections

The major direct afferent projections to the motor trigeminal nucleus (Mo5) originate from cells in the pons and medulla. In the rat there is no direct cortical input (Travers and Norgren, 1983; Vornov and Sutin, 1983). Midbrain projections consist almost entirely of cells from the rostral extension of the ipsilateral mesencephalic trigeminal nucleus (Me5). Anatomic studies show the Me5 projecting throughout the Mo5 in the rat (Matesz, 1981), although physiologically, in the cat, they exert their greatest effect on masseteric motoneurons (Jerge, 1963). In the cat, the red nucleus may influence the Mo5 via the juxtatrigeminal area, that part of the lateral reticular formation interposed between Pr5 and the ventrolateral aspect of Mo5 (Mizuno *et al.*, 1974). This projection has been difficult to analyze because the rubrospinal tract courses ventrolateral to the Mo5 (Courville, 1966b; Martin and Dom, 1970; Miller and Strominger, 1973) such that injections of HRP which damage this area label the contralateral red nucleus in the rat (Travers and Norgren, 1983; Vornov and Sutin, 1983).

Pontine projections to the motor trigeminal nucleus begin rostrally at the level of the decussation of the superior cerebellar peduncle in an area just medial to the lateral lemniscus (Vornov and Sutin, 1983). These projections constitute the primary source of the dense noradrenaline (NE) input to the Mo5 (Fuxe, 1965; Levitt and Moore, 1979; Vornov and Sutin, 1983). This lateral distribution of cells projecting to Mo5 continues further caudally in the pons but the neurotransmitter concerned is different. Bilaterally situated cells in the supratrigeminal and intertrigeminal nuclei (Travers and Norgren, 1983; Vornov and Sutin, 1983) may provide cholinergic input into the Mo5, because immunohistochemical localization of muscarinic cholinergic receptors labeled this perinuclear region (Wamsley *et al.*, 1981). The principal trigeminal nucleus (Pr5), primarily the dorsomedial region, also projects bilaterally to the Mo5. In the cat, this part of Pr5 receives cutaneous afferent projections from the oral region (Eisenman *et al.*, 1963) and is part of a jaw opening reflex pathway (reviewed in Nakamura, 1980). Inhibition of masseteric (jaw closing muscle) motoneurons during jaw opening may be mediated by cells in the supratrigeminal area (Kidokoro *et al.*, 1968) which receives projections directly from primary trigeminal afferents and indirectly via the Pr5 (Mizuno, 1970; Torvik, 1956). A projection from the pontine raphe (Vornov and Sutin, 1983) may be the source of serotonergic input to Mo5 (Palkovits *et al.*, 1974; Schaffer *et al.*, 1982; Steinbusch, 1981).

Projections to the motor trigeminal nucleus from the medulla originate from groups of cells continuous with those in the pons (Travers and Norgren, 1983; Vornov and Sutin, 1983). Clusters of cells projecting to the Mo5 from the medullary reticular formation are retrogradely labeled following HRP deposits into the motor nucleus. Clusters of small sized neurons in the parvocellular (lateral) reticular formation (PCRt) project primarily ipsilaterally to the Mo5; clusters of larger neurons, more medially situated in the reticular formation, project bilaterally. At the level of the hypoglossal nucleus, however, the medial distribution is primarily contralateral in the dorsal and ventral parts of the reticular nucleus of the medulla (MdD, MdV) (Travers and Norgren, 1983). These medullary reticular distributions correspond to projections from the medial part of the lateral propriobulbar zone described in the cat (Holstege *et al.*, 1977). In the cat, cells in this medial zone project monosynaptically to the Mo5 and produce differential responses in masseteric (jaw closer) and anterior digastric (jaw opener) motoneurons (Takatori *et al.*, 1981). These cells are thought to be the final interneurons in the orbital cortex control of masticatory movement (Nakamura, 1980). Within the spinal trigeminal complex, cells in the ipsilateral oral part (Sp5O) project to Mo5 (Travers and Norgren, 1983; Vornov and Sutin, 1983), but a projection from the caudal part (Sp5C) is disputed, both in the rat (Travers and Norgren, 1983; Vornov and Sutin, 1983) and the cat (Burton *et al.*, 1979; Stewart and King, 1963).

III. FACIAL NUCLEUS

A. Cytoarchitecture and functional organization

The facial nucleus (7) of the rat consists of approximately 5000 to 6000 stellate shaped neurons that innervate the superficial muscles of the head and neck.

Cells within the nucleus range from 15 μm–55 μm in diameter and are segregated into several subdivisions (Martin *et al.*, 1977; Szekely and Matesz, 1982; Watson *et al.*, 1982). Papez (1927) described four major subdivisions (medial, intermediate, lateral and dorsal) and two minor subdivisions (ventromedial and ventrolateral) in the rat. Recent descriptions based on Nissl stained sections differ only slightly from the original work. Martin and Lodge (1977) do not delineate a distinct ventrolateral group and Watson *et al.* (1982) specify neither a ventromedial or ventrolateral group. Further, Watson *et al.* (1982) describe a dorsal subdivision, (primarily over the intermediate division) in contrast to the dorsal division described by both Papez (1927) and Martin and Lodge (1977) which is dorsolateral and designated as such by Watson *et al.* (1982). There is general agreement, however, that these subdivisions consist largely of cell groupings separated by white matter, but without obvious cytoarchitectural differences. The divisions designed in the *Atlas* (Fig. 37) correspond both to aggregates of neurons evident in Nissl stained sections and myotopic representations derived from experimental studies (see below).

The organization of the rat facial nucleus has been mapped both for nerve representation within the nucleus (Martin and Lodge, 1977; Papez, 1927; Szekely and Matesz, 1982) and muscle representation (Shohara and Sakai, 1983; Watson *et al.*, 1982). Minor discrepancies among these studies may be accounted for by the multiple innervation of a single muscle by different nerve branches and innervation of several muscles by one nerve branch (Watson *et al.*, 1982; Table 1). In general, however, a common pattern characterizes a number of species (Courville, 1966a; Dom, 1982; Kume *et al.*, 1978; Martin and Lodge, 1977; Papez, 1927; Provis, 1977): in which dorsal muscles are represented dorsally, ventral muscles ventrally and the anterior–posterior axis lateromedially (Provis, 1977). This topographic organization for the rat is shown in Fig. 1(B) where it can be seen that motoneurons innervating the proboscis are lateral within the facial nucleus, oral and orbital motoneurons are intermediate, and facial motoneurons innervating the ear and neck are medial. Table 1 further specifies the innervation of facial musculature from the subdivisions of the 7.

Other muscles innervated by the facial nerve include the posterior digastric, stylohyoid, and stapedius (Huber and Hughson, 1926–27). These deep skeletal muscles, associated with the hyoid arch, are not involved in facial expression, and are innervated by motoneurons outside the main motor nucleus (Huber and Hughson, 1926–27; Provis, 1977). Motoneurons innervating both the posterior belly of the digastric and stylohyoid (active primarily during jaw opening) are found dorsal to the anterior part of the 7 along the existing facial root and form the accessory facial nucleus (Su7) (Hutson *et al.*, 1979; Shohara and Sakai, 1983; Szekely and Matesz, 1982; Watson *et al.*, 1982). Within Su7, motoneurons innervating the stylohyoid are ventral to those innervating the posterior belly of the digastric (Shohara and Sakai, 1983). The size (approximately 35 μm) and

Table 1: Innervation of the facial musculature*

7 Subdivision	Nerve-branch	Muscle
lateral	superior buccolabial	superior labial naris dilator
dorsolateral	lower zygomatic	buccinator
ventral intermediate	cervical marginal mandibular	platysma mentalis
dorsal intermediate	superior buccolabial lower zygomatic	 zygomatic
	temporal upper zygomatic	 orbicularis oris
ventral medial	posterior auricular	posterior auricular
dorsal medial	temporal	anterior auricular

*From Greene, 1963; Martin and Lodge, 1977; Papez, 1927; Watson *et al.*, 1982.

shape of cells in Su7 are similar to those in the facial nucleus although there is a tendency for the soma and dendrites to be elongated in the dorsoventral direction (Szekely and Matesz, 1982). The axonal trajectory of Su7 neurons is similar to other facial motoneurons, but the axons form a distinct bundle parallel to the main existing root. The location of Su7, dorsal to the rostral end of the 7, places these cells nearly continuous with trigeminal motoneurons innervating the anterior belly of the digastric with which they share both common morphologic (Szekely and Matesz, 1982) and functional characteristics (Basmajian, 1979: 386). Facial motoneurons innervating the stapedius muscle of the middle ear have not been studied in the rat, but have been located anterior to the facial nucleus, adjacent to the superior olive in the cat (Lyon, 1978).

B. Afferent projections

Different divisions of the rat facial nucleus receive substantial and differential input from nuclei of the midbrain, pons, and medulla (Erzurumlu and Killackey, 1979; Hinrichsen and Watson, 1983; Senba and Tohyama, 1983a, b; Travers and Norgren, 1983). These projections are generally consistent with those described in cats (Courville, 1966b; Edwards, 1972; Henkel and Edwards, 1978; Holstege *et al.*, 1977; Takeuchi *et al.*, 1979) and opossums (Dom *et al.*, 1973; Panneton and Martin, 1978, 1979, 1983). In general, well defined central afferent projections differentiate between subdivisions of the facial nucleus that innervate muscles of the orofacial region and those that innervate facial muscles of the ear and eye. In the midbrain, cells in and around the anterior part of the contralateral nucleus of the lateral lemniscus (*Atlas* Fig. 30)—the paralemniscal zone first described in the cat(Henkel and Edwards, 1978)—project specifically to the medial subdivision of 7 (Hinrichsen and Watson, 1983; Travers and Norgren, 1983). In the cat, this region of the midbrain receives input from the superior colliculus and has been implicated in pinna orientation to sound (Henkel and Edwards, 1978). The lateral and intermediate subdivisions of 7 receive projections from the (contralateral) red nucleus, providing a relay for cortical and cerebellar input to the 7 (Hinrichsen and Watson, 1983; Panneton and Martin, 1983). Scattered cells in the central gray, periocular nuclei (interstitial nucleus of Cajal and Darkschewitch nucleus) and the midbrain reticular formation also project to the facial nucleus and may mediate vocal–facial behavior (Hinrichsen and Watson, 1983; Panneton and Martin, 1983).

Differential projections to the subdivisions of facial nucleus are evident in the pons as well. More cells in the ipsilateral Kölliker-Fuse (KF) supratrigeminal, and ventrolateral parabrachial (VPB) nuclei project to the lateral and intermediate divisions of 7 than to the medial (Hinrichsen and Watson, 1983; Travers and Norgren, 1983). The ventral parabrachial nucleus which receives second order vagal input (Norgren, 1978; Ricardo and Koh, 1978) may be involved in exploratory sniffing and nasal breathing, activities both requiring coordination between respiratory and facial motoneurons (Hinrichsen and Watson, 1983; Panneton and Martin, 1983). Likewise, the ventral parabrachial nucleus which lies adjacent to gustatory responsive cells in the pons (Norgren and Pfaffman, 1975) may mediate gustatory evoked facial responses (Panneton and Martin, 1983). In contrast to the lateral divisions of the facial nucleus, more cells in the reticular formation dorsal to the superior olive at the level of the Mo5 project to the medial division. Other regions of the pons projecting to 7 include the Kölliker-Fuse nucleus which may be a source of norepinephrine in 7 (Levitt and Moore, 1979). Cells surrounding the Mo5 may provide cholinergic input to the 7 (Wamsley *et al.*, 1981). Projections from the Pr5 to 7 are relatively sparse and are mainly to the lateral divisions (Erzurumlu and Killackey, 1979; Travers and Norgren, 1983).

In the medulla, most of the projections to the facial nucleus originate from cells in the reticular formation, bilaterally distributed in the parvocellular reticular nucleus at the level of 7 and MdD and MdV more caudally (Travers and Norgren, 1983). Different regions of the medulla project to different divisions of the facial nucleus, although these differences are not as pronounced as projections from the midbrain and pons. At the level of the hypoglossal nucleus (12) many reticular cells, bilaterally distributed in the MdD, MdV and along the medial border of the ipsilateral Sp5C, project to the intermediate and lateral divisions of 7 (Hinrichsen and Watson, 1983; Travers and Norgren, 1983). Cells in the caudal medulla that project to the medial subdivisions of 7 are fewer in number and tend to be ventrally located, dorsal to the lateral reticular nucleus and medial to the Sp5C (Senba and Toyhama, 1983b; Travers and Norgren,

983). The presence of immunoreactive enkephalinergic cells in this same ventral reticular region (Senba and Tohyama, 1983b) suggests a source of enkephalinergic fibers (Finley *et al.*, 1981) that are concentrated primarily within the medial division of 7 (Senba and Tohyama, 1983a). Reticular neurons that project preferentially to the medial division of 7 also extend into the cervical levels of the spinal cord, originating in the intermediolateral gray column and lateral part of the ventral gray (Hinrichsen and Watson, 1983; Senba and Tohyama, 1983b).

Projections from the spinal trigeminal complex (Sp5) to the facial nucleus include ipsilateral projections from the Sp50 to the lateral and intermediate divisions (Erzurumlu and Killackey, 1979; Hinrichsen and Watson, 1983; Travers and Norgren, 1983) and an ipsilateral projection from the medial lamina of Sp5 at the level of C1 and C2 to the medial division of 7 (Hinrichsen and Watson, 983; Senba and Tohyama, 1983b). Although a weak projection to both medial and lateral divisions of 7 from the Sp5C is suggested by the results of HRP experiments (Hinrichsen and Watson, 1983; Travers and Norgren, 1983), the dense projection described in the rat using degeneration techniques (Erzurumlu and Killackey, 1979) probably originates from the distribution of cells along the medial border of the Sp5C (Hinrichsen and Watson, 1983; Travers and Norgren, 983). These trigeminofacial and spinal dorsal horn projections appear topographically organized such that facial motoneurons receive cutaneous input from areas overlying the muscles they innervate (Panneton and Martin, 1983). In the cat it has been proposed that interneurons in the Sp5 also mediate a cortical influence on facial nucleus motoneurons (Tanaka, 1976).

Minor projections to the medial division of the facial nucleus from the medial vestibular nuclei have also been reported in the rat (Hinrichsen and Watson, 983), consistent with investigations in the cat (Shaw and Baker, 1983) and opossum (Panneton and Martin, 1983). Serotonergic fibers have been identified in the facial nucleus (Aghajanian and McCall, 1980; Palkovits *et al.*, 1974; Steinbusch, 1981), but their source is unclear. HRP studies in both the rat Hinrichsen and Watson, 1983; Travers and Norgren, 1983) and the opossum Panneton and Martin, 1983) have not labeled raphe neurons, a source of serotonergic projections (McCall and Aghajanian, 1979; Palkovits *et al.*, 1974). Although iontophoretic application of serotonin does not directly activate facial motor neurons, it facilitates excitatory responses induced by other means McCall and Aghajanian, 1979).

V. HYPOGLOSSAL NUCLEUS

A. Cytoarchitecture and functional organization

The hypoglossal nucleus (12) of the rat consists of several longitudinally oriented cell columns stacked upon one another in the dorsoventral plane (Barnard, 1940; Krammer *et al.*, 1979; Lewis *et al.*, 1971; Odutola, 1976). Axons from cells of the dorsal division innervate tongue retractor muscles (the styloglossus and hyoglossus) and travel in the lateral branch of the hypoglossal nerve (distal to the bifurcation of the nerve at the hyoid bone). Axons from cells in the ventral division innervate tongue protruders (the genioglossus and geniohyoid) and travel in the medial branch of the nerve. Although the geniohyoid is not a true lingual muscle (Chibuzo and Cummings, 1982), it has a common embryologic origin with lingual muscles (Edgeworth, 1907) and often functions together with the genioglossus (Cunningham and Basmajian, 1969; Krammer *et al.*, 1979).

The main cell groupings of the hypoglossal nucleus are schematically depicted in Fig. 1(C) in a sagittal plane. The ventrolateral subdivision, seen only in the caudal third of the nucleus (Krammer *et al.*, 1979) is continuous with cells in the dorsomedial part of the ventral horn (Kitamura *et al.*, 1983) and innervates primarily the geniohyoid muscle. Cells in the ventromedial subdivision supply the genioglossus muscle (Krammer *et al.*, 1979) and possibly the intrinsic muscles of the tongue (Lewis *et al.*, 1971). The dorsal division occupying the rostral two thirds of the nucleus can be further subdivided myotopically: the rostral region primarily innervates the styloglossus muscle; and the caudal part the hyoglossus (Krammer *et al.*, 1979).

Based on soma and axon counts, each hypoglossal nucleus contains approximately 3500 motoneurons (Lewis *et al.*, 1971), ranging from 17 μm–40 μm in diameter (Cooper, 1981; Kitamura *et al.*, 1983; Odutola, 1976). Cytoarchitectural characteristics of size, shape and dendritic aborization further differentiate the myotopically defined subdivisions (Cooper, 1981; Kitamura *et al.*, 1983; Odutola, 1976). Nissl stained cells in the dorsal division appear fusiform, oriented along the mediolateral axis and range from 17 μm–27 μm in diameter (Odutola, 1976). These contrast with cells in the central regions of the nucleus that are more globular in shape (17 μm–30 μm diameter) and those of the ventral division that are multipolar shaped and include small (10 μm–12 μm diameter) neurons (Odutola, 1976). Cell diameters average 28.7 μm in the ventrolateral division in a unimodal distribution, suggesting a lack of γ efferents (Kitamura *et al.*, 1983).

Neurons can be further classified on the basis of dendritic morphology. Using Golgi stained material, Cajal (1972) designated hypoglossal cells "external" with dendrites extending beyond the borders of the nucleus and "internal" those with dendrites confined to the nuclear region. In addition, some dendrites extend into the contralateral nucleus (Wan *et al.*, 1982). Recent work using the Golgi technique (Odutola, 1976) and retrograde cholera toxin HRP (Wan *et al.*, 1982) further differentiates between cells of the various subdivisions based on dendritic pattern. Dendrites of cells of the dorsal division can be either internal or external (Odutola, 1976). External dendrites may reach 1 mm in length, extending laterally into the adjacent reticular formation (MdD and MdV) to reach the spinal trigeminal nucleus where they may ramify into tufted ends (Wan *et al.*, 1982). Dendrites from these cells also extend into the ipsilateral nucleus of the solitary tract (Sol) and a very few traverse the dorsal motor nucleus of the vagus (10) to reach the contralateral Sol (Wan *et al.*, 1982). Dendritic fields from cells in the ventral division include an apical dendrite that ramifies within the dorsal divisions of 12 and basal dendrites that ramify laterally within 12 and the adjacent reticular formation (Odutola, 1976), extending ventrally into the medial longitudinal fasciculus and raphe obscurus nucleus (Wan *et al.*, 1982). Dendrites from multipolar cells in the center of 12 fan out to ramify throughout the nucleus (Odutola, 1976).

B. Afferent projections

Afferent projections reach the hypoglossal nucleus of the rat from cells in the midbrain, pons and medulla (Aldes, 1980; Borke *et al.*, 1983; Cooper and Fritz, 1981; Travers and Norgren, 1983). These projections originate from both highly specific aggregates of neurons in identified brain stem structures and from more widely distributed, poorly defined regions of the reticular formation. Many of the projections to 12, both specific and diffuse, have been closely identified with specific lingual reflexes evoked from electrical stimulation of somatic and visceral sensory afferents or from cortical stimulation.

In the midbrain, a few cells in the contralateral reticular formation project to the hypoglossal nucleus as demonstrated with HRP in the rat (Aldes, 1980; Travers and Norgren, 1983). However a similar study in the opossum (Panneton and Martin, 1979) localized midbrain projections primarily within the central gray. Some controversy exists as to whether the mesencephalic nucleus of the trigeminal (Me5) projects directly to 12 via Probst's tract (see Lowe, 1981). Although this projection is evident in the cat using degeneration techniques (Mizuno and Sauerland, 1970; Szentagothai, 1948), relatively few Me5 cells have been labeled in retrograde HRP experiments in the opossum (Panneton and Martin, 1979), or the rat (Travers and Norgren, 1983). This projection may be to dendrites of 12 motoneurons extending laterally in the adjacent reticular formation (Matesz, 1981; Ruggiero *et al.*, 1982). Nevertheless, physiologic data indicate that there is a proprioceptive interaction between the muscles of mastication and lingual muscles (Lowe, 1981). Proprioceptive information from jaw muscles may also reach 12 through the supratrigeminal region in the pons (Mizuno, 1970). This region receives monosynaptic input from the Me5 (Matesz, 1981) and projects to 12 in the rat (Travers and Norgren, 1983).

Neurons in proximity to the Mo5 also project to the hypoglossal nucleus. These include the ventral parabrachial intertrigeminal nuclei (bilaterally), and Kölliker-Fuse nucleus (ipsilaterally) (Aldes, 1980; Borke *et al.*, 1983; Saper and Loewy, 1980; Travers and Norgren, 1983). The Kölliker-Fuse nucleus, part of the lateral tegmental group of catecholamine containing cells (Stevens *et al.*, 1982) may be a source of norepinephrine innervation of 12 (Levitt and Moore, 1979) and intertrigeminal cells may be cholinergic (Wamsley *et al.*, 1981). Cells in the dorsal part of the Pr5 project to 12 though descriptions vary as to whether the projection is bilateral (Borke *et al.*, 1983) or ipsilateral (Aldes, 1980; Travers and Norgren, 1983). In the cat, the Pr5 is topographically organized such that somatosensory afferent axons from the tongue project to the dorsomedial region (Eisenman *et al.*, 1963) and there is reason to believe that the rat is similarly organized (Nord, 1967). Thus, somatosensory input from the tongue can influence hypoglossal motoneurons through disynaptic pathways.

The majority of cells projecting to hypoglossal nucleus originate from the medullary reticular formation (Aldes, 1980; Borke *et al.*, 1983; Cooper and Fritz, 1981; Travers and Norgren, 1983). These cells are bilaterally distributed, primarily rostral to 12 in the parvocellular reticular nucleus, with a few cells in the gigantocellular nucleus (Gi). At caudal levels of the medulla, reticular cells projecting to 12 are primarily in the MdD rather than in MdV. This reticular distribution of cells, visualized with retrograde deposits of horseradish peroxidase in 12 (Aldes, 1980; Borke *et al.*, 1983; Travers and Norgren, 1983) overlaps the dendritic fields of 12 motoneurons visualized with either horseradish peroxidase or Golgi techniques. Many of these reticular projections may be cholinergic as suggested by the overlap of label following cholinergic autoradiography in areas that project to 12, including the parvocellular reticular nucleus at the level of 7 and MdD and MdV at the level of 12 (Wamsley *et al.*, 1981). The complexities of this synaptic substrate have not been closely explored, but correspond to the location of swallowing "centers" (Doty *et al.*, 1967) and other regions organizing complex synergies involving orofacial–pharyngeal afferent inputs and their corresponding motor efferents.

Medullary reticular neurons projecting to the hypoglossal nucleus form a continuous column along the rostrocaudal axis, beginning lateral to 12 and extending rostral through the medulla. Though these cells are primarily dorsally located along the caudal extent of the distribution in MdD, they occupy increasingly ventral locations in the parvocellular reticular nucleus further rostrally (Travers and Norgren, 1983), thus assuming positions adjacent to the ambiguus nucleus and 7. The central core of the medullary brain stem, encompassing most of Gi and the raphe, are not involved in these direct projections, consistent with observations in the cat (Basbaum *et al.*, 1978; Holstege *et al.*, 1977).

A second bilateral distribution of medullary cells projecting to the hypoglossal nucleus is from the dorsal part of the Sp50 and the nucleus of the spinal tract of the trigeminal, interpolar part (Sp5I) (Aldes, 1980; Borke *et al.*, 1983; Travers and Norgren, 1983). In the cat, these neurons are associated with polysynaptic trigeminolingual reflexes elicited from electric stimulation of the trigeminal nerve (reviewed by Lowe, 1981). Multiple routes from the Sp5 to 12 include projections via short axon interneurons in the adjacent reticular formation (Scheibel, A. B., 1955; Scheibel, M. E., 1955; Valverde, 1961) and by the extension of hypoglossal motoneuron dendrites into this area (Odutola, 1976; Wan *et al.*, 1982). Cortical projections to the Sp5 have also been proposed as part of a corticohypoglossal pathway in the cat (Porter, 1967). Inhibitory interneurons adjacent to 12 have also been implicated in trigeminohypoglossal reflexes (Sumino and Nakamura, 1974).

Relatively few cells from the nucleus of the solitary tract (Sol) project to the hypoglossal nucleus. Those that do are ipsilateral, located primarily in the caudal half of the nucleus (Borke *et al.*, 1983; Norgren, 1978; Travers and Norgren, 1983) in an area that receives afferents via the glossopharyngeal and vagus nerves (Contreras *et al.*, 1982); Hamilton and Norgren, 1984; Torvik, 1956). Hypoglossal reflexes induced by stimulation of these nerves (reviewed in Lowe, 1981) may be mediated by interneurons in the Sol or by direct synaptic contact via hypoglossal dendrites that extend into the Sol (Odutola, 1976; Wan *et al.*, 1982). The schema of both somatic and visceral afferent nerves interacting with 12 via the Sp5 and solitary nucleus remains the same as proposed by Cajal (1972: 71; Fig. 294).

Finally, a set of afferent projections to the hypoglossal nucleus are proprioceptive. Though some uncertainty surrounds the nature and pathway of lingual proprioceptors in subprimate species (Law, 1954), and the route by which they reach the central nervous system, a lingual-proprioceptive system seems likely to exist (Lowe, 1981). Recently, lingual proprioceptive afferents have been shown to be carried in the hypoglossal nerve of the dog via the cervical spinal and jugular ganglia (Chibuzo and Cummings, 1981), though electric stimulation of the hypoglossal nerve in the rat failed to evoke responses in neurons of the hypoglossal nucleus that could be attributed to afferent stimulation (Lodge *et al.*, 1973). Lingual proprioceptive input also influences other related oral motor nuclei. Electric stimulation of the twelfth nerve elicits responses in both facial (Tanaka, 1975) and trigeminal motoneurons in the cat (for example, Nakamura *et al.*, 1978; Takata *et al.*, 1979). Injections of HRP into 12 of the rat (Travers and Norgren, 1983) shows evidence of a projection to the intermediate division of the facial nucleus.

The projections considered do not constitute a complete description of all the direct influences on hypoglossal motoneurons. Extensive dendritic arborizations of hypoglossal motoneurons throughout the adjacent reticular formation, the raphe magnus nucleus and the medial longitudinal fasciculus allow further monosynaptic influence from other regions of the central nervous system. Cerebellar influences on 12, for example, may be mediated by 12 motoneuron dendrites in the medial longitudinal fasciculus (Wan *et al.*, 1982).

V. SUMMARY AND CONCLUSIONS

Each of the orofacial motor nuclei comprises several subdivisions which can be defined both myotopically and cytoarchitecturally. In the case of the motor trigeminal and hypoglossal nuclei, muscle antagonists are segregated in cytoarchitecturally distinct subdivisions in which motor agonists have adjacent representation. In the facial nucleus, a pronounced topography with peripheral musculature is evident. Myotopic subdivisions are further defined by the specificity of the central projections they receive. In the motor trigeminal nucleus, the Me5 and the supratrigeminal region project preferentially to jaw closer motoneurons in the dorsolateral division. In the facial nucleus, aural musculature motoneurons in the medial subdivision receive highly specific projections from the paralemniscal zone in the midbrain and an enkephalinergic projection from the ventral region of the caudal medulla. These projections contrast with those to the intermediate and lateral divisions of 7 which originate in the Kölliker-Fuse and ventral parabrachial nuclei in the pons, and MdD in the medulla.

Though these central afferent projections further differentiate between motor nuclei subdivisions, the overlap of central projections to the orofacial motor nuclei unifies them functionally. Neurons from the dorsal region of the sensory trigeminal complex (which receives intraoral somatosensory input) project to each of the orofacial motor nuclei. Likewise, proprioceptive information from jaw closing muscles can influence each of the orofacial motor nuclei via the supratrigeminal region. Complex synergies involving orofacial musculature are, no doubt, mediated by regions of the medullary reticular formation that also project to each of the motor nuclei.

ACKNOWLEDGEMENTS

I would like to thank Drs R. Norgren, T. Pritchard, R. Hamilton and S. Travers for critical readings of the manuscript. This work was supported by NS 20477.

REFERENCES

Aghajanian, G. K. and McCall, R. B. (1980). Serotonergic synaptic input to facial motoneurons: Localization by electron-microscopic autoradiography. *Neurosci.* 5, 2155–2162.

Aldes, L. D. (1980). Afferent projections to the hypoglossal nuclei in the rat and cat. *Anat. Rec.* 196, 7A.

Barnard, J. W. (1940). The hypoglossal complex of vertebrates. *J. Comp. Neurol.* 72, 489–524.

Basbaum, A. I., Clanton, C. H. and Fields, H. L. (1978). Three bulbospinal pathways from the rostral medulla of the cat: An autoradiographic study of pain modulating systems. *J. Comp. Neurol.* 178, 209–224.

Basmajian, J. V. (1979). *Muscles alive: Their functions revealed by electromyography.* Williams & Wilkins, Baltimore.

Berman, A. L. (1968). *The brain stem of the cat: A cytoarchitectonic atlas with stereotaxic coordinates.* University of Wisconsin Press, Madison, Wis.

Borke, R. C., Nau, M. E. and Ringler, R. L. (1983). Brain stem afferents of hypoglossal neurons in the rat. *Brain. Res.* 269, 47–55.

Burton, H., Craig, A. D., Jr., Poulos, D. A. and Molt, J. T. (1979). Efferent projections from temperature sensitive recording loci within the marginal zone of the nucleus caudalis of the spinal trigeminal complex in the cat. *J. Comp. Neurol.* 183, 753–778.

Cajal, S. Ramón y (1972). *Histologie du systeme nerveux de l'homme et des vertebres.* Tome I., Instituto Ramon y Cajal, Madrid.

Chibuzo, G. A. and Cummings, J. F. (1981). The origins of the afferent fibers to the lingual muscles of the dog, a retrograde labeling study with horseradish peroxidase. *Anat. Rec.* 200, 95–101.

Chibuzo, G. A. and Cummings, J. F. (1982). An enzyme tracer study of the organization of the somatic motor center for the innervation of different muscles of the tongue: Evidence for two sources. *J. Comp. Neurol.* 205, 273–281.

Contreras, R. J., Beckstead, R. M. and Norgren, R. (1982). The central projections of the trigeminal, facial, glossopharyngeal and vagus nerves: An autoradiographic study in the rat. *J. Auton. Nerv. Sys.* 6, 303–322.

Cooper, M. H. (1981). Neurons of the hypoglossal nucleus of the rat. *Otol. Head Neck Surg.* 89, 10–15.

Cooper, M. H. and Fritz, H. P. (1981). Input of the rat hypoglossal nucleus from the reticular formation. *Anat. Rec.* 199, 58A.

Courville, J. (1966a). The nucleus of the facial nerve: The relation between cellular groups and peripheral branches of the nerve. *Brain Res.* 1, 3338–354.

Courville, J. (1966b). Rubrobulbar fibres to the facial nucleus and the lateral reticular nucleus (nucleus of the lateral funiculus): An experimental study in the cat with silver impregnation methods. *Brain Res.* 1, 317–337.

Cunningham, D. P. and Basmajian, J. V. (1969). Electromyography of genioglossus and geniohyoid muscles during deglutition. *Anat. Rec.* 165, 401–410.

Dom, R. M. (1982). Topographical representation of the peripheral nerve branches of the facial nucleus of the opossum: A study utilizing horseradish peroxidase. *Brain Res.* 246, 281–284.

Dom, R., Falls, W. and Martin, G. F. (1973). The motor nucleus of the facial nerve in the opossum (*Didelphis marsupialis virginiana*): Its organization and connections. *J. Comp. Neurol.* 152, 373–402.

Doty, R. W., Richmond, W. H. and Storey, A. T. (1967). Effect of medullary lesions on coordination of deglutition. *Exp. Neurol.* 17, 91–106.

Edgeworth, F. H. (1907). On the development and morphology of the mandibular and hyoid muscles of mammals. *Quart. J. Micro. Sci.* 51, 511–556.

Edwards, S. B. (1972). The ascending and descending projections of the red nucleus in the cat: An experimental study using an autoradiographic tracing method. *Brain Res.* 48, 45–63.

Eisenman, J., Landgren, S. and Novin, D. (1963). Functional organization in the main sensory trigeminal nucleus and in the rostral subdivision of the nucleus of the spinal trigeminal tract in the cat. *Acta Physiol. Scand. Suppl.* 214, 3–44.

Erzurumlu, R. S. and Killackey, H. P. (1979). Efferent connections of the brainstem trigeminal complex with the facial nucleus of the rat. *J. Comp. Neurol.* 188, 75–86.

Finley, J. C. W., Maderdrut, J. L. and Petrusz, P. (1981). The immunocytochemical localization of enkephalin in the central nervous system of the rat. *J. Comp. Neurol.* 198, 541–565.

Fuxe, K. (1965). Evidence for the existence of monoamine neurons in the central nervous system. IV. The distribution of monoamine terminals in the central nervous system. *Acta Physiol. Scand.* 64(suppl.), 38–120.

Greene, E. C. (1963). *Anatomy of the rat.* Hafner, New York.

Hamilton, R. B. and Norgren, R. (1984). Central projections of gustatory nerves in the rat. *J. Comp. Neurol.* 222, 560–577.

Henkel, C. K. and Edwards, S. B. (1978). The superior colliculus control of pinna movements in the cat: Possible anatomical connections. *J. Comp. Neurol.* 182, 763–776.

Hinrichsen, C. F. L. and Watson, C. D. (1983). Brain stem projections to the facial nucleus of the rat. *Brain Behav. Evol.* 22, 153–163.

Holstege, G., Kuypers, H. G. J. M. and Dekker, J. J. (1977). The organization of the bulbar fibre connections to the trigeminal, facial, and hypoglossal motor nuclei II. An autoradiographic tracing study in cat. *Brain* 100, 265–286.

Huber, E. and Hughson, W. (1926–27). Experimental studies on the voluntary motor innervation of the facial musculature. *J. Comp. Neurol.* 42, 113–163.

Hutson, K. A., Glendenning, K. K. and Masterson, R. B. (1979). Accessory abducens nucleus and its relationship to the accessory facial and posterior trigeminal nuclei in cat. *J. Comp. Neurol.* 188, 1–16.

Jacquin, M. F., Rhoades, R. W., Enfiejian, H. L. and Egger, M. D. (1983). Organization and morphology of masticatory neurons in the rat: A retrograde HRP study. *J. Comp. Neurol.* 218, 239–256.

Jerge, C. R. (1963). Organization and function of the trigeminal mesencephalic nucleus. *J. Neurophysiol.* 26, 379–392.

Karlsen, K. (1965). The location of motor end plates and the distribution and histological structure of muscle spindles in jaw muscles of the rat. *Acta Odontol. Scand.* 23, 521–547.

Keller, J. T., Saunders, M. C., Ongkiko, C. M., Johnson, J., Frank, E., Van Loveren, H. and Tew, J. M. (1983). Identification of motoneurons innervating the tensor tympani and tensor veli palatini muscles in the cat. *Brain Res.* 270, 209–215.

Kemplay, S. and Cavanagh, J. B. (1983). Bilateral innervation of the anterior digastric muscle by trigeminal motor neurons. *J. Anat.* 136, 417–423.

Kidokoro, Y., Kubota, K., Shuto, S. and Sumino, R. (1968). Possible interneurons responsible for reflex inhibition of motorneurons of jaw-closing muscles from the inferior dental nerve. *J. Neurophysiol.* 31, 709–716.

Kitamura, S., Nishiguchi, T. and Sakai, A. (1983). Location of cell somata and the peripheral course of axons of the geniohyoid and thyrohyoid motoneurons: A horseradish peroxidase study in the rat. *Ex. Neurol.* 79, 87–96.

Krammer, E. B., Rath, T. and Lischka, M. F. (1979). Somatotopic organization of the hypoglossal nucleus: A HRP study in the rat. *Brain Res.* 170, 533–537.

Kume, M., Uemura, M., Matsuda, K., Matsushima, R. and Mizuno, N. (1978). Topographical representation of peripheral branches of the facial nerve within the facial nucleus: A HRP study in the cat. *Neurosci. Lett.* 8, 5–8.

Law, M. E. (1954). Lingual proprioception in pig, dog and cat. *Nature*, 174, 1107–1108.

Levitt, P. and Moore, R. Y. (1979). Origin and organization of brainstem catecholamine innervation in the rat. *J. Comp. Neurol.* 186, 505–528.

Lewis, P. R., Flumerfelt, B. A. and Shute, C. C. D. (1971). The use of cholinesterase techniques to study topographical localization in the hypoglossal nucleus of the rat. *J. Anat.* 110, 203–213.

Limwongse, V. and DeSantis, M. (1977). Cell body locations and axonal pathways of neurons innervating muscles of mastication in the rat. *Am. J. Anat.* 149, 477–488.

Lodge, D., Duggan, A. W., Biscoe, T. J. and Caddy, K. W. T. (1973). Concerning recurrent collaterals and afferent fibers in the hypoglossal nerve of the rat. *Exp. Neurol.* 41, 63–75.

Lowe, A. A. (1981). The neural regulation of tongue movements. *Prog. Neurobiol.* 15, 295–344.

Lyon, M. J. (1978). The central location of the motor neurons to the stapedius muscle in the cat. *Brain Res.* 143, 437–444.

McCall, R. B. and Aghajanian, G. K. (1979). Serotonergic facilitation of facial motoneuron excitation, *Brain Res.* 169, 11–27.

Martin, G. F. and Dom, R. (1970). Rubrobulbar projections of the opossum (*Didelphis virginiana*). *J. Comp. Neurol.* 139, 199–214.

Martin, M. R. and Lodge, D. (1977). Morphology of the facial nucleus of the rat. *Brain Res.* 123, 1–12.

Martin, M. R., Caddy, K. W. T. and Biscoe, T. J. (1977). Numbers and diameters of motoneurons and myelinated axons in the facial nucleus and nerve of the albino rat. *J. Anat.* 123, 579–587.

Matesz, C. (1981). Peripheral and central distribution of fibres of the mesencephalic trigeminal root in the rat. *Neurosci. Lett.* 27, 13–17.

Matsuda, K., Uemura, M., Kume, M., Matsushima, R. and Mizuno, N. (1978). Topographical representation of masticatory muscles in the motor trigeminal nucleus in the rabbit: A HRP study. *Neurosci. Lett.* 8, 1–4.

Meessen, H. and Olszewski, J. (1949). *A cytoarchitectonic atlas of the rhombencephalon of the rabbit.* S. Karger, Basel.

Miller, R. A. and Strominger, N. L. (1973). Efferent connections of the red nucleus in the brainstem and spinal cord of the rhesus monkey. *J. Comp. Neurol.* 152, 327–346.

Mizuno, N. (1970). Projection fibers from the main sensory trigeminal nucleus and the supratrigeminal region. *J. Comp. Neurol.* 139, 457–472.

Mizuno, N. and Sauerland, E. K. (1970). Trigeminal proprioceptive projections to the hypoglossal nucleus and the cervical ventral gray column. *J. Comp. Neurol.* 139, 215–226.
Mizuno, N., Nakamura, Y. and Iwahori, N. (1974). Central afferent fibers to trigeminal motor system. *Bull. Tokyo Med. Dent. Univ.* 21 (Suppl.) 19–21.
Mizuno, N., Konishi, A. and Sato, M. (1975). Localization of masticatory motoneurons in the cat and rat by means of retrograde axonal transport of horseradish peroxidase. *J. Comp. Neurol.* 164, 105–116.
Nakamura, Y. (1980). Brainstem neuronal mechanisms controlling the trigeminal motoneuron activity. *Prog. Clin. Neurophysiol.* 8, 181–202.
Nakamura, Y., Goldberg, L. J., Mizuno, N. and Clemente, C. D. (1978). Effects of hypoglossal afferent stimulation on masseteric motoneurons in cats. *Exp. Neurol.* 61, 1–14.
Nord, S. G. (1967). Somatotopic organization in the spinal trigeminal nucleus, the dorsal column nuclei and related structures in the rat. *J. Comp. Neurol.* 130, 343–356.
Norgren, R. (1978). Projections from the nucleus of the solitary tract in the rat. *Neurosci.* 3, 207–218.
Norgren, R. and Pfaffmann, C. (1975). The pontine taste area in the rat. *Brain Res.* 91, 99–117.
Odutola, A. B. (1976). Cell grouping and golgi architecture of the hypoglossal nucleus of the rat. *Exp. Neurol.* 52, 356–371.
Palkovits, M., Brownstein, M. and Saavedra, J. M. (1974). Serototin content of the brain stem nuclei in the rat. *Brain Res.* 80, 237–249.
Panneton, W. M. and Martin, G. F. (1978). Midbrain projections to the facial nucleus in the opossum. *Brain Res.* 145, 355–359.
Panneton, W. M. and Martin, G. F. (1979). Midbrain projections to the trigeminal, facial and hypoglossal nuclei in the opossum. A study using axonal transport techniques. *Brain Res.* 168, 493–511.
Panneton, W. M. and Martin, G. F. (1983). Brainstem projections to the facial' nucleus of the opossum: A study using axonal transport techniques. *Brain Res.* 267, 19–33.
Papez, J. W. (1927). Subdivisions of the facial nucleus. *J. Comp. Neurol.* 43, 159–191.
Paxinos, G. and Watson, C. R. R. (1982). *The rat brain in stereotaxic coordinates.* Academic Press, Sydney.
Porter, R. (1967). Cortical actions on hypoglossal motoneurones in cats: A proposed role for a common internuncial cell. *J. Physiol.* 193, 295–308.
Provis, J. (1977). The organization of the facial nucleus of the brush-tailed possum (*Trichosurus vulpecula*). *J. Comp. Neurol.* 172, 177–188.
Ricardo, J. A. and Koh, E. T. (1978). Anatomical evidence of direct projections from the nucleus of the solitary tract to the hypothalamus, amygdala, and other forebrain structures in the rat. *Brain Res.* 153, 1–26.
Ruggiero, D. A., Ross, C. A., Kumada, M. and Reis, D. J. (1982). Reevaluation of projections from the mesencephalic trigeminal nucleus to the medulla and spinal cord: New projections. A combined retrograde and anterograde horseradish peroxidase study. *J. Comp. Neurol.* 206, 278–292.
Saper, C. B. and Loewy, A. D. (1980). Efferent connections of the parabrachial nucleus in the rat. *Brain Res.* 197, 291–317.
Sasamoto, K. (1979). Motor nuclear representation of masticatory muscles in the rat. *Jap. J. Physiol.* 29, 739–747.
Schaffer, N., Jean, A. and Calas, A. (1982). Identification of serotoninergic axon terminals in the rat trigeminal motor nucleus: A radioautographic study. *Neurosci. Lett. Suppl.* 10, S433.
Scheibel, A. B. (1955). Axonal afferent patterns in the bulbar reticular formation. *Anat. Rec.* 121, 361.
Scheibel, M. E. (1955). Axonal efferent patterns in the bulbar reticular formation. *Anat. Rec.* 121, 362.
Senba, E. and Tohyama, M. (1983a). Leucine-enkephalin-containing neuron system in the facial nucleus of the rat with special reference to its fine structure. *Brain Res.* 274, 17–23.
Senba, E. and Tohyama, M. (1983b). Reticulo-facial enkephalinergic pathway in the rat: An experimental immunohistochemical study. *Neurosci.* 10, 831–839.
Shaw, M. D. and Baker, R. (1983). Direct projections from vestibular nuclei to facial nucleus in cats. *J. Neurophysiol.* 50, 1265–1280.

Shohara, E. and Sakai, A. (1983). Localization of motorneurons innervating deep and superficial facial muscles in the rat: A horseradish peroxidase and electrophysiologic study. *Exp. Neurol.* 81, 14–33.

Spangler, K. M., Henkel, C. K. and Miller, I. J., Jr. (1982). Localization of the motor neurons to the tensor tympani muscle. *Neurosci. Lett.* 32, 23–27.

Steinbusch, H. W. M. (1981). Distribution of serotinin-immunoreactivity in the central nervous system of the rat-cell bodies and terminals. *Neurosci.* 6, 557–618.

Stevens, R. T., Hodge, C. J., Jr. and Apkarian, A. V. (1982). Kölliker-Fuse nucleus: The principal source of pontine catecholaminergic cells projecting to the lumbar spinal cord of cat. *Brain Res.* 239, 589–594.

Stewart, W. A. and King, R. B. (1963). Fiber projections from the nucleus caudalis of the spinal trigeminal nucleus. *J. Comp. Neurol.* 121, 271–286.

Sumino, R. and Nakamura, Y. (1974). Synaptic potentials of hypoglossal motoneurons and a common inhibitory interneuron in the trigemino-hypoglossal reflex. *Brain Res.* 73, 439–454.

Szekély, G. and Matesz, C. (1982). The accessory motor nuclei of the trigeminal, facial, and abducens nerves in the rat. *J. Comp. Neurol.* 210, 258–264.

Szentágothai, J. (1948). Anatomical considerations of monosynaptic reflex arcs. *J. Neurophysiol.* 11, 445–454.

Taber, E. (1961). The cytoarchitecture of the brain stem of the cat. I. Brain stem nuclei of cat. *J. Comp. Neurol.* 116, 27–70.

Takata, M., Fujita, S. and Shohara, E. (1979). Postsynaptic potentials in the hypoglossal motoneurons set up by hypoglossal nerve stimulation. *Jap. J. Physiol.* 29, 49–60.

Takatori, M., Nozaki, S. and Nakaura, Y. (1981). Control of trigeminal motoneurons exerted from bulbar reticular formation in the cat. *Exp. Neurol.* 72, 122–140.

Tanaka, T. (1975). Afferent projections in the hypoglossal nerve to the facial neurons of the cat. *Brain Res.* 99, 140–144.

Tanaka, T., (1976). Pyramidal activation of the facial nucleus in the cat. *Brain Res.* 103, 389–393.

Torvik, A. (1956). Afferent connections to the sensory trigeminal nuclei, the nucleus of the solitary tract and adjacent structures. *J. Comp. Neurol.* 106, 51–141.

Travers, J. B. and Norgren, R. (1983). Afferent projections to the oral motor nuclei in the rat. *J. Comp. Neurol.* 220, 280–298.

Uemura-Sumi, M., Takahashi, O., Matsushima, R., Takata, M., Yasui, Y. and Mizuno, N. (1982). Localization of masticatory motoneurons in the trigeminal motor nucleus of the guinea pig. *Neurosci. Lett.* 29, 219–224.

Valverde, F. (1961). A new type of cell in the lateral reticular formation of the brain stem. *J. Comp. Neurol.* 117, 189–195.

Vornov, J. J. and Sutin, J. (1983). Brainstem projections to the normal and noradrenergically hyperinnervated trigeminal motor nucleus. *J. Comp. Neurol.* 214, 198–208.

Wamsley, J. K., Lewis, M. S., Young, W. S. and Kuhar, M. J. (1981). Autoradiographic localization of muscarinic cholinergic receptors in rat brainstem. *J. Neurosci.* 1, 176–191.

Wan, X. S. T., Trojanowski, J. Q., Gonatas, J. O. and Liu, C. N. (1982). Cytoarchitecture of the extranuclear and commissural dendrites of hypoglossal nucleus neurons as revealed by conjugates of horseradish peroxidase with cholera toxin. *Exp. Neurol.* 78, 167–175.

Watson, C. R. R., Sakai, S. and Armstrong, W. (1982). Organization of the facial nucleus in the rat. *Brain Behav. Evol.* 20, 19–28.

7
Somatosensory system

DAVID J. TRACEY

University of New South Wales
Kensington, NSW, Australia

I. INTRODUCTION

This chapter is intended as an overview of the neuroanatomy of the somatosensory system, with particular reference to the rat. The somatosensory system is regarded here as those parts of the peripheral and central nervous systems which provide information from somatosensory receptors to the cerebral cortex. Some parts of the somatosensory system, such as the ventrobasal thalamus and somatosensory cortex, will also be dealt with, to some extent, in other chapters.

A. Somatosensory receptors

These receptors sense the condition of the skin, muscles and joints of the head and body, and their fibers transmit information along the peripheral nerves to the central nervous system. They are the terminals of primary afferent fibers, with cell bodies in the spinal ganglia, or in the trigeminal ganglion (see Bannister, 1976, for review).

B. Mechanoreceptors

Somatosensory receptors may be classified as mechanoreceptors, thermoreceptors, or nociceptors. The sensory terminals of most mechanoreceptors are associated with accessory cells or structures which may affect the rate of adaptation of the receptor to a constant stimulus. The terminals together with their accessory structure are then spoken of as the sensory receptor.

Mechanoreceptors are often classified according to their location, as receptors in skin, muscles or joints. However, many mechanoreceptor types such as Pacinian corpuscles and Ruffini endings are found in more than one of these tissues, and so they will be treated here according to their morphology (Bannister, 1976).

THE RAT NERVOUS SYSTEM
ISBN 0 12 547632 9

1. Mechanoreceptors with complex lamellar capsules

The Pacinian corpuscle is a large receptor structure, in which the unmyelinated terminal of the primary afferent is encapsulated by a number of lamellae arranged rather like the layers of an onion. These lamellae constitute an inner core, derived from Schwann cells, and an outer core whose cells have ultrastructural features similar to fibroblasts. Pacinian corpuscles are rapidly adapting, extremely sensitive to vibration, and have very large receptive fields. They are found in the subcutaneous tissue of all types of skin and in particular beneath glabrous skin.

Paciniform endings are similar in structure, but are much smaller, since they have only "inner core" lamellae and no outer core. They occur in sinus hair follicles, and in various types of connective tissue, such as fascia, periosteum and joint capsules, and are rapidly adapting (Poláček, 1966).

Krause's spherical end bulbs are found around the mouth and in the mammalian penis and clitoris (genital end bulbs). Genital end bulbs in the rat are morphologically similar to those found in humans (Patrizi and Munger, 1965). The signaling properties of the end bulbs have not yet been established; however, an electrophysiologic study of the rat penis found both slowly adapting and rapidly adapting mechanoreceptors (Kitchell *et al.*, 1982).

Meissner corpuscles may be present in the hairless skin of the snout in the rat (Cauna, 1976) while Krause corpuscles may be found in the hairy skin (Cunningham and Fitzgerald, 1972).

2. Lanceolate endings of hair follicles

The lanceolate endings of hair follicles are the fine, flattened endings of myelinated fibers which run longitudinally in the follicles of down, guard and sinus hairs or vibrissae (Cauna, 1969; Andres, 1966). They are apparently coupled mechanically to the follicle by fine collagen fibers (Fig. 1a). They are rapidly adapting, and the response properties of those associated with down hairs (type D) and guard hairs (type G) have been studied in the rat (Lynn and Carpenter, 1982). Type D units have relatively large receptive fields and respond to slow movements of down hairs, while type G units have relatively small receptive fields, and are excited only by fast movement of guard hairs.

3. Mechanoreceptors associated with specialized cells

Merkel disc endings are expanded terminals which are associated with specialized epidermal cells, the Merkel cells (Fig. 1b). They are generally located in groups in the lowermost layers of the epidermis, in both hairy and glabrous skin. In hairy skin they occur as compact clusters of 50–70 cells, associated with

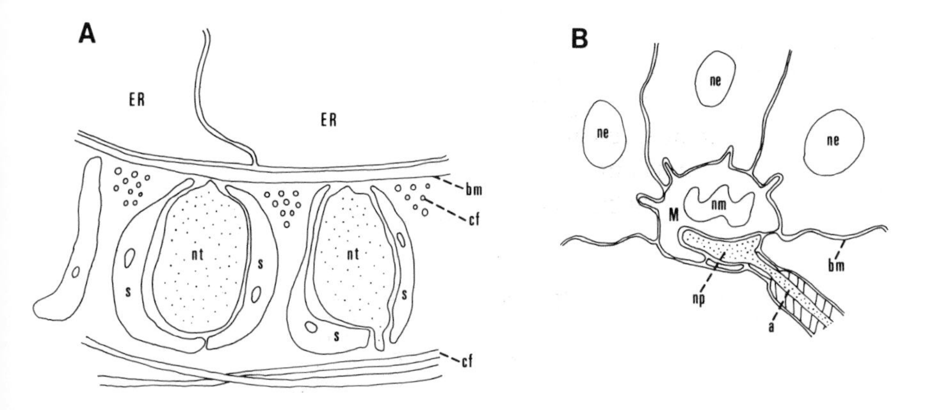

Fig. 1: A: *Lanceolate endings of a fine hair.* The section has cut the hair and its epithelial root sheath (ER) transversely, so that the lanceolate endings (nt) are also seen in transverse section. The shaft of the hair is not shown; it is off the top edge of the diagram. Note that the array of lanceolate endings may not completely surround the hair shaft (modified from Cauna, 1969). B: *Merkel ending.* The nerve ending or nerve plate is closely associated with a specialized epidermal cell, the Merkel cell. The Merkel disc ending is a slowly adapting mechanoreceptor found in glabrous and hairy skin, and in the follicles of the vibrissae. a, myelinated axon; bm, basement membrane; cf, collagen fiber; M, Merkel cell; ne, epithelial cell nucleus; nm, Merkel cell nucleus; np, nerve ending or nerve plate; nt, unmyelinated nerve terminal; s, Schwann cell process.

the terminals of a single myelinated axon. These clusters are often found associated with the follicles of large tylotrich hairs, in the rat as in other mammals (Straile, 1960); the clusters are generally located beneath an elevation in the skin, referred to as a Haarscheibe or touch corpuscle. They are the slowly adapting (type I) mechanoreceptors and have small receptive fields. In some ways the snout of the rat is comparable with the fingertip of the primate: both bear ridged glabrous skin important in tactile discrimination. In both cases Merkel discs are found at the base of rete ridges or pegs (Mackintosh, 1975).

Merkel disc endings are also found in the follicles of vibrissae or sinus hairs, where they provide only one of a range of different receptor types including paciniform endings, lanceolate terminals and Ruffini endings (Andres, 1966). The relationship between the Merkel disc or neurite and the Merkel cell has been described in the rat by Patrizi and Munger (1966).

4. Mechanoreceptors with capsules which enclose other tissues

The Ruffini endings are found in the connective tissue of skin and joint capsules. The unmyelinated terminal branches are intertwined with bundles of collagen fibers, and the innervated bundle is generally surrounded by a capsule. Ruffini endings are slowly adapting, and correspond with the slowly adapting (type II) receptors of hairy skin, where they have relatively large receptive fields. They occur in human glabrous skin, where they were first identified. Ruffini endings also occur as the slowly adapting "spray-like" endings of the capsule and ligaments of joints. In the rat joint, they lack the capsule which is usually found in the spray-like joint receptors of cats and primates (Poláček, 1966).

Golgi tendon organs and muscle spindles are further examples of mechanoreceptors in which primary afferents terminate on accessory structures enclosed within a capsule.

C. Thermoreceptors and nociceptors

These receptors are generally thought to be afferents with unencapsulated endings and unmyelinated or thin myelinated axons. Cold receptors have axons belonging to both groups, and have been identified histologically. Warm receptors have unmyelinated axons, but their receptor structure has yet to be characterized. Nociceptors are thought to correspond with free nerve endings, such as the papillary endings found in the skin of the rat (Cauna, 1969) or the penicillate endings of unmyelinated fibers in the corium, just under the epidermis (Cauna, 1976). Some unmyelinated or C fibers also function as mechanoreceptors.

II. PROJECTIONS OF SOMATOSENSORY RECEPTORS TO THE SPINAL CORD

The cell body of a primary afferent neuron has a peripheral process which carries the receptor ending and travels in a peripheral nerve towards the spinal ganglion. It also has a central process which is carried by the dorsal root towards the spinal cord. The dorsal roots enter the spinal cord in the dorsolateral sulcus, between the dorsal columns medially and the dorsolateral fasciculus laterally. On entering the cord, the primary afferent splits into an ascending branch and a descending branch. Some of the ascending branches are long, and contribute to the dorsal column pathway, one of the main ascending somatosensory pathways. Other branches, both rostral and caudal, are relatively short and give rise to collateral branches which end as synaptic terminals on interneurons or motoneurons. Some of the recipient interneurons are part of the neuronal circuitry which gives rise to reflexes at the level of the spinal cord. Others are the cells of origin of ascending somatosensory pathways such as the spinothalamic pathway.

A. Laminar terminations and somatotopy

Since the introduction of intracellular staining of neurons with horseradish peroxidase through a microelectrode, it has been possible to examine the detailed morphology of the central processes of identified primary afferents. The greater part of this work has been done in the cat, and has been reviewed recently by Brown (1982). From this work it appears that the pattern of termination is quite specific, and depends on the receptor type of the primary afferent. Thus low threshold mechanoreceptors such as Krause corpuscles, Pacinian corpuscles, and the slowly adapting Merkel and Ruffini endings have central processes with characteristic terminal arbors extending from layer 3 to 6 of the dorsal horn. The terminal arbors of afferents from hair follicles are restricted to layer 3.

The group Ia and group II afferents from muscle spindles have terminations in layers 7 and 9 of the ventral horn (layer 9 containing the motoneurons), while the group Ib fibers of the Golgi tendon organs terminate in layers 5 to 7. The sites of termination of afferents from the joint capsules have not yet been reported.

Unmyelinated fibers terminate in layer 2 (substantia gelatinosa) while thin myelinated fibers, some of which belong to nociceptors, terminate in layer 1 (marginal zone) as well as in layer 2 and 3 (Light and Perl, 1979).

Superimposed on the laminar organization of primary afferent terminals is a somatotopic organization, which is revealed by recording from second order neurons. Such recordings reveal a rostrocaudal organization (due to the segmental representation of dermatomes) and a mediolateral organization. Thus, in the substantia gelatinosa of the rat, the foot is represented medially while the proximal thigh is represented laterally (Devor and Claman, 1980).

B. Ascending pathways

There are several ascending pathways which carry somatosensory information from peripheral receptors (see Willis and Coggeshall, 1978, for review). Two of these have been studied in the rat: the dorsal column pathway, and the spinothalamic pathway. Other somatosensory pathways have been described in the cat, but for the rat less information is available on their existence or nature. Pathways described in the cat include the spinocervical pathway and the nucleus Z pathway.

1. Dorsal column pathway

The ascending branches of some primary afferent fibers travel up the dorsal columns and terminate on the cells of the gracile and cuneate nuclei. Cells in the dorsal column nuclei project via the medial lemniscus to the ventrobasal thalamus. Neurons in this region of the thalamus then send their axons to the somatosensory cortex (Fig. 2a).

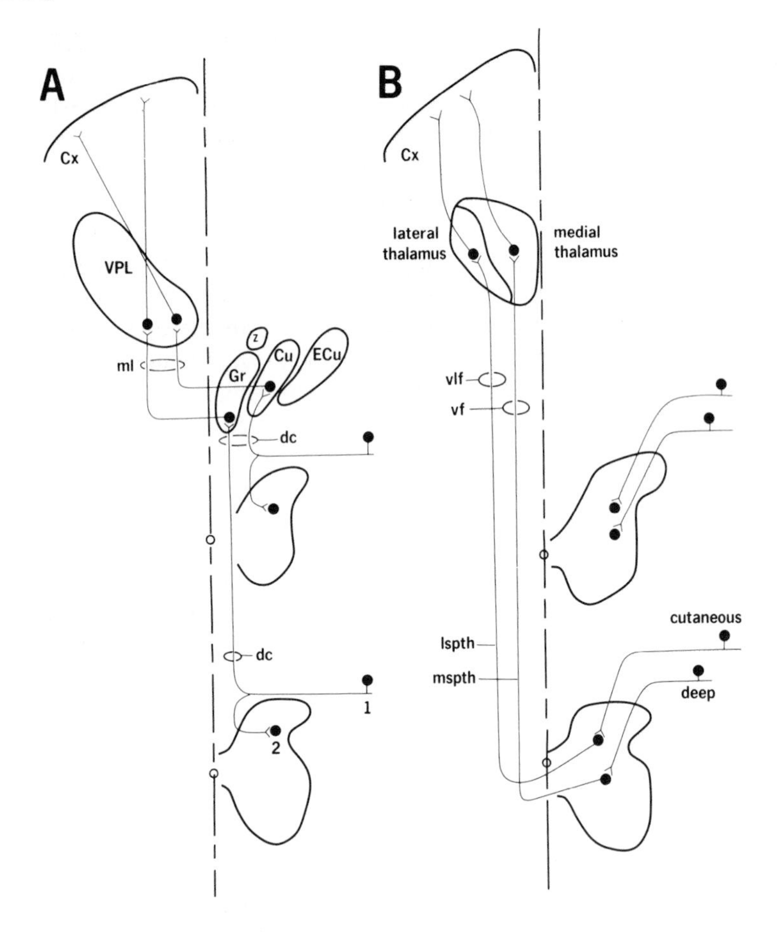

Fig. 2: A. *Dorsal column pathway in the rat.* A primary afferent with its cell body (1) in the dorsal root ganglion gives rise to an ascending collateral, which travels in the dorsal columns (dc) and terminates in the dorsal column nuclei. The primary afferent may also terminate on a second-order dorsal column neuron (2) whose axon also travels in the dorsal columns and terminates in the gracile or cuneate nucleus. Primary afferents also send collaterals to neurons in Clarke's column and to the external cuneate nucleus (not shown). The axons of postsynaptic dorsal column neurons are omitted for the sake of clarity (after Brown and Gordon, 1977). Cx, cortex; Cu, cuneate nucleus; dc, dorsal columns; ECu, external cuneate nucleus; Gr, gracile nucleus; ml, medial lemniscus; VPL, ventroposterior thalamic nucleus, lateral part; Z, nucleus Z. B. *Spinothalamic (spth) pathway in the rat.* A primary afferent terminates on a spinothalamic neuron. These neurons are located in the ventral part of the dorsal horn and intermediate gray matter (deep inputs) or in the nucleus proprius and marginal zone (cutaneous inputs). Spinothalamic neurons receiving deep inputs belong to the 'medial' spth (mspth) and project in the ventral funiculus (vf) to the medial thalamus, including the intralaminar nuclei. Spinothalamic neurons receiving cutaneous inputs belong to the 'lateral' spth (lspth) and project in the ventrolateral funiculus (vlf) to the lateral thalamus, including VPL (based on Giesler *et al.*, 1979a). Note that the 'medial thalamus' includes nucleus submedius, which receives nociceptive inputs from layer 1. Cx, cortex; lspth, lateral spinothalamic tract; mspth, medial spinothalamic tract; vf, ventral funiculus of the spinal cord; vlf, ventrolateral funiculus of the spinal cord.

The ascending fibers of the dorsal columns are somatotopically organized, so that those joining the cord most caudally run next to the midline. Primary afferents from spinal ganglia at increasingly rostral levels of the spinal cord are added laterally, so that a cross-section of the dorsal columns at the upper cervical level will contain a mediolateral sequence of primary afferents from the tail, hindlimb, trunk and forelimb.

The dorsal columns do not consist only of primary afferent collaterals ascending to the dorsal column nuclei. In the cat, about 80% of primary afferents entering the dorsal columns leave again without reaching the gracile and cuneate nuclei—for example, many leave to terminate on cells in Clarke's column (dorsal nucleus). Some of the fibers ascending the dorsal columns are not collaterals of primary afferents, but are the axons of second order neurons, which also carry somatosensory information. Finally, there are descending axons in the dorsal columns. Some of these originate in the dorsal column nuclei; and, in the rat, the corticospinal tract also descends in the dorsal column.

The dorsal column system has received relatively little attention in the rat, but it is known that the primary afferents of the gracile fasciculus terminate at all levels of the gracile nucleus, in Clarke's column (dorsal nucleus), and in other regions such as the medullary reticular formation (Ganchrow and Bernstein, 1981). In the rat (as in humans) the majority of axons in the dorsal columns are unmyelinated (Langford and Coggeshall, 1981).

2. *Spinothalamic pathway*

Primary afferents terminate on spinothalamic neurons located just rostral to the entry of the dorsal root. The axons of the spinothalamic neurons cross to the contralateral side of the spinal cord and ascend in its ventrolateral quadrant. After reaching the thalamus, they terminate in four regions: the rostral part of the ventrobasal complex; the intralaminar nuclei; the posterior group; and the nucleus submedius or gelatinosus (Craig and Burton, 1981; Lund and Webster, 1967). Neurons of the ventrobasal thalamus which receive spinothalamic inputs probably project to the somatosensory cortex.

In the rat, a medial and a lateral spinothalamic tract (spth) have recently been distinguished, based on their terminations in medial or lateral thalamic nuclei. The neurons of the medial spth are located in the ventral part of the dorsal horn and "intermediate gray zone", and are activated by stimulating deep tissues such as muscle and tendon (Fig. 2b). The neurons of the lateral spth are located in the marginal zone and nucleus proprius, and are activated primarily by cutaneous stimuli, both noxious and innocuous (Giesler *et al.*, 1979a). In other animals, such as the monkey, some spinothalamic axons also carry proprioceptive information, but do not seem to terminate differently in the thalamus compared with other spinothalamic axons. Spinothalamic neurons in the rat may also send collateral branches to the medullary reticular formation (Kevetter and Willis, 1983) and to the cerebellum (Yezierski and Bowker, 1981).

III. MEDULLARY NUCLEI

The nuclei which will be discussed here are the dorsal column nuclei, nucleus Z, and the lateral cervical nucleus. The dorsal column nuclei are positioned at the summit of the dorsal columns and include the gracile, cuneate and external cuneate. Some of the neurons in the dorsal column nuclei send their axons to the thalamus. Their axons arch ventrally as internal arcuate fibers to the region of the inferior olive, and then cross the midline and gather into a bundle called the medial lemniscus, which ascends to the thalamus.

A. Gracile nucleus

The gracile nucleus (Gr) is divided into caudal and rostral parts which probably correspond with the caudal and rostral parts of the cuneate nucleus (see below). Information on the gracile nucleus of the rat is based largely on Golgi material. The gracile nucleus is said to contain seven classes of cell, distinguishable on morphological grounds. Two of these neuronal types, the medium sized fusiform neuron and the large stellate neuron, have axons which appear to enter the medial lemniscus. More recently, the cells projecting to the lateral part of the ventroposterior thalamic nucleus (VPL) have been identified using the retrograde transport of HRP (Tan and Lieberman, 1978). The labeled cells were of two types: some were round with a spherical nucleus, while others were oval with a flattened nucleus. These are probably the large stellate and medium fusiform neurons described in the Golgi material.

In the gracile nucleus, vertical columns of intertwined dendrites belonging to small and medium sized cells were observed, but the cell bodies were not aggregated into any discernible arrangement. The absence of cell nests agrees with observations on the rat cuneate, but the absence of "slabs" like those found in the caudal region of the rat cuneate is surprising. The caudal region of the gracile nucleus is traversed by large primary afferents of the dorsal columns, which travel in a median plane giving off vertical collaterals. On reaching the rostral region, most of the primary axons bend sharply to traverse the nucleus in a transverse and slightly rostral direction (Gulley, 1973). The arrangement of the primary afferent fibers in the caudal region of the nucleus no doubt underlies the somatotopic arrangement of paramedian strips reported in physiologic experiments.

B. Cuneate nucleus

The cuneate nucleus (Cu) of the rat can be divided into two parts—a caudal region with distinct aggregates of cells arranged as vertical "slabs", and a rostral region in which the cells are not organized into focal groups. Each slab of cells in the caudal region appears to receive a discrete terminal field projection from a single dorsal root, or perhaps even a single primary afferent (Basbaum and Hand, 1973). The caudal region projects primarily to VPL.

The rostral region receives different overlapping terminal fields from the dorsal roots, and projects not only to the VPL but also to other regions including the tectum, pretectum and zona incerta. A comparable organization is found in the cat.

C. External cuneate nucleus

The external cuneate nucleus (ECu) is located just lateral to the main cuneate, and the two are partly connected by bridges of cells. Like the main cuneate nucleus, the external cuneate receives the terminations of primary afferents from the forelimb. Unlike the main cuneate, the external cuneate projects primarily to the cerebellum, although a projection to the thalamus has been reported in the rat and the primate. The ECu is the forelimb equivalent of Clarke's column, which receives primary afferent terminations from the hindlimb and also projects to the cerebellum. Since the external cuneate has been shown to project to the thalamus in the rat (Fukushima and Kerr, 1979) it can be regarded as a proper part of the somatosensory system.

The ECu is composed mainly of large globular cells which tend to be grouped in small clusters. The terminations of primary afferents in the feline nucleus are somatotopically organized into spiral oblique lamellae surrounding a core provided by the seventh cervical dorsal root (C7). This organization is probably similar in the rat. Unfortunately it is difficult to correlate the anatomic work on projections of individual dorsal roots in the cat with the elegant electrophysiologic work on the somatotopic organization of neurons in the rat ECu activated by muscle afferents. This work showed that the cells activated by forearm muscle afferents are situated medially, while cells activated by afferents from shoulder and thorax are found progressively more laterally (Campbell *et al.*, 1974).

D. Nucleus Z

At the rostral end of the gracile nucleus is a cell group called nucleus Z, which was originally described as being associated with the vestibular nuclei. There is now good evidence from the cat that nucleus Z is that subdivision of the dorsal column nuclei which relays information to the thalamus from proprioceptors in the hindlimb. While the cuneate and gracile nuclei receive ascending terminations almost exclusively from primary and second order afferents in the dorsal columns, nucleus Z receives its projections from collaterals of axons which travel in the dorsolateral fasciculus and belong to the dorsal spinocerebellar tract.

Nucleus Z is difficult to distinguish from the rostral part of the gracile nucleus in Nissl stained sections. However, there is evidence that nucleus Z is present in the rat, since axons in the dorsolateral fasciculus project to the appropriate region near the rostral end of the gracile nucleus (Zemlan *et al.*, 1978). Recent experiments using double retrograde labeling from nucleus Z and the cerebellum have confirmed that spinal afferents to nucleus Z are collaterals of dorsal spinocerebellar tract axons, with cell bodies in Clarke's column (Low and Tracey, 1984).

E. Lateral cervical nucleus

The lateral cervical nucleus (LatC) of the rat is a somewhat diffuse group of cells which is located close to the ventrolateral border of the dorsal horn in the upper three segments of the spinal cord (Giesler *et al.*, 1979b). It has been shown that the lateral cervical nucleus of the rat receives afferents from all levels of the spinal cord (Lund and Webster, 1967) and projects to the thalamus (Giesler *et al.*, 1979a). This is consistent with extensive findings in the cat, where the spinocervical neurons are located mainly in layer 4 of the spinal cord, and LatC projects to VPL. In the rat, LatC also projects to the midbrain and tectum via the medial lemniscus (M. Björklund, personal communication).

F. Cortical inputs to sensory nuclei of the medulla

Cortical inputs from the pyramidal tract to most of the sensory nuclei of the medulla have been demonstrated in the rat (Valverde, 1966). They apparently terminate primarily on the cell bodies of neurons, in contrast to the primary afferent fibers, which terminate on the dendrites. Cortical fiber terminals are concentrated in the ventral part of the gracile and cuneate nuclei (called the pars compacta by Valverde), but it is not clear whether they avoid the cell "slab" region in the rat in the same way as they avoid the cell "cluster" region in the cat.

IV. SENSORY TRIGEMINAL NUCLEI

The sensory nuclei of the trigeminal nerve receive afferent input from the skin of the face and oral mucosa, as well as from receptors in the facial musculature and the temporomandibular joint. The nuclei extend from the mesencephalon down to the upper cervical region of the spinal cord, and are divided into three main parts: the mesencephalic sensory nucleus; the principal sensory nucleus; and the nucleus of the spinal tract. The nucleus of the spinal tract is further subdivided into three parts: oral subnucleus; interpolar subnucleus; and caudal subnucleus (Fig. 3). For review see Darian-Smith, 1973.

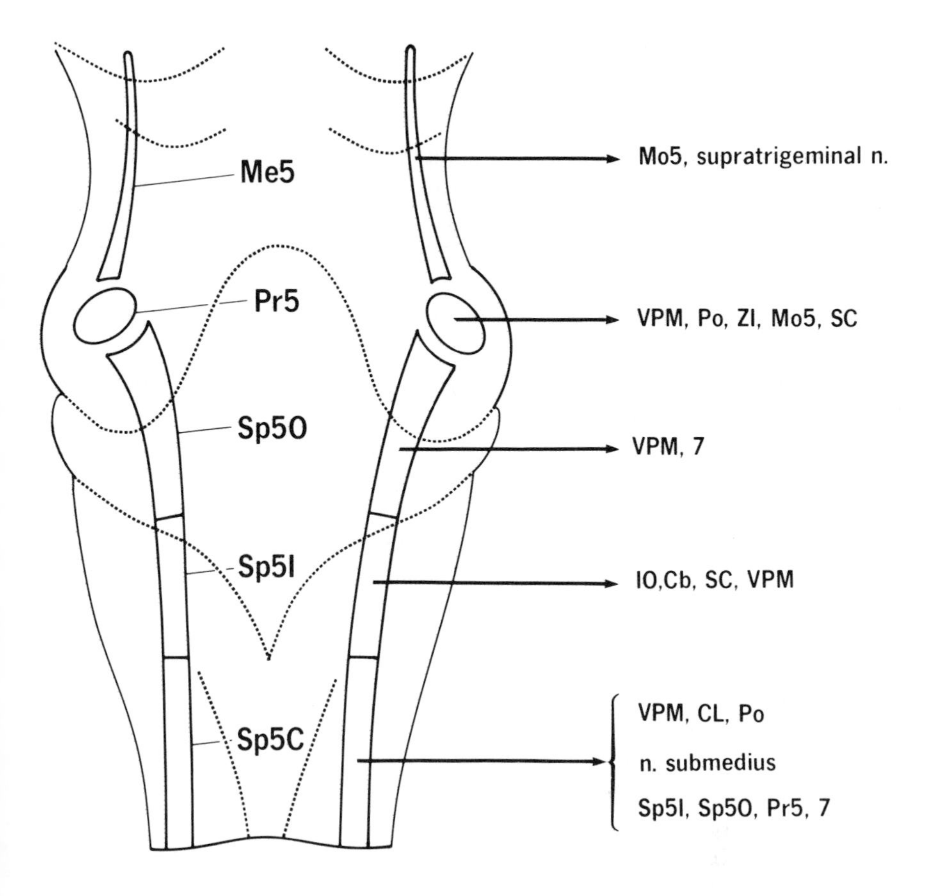

Fig. 3: Projections of the sensory trigeminal nuclei in the rat (outline based on Olszewski, 1950). 7, facial nucleus; Cb, cerebellum; CL, centrolateral thalamic nucleus; IO, inferior olive; Me5, nucleus of the mesencephalic tract of the trigeminal nerve; Mo5, motor trigeminal nucleus; Po, posterior thalamic nuclear group; Pr5, principal sensory trigeminal nucleus; SC, superior colliculus; Sp5C, nucleus of the spinal tract of the trigeminal nerve, caudal subnucleus; Sp5I, nucleus of the spinal tract of the trigeminal nerve, interpolar subnucleus; Sp5O, nucleus of the spinal tract of the trigeminal nerve, oral subnucleus; VPM, ventroposterior thalamic nucleus, medial part; ZI, zona incerta.

A. Primary afferents

The primary afferents of the trigeminal nerve have cell bodies located mainly in the trigeminal ganglion, where they are organized in a somatotopic fashion (Gregg and Dixon, 1973; see also Arvidsson, 1982). However, the cell bodies of some trigeminal afferents are located in the mesencephalic trigeminal nucleus (see below). On entering the pons, most myelinated fibers divide into an ascending and a descending branch. The ascending branches terminate in the principal sensory nucleus, while the descending branches contribute to the spinal tract of the trigeminal nerve. Unmyelinated fibers of the trigeminal nerve do not branch, but descend to terminate in subnucleus caudalis. As more caudal levels of the tract are reached, the fiber diameter becomes significantly smaller, due partly to termination of the larger fibers at more rostral levels, and partly to a decrease in the fiber diameter of remaining fibers as they give off collaterals. Below the level of the obex, the majority of fibers are unmyelinated. The small diameter of these fibers agrees with the idea that the caudal part of the spinal tract, like the caudal subnucleus, is the region concerned most with noxious and thermal stimuli. However, the caudal nucleus also receives input from mechanoreceptors.

Recordings of primary afferents in the trigeminal ganglion of the rat by Zucker and Welker (1969) revealed five classes of primary afferents from the vibrissae. Two classes were slowly adapting and responded to low velocities of vibrissal deflection: these probably correspond to the Merkel and Ruffini endings. Another two classes were rapidly adapting and responded only to higher velocities of vibrissal deflection. Presumably these were the paciniform and lanceolate terminals.

B. Laminar organization and somatotopy

As in the spinal cord, mechanoreceptors terminate in a way which is characteristic of the receptor type. For example, in the caudal subnucleus of the spinal tract of the trigeminal nerve (Sp5C), vibration-sensitive afferents (possibly belonging to Pacinian corpuscles) have widely distributed arborizations in layers 2 to 5, while rapidly adapting and slowly adapting vibrissal afferents have circumscribed terminations in layers 3 or 4 (Jacquin *et al.*, 1983). In the more rostral parts of the sensory trigeminal nuclei, terminal arbors are located more superficially, and are restricted to the outer half of the subnucleus interpolaris and oralis as well as the principal sensory nucleus (Hayashi, 1983). In some cases a single afferent may give off terminal arbors to the principal sensory trigeminal nucleus (Pr5) and all subnuclei of Sp5 (Hayashi, 1980).

Superimposed on this laminar organization is a somatotopic organization. This has been best studied for vibrissal afferents, where it is found that the most dorsal rows of vibrissae are represented most ventrally in the sensory trigeminal nuclei (Arvidsson, 1982). However, the somatotopic patterns defined by anterograde transport of HRP are somewhat blurred, since they include terminations from primary afferents of different types, which have different laminar terminations. Nevertheless, there is some evidence from such studies that the primary afferents from a single vibrissa may have terminal arbors corresponding with a single cluster of cells, analogous to a cortical barrel; such clusters can be distinguished in the sensory trigeminal complex of neonatal rats (Bates and Killackey, 1983; Belford and Killackey, 1979).

C. Mesencephalic trigeminal nucleus

The nucleus of the mesencephalic tract of the trigeminal nerve (Me5) forms a narrow layer of scattered cells at the lateral margins of the periaqueductal gray matter. These cells extend from the level of the superior colliculus rostrally, to the level of the trigeminal motor nucleus caudally. The neurons are the cell bodies of primary afferents, and are the only such cells to be found within the central nervous system. Most of the somata of trigeminal sensory receptors are located in the trigeminal ganglion; those which are found in the mesencephalic nucleus belong to proprioceptors, in particular, to muscle spindles of the muscles of mastication (Alvarado-Mallart *et al.*, 1975). The somata of mechanoreceptors associated with the teeth are also located in the mesencephalic nucleus (Gonzalo-Sanz and Insausti, 1980). However, not all proprioceptive afferents have their cell bodies in the mesencephalic nucleus: some are located in the trigeminal ganglion (Alvarado-Mallart *et al.*, 1975; Romfh *et al.*, 1979). In particular, the somata of tendon organs are located in the trigeminal ganglion (Cody *et al.*, 1972).

Neurons in the mesencephalic nucleus project to several regions, including the cerebellum (Somana *et al.*, 1980). In the rat the main terminations are in the trigeminal motor nucleus, the supratrigeminal nucleus, and in a region between the trigeminal and facial motor nuclei (Matesz, 1981; Travers and Norgren, 1983). It appears that primary afferents of the trigeminal ganglion do not penetrate the motor nucleus of the trigeminal, which corroborates the idea that these afferents are from tendon organs rather than from muscle spindles.

D. Principal sensory nucleus

The principal sensory trigeminal nucleus (Pr5) has neurons which are loosely distributed and may be fusiform or multipolar. Relatively few Nissl bodies are found. Primary afferents run longitudinally alongside the nucleus, and give off side branches which run medially. These side branches bear dilations which appear to be synaptic boutons, and each bouton forms the core of a synaptic glomerulus. Forming part of the glomerulus and surrounding the central bouton of the primary afferent are the dendrites of postsynaptic neurons, and peripheral boutons which are probably terminations of axons projecting from the nucleus of the spinal tract and from the cerebral cortex (Gobel and Dubner, 1969).

Representations of the vibrissae occupy a large part of the trigeminal sensory nuclei; in fact, the vibrissae are represented in three separate maps. One of these is in the principal sensory nucleus, while the other two are in the spinal nucleus.

These representations have been demonstrated histologically in the neonatal rat using the succinic dehydrogenase method, which probably shows the pattern of afferent terminations. Similar patterns have been demonstrated using anterograde transport of horseradish peroxidase. These patterns suggest the presence of organized clusters of neurons comparable with the barrels found in the somatosensory cortex (Arvidsson, 1982; Belford and Killackey, 1979).

The principal sensory nucleus sends the strongest projections of any region of the trigeminal complex to the contralateral ventrobasal thalamus. The projection is from small and medium sized neurons, not from the larger neurons, and runs in the ventral trigeminothalamic tract. There is also a projection to the ipsilateral ventrobasal complex from the dorsal third of the rostral pole of the main sensory nucleus, via the small dorsal trigeminothalamic tract (Fukushima and Kerr, 1979). In addition, the principal sensory nucleus projects to the posterior group of thalamic nuclei and the ventral part of the zona incerta (Smith, 1973), to the motor nucleus of the trigeminal (Travers and Norgren, 1983), and to the superior colliculus (Huerta *et al.*, 1983).

E. Nucleus of the spinal tract

The nucleus of the spinal tract of the trigeminal nerve (Sp5) is continuous with the substantia gelatinosa of the dorsal horn of the spinal cord caudally, and with the principal or main sensory nucleus rostrally. In fact, it is plausible to regard the nuclei of the spinal tract as being homologous with the four most superficial layers of the dorsal horn. They receive information from thermoreceptors and nociceptors as well as from mechanoreceptors. On the basis of cytoarchitecture the nucleus of the spinal tract is subdivided into three regions: subnucleus oralis; subnucleus interpolaris; and subnucleus caudalis (Olszewski, 1950).

The terminations of primary afferents from the trigeminal nerve are somatotopically organized within the nucleus of the spinal tract so that the afferents from the mandibular division terminate most dorsally, whilst those from the ophthalmic division terminate most ventrally. This organization has been confirmed in the rat (Torvik, 1956). Unlike the subnucleus interpolaris and caudalis, the oral part of the spinal nucleus does not appear to receive a significant projection from the vibrissae. Instead, it appears to receive projections from the nasal and oral cavities, at least in the cat (Wall and Taub, 1962).

The synaptic organization of the nucleus of the spinal tract is similar to that of the principal sensory nucleus (see above) with the exception of the superficial layers of the subnucleus caudalis. Synaptic glomeruli, whose central element is a bouton or dilation of a primary afferent, are found. Grouped around this bouton are dendrites and synaptic terminals with flattenable synaptic vesicles (Ide and Killackey, 1983).

As in the principal sensory nucleus, there is a suggestion that some of the neurons of the spinal nucleus may be arranged as clusters comparable to cortical barrels. This idea is based mainly on the pattern of afferent terminations, which show that there are two separate representations of the mystacial vibrissae in the nucleus of the spinal tract—one in the interpolar subnucleus and a more complex one in the caudal subnucleus. Within the interpolar subnucleus, each vibrissa is represented at all rostrocaudal levels, while in the caudal subnucleus, more rostral dermatomes are represented at more rostral levels of the subnucleus, and more posterior dermatomes are represented at successively more caudal levels (Arvidsson, 1982; Belford and Killackey, 1979).

1. Oral subnucleus

Most of the neurons are small and similar in morphology to those found in the principal sensory nucleus; that is, they are loosely distributed throughout the nucleus individually or in clusters (Gobel and Dubner, 1969).

Some neurons in the oral subnucleus project to the contralateral medial part of the ventroposterior thalamic nucleus (VPM) via the trigeminal lemniscus, although the projection is sparse (Fukushima and Kerr, 1979; Kruger *et al.*, 1977). In addition there are projections to other regions of the brain stem such as the facial nucleus (Erzurumlu and Killackey, 1979).

2. Interpolar subnucleus

This nucleus extends from the obex to the rostral third of the inferior olivary nucleus. In the mouse, there is a marginal plexus with relatively large cells, whose dendrites are oriented radially. The rest of the nucleus consists of smaller neurons with spherical perikarya and compact dendritic branches (Astrom, 1953). A recent Golgi study in the rat showed that the largest neurons were concentrated in the rostral half of the nucleus; these may be neurons projecting to other regions. Several types of small and medium sized neurons (10 μm–30 μm in diameter) were located throughout the interpolar nucleus (Phelan and Walls, 1983).

The interpolar subnucleus is the main site of neurons projecting to the cerebellum. About 70% of neurons in this subnucleus project directly to the cerebellum, mainly ipsilaterally to the central region of the vermis (Watson and Switzer, 1978). There is also an indirect projection to crura I and II of the cerebellar cortex via the inferior olive, and a projection to the superior colliculus (Huerta *et al.*, 1983). Some neurons in the interpolar subnucleus project to the thalamus, including the VPM and the posterior group (Erzurumlu and Killackey, 1980; Fukushima and Kerr, 1979; Kruger *et al.*, 1977).

3. Caudal subnucleus

This nucleus is made up of three layers, corresponding with layers 1 to 4 of the spinal cord. The most peripheral is the marginal zone, corresponding to layer 1 of the dorsal horn; it is narrow with a few large "marginal" cells and many horizontally oriented smaller cells. It is this region which contains neurons responding specifically to noxious and thermal stimuli. Next is the gelatinous layer, corresponding to layers 2 and 3. This is broader, and contains a feltwork of unmyelinated fibers in the outer part (layer 2). The neurons correspond in appearance with the small spindle shaped cells of the substantia gelatinosa of the spinal cord. The deepest part is the magnocellular part, corresponding to layer 4 and containing neurons variable in shape and size. Some authors include a layer 5 in the caudal subnucleus.

Neurons in the marginal and gelatinous layers of the caudal subnucleus project to more rostral trigeminal nuclei and to the thalamus. The thalamic projections are comparable to those from the spinal cord, so that, in the rat, there are projections not only to the ventrobasal complex (Fukushima and Kerr, 1979) but also to intralaminar nuclei such as the centrolateral thalamic nucleus (CL) (Kruger *et al.*, 1977; Mehler, 1969; Stewart and King, 1963) and the posterior group (Lund and Webster, 1967; Shigenaga *et al.*, 1979). In addition, there is a recently described projection to nucleus submedius or gelatinosus of the thalamus (Craig and Burton, 1981). The majority of these projections, at least in the cat, are from the marginal layer (Hockfield and Gobel, 1978). This region projects to the dorsomedial part of the VPM bilaterally (the locus of projections from thermoreceptors) and contralaterally to the main part of the VPM and the posterior group (Burton *et al.*, 1979).

The magnocellular region of the caudal nucleus has a strong projection to the lateral part of the facial nucleus (Erzurumlu and Killackey, 1979).

V. THALAMUS

The regions of the mammalian thalamus which receive somatosensory input include the ventrobasal complex, the intralaminar nuclei, the posterior group of nuclei, and the nucleus submedius (see Volume 1, Chapter 5). The ventrobasal nuclei project specifically to the somatosensory cortex. Their neurons generally respond only to one submodality of somatosensory receptor, and to receptor stimulation with a relatively short latency; they have relatively restricted receptive fields. These properties have been termed "lemniscal". Nucleus submedius receives a projection from the marginal zone of the spinal cord and the caudal part of the spinal trigeminal nucleus, and probably has an important role in nociception. Somatosensory inputs to the thalamus and the appropriate projections to the cortex are outlined in Fig. 4.

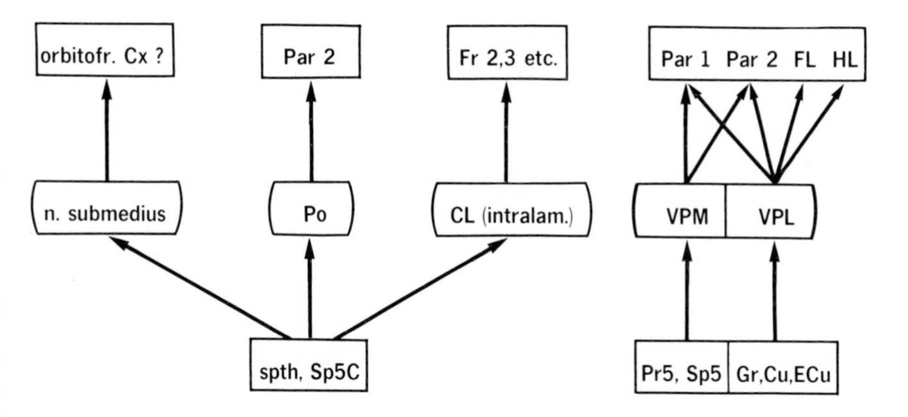

Fig. 4: Somatosensory pathways through the thalamus in the rat. Cortical areas are in the upper row, thalamic nuclei in the middle row; afferents to these nuclei are in the bottom row. The main somatosensory projection of mechanoreceptors is shown on the right, from the dorsal column nuclei and sensory trigeminal to the ventrobasal thalamus, and then to somatosensory areas 1 and 2 of the cortex. Projections of high threshold receptors such as nociceptors and thermoreceptors are shown on the left. CL, centrolateral thalamic nucleus; Cu, cuneate nucleus; ECu, external cuneate nucleus; FL, forelimb part of parietal cortex; Fr 2,3, frontal cortical areas 2 and 3 (primary motor cortex); Gr, gracile nucleus; HL, hindlimb part of parietal cortex; Par 1, area 1 of parietal cortex (SI); Par 2, area 2 of parietal cortex (SII); Po, posterior thalamic nuclear group; Pr5, principal sensory trigeminal nucleus; Sp5, nucleus of the spinal tract of the trigeminal nerve; Sp5C, nucleus of the spinal tract of the trigeminal nerve, caudal subnucleus; spth, spinothalamic tract; VPL, ventroposterior thalamic nucleus, lateral part; VPM, ventroposterior thalamic nucleus, medial part.

A. Ventrobasal complex

The term "Ventrobasal complex" was introduced to describe the thalamic region which receives tactile inputs. It is divided into two distinct parts: one receives projections from the cuneate and gracile nuclei, the ventroposterolateral nucleus (VPL); and the other receives projections from the sensory trigeminal nuclei, the ventroposteromedial nucleus (VPM). Thus the VPL receives somatosensory input from the trunk and limbs, while the VPM receives somatosensory input from the face.

In the rat, all neurons in the ventrobasal thalamus appear to send their axons to the somatosensory cortex. These are the thalamocortical relay neurons, some of which send collaterals to both area I (SI) and area II (SII) of the somatosensory cortex. In other mammals, in addition there are local interneurons, which play some role, as yet unclear, in the processing of somatosensory information in the thalamus. However, the rat VPL does not seem to have such intrinsic or local circuit neurons (McAllister and Wells, 1981).

The rat VPL is composed exclusively of medium sized, multipolar neurons, arranged in a series of concentric layers running parallel to the external medullary lamina. As in other mammals, the VPL in the rat is somatotopically organized, with the forelimb represented medially and the hindlimb laterally (Angel and Clarke, 1975; Davidson, 1965).

The rat VPM is an ovoid nucleus just medial to the VPL, and somewhat similar in cytoarchitecture. In fact, the part of the somatosensory system devoted to a representation of the face is particularly interesting. It is primarily occupied by representations of the vibrissae at all levels, including the sensory trigeminal nuclei, the VPM and the face area of the somatosensory cortex. At the cortical level, each vibrissa is represented by a morphologically distinct group of cells, known as a "barrel". The basis for this cortical representation must be provided by an underlying organization of the corresponding thalamocortical relay cells. In neonatal rats, such an organization can be seen by special staining of the vibrissal region of the VPM; a similar pattern can also be found in the appropriate parts of the sensory trigeminal nuclei (Belford and Killackey, 1979). In the mouse VPM it is possible to see neurons organized as "barreloids" in appropriately cut horizontal sections. These barreloids are organized into a two dimensional array similar to the array of vibrissal follices on the snout, and to the array of barrels in the appropriate region of the somatosensory cortex (Van der Loos, 1976). The system has provided a fertile field for study of the way in which primary afferents determine the organization of second and higher order neurons at successive stages of the somatosensory pathway to the cortex.

B. Intralaminar nuclei

The intralaminar nuclei project widely and diffusely to the cortex; unlike the ventrobasal complex, their projection is not confined to somatosensory cortex (Jones and Leavitt, 1974). Some of the intralaminar nuclei have somatosensory inputs, especially the centrolateral nucleus (CL), and these inputs are derived from the spinothalamic tract. In most mammals, the spinothalamic tract has a set of terminations on the large cells in the caudal part of the CL; these cells project in turn to the motor cortex. In the rat, neurons in the posterior intralaminar region are activated by rather nonspecific stimuli such as brisk taps, and are most effectively activated by combined cutaneous and deep input (Peschanski *et al.*, 1981). The afferent pathway appears to be a subdivision of the spinothalamic tract which has been termed the medial spinothalamic tract (Giesler *et al.*, 1979a).

C. Posterior group

The posterior group is a complex of several nuclei (see Volume 1, Chapter 5). Only the medial division, which embraces the caudal pole of the ventrobasal complex, is considered a part of the somatosensory system (Jones, 1981). This region receives fibers from the spinal cord and sensory trigeminal nuclei (Lund and Webster, 1967; Shigenaga *et al.*, 1979). It can be identified in the rat (Jones and Leavitt, 1974) and probably corresponds with the caudal transitional zone of McAllister and Wells (1981).

VI. SOMATOSENSORY CORTEX

The somatosensory cortex of mammals generally includes a first and second somatic sensory area (SI and SII).

A. First somatic sensory area

In humans, as well as in other primates and cats, the SI can be divided into four mediolateral strips which are distinguishable on the basis of their cytoarchitecture. From anterior to posterior these are areas 3a, 3b, 1 and 2. Each of these mediolateral strips receives a projection from the whole surface of the body, but from different submodalities of mechanoreceptor; for this reason each strip has come to be regarded as a separate representation of the body surface.

In the rat, the SI does not appear to be subdivided into areas homologous with Brodmann's areas 3, 1 and 2 (Zilles *et al.*, 1980) and there is so far no evidence of multiple sensory representation within SI. Rather the SI appears to be divided into three parts: Par1, FL and HL. Par1 is a distinct cytoarchitectural subarea within the SI largely devoted to the representation of the mystacial vibrissae and sinus hairs, but including in its caudal part the representation of the trunk and tail. Par1 contains discrete aggregations of neurons, or "barrels", and there is one barrel for each contralateral whisker (Fig. 5). The barrels of the SI are located in layer 4, in which the density of terminations of thalamocortical neurons is greatest (Welker and Woolsey, 1974).

Regions of layer 4 within the SI, but outside the representation of the face, contain few or no barrels, and display several distinct areas of high cell density representing the forelimb (area FL) and hindlimb (area HL) (Volume 1, Chapter 10). These two regions constitute a zone of overlap between sensory and motor cortices in the sense that they receive projections from both sensory and motor nuclei of the thalamus, that is, from VPL and VL (Donoghue *et al.*, 1979). Such overlap of sensory and motor areas appears to be a primitive feature of mammalian cortex: it is found to a considerable extent in marsupials and some primitive placental mammals (Haight and Neylon, 1979).

Since the rat cortex is lissencephalic, the somatosensory representation of the whole of one side of the head and body can be visualized in a section of the flattened neocortex which passes through layer 4 (Welker, 1976). Superficial to layer 4, layers 2 and 3 are difficult to distinguish from one another, and both consist of medium sized pyramidal and nonpyramidal cells. Deep to layer 4 is

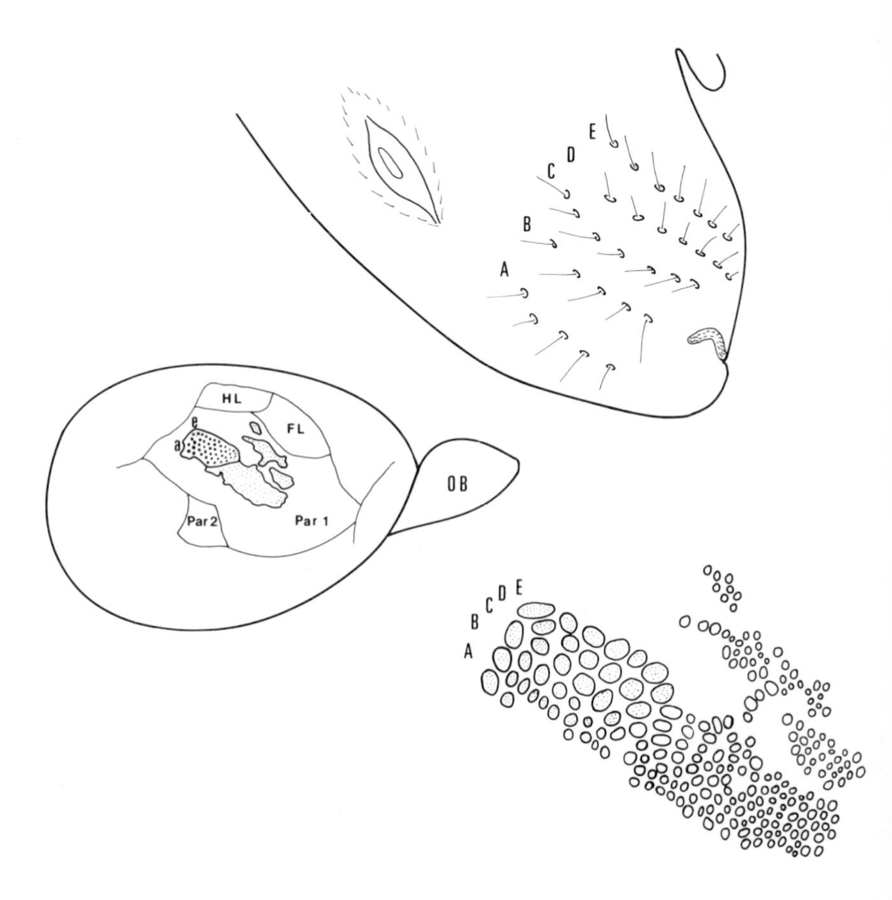

Fig. 5: Somatotopic representation of the mystacial vibrissae in somatosensory cortex of the rat. There are typically five rows of vibrissae, lettered A to E from dorsal to ventral on the face of the rat (top panel). These vibrissae project to an array of barrels with a similar topology; this array is the posteromedial barrel subfield (PMBSF). The middle panel shows the location of the PMBSF within Par 1 of the parietal cortex (heavy dots). The bottom panel shows the PMBSF in more detail (stippled barrels). The barrels which are not a part of the PMBSF correspond to sinus hairs on the face (these sinus hairs are not shown) (modified from Woolsey *et al.*, 1975). FL, forelimb part of parietal cortex; HL, hindlimb part of parietal cortex; OB, olfactory bulb; Par 1, area 1 of parietal cortex (SI); Par 2, area 2 of parietal cortex (SII).

layer 5, which can be subdivided into two laminae: a superficial lamina (5a) with a relatively low density of pyramidal cells; and a deep lamina (5b) with the largest pyramidal cells of the SI cortex. The cortical neurons whose axons project to subcortical structures are found only in layers 5 and 6 (Wise and Jones, 1977).

B. Second somatic sensory area

This region of the somatosensory cortex has been examined extensively in cats and primates. In these animals, each SII cortex has a largely bilateral representation of the body surface (Dykes, 1978), although footpad receptors only project contralaterally. The SII receives projections from the ventrobasal thalamus; some of these projections are from collaterals of the thalamic neurons which project to the SI. In addition it receives terminations from the ipsilateral and contralateral SI: the callosal connections between the SI and SII of the two hemispheres presumably account for the bilateral receptive fields found in cats and primates.

The cytoarchitecture of the SII in the rat is comparable with that of the SI. However, in the SII there are no aggregates of granule cells, the supragranular and granular layers are thicker than in the SI, and the lamination if less clear than that of SI (Wise and Jones, 1977).

The somatotopic organization of the SII in the rat apparently differs from that of other animals in that physiologic recordings show that receptive fields of SII neurons are contralateral rather than bilateral for all parts of the body except for the teeth and face. However, callosal connections are present between ipsilateral and contralateral somatosensory cortex in the rat (see Volume 1, Chapter 10), so the physiologic results are probably artefacts related to the use of anaesthetic agents in these experiments. As in the SI, the rat SII shows no evidence of organization into mediolateral dermatomal trajectories, but bears a simple two dimensional map of the rat's body. This map is upside down, with the distal parts of the limbs represented medially and the trunk represented laterally (Welker and Sinha, 1972).

ACKNOWLEDGEMENTS

I should like to thank Dr Saw Kin Loo and Professor Mark Rowe for constructive criticism of the manuscript.

REFERENCES

Alvarado-Mallart, M. R., Batini, C., Buisseret-Delmas, C. and Corvisier, J. (1975). Trigeminal representations of the masticatory and extraocular proprioceptors as revealed by horseradish peroxidase retrograde transport. *Exp. Brain Res.* 23, 167–179.

Andres, K. H. (1966). Ueber die Feinstruktur der Rezeptoren an Sinushaaren. *Z. Zellforsch.* 75, 339–365.

Angel, A. and Clarke, K. A. (1975). An analysis of the representation of the forelimb in the ventrobasal thalamic complex of the albino rat. *J. Physiol. (London)* 249, 339–423.

Arvidsson, J. (1982). Somatotopic organization of vibrissae afferents in the trigeminal sensory nuclei of the rat studied by transganglionic transport of HRP. *J. Comp. Neurol.* 211, 84–92.

Aström, K. E. (1953). The central course of afferent fibres in the trigeminal, facial, glossopharyngeal and vagal nerves and their nuclei in the mouse. *Acta Physiol. Scand. Supp.* 106, 209–320.
Bannister, L. H. (1976). Sensory terminals of peripheral nerves. In D. N. Landon (Ed.). *The peripheral nerve*, pp. 396–463. Chapman and Hall, London.
Basbaum, A.I. and Hand, P. J. (1973). Projections of cervicothoracic dorsal roots to the cuneate nucleus of the rat, with observations on cellular "bricks". *J. Comp. Neurol.* 148, 347–360.
Bates, C. A. and Killackey, H. P. (1983). Pattern of trigeminal afferents in the brainstem trigeminal complex of the rat. *Soc. Neurosci. Abst.* 9, 245.
Belford, G. and Killackey, H. P. (1979). Vibrissae representation in subcortical trigeminal centers of the neonatal rat. *J. Comp. Neurol.* 183, 305–322.
Brown, A. G. (1982). The dorsal horn of the spinal cord. *Quart. J. Exp. Physiol.* 67, 193–212.
Brown, A. G. and Gordon, G. (1977). Subcortical mechanisms concerned in somatic sensation. *Brit. Med. Bull.* 33, 121–128.
Burton, H., Craig, A. D., Poulos, D. A. and Molt, J. T. (1979). Efferent projections from temperature sensitive recording loci within the marginal zone of the nucleus caudalis of the spinal trigeminal complex in the cat. *J. Comp. Neurol.* 183, 753–778.
Campbell, S. K., Parker, T. D. and Welker, W. (1974). Somatotopic organization of the external cuneate nucleus in albino rats. *Brain Res.* 77, 1–23.
Cauna, N. (1969). The fine morphology of the sensory receptor organs in the auricle of the rat. *J. Comp. Neurol.* 136, 81–98.
Cauna, N. (1976). Morphological basis of sensation in hairy skin. *Prog. Brain Res.* 43, 35–45.
Cody, F. W. J., Lee, R. W. H. and Taylor, A. (1972). A functional analysis of the components of the mesencephalic nucleus of the fifth nerve in the cat. *J. Physiol. (London)* 226, 249–261.
Craig, A. D. and Burton, H. (1981). Spinal and medullary lamina I projection to nucleus submedius in medial thalamus: A possible pain center. *J. Neurophysiol.* 45, 443–466.
Cunningham, F. O. and Fitzgerald, M. J. T. (1972). Encapsulated nerve endings in hairy skin. *J. Anat.* 112, 93–97.
Darian-Smith, I. (1973). The trigeminal system. *In* A. Iggo (ed.), *Handbook of Sensory Physiology*, vol. 2, pp. 271–314. Springer, Berlin.
Davidson, N. (1965). The projection of afferent pathways on the thalamus of the rat. *J. Comp. Neurol.* 124, 377–390.
Devor, M. and Claman, D. (1980). Mapping and plasticity of acid phosphatase afferents in the rat dorsal horn. *Brain Res.* 190, 17–28.
Donoghue, J. P., Kerman, K. L. and Ebner, F. F. (1979). Evidence for two organizational plans within the somatic sensory-motor cortex of the rat. *J. Comp. Neurol.* 183, 647–663.
Dykes, R. W. (1978). The anatomy and physiology of the somatic sensory cortical regions. *Prog. Neurobiol.* 10, 33–88.
Erzurumlu, R. S. and Killackey, H. P. (1979). Efferent connections of the brainstem trigeminal complex with the facial nucleus of the rat. *J. Comp. Neurol.* 188, 75–86.
Erzurumlu, R. S. and Killackey, H. P. (1980). Diencephalic projections of the subnucleus interpolaris of the brainstem trigeminal complex in the rat. *Neurosci.* 5, 1891–1901.
Fukushima, T. and Kerr, F. W. L. (1979). Organization of trigeminothalamic tracts and other thalamic afferent systems of the brainstem in the rat: Presence of gelatinosa neurons with thalamic connections. *J. Comp. Neurol.* 183, 169–184.
Ganchrow, D. and Bernstein, J. J. (1981). Projections of caudal fasciculus gracilis to nucleus gracilis and other medullary structures, and Clarke's nucleus in the rat. *Brain Res.* 205, 383–390.
Giesler, G. J., Menétrey, D. and Basbaum, A. I. (1979a). Differential origins of spinothalamic tract projections to medial and lateral thalamus in the rat. *J. Comp. Neurol.* 184, 107–126.
Giesler, G. J., Urca, G., Cannon, T. and Liebeskind, J. C. (1979b). Response properties of neurons of the lateral cervical nucleus in the rat. *J. Comp. Neurol.* 186, 65–78.
Gobel, S. and Dubner, R. (1969). Fine structural studies of the main sensory trigeminal nucleus in the cat and rat. *J. Comp. Neurol.* 137, 459–493.
Gonzalo-Sanz, L. M. and Insausti, R. (1980). Fibers of trigeminal mesencephalic neurons in the maxillary nerve of the rat. *Neurosci. Lett.* 16, 137–141.
Gregg, J. M. and Dixon, A. D. (1973). Somatotopic organization of the trigeminal ganglion in the rat. *Arch. Oral Biol.* 18, 487–498.

Gulley, R. L. (1973). Golgi studies of the nucleus gracilis in the rat. *Anat. Rec.* 177, 325–342.

Haight, J. R. and Neylon, L. (1979). The organization of neocortical projections from the ventrolateral thalamic nucleus in the brush-tailed possum, *Trichosurus vulpecula*, and the problem of motor and somatic sensory convergence within the mammalian brain. *J. Anat.* 129, 673–694.

Hayashi, H. (1980). Distributions of vibrissae afferent fiber collaterals in the trigeminal nuclei as revealed by intra-axonal injection of horseradish peroxidase. *Brain Res.* 183, 442–446.

Hayashi, H. (1983). Arrangement of terminal arbors of physiologically identified vibrissa afferents in the rat trigeminal system. *Soc. Neurosci. Abst.* 9, 245.

Hockfield, S. and Gobel, S. (1978). Neurons in and near nucleus caudalis with long ascending projection axons demonstrated by retrograde labeling with horseradish peroxidase. *Brain Res.* 139, 333–339.

Huerta, M. F., Frankfurter, A. and Harting, J. K. (1983). Studies of the principal sensory and spinal trigeminal nuclei of the rat: Projections to the superior colliculus, inferior olive, and cerebellum. *J. Comp. Neurol.* 220, 147–167.

Ide, L. S. and Killackey, H. P. (1983). Fine structural organization of the brainstem sensory trigeminal complex in the rat. *Soc. Neurosci. Abs.* 9, 245.

Jacquin, M. F., Mooney, R. D. and Rhoades, R. W. (1983). Morphology and topographic organization of functionally identified trigeminal (V) primary afferent fibers in rat. *Soc. Neurosci. Abs.* 9, 244.

Jones, E. G. (1981). Organization of the thalamocortical complex and its relation to sensory processes. In J. M. Brookhart, V. B. Mountcastle and I. Darian Smith (Eds). *Handbook of Physiology.* Section I: The nervous system. Vol. 2: Sensory Processes. American Physiological Society, Bethesda, Md.

Jones, E. G. and Leavitt, R. Y. (1974). Retrograde axonal transport and the demonstration of non-specific projections to the cerebral cortex and striatum from thalamic intralaminar nuclei in the rat, cat and monkey. *J. Comp. Neurol.* 154, 349–378.

Kevetter, G. A. and Willis, W. D. (1983). Collaterals of spinothalamic cells in the rat. *J. Comp. Neurol.* 215, 453–464.

Kitchell, R. L., Gilanpour, H. and Johnson, R. D. (1982). Electrophysiologic studies of penile mechanoreceptors in the rat. *Exp. Neurol.* 75, 229–244.

Kruger, L., Saporta, S. and Feldman, S. G. (1977). Axonal transport studies of the sensory trigeminal complex. *In* D. J. Anderson and B. Matthews (eds), *Pain in the Trigeminal Region*, pp. 191–202. Elsevier, Amsterdam.

Langford, L. A. and Coggeshall, R. E. (1981). Unmyelinated axons in the posterior funiculi. *Science* 211, 176–177.

Light, A. R. and Perl, E. R. (1979). Spinal termination of functionally identified primary afferent neurons with slowly conducting myelinated fibers. *J. Comp. Neurol.* 186, 133–150.

Low, J. S. T. and Tracey, D. J. (1984). Spinal afferents to nucleus z are collaterals of dorsal spinocerebellar tract neurones. *Proc. Aust. Physiol. Pharmacol. Soc.* 15, 78P.

Lund, R. D. and Webster, K. E. (1967). Thalamic afferents from the spinal cord and trigeminal nuclei. An experimental anatomical study in the rat. *J. Comp. Neurol.* 130, 313–328.

Lynn, B. and Carpenter, S. E. (1982). Primary afferent units from the hairy skin of the rat hind limb. *Brain Res.* 238, 29–43.

McAllister, J. P. and Wells, J. (1981). The structural organization of the ventral posterolateral nucleus in the rat. *J. Comp. Neurol.* 197, 271–301.

Mackintosh, S. R. (1975). Observations on the structure and innervation of the rat snout. *J. Anat.* 119, 537–546.

Matesz, C. (1981). Peripheral and central distribution of fibres of the mesencephalic trigeminal root in the rat. *Neurosci. Lett.* 27, 13–17.

Mehler, W. R. (1969). Some neurological species differences—*a posteriori. Ann. N.Y. Acad. Sci.* 167, 424–468.

Olszewski, J. (1950). On the anatomical and functional organization of the spinal trigeminal nucleus. *J. Comp. Neurol.* 92, 401–413.

Patrizi, G. and Munger, B. L. (1965). The cytology of encapsulated nerve endings in the rat penis. *J. Ultrastruct. Res.* 13, 500–515.

Patrizi, G. and Munger, B. L. (1966). The ultrastructure and innervation of rat vibrissae. *J. Comp. Neurol.* 126, 423–435.

Peschanski, M., Guilbaud, G. and Gautron, M. (1981). Posterior intralaminar region in rat: Neuronal responses to noxious and nonnoxious cutaneous stimuli. *Exp. Neurol.* 72, 226–238.

Phelan, K. D. and Walls, W. M. (1983). A Golgi analysis of trigeminal nucleus interpolaris in the adult rat. *Soc. Neurosci. Abst.* 9, 246.

Poláček, P. (1966). Receptors of the joints. Their structure, variability and classification. *Acta Facul. Med. Univ. Brunensis* 23.

Romfh, J. H., Capra, N. F. and Gatipon, G. B. (1979). Trigeminal nerve and temporomandibular joint of the cat: A horseradish peroxidase study. *Exp. Neurol.* 65, 99–106.

Shigenaga, Y., Takabatake, M., Sugimoto, T. and Sakai, A. (1979). Neurons in marginal layer of trigeminal nucleus caudalis projecting to ventrobasal complex (VB) and posterior nucleus group (PO) demonstrated by retrograde labelling with horseradish peroxidase. *Brain Res.* 166, 391–396.

Smith, R. L. (1973). The ascending fiber projections from the principal sensory trigeminal nucleus in the rat. *J. Comp. Neurol.* 148, 423–446.

Somana, R., Kotchabhakdi, N. and Walberg, F. (1980). Cerebellar afferents from the trigeminal sensory nuclei in the cat. *Exp. Brain Res.* 38, 57–64.

Stewart, W. A. and King, R. B. (1963). Fiber projections from the nucleus caudalis of the spinal trigeminal nucleus. *J. Comp. Neurol.* 121, 271–286.

Straile, W. E. (1960). Sensory hair follicles in mammalian skin: The tylotrich follicle. *Am. J. Anat.* 106, 133–147.

Tan, C. K. and Lieberman, A. R. (1978). Identification of thalamic projection cells in the rat cuneate nucleus: A light and electron microscopic study using horseradish peroxidase. *Neurosci. Lett.* 10, 19–22.

Torvik, A. (1956). Afferent connections to the sensory trigeminal nuclei, the nucleus of the solitary tract and adjacent structures: An experimental study in the rat. *J. Comp. Neurol.* 106, 51–142.

Travers, J. B. and Norgren, R. (1983). Afferent projections to the oral motor nuclei in the rat. *J. Comp. Neurol.* 220, 280–298.

Valverde, F. (1966). The pyramidal tract in rodents: A study of its relations with the posterior column nuclei, dorsolateral reticular formation of the medulla oblongata, and cervical spinal cord. *Z. Zellforsch.* 71, 297–363.

Van der Loos, H. (1976). Barreloids in mouse somatosensory thalamus. *Neurosci. Lett.* 2, 1–6.

Wall, P. D. and Taub, A. (1962). Four aspects of trigeminal nucleus and a paradox. *J. Neurophysiol.* 25, 110–126.

Watson, C. R. R. and Switzer, R. C. (1978). Trigeminal projections to cerebellar tactile areas in the rat—origin mainly from n. interpolaris and n. principalis. *Neurosci. Lett.* 10, 77–82.

Welker, C. (1976). Receptive fields of barrels in the somatosensory neocortex of the rat. *J. Comp. Neurol.* 166, 173–190.

Welker, C. and Sinha, M. M. (1972). Somatotopic organization of SmII cerebral neocortex in albino rat. *Brain Res.* 37, 132–136.

Welker, C. and Woolsey, T. A. (1974). Structure of layer IV in the somatosensory neocortex of the rat: Description and comparison with the mouse. *J. Comp. Neurol.* 158, 437–454.

Willis, W. D. and Coggeshall, R. E. (1978). *Sensory mechanisms of the spinal cord.* Wiley, New York.

Wise, S. P. and Jones, E. G. (1977). Cells of origin and terminal distribution of descending projections of the rat somatic sensory cortex. *J. Comp. Neurol.* 175, 129–158.

Yezierski, R. P. and Bowker, R. M. (1981). A retrograde double label tracing technique using horseradish peroxidase and the fluorescent dye 4′,6-diamidino-2-phenylindole 2HCl (DAPI). *J. Neurosci. Meth.* 4, 53–62.

Zemlan, F. P., Leonard, C. M., Kow, L.-M. and Pfaff, D. W. (1978). Ascending tracts of the lateral columns of the rat spinal cord: A study using the silver impregnation and horseradish peroxidase techniques. *Exp. Neurol.* 62, 298–334.

Zilles, K., Zilles, B. and Schleicher, A. (1980). A quantitative approach to cytoarchitectonics. VI. The areal patterns of the cortex of the albino rat. *Anat. Embryol.* 159, 335–360.

Zucker, E. and Welker, W. I. (1969). Coding of somatic sensory input by vibrissae neurons in the rat's trigeminal ganglion. *Brain Res.* 12, 138–156.

8
Auditory system

WILLIAM R. WEBSTER

Monash University
Melbourne, Victoria, Australia

1. INTRODUCTION

Compared with other species, the central auditory pathway of the rat has not been intensively studied apart from the work on the cochlear nucleus by Harrison and his colleagues (Harrison, 1978; Harrison and Feldman, 1970). A possible reason for the lack of research is the prevalence of chronic otitis media in many laboratory rats caused by *Mycoplasma pulmonis* (Crowley, 1974; Daniel *et al.*, 1973). Since the rat is one of the most commonly used laboratory animals, it is worth reviewing what we know of the structure of its auditory system. Throughout this chapter, one of the main themes will be the correlation of structure with function. However, there are problems associated with such an approach. First, some parts of the auditory pathway have not received detailed anatomic or physiologic examination. Indeed, some parts have not been examined at all with physiologic techniques. To overcome the dearth of functional data for the rat, where possible, comparisons will be made with anatomic and physiologic data from the cat. It is fortunate that the overall structure of the rat auditory pathway is very similar to that of the cat. A block diagram of a schematic auditory pathway, would apply reasonably well to both species (Fig. 1). In the schematic pathway of Fig. 1, the tonotopic organization for frequency shown is based largely on data from the cat. The cortical representation is based on an interpretation of the work of Ryugo (1976).

Another serious problem for a correlation between structure and function in the rat is that there is a surprising lack of behavioral data about the rat auditory pathway (Neff *et al.*, 1975). While there is no dearth of behavioral studies using auditory stimuli, there are few experiments which test specific hypotheses about the rat's auditory performance in relation to the structure of the auditory pathway. It is also not clear exactly what items should be considered indicators of function in the auditory pathway. For the purpose of this chapter, function will be restricted in scope and defined in terms of a limited set of experimental

THE RAT NERVOUS SYSTEM
ISBN 0 12 547632 9

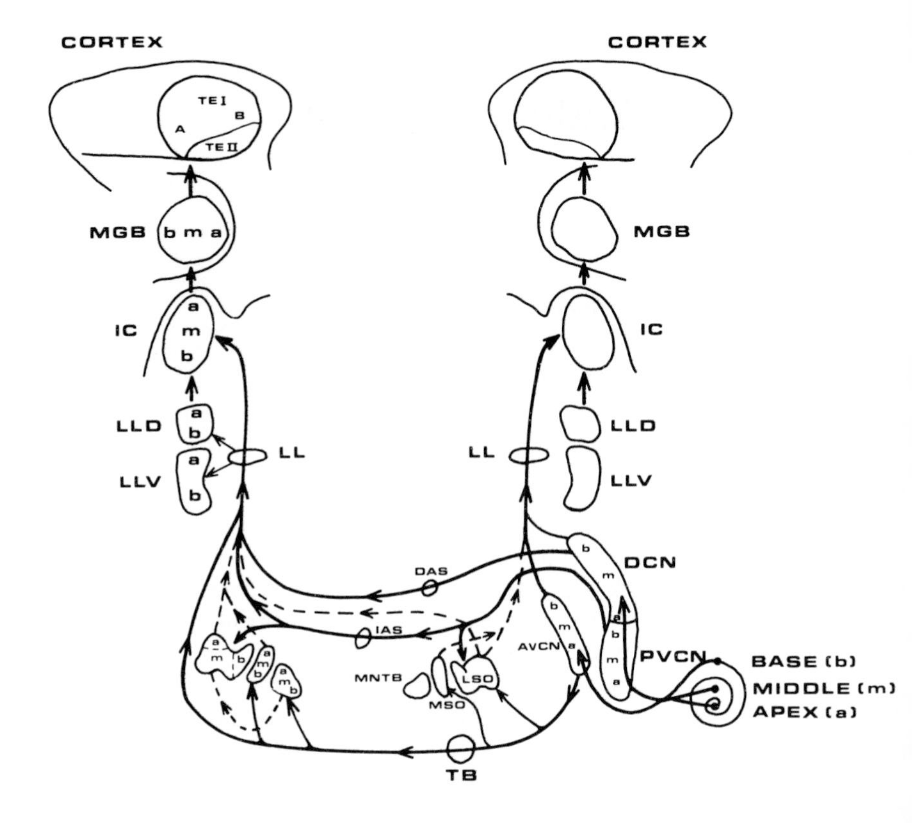

Fig. 1: The auditory pathway of the rat showing the tonotopic organization as a function of position of the basilar membrane. Most of the information about tonotopic organization comes from the data from the cat except for the inferior colliculus (IC) and the cortex (TEI). The cortical data are estimations based on the work of Ryugo (1976). The secondary projection from the cochlear nuclei are shown in solid lines and the tertiary projections in dotted lines. There is some dispute as to the size of the ipsilateral projection from the cochlear nuclei. At the level of MG, the lateral part of the brain is located medially in the diagram. At the level of the cortex, the anterior part of cortex is located laterally in the figure for each cortex. AVCN, anterior ventral cochlear nucleus; DAS, dorsal acoustic stria; DCN, dorsal cochlear nucleus; IAS, intermediate acoustic stria; IC, inferior colliculus; LLD, dorsal nucleus of the lateral lemniscus; LLV, ventral nucleus of the lateral lemniscus; LSO, lateral superior olive; MGB, medial geniculate body; MSO, medial superior olive; TEI, presumed primary auditory cortex; TEII, presumed secondary auditory cortex.

paradigms: (i) single unit and neural tracing studies of auditory centers; (ii) behavioral studies that test for necessary conditions in removing auditory areas by lesions; and (iii) behavioral studies which examine function with animal psychophysics. No attempt will be made to incorporate a vast literature that relates either auditory multiunit or evoked potential recordings to conditioning procedures (Olds, 1973). It is a prejudice of the author that these types of recordings from the brain lack the precision necessary to bring out many of the detailed functional relationships of a sensory system.

II. COCHLEA

A. Structure

The cochlea of the rat has two and a half turns (Wada, 1923) and, in young rats, contains 15 800 spiral ganglion cells (Keithley and Feldman, 1979). The rat cochlea is a typical mammalian one with one row of inner hair cells (IHCs) and three rows of outer hair cells (OHCs) on the basilar membrane (Fig. 2). The OHCs have the typical "W" shaped hairs of a mammalian cochlea (Ross, 1977). Both afferent and efferent synapses make contact with each type of hair cell. Pujol *et al.* (1980)·have shown remarkable changes in innervation pattern during hair cell synaptogenesis in the rat. At the onset of hearing, the IHCs are innervated directly by many afferent and efferent synapses, whereas the OHCs have only afferent synapses. When adult like hearing is achieved, the IHCs still have many afferent synapses on the soma, but the efferent synapses now contact the afferent fibers. At this stage, the OHCs have one or two large efferent synapses which compete with one small afferent synapse to occupy the hair cell synaptic membrane (Lenoir *et al.*, 1980; Pujol *et al.*, 1980).

In the cat, all afferent fibers from large ganglion cells (95% of the total number of afferent fibers) innervate the IHCs, whereas the remaining 5%, from small ganglion cells, terminate on the OHCs (Spoendlin, 1972, 1978). A small number of giant fibers have been reported to innervate many IHCs. These data are based on inferences from thin sections of normal and partly denervated cochleas. More direct observations of fibers, using Golgi material, have confirmed Spoendlin's general observation in both young rats and cats (Ginzberg and Morest, 1983; Perkins and Morest, 1975). Two populations of spiral ganglion cells are seen: radial fiber neurons innervating only IHCs; and spiral fiber neurons innervating OHCs. The radial fibers contact one or two IHCs to establish nearly point to point connections between the auditory nerve and the organ of Corti. In contrast, each spiral fiber can contact from 10–60 OHCs. The largest group of spiral fibers contacts only one row of OHCs and another large group contacts two rows. A small group of spiral fibers contacts all three rows of IHCs (Perkins and Morest, 1975). Each IHC is connected to up to 20 afferent neurons by means of unbranched radial fibers and each OHC is innervated by one or two spiral fibers.

In the rat, there are also two patterns of efferent innervation of the hair cells (Perkins and Morest, 1975). In the simpler pattern, an efferent fiber innervates

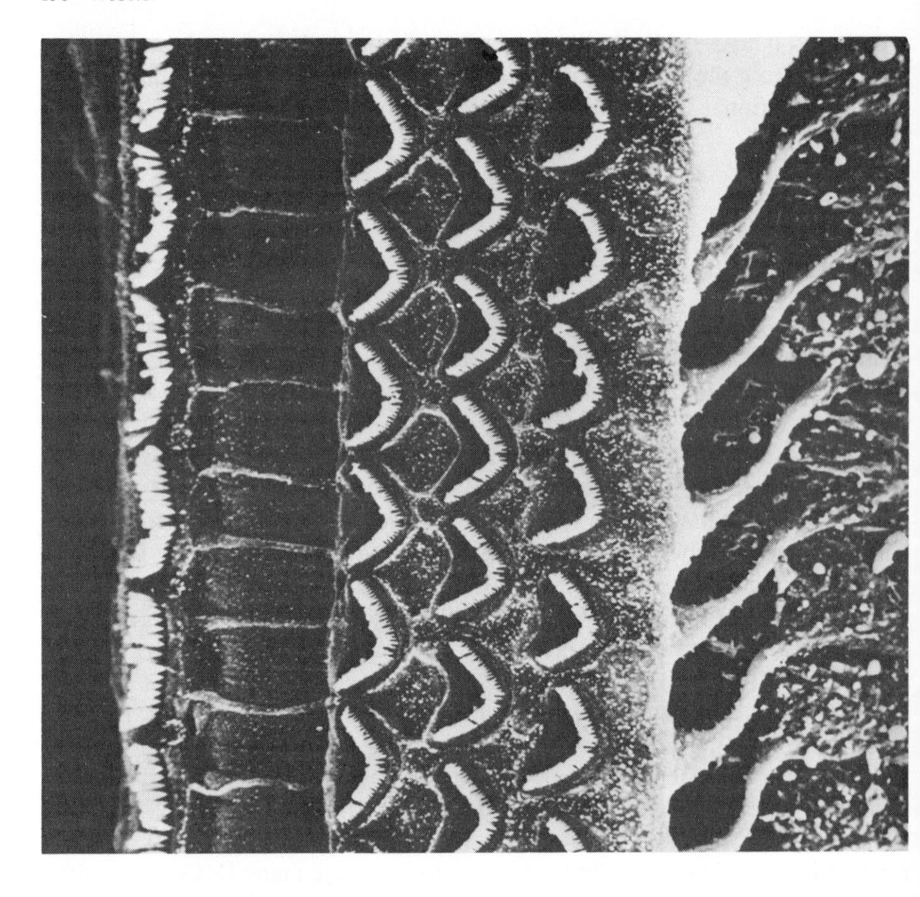

Fig. 2: Scanning electron microscope picture of the hair cells in the rat cochlea. Three rows of outer hair cells with typical W shaped hairs are clearly seen. There is one row of inner hair cells (from work by Lenoir and Pujol available to the writer).

an IHC and then crosses the middle of the tunnel and innervates one OHC. The complex patterns are formed by larger axons which branch and innervate up to 12 IHCs and then innervate up to six OHCs.

B. Function

The most obvious and interesting question is whether the two types of hair cells have different functions. Unfortunately, there are few direct data from the rat cochlea on the differential roles of the hair cells (Pujol *et al.*, 1980). Most of the functional data come from the cat and guinea pig (Dallos, 1973, 1981). There have been many theories put forward to explain the role of the two types of cells (See Dallos, 1981, and Pickels, 1982, for reviews of the evidence and theories).

Briefly, measurements using either the Mössabauer (Johnstone *et al.*, 1970) or the capacitance probe techniques (Wilson and Johnstone, 1975) have indicated that the basilar membrane was broadly tuned for frequency. But the tuning of auditory nerve fibers was much sharper (Evans, 1975). This discrepancy led Evans to postulate a physiologically vulnerable second filter. It was also shown that ototoxic drugs removed OHCs only, and that there was concurrent loss of sensitivity of auditory fibers (Evans, 1975). While it was not postulated that the OHCs were the second filter, it was argued that the IHCs needed the integrity of the OHCs (Evans, 1975). The theory of the second filter was dealt a death blow by evidence from intracellular recordings from IHCs (Russell and Sellick, 1978). It was shown that receptor potentials were as sharply tuned as fibers, thus there was no need to postulate a second filter. It was then shown that the broadness of tuning of the basilar membrane, reported for so many years with the Mössbauer technique, was as sharp as that of auditory nerve fibers (Sellick *et al.*, 1982). This still left the general problem of vulnerability and the role of the OHCs. It was found that the mammalian ear could produce emissions or echos (Kemp, 1978) and that these emissions could be influenced by stimulation of the efferents (Brownell, 1982, 1983). It has been suggested that this evidence indicates that an active mechanical force acting as a feedback may be responsible for the sharp tuning and the vulnerability. Brownell (1983) has some evidence that OHCs may act as an effector to alter the micromechanics of IHCs and so alter tuning.

Another obvious question is whether there are two types of auditory nerve fibers correlating with two types of hair cells (Dallos, 1981). The answer to this question has generally been in the negative (Dallos, 1981; Evans, 1975), but recent intracellular labeling of auditory nerve fibers has suggested why. Elegant experiments by Liberman (1982) and Kiang *et al.* (1982) have found that fibers going to the OHCs were always less than 1.0 μm in diameter. Current physiologic techniques would almost surely fail to record from such fibers.

The rat is sensitive over an extraordinary range of frequencies (0.10 kHz to 100.0 kHz) as indicated by evidence from cochlear potentials (Crowley *et al.*, 1965) and animal psychophysics (Gourevitch and Hack, 1966). The early threshold behavioral data indicated that the rat was more sensitive for an octave around 40.0 kHz (Gourevitch and Hack, 1966). However, Kelly and Masterton (1977) found the minimum threshold was around 16.0 kHz, but in their data there is also a sensitive area around 38.0 kHz. Compared with the cat, there have been few single unit recordings from rat auditory nerve fibers (Møller, 1976b; 1983) and the amount of data reported is not extensive. In general, the results are similar to that obtained from the cat (Evans, 1975; Kiang *et al.*, 1965). The dynamic range of auditory nerve fibers in cats and rats is limited to 30 dB–40 dB (Møller, 1983), a finding which provides difficulties for all theories of auditory coding (Evans, 1975).

III. COCHLEAR NUCLEI

A. General structure

The cochlear nuclei of mammals are located at the dorsolateral border of the junction of the pons and the medulla. The cochlear nuclei have been traditionally divided into a ventral cochlear nucleus (VCo) with an anterior and posterior subdivision, a dorsal cochlear nucleus (DCo) and an interstitial nucleus (I8) (Cajal, 1955; Lorento de Nó, 1933). More recent Golgi studies of the rat (Harrison and Feldman, 1970) and the cat (Brawer *et al.* 1974) have revealed more complex subdivisions and these will be outlined later.

The axons of the auditory nerve enter the cochlear nucleus at the I8 where they bifurate into an ascending and descending branch (Feldman and Harrison, 1969; Lorente de Nó, 1933), giving off fine collaterals prior to bifurcation (Harrison and Feldman, 1970). The ascending branch enters the anterior ventral cochlear nucleus (VCoA) where it runs in parallel fascicles which maintain fixed positions relative to one another. The descending branch enters the posterior ventral cochlear nucleus (VCoP) in parallel fascicles and then it gathers into a bundle to enter the DCo and gives terminal ramifications (Harrison and Feldman, 1970).

In the cat, destruction of the basal end (high frequency) of the cochlea leads to degeneration of the dorsal parts of the cochlear nuclei, whereas destruction of the apical end (low frequency) leads to degeneration of the ventral parts of the cochlear nuclei (Powell and Cowan, 1962; Sando, 1965).

B. Anterior ventral cochlear nucleus

In the cat, there have been two very influential schemes which have delineated subdivisions in the VCoA (Table 1). Using Nissl and Glees preparations, Osen (1969) divided the VCoA into five subdivisions, each containing only one major type of cell. Using Nissl and Golgi preparations, Brawer *et al.*, (1974) divided the VCoA into six subdivisions (Table 1) and they described five types of cells, but the two main types, bushy and stellate, were present throughout the VCoA. In the rat the VCoA has been divided into four regions containing seven main types of cells (Harrison and Feldman, 1970; Harrison and Irving, 1965). This classification was based on Nissl, Protargol and some Golgi preparations. Since the cat classifications have been both helpful and influential in guiding functional studies of the VCoA, even in studies of the rat, the scheme for the rat will be compared with it in some detail (Table 1). Both the schemes of Osen (1969) and Brawer *et al.* (1974) are comparable at the level of Nissl staining, but the finer grain revealed by detailed Golgi analysis makes the latter scheme more useful.

Table 1: Correspondence between rat and cat anterior ventral cochlear nucleus*

Study				
Harrison and Feldman (1970)		Osen (1969)	Brawer *et al.* (1974)	
Area	Cell type	Area and Cell type	Area	Cell type
III	c round cells (large end bulbs)	large spherical area (large end bulbs)	AA	bushy (large end bulbs) stellate (Boutons)
—	—	—	APD	bushy (end bulbs) stellate (boutons)
I	i round cells (pale end bulbs) h large multipolar	small spherical cell area (small end bulbs)	AP	bushy (small end bulbs) stellate (boutons)
II	g oval cells (modified end bulbs) d fusiform cells e round cells f long round cells	multipolar cell area	PD	bushy (modified end bulbs) stellate (boutons) giant (boutons)
II	g d, e, f (interstitial nucleus)	globular cell area	PV	bushy (modified end bulbs)
G	granular cell area	granular cell area		granular cell area

*Note there are small cells present throughout the anterior ventral cochlear nucleus according to all three schemata.

In the cat and rat, the fibers of ascending branch of the eighth nerve give rise to large synaptic endings called end bulbs of Held. Electromicroscopy of the rat VCoA has shown that each bulb makes multiple synaptic contacts with the soma of VCoA cells (Lenn and Reese, 1966; Neises *et al.*, 1982). The largest end bulbs are found in region III of the rat or area AA of the cat (Table 1). Harrison and Feldman (1970) described the cells receiving large bulbs as large and round and they designated them as type c. These cells are the equivalent of the bushy cells described in the anterior part of the cat VCoA (Table 1) and the latter terminology will be preferred. It has been asserted that each bushy cell in AA (Lorente de Nó, 1933) and region III (Neises *et al.*, 1982) receives only one large end bulb. However, Brawer and Morest (1975) and Harrison and Feldman (1970) have found multiple end bulbs on cells in, respectively, AA and region III. As in the cat, most of the cells in area III are bushy cells with end bulbs. In addition, a few stellate cells are also found in peripheral parts of AA (Cant and Morest, 1979a). Region I of the rat VCoA is the equivalent of area AP of the cat (Table 1) and it contains bushy cells (type i) innervated with small end bulbs and few large multipolar cells (type h). In the cat, stellate cells are also present in area AP. Region II of rat is shared by both the VCoA and VCoP. The part of region

II in VCoA is the equivalent of areas of PD and PV in the cat (Table 1). This area includes the area of bifurcation (I8). Harrison and Feldman (1970) reported that the area contains four types of cells of which type g is by far the most numerous. The type g cells are equivalent to bushy cells in the cat and are described in the rat as receiving modified end bulbs of Held (Harrison and Feldman, 1970). Harrison and Feldman (1970) found a nucleus in the auditory nerve between the main body of the cochlear nucleus and the Schwann–glial boundary of the nerve; they called it the acoustic nerve nucleus (AN). This structure has not been reported in the cat. The acoustic nerve nucleus consists of very large cells thinly scattered in the acoustic nerve. The cell bodies and the base of the dendrites are innervated by small synaptic endings which are abolished by destruction of the auditory nerve (Harrison and Feldman, 1970). These cells give rise to large axons which appear to enter the trapezoid body (tz).

In all areas of the anterior ventral cochlear nucleus (VCoA), destruction of the auditory nerve leads to the disappearance of all end bulbs (Gentschev and Sotelo, 1973) and the terminals on stellate cells. However, it should not be thought that bushy cells receive only end bulb innervation. Small boutons also disappear in the rat (Gentschev and Sotelo, 1973) and in the cat (Cant and Morest, 1979b). There are also terminals on bushy cells which do not disappear after destruction of the auditory nerve, but the origins of the fibers of these synapses are not known (Cant and Morest, 1979b).

Harrison and Feldman (1970) have also classified rat VCoA cells into two other classes. The first class of cells has dendrites running parallel with the auditory nerve fibers. This class includes the cells receiving end bulbs. The other class has dendrites which extend across the fibers. Harrison and Feldman (1970) want to argue that the first class are cells that reflect tonotopic organization of the auditory nerve, while the second class does not. However, bushy cells in other species do not show this strict organization (Cant and Morest, 1979a, b).

C. Posterior ventral cochlear nucleus

Harrison and Feldman (1970) have divided the posterior ventral cochlear nucleus (VCoP) into three regions (part of region II, region IV and region V). In the cat, seven areas have been defined by Brawer *et al.* (1974). Because detailed light and electronmicroscopic studies are only available for the octopus cell region in the cat VCoP (Kane, 1973), most comparisons will be restricted to region IV which is the equivalent area. Kane (1973) has shown that this area in the cat contains only one type of cell, the octopus cell. A similar cell, type k, has been observed in the rat VCoP. Kane (1973) has also shown that this cell in the cat receives two types of synapse from the same auditory nerve. Each type of synapse has a different vesicle, suggesting the hypothesis that one nerve can give rise to both inhibitory and excitatory synapses. It is not known if such synapses exist in the rat. One striking difference between rat and cat VCoP is that end bulbs have been observed innervating some cells in the rat VCoP. In the cat VCoP, most of the neurons are innervated by small synaptic boutons.

D. Dorsal cochlear nucleus

In both the rat and the cat, the dorsal cochlear nucleus (DCo) consists of a considerably more complex arrangement of fibers and cells, which are divided into either four or five layers (Brawer *et al.*, 1974; Harrison and Warr, 1962; Godfrey *et al.*, 1975; Mugnaini *et al.*, 1980). In the rat, the three outer layers are related to specific segments of the fusiform (pyramidal) cells (Mugnaini *et al.*, 1980). The first layer (molecular) contains the apical dendrites of the fusiform cells, small cells and granule cells. The second layer contains the somata of the fusiform cells and the third layer contains the basal dendrites of the fusiform cells. The fourth layer contains a dense network of afferent and efferent fibers and is situated deep to the basal dendrites. The fifth layer is the deepest portion of the DCo and is defined by Lorente de Nó (1933) as the central nucleus. The last two layers contain giant cells as well as small cells.

Work in the cat (Jones and Casseday, 1979) has shown considerable labeling on the basal dendrites of fusiform cells after injection of tritiated amino acids into the cochlea. While the overall labeling is less than in the VCo, these data, together with the Golgi work of Kane (1974) on the cat, indicate that the DCo does receive considerable auditory nerve input and can be considered a secondary relay station. Wenthold (1978) has shown a similar pattern in rat cochlear nucleus after injection of [^{3}H]-proline in the cochlea, but there is less labeling in the rat DCo compared with that in the cat.

E. Function

There have been extensive physiologic studies of cochlear nucleus neurons in the cat (see Aitkin *et al.*, 1984, for a detailed review). Most of these studies have tested the hypothesis that the varied neuronal response patterns found in different parts of the cochlear nucleus are a product of the diverse innervation patterns (Fig. 3). That is, the uniform responses of auditory nerve fibers are transformed by cochlear nucleus neurons. Given these findings, it is somewhat surprising to find that all the physiologic studies of rat cochlear nucleus neurons have not carried out the essential histologic controls. Møller (1969a,b; 1970; 1972; 1976a,b; 1983) has carried out some extremely interesting experiments on rat cochlear nucleus neurons, but in each case he has failed to determine where the neurons are located within the cochlear nucleus. This deficiency has important implications for Møller's work. For example he has reported that all rat cochlear nucleus neurons are more broadly tuned for frequency than auditory

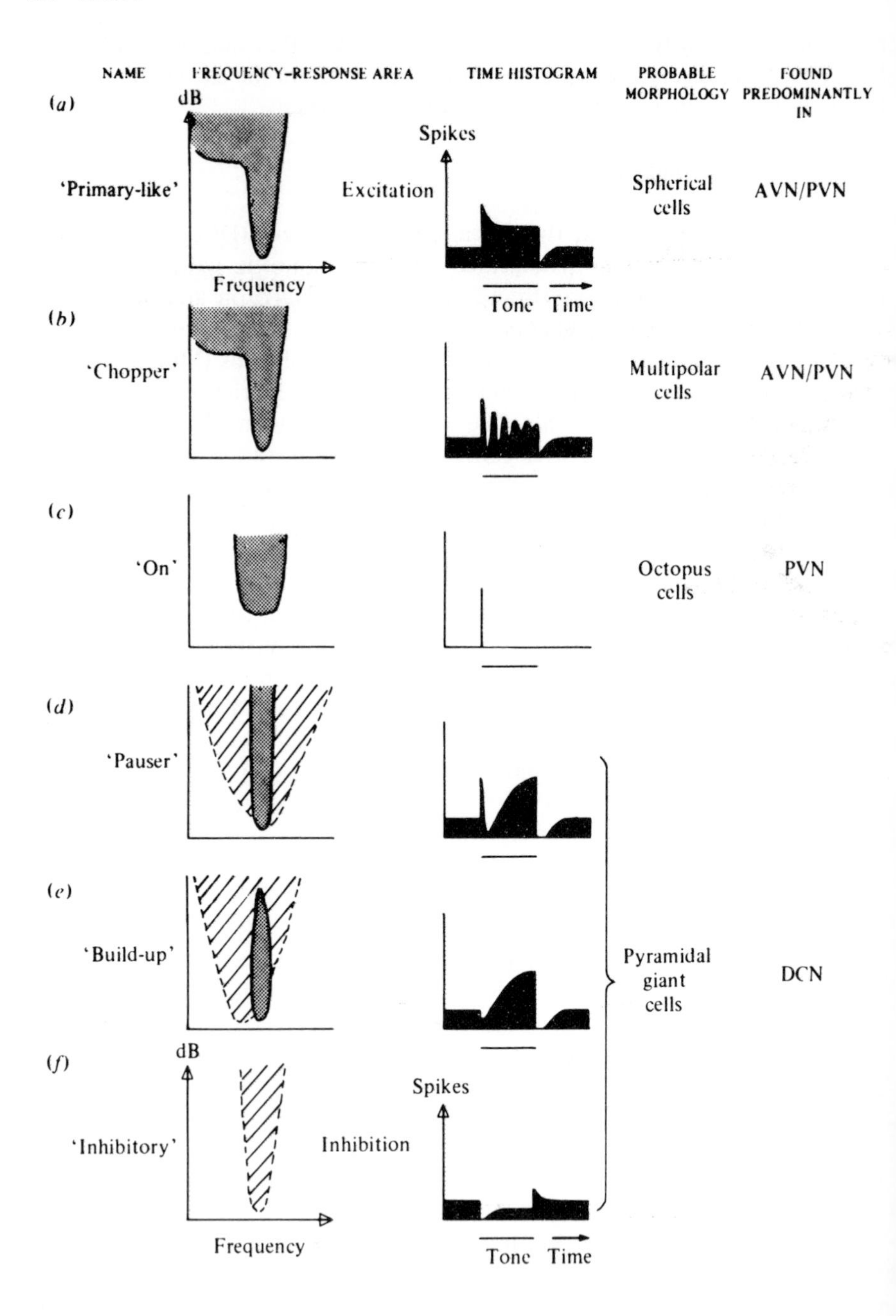

Fig. 3: A–F: response characteristics of units in the cochlear nuclei of the cat defined by time histogram shape. Probable morphology and stimulus location are also shown. Excitatory region is shown by stipple and inhibitory regions by hatched lines (Evans, 1982). AVN anterior ventral cochlear nucleus; PVN posterior cochlear nucleus; DCN dorsal cochlear nucleus.

nerve fibers (Møller, 1976b). However, in the cat VCo A, Bourk (1976) has shown that the neurons innervated by end bulbs are as sharply tuned as auditory nerve fibers. It is possible that Møller's recordings were made in areas which did not contain end bulbs.

The different physiologic types found in cat cochlear nucleus and their relation to cell types and position in the nucleus is illustrated in Fig. 3. Similar data are not available for the rat. It should be noted that octopus cells are "on" responders and DCo neurons have considerable inhibitory inputs. All major regions of the cat cochlear nucleus are tonotopically organized, but, once again, these data are not available for the rat.

IV. SUPERIOR OLIVARY COMPLEX

A. General structure

The superior olivary complex (SOC) of the rat consists of a group of auditory nuclei located in the ventral region of the pons. There has been some dispute about the structure of the SOC in the rat, especially in regard to the size and the location of the medial superior olive (MSO). The rat SOC is much less complex than the cat SOC, which contains nine separate divisions. It is unfortunate that recent Golgi studies of rat MSO failed to give an adequate description of its location (Feng and Rogowski, 1980; Rogowski and Feng, 1981). Its mediolateral extent was overestimated in the Paxinos and Watson (1982) *Atlas*. A favorable section through the MSO reveals it to be a vertical column of cells (see Fig. 3E in Volume 1, Chapter 14).

B. Connections from the cochlear nucleus

Fibers leave the cochlear nucleus in three separate pathways: the trapezoid body (tz), the intermediate acoustic stria (stria of Held) (ias) and the dorsal acoustic stria (Stria of Monakow) (das) (Fig. 1). These pathways innervate the SOC, the nuclei of the lateral lemniscus and the inferior colliculus (IC). Most of the bushy cells project via the tz to the MSO (bilaterally) and LSO (ipsilaterally). The octopus cells project via the intermediate acoustic stria into SOC and the fusiform and giant cells of DCo project to IC via the dorsal acoustic stria (Beyerl, 1978).

C. Medial superior olive

Whenever present in mammals, the medial superior olive (MSO) in frontal sections consists of a column or columns of fusiform cells arranged with their long axes in the mediolateral direction. In the rat, one or more long dendrites originate from each pole of the cell body (Rogowski and Feng, 1981). Often a third dendrite originates from the middle of the cell (Harrison and Feldman,

1970). There is a fine dense neuropil from which boutons arise. These terminate on the dendrites, but never on the cell body. A lesion of the cochlear nucleus in the cat on one side leads to degeneration in the trapezoid body (tz) and the disappearance of boutons on MSO dendrites pointing towards the side of the lesion. No end bulbs have been reported in the MSO of the rat (Harrison and Feldman, 1970). Both area III in the rat and area AA in the cat provide the input to MSO from each side.

D. Superior paraolivary nucleus

Harrison and Warr (1962) initially reported that they could not find an MSO, but only a superior paraolivary nucleus (SPO) which is a region of dense neuropil containing large and small multipolar cells. The large cells are the largest found in the SOC and have conspicuous nuclei. Bouton type endings are found. More detailed Golgi and lesion studies are required to determine the nature and extent of the SPO. Harrison and Warr (1962) point out that the structure of the rat SOC is closer to that of the rabbit and bat than the cat. Both these species do not have a prominent MSO. This nucleus is not shown in the Paxinos and Watson (1982) *Atlas*.

E. Lateral superior olive

In the cat, the lateral superior olive (LSO) has been described as an S shaped structure, but in the rat, Harrison and Feldman (1970) have described it as an N shaped structure. The LSO of the cat consists primarily of fusiform or tufted cells from the poles of which one or more dendrites take origin. The cells are oriented such that in transverse sections their long axis is approximately perpendicular to the surface of the nucleus. There is a dense neuropil and boutons are found upon the dendrites and cell bodies (Harrison and Feldman, 1970). A small number of widely scattered multipolar cells are also found in the LSO (Harrison and Feldman, 1970). The LSO receives ipsilateral input from the cochlear nucleus via the trapezoid body and contralateral input from the medial nucleus of the trapezoid body (MTz) (Fig. 1).

F. Medial nucleus of the trapezoid body

The medial nucleus of the trapezoid body (MTz) in the rat consists of a prominent cluster of globular shaped cells, between 15 μm–18.5 μm in diameter. The nucleus is located on the medial side of the SPO and extends dorsally over the pyramidal tract (Harrison and Warr, 1962). The cells have only one or two short dendrites. The synaptic endings consist of a single chalice of Held on the

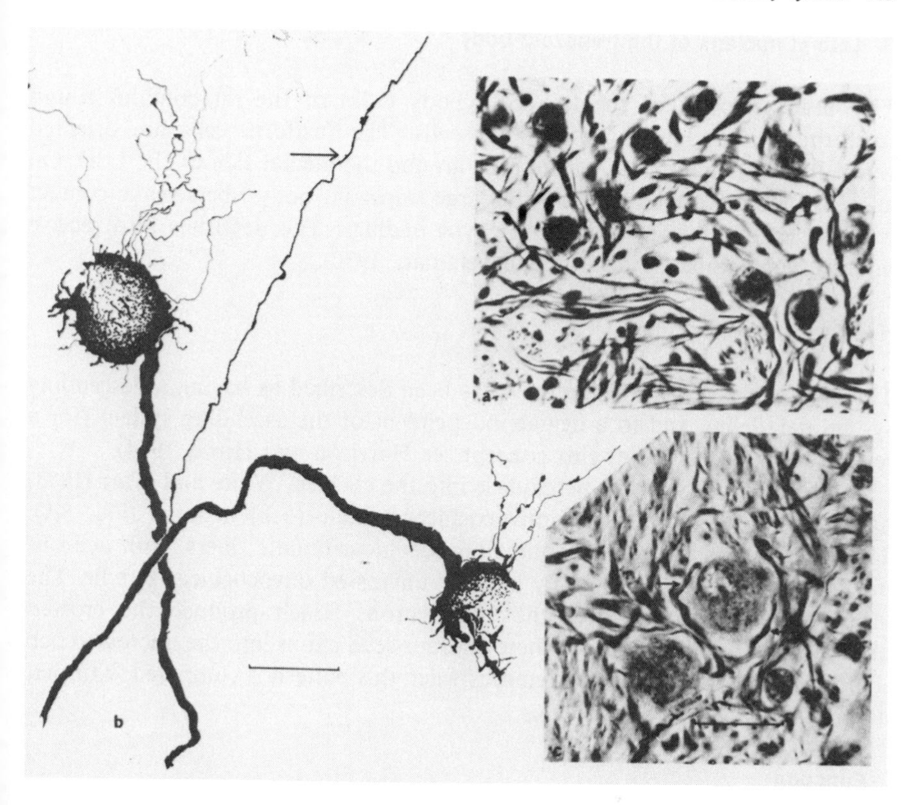

Fig. 4: B Golgi drawings of two chalices of Held in the rat MTz. Inserts: protargol stains showing end bulbs. Calibration 20 μm (Harrison and Feldman, 1970).

body of each cell (Fig. 4). Very infrequently, a bouton ending is seen on the base of a dendrite. The chalice arises from large presynaptic fibers of 2.75 μm–3.5 μm diameter. A lesion of the trapezoid body abolished the chalices of Held in the MTz contralateral to the lesion (Harrison and Warr, 1962). The axons of these MTz cells go to the base of the LSO on the same side. A lesion placed under the LSO led to terminal degeneration in the MTz. Lesioning of the lateral lemniscus in the rat did not produce degeneration in the MTz on the same side, supporting the notion that these cells project only to the LSO, although Glendenning *et al.* (1981) have reported that the MTz in the cat projects to the intermediate nucleus of the lateral lemniscus.

In the rat, Harrison and Irving (1964) have shown that the VCoA projects to the anterior part of the MTz, and the VCoP projects to the posterior part of the MTz. Thus the cochlear nucleus projects twice onto the MTz.

G. Lateral nucleus of the trapezoid body

The lateral nucleus of the trapezoid body (LTz) of the rat contains round fusiform cells and a few multipolar cells. The fusiform cells are oriented vertically and the dorsal dendrite runs around the medial side of the LSO. On all cells in LTz, collaterals from the large trapezoid body fibers make contact with large club shaped and chalice type endings. The dendrites also receive bouton type endings (Harrison and Feldman, 1970).

H. Olivocochlear bundle

The auditory pathway of mammals has been described as having a descending system paralleling and to a degree independent of the ascending system (for a review of the evidence for this concept see Harrison and Howe, 1974).

By injecting horseradish peroxidase into the cochlea, White and Warr (1983) have revealed the location of olivocochlear bundle (ocb) neurons (Fig. 5A). There is a dual origin of the source olivocochlear bundle fibers, with neurons in LSO sending their fibers only into the uncrossed olivocochlear bundle. The other source is confined to the LTz neurons which produce the crossed olivocochlear bundle. A few of these neurons send axons into the uncrossed ocb (Fig. 5A). There are drastic differences when this pattern is compared with that of the cat (Fig. 5B).

I. Function

There has been only one physiologic study of the SOC of the rat (Inbody and Feng, 1981). Before discussing this study, a brief review of the possible function of the SOC will be given based on data from the cat and the beagle (for a detailed review see Aitkin *et al.*, 1984; Webster, 1977). Because the SOC is the site of the first binaural interaction in the auditory pathway, most investigations have been designed to reveal binaural integration and its relationship with sound localization (Webster, 1977). Most studies have been carried out in the context of a duplex model of sound localization (Moore, 1977; Webster, 1977), in which low frequency sounds are localized by cues of interaural phase differences and high frequency sounds are localized by cues of interaural intensity differences. In the cat and the beagle, anatomic and physiologic correlates of this model have been found (Goldberg and Brown, 1968, 1969; Guinan *et al.*, 1972a,b). The MSO has been found to have many neurons sensitive to low frequency cues and the LSO has many neurons sensitive to high frequency cues. Most MSO neurons have characteristic frequencies below 12.0 kHz and most LSO neurons have high frequency characteristic frequencies (greater than 4.0 kHz). While most MSO and LSO neurons are binaurally influenced, two major classes have been

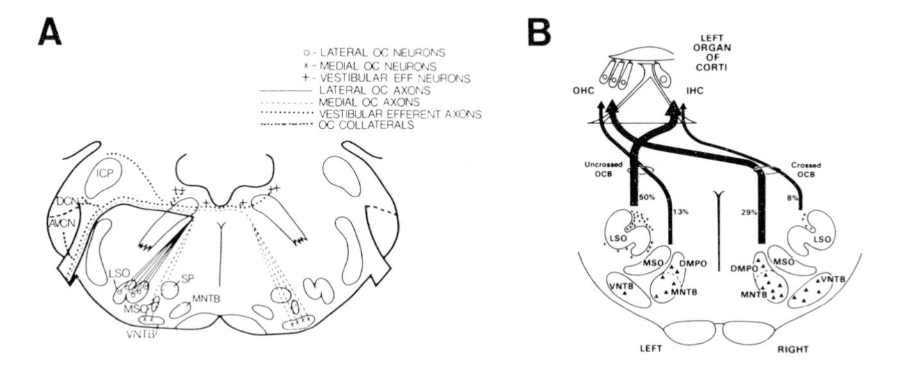

Fig. 5: A: The origin of the cells giving rise to the olivocochlear bundle (OC) in the rat (White and Warr, 1983). B: The cells of origin of the OC in the cat (Warr, 1978). AVCN, anterior ventral cochlear nucleus; DCN, dorsal cochlear nucleus; DMPO, dorsomedial paraolivary nucleus; ICP, inferior cerebellar peduncle; IHC, inner hair cells; LSO, lateral superior olive; MNTB, medial nucleus of the trapezoid body; MSO, medial superior olive; OCB, olivocochlear bundle; OHC, outer hair cells; SP, superior paraolivary nucleus; VNTB, ventral nucleus of the trapezoid body.

distinguished: cells exhibiting predominantly excitatory responses to either ear (E/E or excitatory/excitatory cells); and cells exhibiting excitation to one ear and inhibition to the other (E/I or excitatory/inhibitory cells). In the cat and the beagle MSO, two thirds of the neurons are E/E. The remaining neurons are either E/I (24%) or E/O monaural (11%). In contrast, very few E/E cells are found in the LSO, in which the majority are E/I. Many of the low frequency E/E cells in the MSO are very sensitive to low frequency cues and most of the E/I cells in the LSO are very sensitive to high frequency cues. It should be stressed that a duplex model sound localization is an oversimplification (Moore, 1977), so some caution is needed in interpreting the MSO and LSO as the only neural sites for such a model.

However, it is interesting to look at the study of rat MSO (Inbody and Feng, 1981) in the light of the above data. The response properties of 52 MSO neurons were identified. Inbody and Feng (1981) found that 52% of the cells were E/E, and 48% were E/I. However, the range of characteristic frequencies for both classes was from 2.2 kHz to 6.6 kHz. Thus while the rat MSO has comparatively low characteristic frequencies compared with the animal's range of sensitivity (0.1 to 100.0 kHz), no low frequency neurons (less than 1.0 kHz) were found. In other species, it is the low frequency MSO neurons which are exquisitively sensitive to low frequency cues. It can be concluded that the rat MSO is not important for the localization of low frequency sounds. Such a result might be expected from the small size of the rat's head which would allow maximal interaural time or phase differences of the order of 150 μs. These data indicate why the MSO is a small structure in the rat, suggesting that there has been no selective pressure to evolve a large MSO. The rat LSO is a much larger structure, perhaps indicating that the rat, like other small mammals, uses interaural intensity cues to localize sounds. Unfortunately, there are no available physiologic studies of the rat LSO, but behavioral data indicate that the rat localizes high frequency sounds better than low frequency sounds (Masterton *et al.*, 1975).

V. NUCLEI OF THE LATERAL LEMNISCUS

Virtually nothing is known about either the dorsal (DLL) or the ventral (VLL) nucleus of the lateral lemniscus (ll) in the rat, except that they form two gray regions in the lemniscal fibers.

VI. INFERIOR COLLICULUS

A. General structure

The auditory midbrain is an obligatory synaptic station in the ascending auditory pathway (Goldberg and Moore, 1967; Morest, 1964a). The inferior colliculus (IC) is not a homogeneous structure. In Nissl (Berman, 1968), the cat IC appears to consist of a large, cell dense, central nucleus (CIC) surrounded by an overlying pericentral nucleus (PCIC) and a lateral and caudal external nucleus (EIC). Detailed Golgi studies of the cat have distinguished a more complex structure for IC (Cajal, 1955; Morest, 1964a, 1966; Morest and Oliver, 1984; Oliver and Morest, 1984; Rockel and Jones, 1973a,b). It has been suggested that the central nucleus is organized into tonotopic laminations made up of the principal cells, their dendrites and the incoming afferent fibers. However, the various cat Golgi models do not agree. Rockel and Jones (1973a,b) have proposed that the laminae have a concentric organization. However, studies with 2-deoxyglucose (Servière and Webster, 1983; Servière *et al.*, 1984) have revealed that functional isofrequency contours are not organized concentrically, but are organized diagonally as suggested by Oliver and Morest (1984). There is no published Golgi data on the rat IC but Ryugo (1976) has carried out a Nissl and Golgi study. Ryugo distinguished a central nucleus (CIC), a pericentral nucleus (PCIC), an external nucleus (EIC), an interstitial nucleus (IN) and a dorsomedial nucleus (CICDM).

B. Central nucleus

Ryugo (1976) has shown that the rat central nucleus (CICVL), like the cat homolog, is organized into laminations in the three standard planes (Fig. 6). The

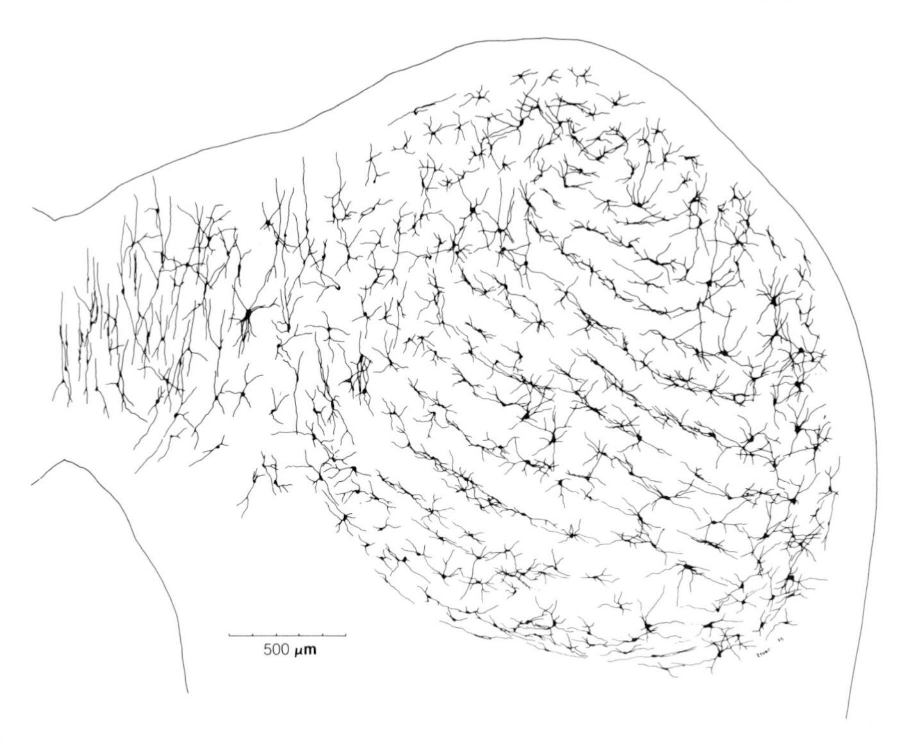

Fig. 6: Golgi drawings of cells in the rat inferior colliculus (IC). Cells in the central nucleus can be seen to line up into laminae. Most of the cells in the central nucleus are tufted neurons. Multipolar neurons are predominant in surrounding regions (Ryugo, 1976).

organization is somewhat similar to the model of Rockel and Jones (1973a,b) for the cat IC, as the laminations are approximated by a series of concentric curves. The principal neurons show a tufted and two dimensional appearance with their longitudinal axes conforming to a lamina in all three standard sections. A few neurons are medium to large multipolars with dendrites crossing several laminae. Some unpublished 2-deoxyglucose data suggest that Ryugo's concentric model is perhaps not the best model for the frequency organization of the rat (CICVL) (Webster, observations).

C. Pericentral nucleus

The pericentral nucleus (PCIC) contains mostly medium and small cells, the majority of which are multipolar with radiate type dendritic fields (Fig. 6) (Ryugo, 1976). In addition, there are fusiform neurons with flattened dendritic fields that are parallel to the tectal surface (Fig. 6). The fields of these multipolar and fusiform cells completely encapsulate the inferior colliculus, except at the dorsomedial aspect through which fibers of the commissure of the inferior colliculus pass. This result contrasts markedly with observations on the cat where the pericentral nucleus has been observed only along the dorsal and posterior edge of the IC (Rockel and Jones, 1973b).

D. External and dorsomedial nuclei

The external nucleus (EIC) is a sheet of primarily large cells intercalated between the PCIC and the lateral edge of the central nucleus of the inferior colliculus (CIC) (Ryugo, 1976). These cells have pyramidal shaped bodies with multipolar, radiate type dendrites reaching lengths up to 400 μm. There is no apparent orientation to their dendritic fields.

The dorsomedial nucleus (CICDM) is readily visible in Golgi preparations. It is perforated by numerous fascicles of commissural fibers and contains medium to large multipolar neurons with radiate dendrites. These do not appear to have any particular orientation (Ryugo, 1976).

E. Afferent connections of the inferior colliculus

The inferior colliculus (IC) in both the rat and cat receives axons from most brain stem auditory nuclei. These fibers gather together to form a large tract, the lateral lemniscus, which proceeds dorsally and rostrally to occupy a progressively more lateral position in the brain stem. In the cat, several studies using retrograde tracers have shown that the IC receives fibers from 13 different brain stem nuclei (Adams, 1979; Brunso-Bechtold *et al.*, 1981; Glendenning and Masterton, 1983; Roth *et al.*, 1978). The predominant ascending projections to IC come from: (i) the contralateral cochlear nucleus (most divisions); (ii) the contralateral and ipsilateral LSO; (iii) the ipsilateral MSO; (iv) the ipsilateral VLL; and (v) the ipsilateral and contralateral DLL. Lesser projections come from: (i) the ipsilateral cochlear nucleus; (ii) periolivary nuclei; and (iii) the contralateral IC. Most cell types in the cochlear nucleus were labeled except the bushy cells of AA and many of the octopus cells.

There has been only one HRP study of the rat IC (Beyerl, 1978) and the data are in general agreement with those in the cat (Fig. 7). Discrete injections into the IC show selective uptake in the cochlear nuclei and LSO, indicating that

---

Fig. 7: Transport of horseradish peroxidase (HRP) after injection in rat inferior colliculus at dorsal (A), middle (B) and ventral (C) parts. Tonotopic uptake of label is seen in anterior ventral (AVCN), posterior ventral (PVCN) and dorsal cochlear nuclei (DCN). Ventral injections have produced labeling in dorsal regions of cochlear nucleus. Dorsal injections produced ventral labeling in cochlear nucleus. *A–*C: show topographic transport of HRP in lateral superior (LSO) and medial superior olive MSO (adapted from Beyerl, 1978). LTB, MTB, VTB, lateral, medial, and ventral nuclei of the trapezoid body.

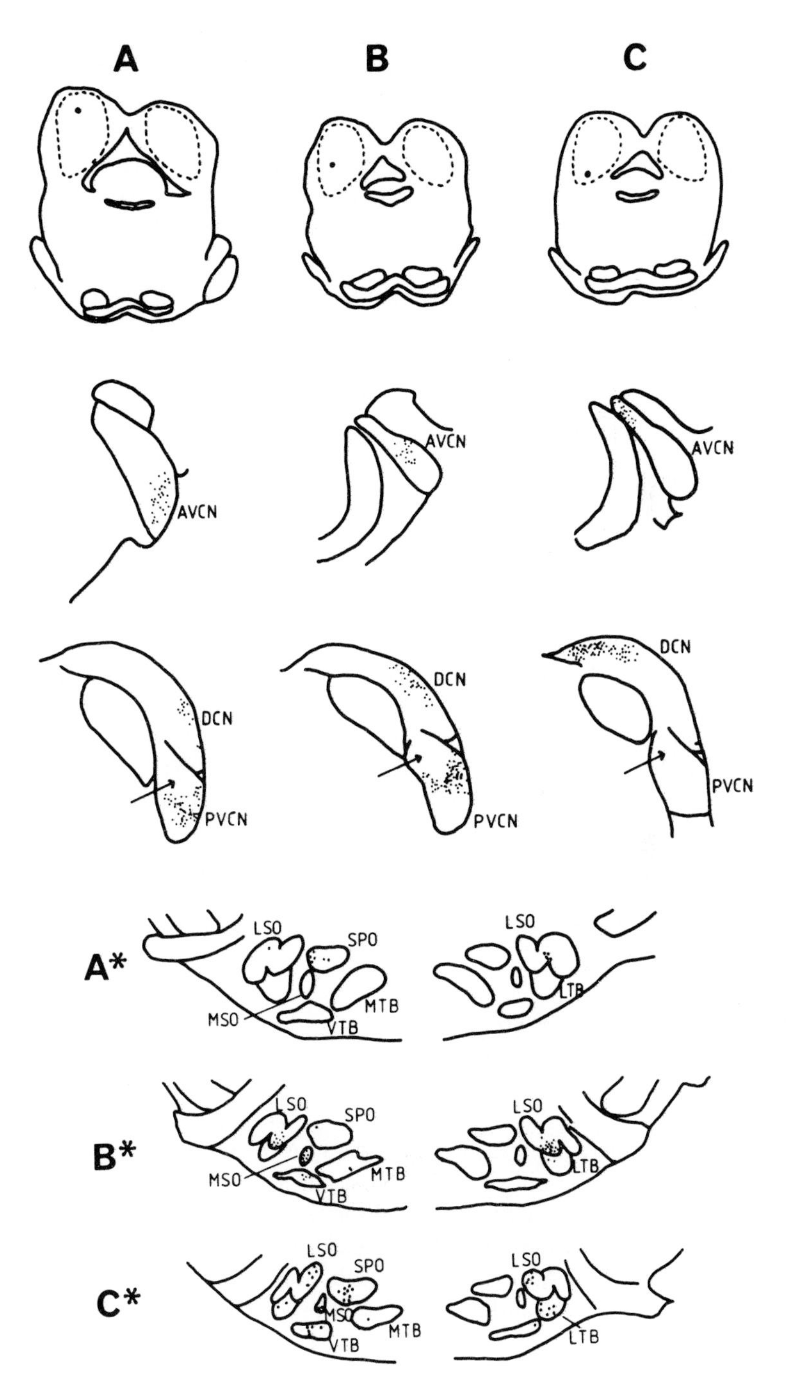
A
B
C
AVCN
AVCN
AVCN
DCN
DCN
DCN
PVCN
PVCN
PVCN
A*
LSO
SPO
MSO
MTB
VTB
LSO
LTB
B*
LSO
SPO
MSO
MTB
VTB
LSO
LTB
C*
LSO
SPO
MSO
MTB
VTB
LSO
LTB

these regions in the rat could be tonotopically organized. However, an injection into a dorsal region of the ICC (presumably a low frequency region) produced labeling in the ventral region of the three main divisions of the cochlear nucleus but failed to produce labeling in the ipsilateral MSO and in what should be low frequency parts of the ipsilateral and contralateral LSO (Fig. 7). This result confirms a previous suggestion that the rat superior olivary complex contains few low frequency cells. More detailed HRP experiments made with concurrent physiologic recordings, as in the cat IC, are needed to support this suggestion (Aitkin *et al.*, 1981). Like the cat, no bushy or type c cells were labeled in Region III and no octopus cells were labeled in Region IV even after extensive HRP injections (Beyerl, 1978).

F. Function

The most obvious functional question concerns the role of the ubiquitous laminae found in mammalian IC. It has been proposed that they are the structural basis for the tonotopic organization found in both the cat (Aitkin *et al.*, 1975; Semple and Aitkin, 1979) and the rat IC (Clopton and Winfield, 1973).

Compared with the cat, there have been relatively few single unit studies of the adult rat IC (Clopton and Winfield, 1973; Flammino and Clopton, 1975; Syka *et al.*, 1981). The meagre data indicate that the central nucleus contains binaural neurons sensitive to interaural intensity and time differences as in the cat. There are no studies which have looked at other divisions of the IC similar to studies carried out in the cat (Aitkin *et al.*, 1975; 1984).

The specific function of IC has not yet been ascertained. Unilateral lesions of the IC produce sound localization deficits confined to contralateral space (Jenkins and Masterton, 1982). However, when similar lesions are made more centrally, the same deficit is produced. The nucleus is a site of convergence of a number of parallel pathways, but the functional role of these pathways remains to be discovered.

VII. MEDIAL GENICULATE BODY

A. General structure

As with the IC, there are few published data on the rat medial geniculate body (MG). Once again, the only detailed Nissl and Golgi study is the unpublished work of Ryugo (1976). The thalamic auditory region receives ascending afferents from the auditory midbrain and the adjacent tegmentum in the cat (Kudo and Niimi, 1980). The cat MG (Morest, 1964b, 1965) has been divided by Golgi methods into a principal and a medial division (MGM). The principal division consists of a dorsal (MGD) and a ventral division (MGV). The ventral division is laminated, like the IC, by tufted neurons. Ryugo (1976) was also able to distinguish a principal and a medial division in the rat MG.

B. Principal division ventral part

Ryugo (1976) found two types of cells in the MGV which were similar to those in the cat. A large principal neuron was tufted and organized into laminations, and these displayed a pronounced concentric arrangement. There were also small Golgi Type II cells which were inconsistently impregnated. The principal cells are homologous with the thalamocortical relay neurons described in the cat MGV by Morest (1964b).

C. Principal division dorsal part

The principal division (dorsal) (MGD) is characterized by a relatively homogeneous population of neurons (small and medium sized) characterized by freely radiating dendrites that are diffusely arranged within a spherical field. The area containing these neurons is marked by very light staining for myelin. There does not appear to be any particular organization of cell bodies or dendrites in this division (Ryugo, 1976).

D. Medial division

Two cell populations are apparent within the MGM with Golgi stained material. There are small neurons with an elaborately branching dendritic system which appears to be very similar to the Golgi Type II cells in the MGV. Like the MGV neurons, these cells impregnate rarely. The main cells are large, flask shaped and multipolar (Ryugo, 1976).

E. Afferent connections of the medial geniculate body

The ascending projections of the IC have been studied by Ryugo (1976) who analyzed fiber degeneration patterns. However, the confounding by commissural fibers of passage did not allow an analysis of interstitial, dorsomedial or pericentral parts of the inferior colliculus. However, the projections of the central nucleus and the external nucleus were revealed.

Discrete lesions in the dorsal part of the central nucleus produced some degenerating fibers which penetrated the dorsomedial (CICDM) and interstitial (IN) parts and entered the commissure of the IC. Most of the fibers swept dorsally and laterally through the pericentral nucleus to enter the brachium of the IC on the same side. These fibers terminated in the dorsal and lateral part of the MGV, in correspondence with the laminar structure of this part of the MGV. No degeneration was seen in either the MGD or MGM.

The pattern of ascending projections of the external nucleus were quite different from that of the CICVL. Most of the fibers enter the brachium of the IC and terminate evenly within the MGM. There is a small focus of degeneration in the dorsal PCIC.

It would be interesting to have some neural tracing studies of the rat MG. Injections of tritiated amino acids into the cat CICVL provided terminal labeling in the MGV oriented in the same plane as the laminae (Anderson *et al.*, 1980). Thus, the cat CICVL appears to project primarily and tonotopically onto the MGV. Also, when HRP is injected into the cat MGV, in small areas specified physiologically, the labeled neurons are almost exclusively confined to the CICVL and are topographically related to the injection site (Aitkin *et al.*, 1981). The dorsal division of the cat MG receives afferents from the PCIC (Aitkin *et al.*, 1981), while the EIC projects mainly to the MGM. In contrast, the CICDM projects to every division of the cat MG. While these projections are ipsilateral, there are also contralateral projections to the thalamus that are tonotopically similar but are less dense.

F. Function

There are no functional data available for the rat medial geniculate (MG). In studies of the cat MGV, penetrations from dorsocaudal to ventrorostral have shown tonotopic organization in that the apical segments of the cochlea are represented more caudally, laterally and ventrally while basal turns are represented more rostrally, medially and dorsally (Calford and Webster, 1981). Tonotopic organization has not been shown for either the MGM or MGD.

In the MGV, responses to binaural tones are similar to those in lower auditory nuclei. Thus, EE and EI cells form the majority and a much smaller proportion are monaural or EO (Aitkin and Webster, 1971, 1972; Calford and Webster, 1981). Low critical frequency (CF) units can respond to ITD's and high frequency units may be sensitive to IID's. Strong inhibitory effects are seen in the MG (Aitkin *et al.*, 1966) and the inhibition has a longer time course than that occurring in IC.

The neurons in the cat MGM are broadly tuned, receive binaural input, fire at the onset of a stimulus (Aitkin, 1973) and about one third are polysensory. These characteristics are shared by the EIC, the main source of afferents to MGM. In the cat MGD, units are excited by tones with latencies as long as 100 ms–250 ms (Calford and Webster, 1981). It is difficult to reconcile these latencies with projections from the PCIC or the auditory cortex.

VIII. AUDITORY CORTEX

A. General structure and connections

The rat auditory cortex and its connections with the thalamus have been studied by early workers in this field (Kreig, 1964a, b; Lashley, 1941; Waller, 1947). However, it is extremely difficult to develop a clear idea about the relationships between auditory cortex and the thalamus from this literature. Kreig (1946a, b) identified three auditory areas in temporal cortex: area 41, area 20 and area 36. Zilles and his colleagues, using a computerized gray level index analysis, have recognized three auditory areas: Tel, Te2 and Te3 (Zilles *et al.*, 1980; see also Volume 1, Chapter 10). Overall, the classification of Zilles and colleagues corresponds to that of Kreig, but within the overall area many topologic discrepancies are found. For example, the primary auditory cortex (area 41) of Kreig, does not match exactly the primary area Tel. It is a pity that the only HRP study of these areas (Patterson, 1977) remains unpublished.

To describe the auditory cortical areas, I am once again dependent on the work of Ryugo (1976) and also Ryugo and Killackey (1974). Ryugo (1976) placed discrete lesions in the three subdivisions of the MG. Lesions placed in the MGD showed that it projected in a topographical manner onto a narrow belt of neocortex situated immediately dorsal to the posterior extension of the rhinal fissure (Fig. 8). Ryugo (1976) argues that this area is comparable with areas 20 and 36 of Krieg (1946a,b).

Discrete lesions placed in the MGV also reveal a topographic connection between thalamus and cortex (Fig. 8). Following a discrete lesion of the MGV, large caliber degenerating fibers can be traced up to and through the internal capsule where they enter the cortical white matter. These fibers transverse back within the cortical white matter and layer 6 before repeatedly branching in layer 5 and terminating densely in a restricted position of layer 4 and the adjacent part of layer 3. There is no evidence of fiber or terminal degeneration in layers 2 and 1. The area extent of this projection conforms to area 41 of Kreig (1946a,b) and therefore does not match entirely Tel of Zilles and Wree (Volume 1, Chapter 10). Lesions placed in various quadrants within MGV exhibit topographically organized thalamocortical projections (Fig. 8).

Lesions placed in the MGM, in contrast to the other two areas, do not reveal a topographic projection pattern (Fig. 8). Instead a comparatively small lesion in the MGM will lead to degeneration in the areas projected to by the MGD and MGV, as well as cortical areas above these two areas (Fig. 8).

B. Architecture of auditory fields

On the basis of the differential patterns of thalamic and callosal input, a division

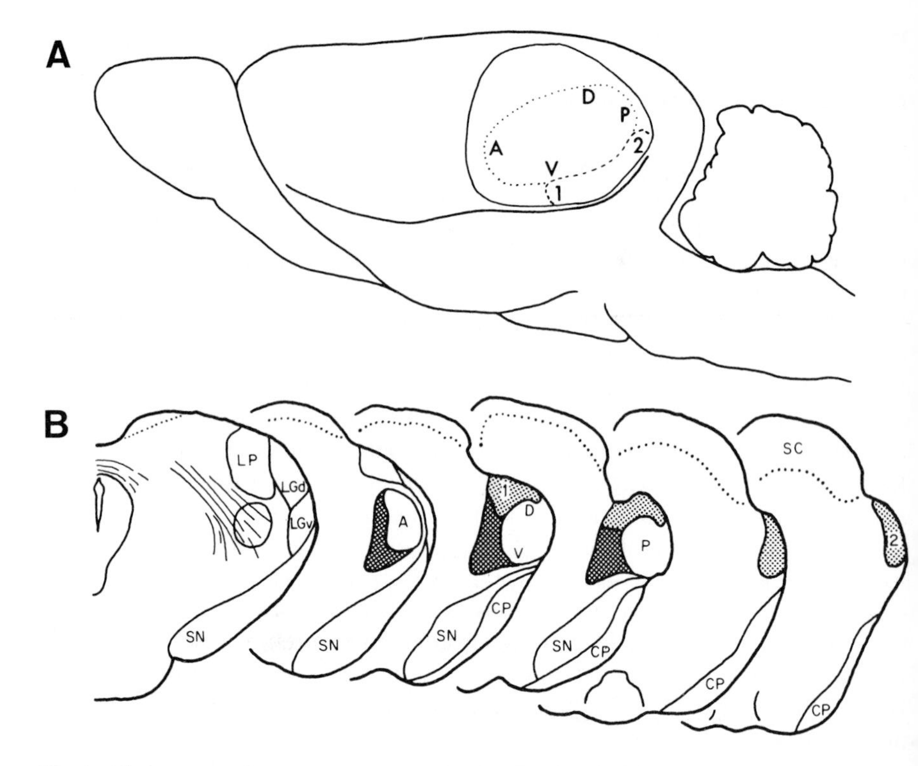

Fig. 8: Thalamocortical projections of medial geniculate body (MGB); A: shows projections of the three divisions of MGB to the auditory cortex. The projection from MGv is topographic. Lesions at D (dorsal) and V (ventral) (located in B) produce medial and lateral degeneration in the cortex. Lesions at A (anterior) and P (posterior) (located in B) produce anterior and posterior degeneration in auditory cortex. MGd (light stipple) also projects topographically to nonprimary cortex. Position of lesions and target marked by 1 and 2. MGm (dark stipple) projects without topography on to a large cortical region marked by a solid line in the cortex in A (Ryugo, 1976).

of rat auditory cortex into a primary cortical field and nonprimary cortical belt has been proposed (Ryugo, 1976). Zilles and Wree (Volume 1, Chapter 10) have reported that the HRP study of Patterson (1977) also supports a notion of a "core cortex" and a "belt cortex". A histologic examination of the cortical field yields additional cytoarchitectural confirmation of the auditory cortical subdivisions (Ryugo, 1976). In Nissl stained material, the primary cortical field is distinguished by its broad and densely packed cell population of layer 4. Myelin stained sections reveal that primary cortex is heavily myelinated. The belt cortex is much less myelinated.

Golgi analysis confirms differences in the Nissl material. There is a dense band of stellate cells in layer 4 of the primary cortex (Te1) which are not present in the "belt" cortex (Ryugo, 1976).

This parcelation of the rat auditory cortex by Ryugo (1976) and Zilles and Wree (Volume 1, Chapter 10) should be regarded as tentative. In the cat, the combination of detailed physiologic mapping and modern tracing methods have revealed cortical fields other than the original proposal of a "core" and a "belt" cortex (Woolsey and Walzl, 1942). For a detailed review of cortical structure in other species refer to Aitkin *et al.* (1984).

The corticofugal system has not been studied extensively in the rat, apart from Beyerl's (1978) HRP study of the IC. Beyerl found that the primary cortex projected to the CIC. This observation agrees with data from the owl monkey but not with observations on the cat (Aitkin *et al.*, 1984).

C. Function

As with other parts of the rat auditory system, there have been no single unit studies of the auditory cortex. There have also been no HRP or autoradiographic studies of these cortical fields. By contrast, in the cat and other species, there have been many detailed studies of the function and connections of auditory fields (Aitkin *et al.*, 1984). These studies have revealed, in the cat, that as well as primary or AI cortex, there are a number of divisions of auditory cortex which are tonotopically organized. Also, AI has been shown to have binaural strips or columns as well as tonotopic strips or columns (Imig and Adrian, 1977). In the cat and other species, the functional significance of these multiple cortical fields remains to be elucidated. These areas are interconnected but the significance of these connections has not been revealed. In the rat, there is obviously a need to define the cortical fields, both by electrophysiologic and histochemical methods, before we can put forward functional hypotheses for testing.

D. Behavioral–adaptation studies

The role of the rat auditory cortex in discriminating spatial location has been examined by Kelly and Glazier (1978). They found that the auditory cortex (and the degenerated MG) were not necessary for discriminating the spatial locations of auditory stimuli. Their lesions removed both primary and "belt" cortical areas. There has been considerable controversy in the literature over the role of the auditory cortex in other species (see Aitkin *et al.*, 1984, for a more detailed review). Most studies appear to show that animals without an auditory cortex can lateralize, but could not localize. That is, animals could indicate that a sound source was on the left, but could not locomote towards such a source (Aitkin *et al.*, 1984). These studies have led to the view that the deficit after lesions is not a sensory or perceptual one, but is a deficit in sensory–motor integration (Heffner, 1978; Heffner and Masterton, 1975). However, deficits in tasks requiring integration of binaural stimuli (for example, interaural time difference cues) indicate a reduction in sensory or perceptual processes. Detailed studies of the rat auditory system are needed which take into account modern procedural issues.

IX. NEUROPHARMACOLOGY OF THE RAT AUDITORY PATHWAY

It is not possible to provide a detailed account of the neuropharmacology of the rat auditory pathway because of space limitations. Most work has been carried out in the cochlea and the cochlear nucleus (see Klinke, 1981, for a detailed review). In the cochlea, there is no clear evidence to indicate what the afferent transmitter is which acts between the junction of the hair cells and the afferent fibers. There is reasonable evidence that ACh is the efferent transmitter (Klinke, 1981).

The importance of the subdivisions of the cochlear nucleus based on the system in the cat (Brawer *et al.*, 1974) is emphasized in work on the rat cochlear nucleus. A computer model from the cat cochlear nucleus has been developed for the rat cochlear nucleus which helps quantitative histochemical mapping of the subdivisions (Godfrey *et al.*, 1978). The method allows a precise correlation between putative transmitter and position in the cochlear nucleus. Godfrey *et al.* (1978) have shown that the inhibitory transmitters γ-aminobutyric acid (GABA) and Glycine are concentrated in dorsal cochlear nucleus (DCo). By using their computer map, they have shown that the molecular and fusiform layers of DCo contain the highest levels of GABA.

There is little information on higher centers of either the rat or the cat auditory pathway (Adams and Wenthold, 1979). Without a detailed anatomical map of the higher centers in the rat, it is not possible to carry out experiments similar to the type carried out in the cochlear nucleus (Godfrey *et al.*, 1978; Godfrey and Matschinsky, 1981).

X. GENERAL CONCLUSION

The rat auditory pathway has many of the characteristics of a typical mammalian pathway. The general structure of the pathway is similar to that of the cat. There are, however, a great many gaps in our knowledge of structure, function and neuro-pharmacology of the rat system. Detailed single unit studies are urgently required at all levels of the auditory pathway, particularly studies which correlate function with structure. It will be necessary to apply modern tracing techniques to study the connections between each level of the pathway. One can look forward to an exciting era of work on the rat auditory system, when modern techniques provide us with specific answers relating structure to function.

ADDENDUM

Since this article was first written, an important paper on the inferior colliculus has been published [Faye-Lund, H. and Osen, K.K. (1985). The anatomy of the inferior colliculus in rat, *Anat. Embryol.*, 171, 1-20]. This paper is the type of paper which is needed in the study of the anatomy of the rat auditory pathway. While confirming the general work of Ryugo (1976), the results show the three-dimensional structure of the inferior colliculus in Nissl and Golgi and confirm that the orientation of the lamina agree with the 2-deoxyglucose data.

ACKNOWLEDGEMENTS

This work was supported by grants from the Australian Research Grant Scheme and from the National Health and Medical Research Council of Australia. The author is grateful to Dr D. Ryugo and Dr R. Pujol for making available their unpublished material. Mr M. Brown is thanked for invaluable help with word processing.

REFERENCES

Adams, J. C. (1979). Ascending projections to the inferior colliculus. *J. Comp. Neurol.* 183, 519–538.

Adams, J. C. and Wenthold, R. J. (1979). Distribution of putative amino acid transmitters, choline acetyltransferase and glutamate decarboxylase in the inferior colliculus. *Neurosci.* 4, 1947–1951.

Aitkin, L. M. (1973). Medial geniculate body of the cat: Responses to tonal stimuli of neurons in the medial division. *J. Neurophysiol.* 36, 275–283.

Aitkin, L. M. and Webster, W. R. (1971). Tonotopic organization in the medial geniculate body of the cat. *Brain Res.* 20, 402–405.

Aitkin, L. M. and Webster, W. R. (1972). Medial geniculate body of the cat: Organization and responses to tonal stimuli of neurons in the ventral division. *J. Neurophysiol.* 35, 365–380.

Aitkin, L. M., Dunlop, C. W. and Webster, W. R. (1966). Click-evoked response patterns of single units in the medial geniculate body of the cat. *J. Neurophysiol.* 29, 109–123.

Aitkin, L. M., Webster, W. R., Veale, J. L. and Crosby, D. (1975). Inferior colliculus. I. Comparison of response properties of neurons in the central, pericentral and external nuclei of adult cat. *J. Neurophysiol.* 38, 1195–1207.

Aitkin, L. M., Calford, M. B., Kenyon, C. E. and Webster, W. R. (1981). Some facets of the organization of the principal division of the cat medial geniculate body. In J. Syka and K. M. Aitkin (Eds). *Neuronal mechanism of hearing*, pp. 163–182. Plenum, New York.

Aitkin, L. M., Irvine, D. R. F. and Webster, W. R. (1984). Central neural mechanisms of hearing. In I. Darian Smith (Ed.). *Handbook of physiology*. American Physiological Society, New York, 675–737.

Anderson, R. A., Roth, G. L., Aitkin, L. M. and Merzenich, M. M. (1980). The efferent projection of the central nucleus and the pericentral nucleus of the inferior colliculus in the cat. *J. Comp. Neurol.* 194, 649–662.

Berman, A. L. (1968). *The brain stem of the cat: A cytoarchitectonic atlas with stereotaxic coordinates.* University of Wisconsin Press, Madison, Wis.

Beyerl, B. D. (1978). Afferent projections to the central nucleus of the inferior colliculus of the rat. *Brain Res.* 145, 209–223.

Bourk, T. R. (1976). Electrical responses of neural units in the anteroventral cochlear nucleus of the cat. Doctoral thesis. Massachusetts Institute of Technology.

Brawer, J. R. and Morest, D. K. (1975). Relations between auditory nerve endings and cell types in the cat's anteroventral cochlear nucleus seen with the Golgi method and Normarski optics. *J. Comp. Neurol.* 160, 491–506.

Brawer, J. R., Morest, D. K. and Kane, E. C. (1974). The neuronal architecture of the cochlear nucleus of the cat. *J. Comp. Neurol.* 155, 251–299.

Brownell, W. E. (1982). Cochlear transduction: An integrative model and review. *Hearing Res.* 6, 335–350.

Brownell, W. R. (1983). Observations on a motile response in isolated outer hair cells. *In* W. R. Webster and L. M. Aitkin (eds), *Mechanisms of hearing*, pp. 5–10. Monash University Press, Melbourne.

Brunso-Bechtold, J. K., Thompson, G. C. and Masterton, R. B. (1981). HRP study of the organization of auditory afferents ascending to central nucleus of inferior colliculus in cat. *J. Comp. Neurol.* 197, 705–722.

Cajal, S. Ramón y (1955). *Histologie du système nerveux de l'homme et des vertébrés.* Madrid, Instituto Ramon y Cajal.

Calford, M. B. and Webster, W. R. (1981). Auditory respresentation within principal division of cat medial geniculate body: An electrophysiological study. *J. Neurophysiol.* 45, 1013–1028.

Cant, N. B. and Morest, D. K. (1979a). Organization of the neurons in the anterior division of the anteroventral cochlear nucleus of the cat: Light-microscopic observation. *Neurosci.* 4, 1909–1923.

Cant, N. B. and Morest, D. K. (1979b). The bushy cells in the anteroventral cochlear nucleus of the cat: A study with the electron microscope. *Neurosci.* 4, 1925–1945.

Clopton, B. M. and Winfield, J. A. (1973). Tonotopic organization in the inferior colliculus of the rat. *Brain Res.* 56, 355–358.

Crowley, D. E. (1974). Comment on "otitis media in labatory rats". *Physiol. Psychol.* 2, 99–100.

Crowley, D. E., Hepp-Raymond, M., Tabowitz, D. and Palin, J. (1965). Cochlear potentials in the albino rat. *J. Aud. Res.* 5, 307–316.

Dallos, P. (1973). *The auditory periphery: Biophysics and physiology.* Academic Press, New York.

Dallos, P. (1981). Cochlear physiology. In M. R. Rosenzweig and L. W. Porter (Eds). *Annual Review of Psychology*, pp. 153–190. Annual Reviews Inc., Palo Alto.

Daniel, H. J., Means, L. W., Dressel, M. E. and Loesche, P. J. (1973). Otitis media in laboratory rats. *Physiol. Psychol.* 1, 7–8.

Evans, E. F. (1975). Cochlear nerve and cochlear nucleus. In W. D. Keidel and W. D. Neff (Eds). *Handbook of sensory physiology*, vol. 5/2, pp. 1–108. Springer, Berlin.

Evans, E. F. (1982). Functional anatomy of the auditory system. In H. B. Barlow and J. D. Mollon (Eds). *The senses*, pp. 256–306. Cambridge University Press, Cambridge.

Feldman, M. L. and Harrison, J. M. (1969). The projection of the acoustic nerve to the ventral cochlear nuclei of the rat: A Golgi study. *J. Comp. Neurol.* 137, 267–294.

Feng, A. S. and Rogowski, B. A. (1980). Effects of monaural and binaural occlusion on the morphology of neurons in the medial olivary nucleus of the rat. *Brain Res.* 189, 530–534.

Flammino, F. and Clopton, B. M. (1975). Neural responses in the inferior colliculus of albino rat to binaural stimuli. *J. Acoust. Soc. Amer.* 57, 692–695.

Gentschev, T. and Sotelo, C. (1973). Degenerative patterns in the ventral cochlear of the rat after primary deafferentiation: An ultrastructural study. *Brain Res.* 62, 37–60.

Ginzberg, R. D. and Morest, D. K. (1983). A study of cochlear innervation in the young cat with the Golgi method. *Hearing Res.* 10, 227–246.

Glendenning, K. K., Brunso-Bechtold, J. K., Thompson, G. C. and Masterton, R. B. (1981). Ascending auditory afferents to the nuclei of the lateral lemniscus. *J. Comp. Neurol.* 197, 673–703.

Glendenning, K. K. and Masterton, R. B. (1983). Acoustic chiasm: Efferent projections of the lateral superior olive. *J. Neurosci.* 3, 1521–1537.

Godfrey, D. A. and Matschinsky, F. M. (1981). Quantitative distribution of choline acetyltransferase and acetylcholinesterase activities in the rat cochlear nucleus. *J. Histochem. Cytochem.* 29, 720–730.

Godfrey, D. A., Kiang, N. -Y. S. and Norris, B. E. (1975). Single unit activity in the dorsal cochlear nucleus of the cat. *J. Comp. Neurol.* 162, 269–284.

Godfrey, D. A., Carter, J. A., Lowry, O. H. and Matschinsky, F. A. (1978). Distribution of γ-aminobutyric acid, glycine, glutamate and aspartate in the cochlear nucleus of the rat. *J. Histochem. Cytochem.* 26, 118–126.

Goldberg, J. M. and Brown, P. B. (1968). Functional organization of the dog superior olivary complex: An anatomical and electrophysiological study. *J. Neurophysiol.* 31, 639–656.

Goldberg, J. M. and Brown, P. B. (1969). Response of binaural neurons of dog superior olivary complex to dichotic tonal stimuli: Some physiological mechanisms of sound localization. *J. Neurophysiol.* 32, 613–636.

Goldberg, J. M. and Moore, R. Y. (1967). Ascending projections of the lateral lemniscus in the cat and monkey. *J. Comp. Neurol.* 129, 143–156.

Gourevitch, G. and Hack, M. H. (1966). Audibility in the rat. *J. Comp. Physiol. Psychol.* 122, 289–291.

Guinan, J. J. Jr., Guinan, S. S. and Norris, B. E. (1972a). Single auditory units in the superior olivary complex. I. Responses to sound and classification based on physiological properties. *Internat. J. Neurosci.* 4, 102–120.

Guinan, J. J. Jr., Norris, B. E. and Guinan, S. S. (1972b). Single auditory units in the superior olivary complex. II. Location of units, categories and tonotopic organization. *Internat. J. Neurosci.* 4, 147–166.

Harrison, J. M. (1978). The auditory system of the brain stem. In R. F. Naunton and C. Fernandez (Eds). *Evoked electrical activity in the auditory nervous system*, pp. 353–372. Academic Press, New York.

Harrison, J. M. and Feldman, M. L. (1970). Anatomical aspects of the cochlear nucleus and superior olivary complex. In W. D. Neff (Ed.). *Contributions to sensory physiology*, vol. 4, pp. 95–142. Academic Press, New York.

Harrison, J. M. and Howe, M. E. (1974). Anatomy of descending auditory system (mammalian). In W. D. Keidel and W. D. Neff (Eds). *Handbook of sensory physiology*, vol. V/1, pp. 363–388. Springer, Berlin.

Harrison, J. M. and Irving, R. (1964). Nucleus of the trapezoid body: Dual afferent innervation. *Science* 143, 473–474.

Harrison, J. M. and Irving, R. (1965). The anterior ventral cochlear nucleus. *J. Comp. Neurol.* 124, 15–42.

Harrison, J. M. and Warr, W. B. (1962). A study of the cochlear nucleus and ascending auditory pathways of the medulla. *J. Comp. Neurol.* 119, 341–380.

Heffner, H. (1978). The effect of auditory cortex ablation on localization and discrimination of brief sounds. *J. Neurophysiol.* 41, 963–976.

Heffner, H. and Masterton, B. (1975). Contribution of auditory cortex to sound localization in the monkey (*Macaca mulatta*). *J. Neurophysiol.* 38, 1340–1358.

Imig, T. J. and Adrian, H. O. (1977). Binaural columns in primary auditory cortex. *Brain Res.* 138, 241–257.

Inbody, S. B. and Feng, A. S. (1981). Binaural response characteristics of single neurons in the medial superior olivary nucleus of the albino rat. *Brain Res.* 210, 361–366.

Jenkins, W. M. and Masterton, R. B. (1982). Sound localization: Effects of unilateral lesions in central auditory system. *J. Neurophysiol.* 47, 987–1016.

Johnstone, B. M., Taylor, K. J. and Boyle, A. J. (1970). Mechanisms of guinea pig cochlea. *J. Acoust. Soc. Amer.* 47, 504–509.

Jones, D. R. and Casseday, J. H. (1979). Projections of auditory nerve in the cat as seen by anterograde transport methods. *Neurosci.* 4, 1299–1313.

Kane, E. C. (1973). Octopus cells in the cochlear nucleus of the cat: heterotypic synapses upon homeotypic neurons. *Int. J. Neurosci.* 5, 251–279.

Kane, E. C. (1974). Patterns of degeneration in the caudal cochlear nucleus of the cat after cochlear ablation. *Anat. Rec.* 179, 67–91.

Keithley, E. M. and Feldman, M. L. (1979). Spiral ganglion cell counts in an age-graded series of rat cochleas. *J. Comp. Neurol.* 136, 429–442.

Kelly, J. B. and Glazier, S. J. (1978). Auditory cortex lesions and discrimination of spatial location by the rat. *Brain. Res.* 145, 315–321.

Kelly, J. B. and Masterton, B. (1977). Auditory sensitivity of the albino rat. *J. Comp. Physiol. Psychol.* 91, 930–936.

Kemp, D. T. (1978). Stimulated acoustic emissions from within the human auditory system. *J. Acoust. Soc. Amer.* 64, 1386–1391.

Kiang, N.Y. S., Watanabe, T., Thomas, E. C. and Clark, L. F. (1965). *Discharge patterns of single fibers in cat's auditory nerve.* MIT Press, Cambridge.

Kiang, N.Y. S., Rho, J. M., Northrop, C. C., Liberman, M. C. and Ryugo, D. K. (1982). Hair-cel innervation by spiral ganglion cells in adult cats. *Science* 217, 175–177.

Klinke, R. (1981). Neurotransmitters in the cochlea and the cochlear nucleus. *Acta Otolaryngol.* 91 541–554.

Kreig, W. J. S. (1964a). Connections of the cerebral cortex. I. The albino rat. A topography of th cortical areas. *J. Comp. Neurol.* 84, 221–275.

Kreig, W. J. S. (1964b). Accurate placement of minute lesions in the brain of the albino rat. *Quart Bull. Northwest. Univ. Med. Sch.* 20, 199–208.

Kudo, M. and Niimi, K. (1980). Ascending projections of the inferior colliculus in the cat: A autoradiographic study. *J. Comp. Neurol.* 191, 545–556.

Lashley, K. S. (1941). Thalamocortical connections of the rat's brain. *J. Comp. Neurol.* 75, 67–121

Lenn, N. J. and Reese, T. S. (1966). The fine structure of nerve endings in the nucleus of the trapezoi body and the ventral cochlear nucleus. *Am. J. Anat.* 118, 375–389.

Lenoir, M., Shernson, A. and Pujol, R. (1980). Cochlear receptor development in the rat wit emphasis on synaptogenesis. *Anat. Embryol.* 160, 253–262.

Liberman, M. C. (1982). Single-neuron labelling in the cat auditory nerve. *Science* 216, 1239–1241

Lorento de Nó, R. (1933). Anatomy of the eighth nerve III. General plan of structure of the primar cochlear nuclei. *Laryngoscope* 43, 327–350.

Masterson, B., Thompson, G. C., Bechtold, J. K. and Robards, M. J. (1975). Neuroanatomical basi of binaural phase-difference analysis for sound localization: A comparative study. *J. Comp Physiol. Psychol.* 89, 379–386.

Møller, A. R. (1969a). Unit responses in the rat cochlear nucleus to repetitive, transient sounds. *Act Physiol. Scand.* 75, 542–551.

Møller, A. R. (1969b). Unit responses in the cochlear nucleus of the rat to sweep tones. *Acta Physio Scand.* 76, 503–512.

Møller, A. R. (1970). Unit responses in the cochlear nucleus of the rat to noise and tones. *Act Physiol. Scand.* 78, 289–298.

Møller, A. R. (1972). Coding of sounds in lower levels of the auditory system. *Quart. Rev. Biophys* 5, 59–155.

Møller, A. R. (1976a). Dynamic properties of excitation and two-tone inhibition in the cochlea nucleus studies using amplitude-modulated tones. *Exp. Brain Res.* 25, 307–321.

Møller, A. R. (1976b). Dynamic properties of auditory fibers compared with cells in the cochlea nucleus. *Acta Physiol. Scand.* 98, 157–167.

Møller, A. R. (1983). *Auditory physiology.* Academic Press, New York.

Moore, B. C. J. (1977). *Introduction to the psychology of hearing.* Macmillan, London.

Morest, D. K. (1964a). The laminar structure of the inferior colliculus of the cat. *Anat. Rec.* 148, 314

Morest, D. K. (1964b). The neuronal architecture of the medial geniculate body of the cat. *J. Anat (London)* 98, 611–630.

Morest, D. K. (1965). The laminar structure of the medial geniculate body of the cat. *J. Anat (London)* 99, 143–160.

Morest, D. K. (1966). The cortical structure of the inferior quadrigeminal lamina in the cat. *Anat Rec.* 154, 389–390.

Morest, D. K. and Oliver, D. L. (1984). The neuronal architecture of the inferior colliculus in th cat. Defining the functional anatomy of the auditory midbrain. *J. Comp. Neurol.* 222, 209–236

Mugnaini, E., Osen, K. K., Dahl, A., Friedrich, V. L., Jr. and Korte, G. (1980). The structure o granular cells and related interneurons (termed Golgi cells) in the cochlear complex of cat, ra and mouse. *J. Neurocytol.* 9, 537–570.

Neff, W. D., Diamond, I. T. and Casseday, J. G. (1975). Behavioural studies of auditory discrimina tion: Central nervous system. In W. D. Keidel and W. D. Neff (Eds). *Handbook of sensor physiology* vol. V/2, pp. 307–400. Springer, New York.

Neises, G. R., Maltoxe, D. E. and Gully, R. L. (1982). The maturation of the end bulb of Held ir the rat anteroventral cochlear nucleus. *Anat. Rec.* 204, 271–279.

Olds, J. (1973). Multiple unit recordings from behaving rats. In R. F. Thompson and M. M Patterson (Eds). *Bioelectric recording techniques: Part A, Cellular processes and brain potentials* pp. 167–201. Academic Press, New York.

Oliver, D. L. and Morest, D. K. (1984). The central nucleus of the inferior colliculus in the cat. *J Comp. Neurol.* 222, 237–264.

Osen, K. K. (1969). Cytoarchitecture of the cochlear nuclei in the cat. *J. Comp. Neurol.* 136, 453–483.
Patterson, H. A. (1977). An anterograde degeneration and retrograde axonal transport study of the cortical projections of rat geniculate body. Doctoral thesis. Boston University, Boston, Mass.
Paxinos, G. and Watson, C. (1982). *The rat brain in stereotaxic coordinates.* Academic Press, Sydney.
Perkins, R. E. and Morest, D. K. (1975). A study of cochlear innervation patterns in cats and rats with the Golgi method and Normaski optics. *J. Comp. Neurol.* 163, 129–158.
Pickels, J. O. (1982). *An introduction to the physiology of hearing.* Academic Press, London.
Powell, T. P. S. and Cowan, W. M. (1962). An experimental study of the projection of the cochlea. *J. Anat. (London)* 916, 269–284.
Pujol, R., Carlier, E. and Lenoir, M. (1980). Ontogenetic approach to inner and outer hair cell function. *Hearing Res.* 2, 223–230.
Rockel, A. J. and Jones, E. G. (1973a). The neuronal organization of the inferior colliculus of the cat. I. The central nucleus. *J. Comp. Neurol.* 147, 11–60.
Rockel, A. J. and Jones, E. G. (1973b). The neuronal organization of the inferior colliculus of the cat. II. The pericentral nucleus. *J. Comp. Neurol.* 149, 301–334.
Rogowski, B. A. and Feng, A. S. (1981). Normal postnatal development of medial superior olivary neurons in the albino rat: Golgi and Nissl study. *J. Comp. Neurol.* 196, 85–97.
Ross, M. D. (1977). The tectorial membrane of the rat. *Am. J. Anat.* 139, 449–482.
Roth, G. L., Aitkin, L. M., Anderson, R. A. and Merzenich, M. M. (1978). Some features of the spatial organization of the central nucleus of the inferior colliculus of the cat. *J. Comp. Neurol.* 182, 661–680.
Russell, I. J. and Sellick, P. M. (1978). Intracellular studies of hair cells in the mammalian cochlea. *J. Physiol. (London)* 284, 261–290.
Ryugo, D. K. (1976). An attempt towards an integration of structure and function in the auditory system. Doctoral thesis. Irvine, University of California.
Ryugo, D. K. and Killackey, H, P. (1974). Differential telencephalic projections of the medial and ventral divisions of the medial geniculate body of the rat. *Brain Res.* 82, 173–177.
Sando, I. (1965). The anatomical interrelationships of the cochlear nerve fibers. *Acta Otolaryngol. (Stockholm)* 59, 417–436.
Sellick, P. M., Patuzzi, R. and Johnstone, B. M. (1982). Measurement of basilar membrane motion in the guinea pig using the Mössbauer technique. *J. Acoust. Soc. Amer.* 72, 131–141.
Semple, M. N. and Aitkin, L. M. (1979). Representation of sound frequency and laterality by units in central nucleus of cat inferior colliculus. *J. Neurophysiol.* 42, 1626–1639.
Servière, J. and Webster, W. R. (1983). Excitatory and inhibitory tonotopic contours in the inferior colliculus of the cat: A 2-[^{14}C]-deoxyglucose study. In W. R. Webster and L. M. Aitkin (Eds). *Mechanisms of hearing*, pp. 77–82. Monash University Press, Melbourne.
Servière, J., Webster, W. R. and Calford, M. B. (1984). Iso-frequency labelling revealed by a combined [^{14}C]-2-deoxyglucose, electrophysiological and horseradish peroxidase study of the inferior colliculus of the cat. *J. Comp. Neurol.* 228, 463–477.
Spoendlin, H. (1972). Innervation densities of the cochlea. *Acta Otolaryngol. (Stockholm)* 73, 235–248.
Spoendlin, H. (1978). The afferent innervation of the cochlea. In R. F. Naunton and C. Fernandez (Eds). *Evoked electrical activity on the auditory nervous system*, pp. 21–39. Academic Press, New York.
Syka, J., Druga, R., Popelar, J. and Kalinova, B. (1981). Functional organization of the inferior colliculus. In J. Syka and L. M. Aitkin (Eds). *Neuronal mechanisms of hearing*, pp. 137–154. Plenum, New York.
Wada, T. (1923). Anatomical and physiological studies on the growth of the inner ear of the albino rat. *Memoirs of the Wistar Institute of Anatomy and Biology* 10.
Waller, W. H. (1947). Topographical relations of cortical lesions to thalamic nuclei in the albino rat. *J. Comp. Neurol.* 60, 237–270.
Warr, W. B. (1978). The olivocochlear bundle: Its origins and terminations in the cat. In R. G. Naunton and C. Fernandez (Eds). *Evoked activity in the auditory nervous system*, pp. 43–63. Academic Press, New York.
Webster, W. R. (1977). Central neural mechanisms of hearing. *Proc. Aust. Physiol. Pharmacol. Soc.* 8, 1–7.

Wenthold, R. J. (1978). Glutamic acid and aspartic acid in subdivisions of the cochlear nucleus after auditory nerve lesion. *Brain Res.* 143, 544–548.
White, J. S. and Warr, W. B. (1983). The dual origins of the olivocochlear bundle in the albino rat. *J. Comp. Neurol.* 219, 203–214.
Wilson, J. P. and Johnstone, J. R. (1975). Basilar membrane and middle ear vibration in guinea pig measured by capacitance probe. *J. Acoust. Soc. Amer.* 57, 705–723.
Woolsey, C. N. and Walzl, E. M. (1942). Topical projections of nerve fibers from local regions of the cochlear to the cerebral cortex of the cat. *Bull. Johns Hopkins Hosp.* 71, 315–344.
Zilles, K., Zilles, B. and Schleicher, A. (1980). A quantitative approach to cytoarchitectonics. VI. The area pattern of the cortex in the albino rat. *Anat. Embryol.* 159, 355–360.

9

Anatomy of the vestibular nucleus complex

WILLIAM R. MEHLER

University of California
San Francisco, California, USA

JOSEPH A. RUBERTONE

Hahnemann University
Philadelphia, Pennsylvania, USA

I. INTRODUCTION

There are four major vestibular nuclei and a number of associated smaller cell groups that are collectively referred to as the vestibular nuclear complex (VNC). It has been postulated that the vestibular nuclei were among the first supraspinal cell groups to evolve out of the reticular formation. There are VNC cell groups in the brain stem of the most primitive vertebrates that receive primary fiber connections from labyrinthine receptors and relay impulses to skeletomotor and oculomotor nuclei (Bangma and Ten Donkelaar, 1983; Mehler, 1972).

Cytoarchitecturally, the exact borders between the four major mammalian subdivisions (the superior, lateral, medial and spinal vestibular nuclei) are not always evident in Nissl stained sections. Golgi method studies have shown that some cells retain a "reticular" character in that their dendrites extend across apparent borders between adjacent nuclei. However, the dendrites of the vast majority of cells in the cat stay within the confines of the subdivision where the soma is located (Hauglie-Hanssen, 1968).

The small, topographically related cell groups included in the vestibular nuclear complex (VNC) were described largely by Brodal and Pompeiano (1957a, b) in the cat and designated X, Y, Z and F. They have been identified in a number of mammalian brains and they also appear to exist in the rat brain. Because it

THE RAT NERVOUS SYSTEM
ISBN 0 12 547632 9

is difficult to discern some of these minor cell groups in the rat by cytoarchitecture alone, comparative studies of various VNC fiber connections have been utilized to establish homologies with connectional patterns described in the cat and other species. Owing to the paucity of published material bearing on most aspects of the rat's VNC, we have employed data derived from a number of (largely unpublished) rat Nauta method fiber degeneration series of eighth nerve (Scarpa's ganglion), cerebellar and spinal lesion experiments and more recent horseradish peroxidase (HRP) injection experiments.

The vestibular nuclear complex (VNC) has been aptly described as a mosaic of minor functional units (Angaut and Brodal, 1967). The present synopsis examines the organization of this mosaic in the rat and it is designed to lead the interested reader to a number of key references and reviews for further study, especially those of Professor Alf Brodal to whom this treatise is dedicated.

II. TOPOGRAPHY AND CYTOARCHITECTURE OF THE VESTIBULAR NUCLEI

A. The interstitial nucleus of the vestibular nerve

Cells intercalated in the vestibular nerve root have been described in all mammals studied, including the rat. In most species the cells of this interstitial nucleus of Cajal appear in patches in the eighth nerve as it penetrates the inferior cerebellar peduncle just dorsal to the descending fibers of the spinotrigeminal tract (In8, Fig. 1b). The somata of these cells are middle sized and they may appear to be round in certain preparations and fusiform in others. Golgi sections in the cat have shown multipolar cells with dendrites curled up within the territory of an intercalated cluster and other In8 cells with straight dendrites which parallel the eighth nerve fibers (Hauglie-Hanssen, 1968).

Rostral and caudal In8 groups, related to different parts of the vestibular (Scarpa's) ganglion have been described in the cat (Gacek, 1969). Primary fiber connections have been long recognized, but until the advent of the retrograde cell labeling method little was known about In8's efferent connections. Retrograde cell changes in some cells were suggested following either spinal or medial longitudinal fasciculus lesions (Pompeiano and Brodal, 1957a, b), but neither spinal cord nor ocular motor nuclei HRP injections appear to label In8. However, both cerebellar and vestibular nucleus complex (VNC) injections label In8 cells. The injection of HRP into the VNC produces bilateral In8 labeling, suggesting possible commissural connections or bilateral VNC input. The function of In8 is unclear.

B. The superior vestibular nucleus

The superior vestibular nucleus (Bechterew, SuVe) lies beneath the crescent shaped emerging brachium conjunctivum and other cerebellar efferent (that is,

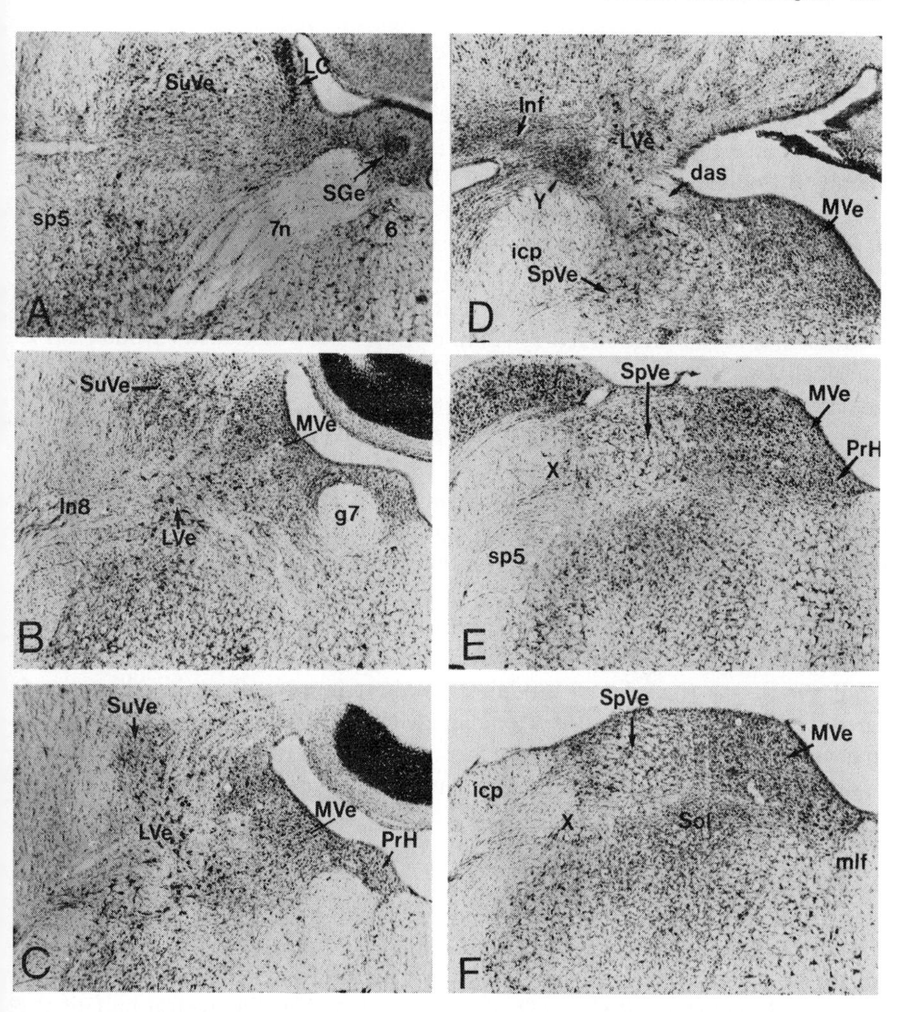

Fig. 1: The vestibular nuclei in the rat. Cresylecht violet stain (X 30). 6, abducens nucleus; 7, facial nucleus; das, dorsal acoustic stria; f, cell group F; g7, genu of the facial nerve; icp, inferior cerebellar peduncle; In8, interstitial nucleus of the eigth nerve; Inf, infracerebellar nucleus; LVe, lateral vestibular nucleus; MVe, medial vestibular nucleus; mlf, medial longitudinal fasciculus; Nod, nodulus; PrH, prepositus hypoglossal nucleus; SGe, suprageniculate nucleus of the pons; Sol, solitary nucleus; sp5, spinal tract of the trigeminal; SpVe, spinal vestibular nucleus; SuVe, superior vestibular nucleus; Uv, uvula; X, nucleus X; Y, nucleus Y.

the uncinate or "hook" bundle) or afferent (ventral spinocerebellar tract) fibers which together constitute the superior cerebellar peduncle (scp). The oral pole abuts the parabrachial nuclei and its dorsomedial edge is slightly invaded by the caudalmost cells of the locus coeruleus (LC, Fig. 1a). The lateral border of SuVe is formed by the middle cerebellar peduncle (mcp) and, except for its extreme oral pole, the medial border of the SuVe is occupied by the medial vestibular nucleus (MVe). The ventral border is also formed chiefly by fiber systems traversing the supratrigeminal region. The ventral part of the caudal pole of SuVe becomes replaced by the ventral part of the anterior portion of the large celled lateral vestibular nucleus (LVe) (*Atlas**; Fig. 1B).

Cytoarchitecturally, Nissl preparations reveal that the SuVe is composed of darker staining medium sized cells and paler small cells. Bundles of myelinated fibers lace through the nucleus in the dorsomedial direction causing some cells to be grouped in strands. In the cat a central core arrangement of predominantly medium sized cells surrounded by small cells has been described (Brodal, 1974; Gacek, 1969). Henkel and Martin (1977a) described a somewhat different organization of SuVe in the American opossum; they noted a lateral reticulated part and a medial compact part which we would consider to be the oral pole of the medial vestibular nucleus (see below). They also tentatively (but prophetically) labeled a cell cluster in the dorsolateral part of SuVe as group Y since it labeled after oculomotor nuclei HRP injection (see below). Such a central core is rarely evident in the rat's SuVe. The fact that SuVe cell axons ascend chiefly in the ipsilateral medial longitudinal fasciculus (mlf) and end on motor cells of the fourth and third nerve nuclei has been known for many years. Commissural connections between the SuVe nuclei, first demonstrated in Nauta method studies in the cat (Ladpli and Brodal, 1968), have been confirmed in the cat and rat with the HRP method (Rubertone and Mehler, 1980) and, as will be detailed in other sections of this Chapter, the mosaic of the internuclear and intranuclear organization of SuVe and other vestibular nuclear complex nuclei is yielding to studies with new methods.

C. The lateral vestibular nucleus

The lateral vestibular nucleus of Deiters (LVe) has been defined classically as that part of the vestibular complex which contains a large number of multipolar, giant sized cells (Fig. 1B, C and D; *Atlas* Figs 35 to 37). The rostral half of the LVe is capped dorsally by the SuVe nucleus and bounded medially by the more cellular MVe. The oral pole of the nucleus of the solitary tract (sol) is subadjacent to the mostly fibrous part of the ventral border of the LVe; fibers of the inferior cerebellar peduncle (icp) form the lateral border of LVe except at that point where the vestibular group Y and the infracerebellar nucleus are juxtaposed (Figs 1D and 2; *Atlas* Fig. 37). A precise border between the LVe and SpVe in many species is difficult to discern (Walberg, 1975) and the rat is no exception. This difficulty is due to the presence of some large neurons in the rostral SpVe that are similar in appearance to those found in LVe which is undercut by SpVe's oral pole. Another controversy exists concerning the border

*References to *Atlas* figures are references to Paxinos and Watson (1982).

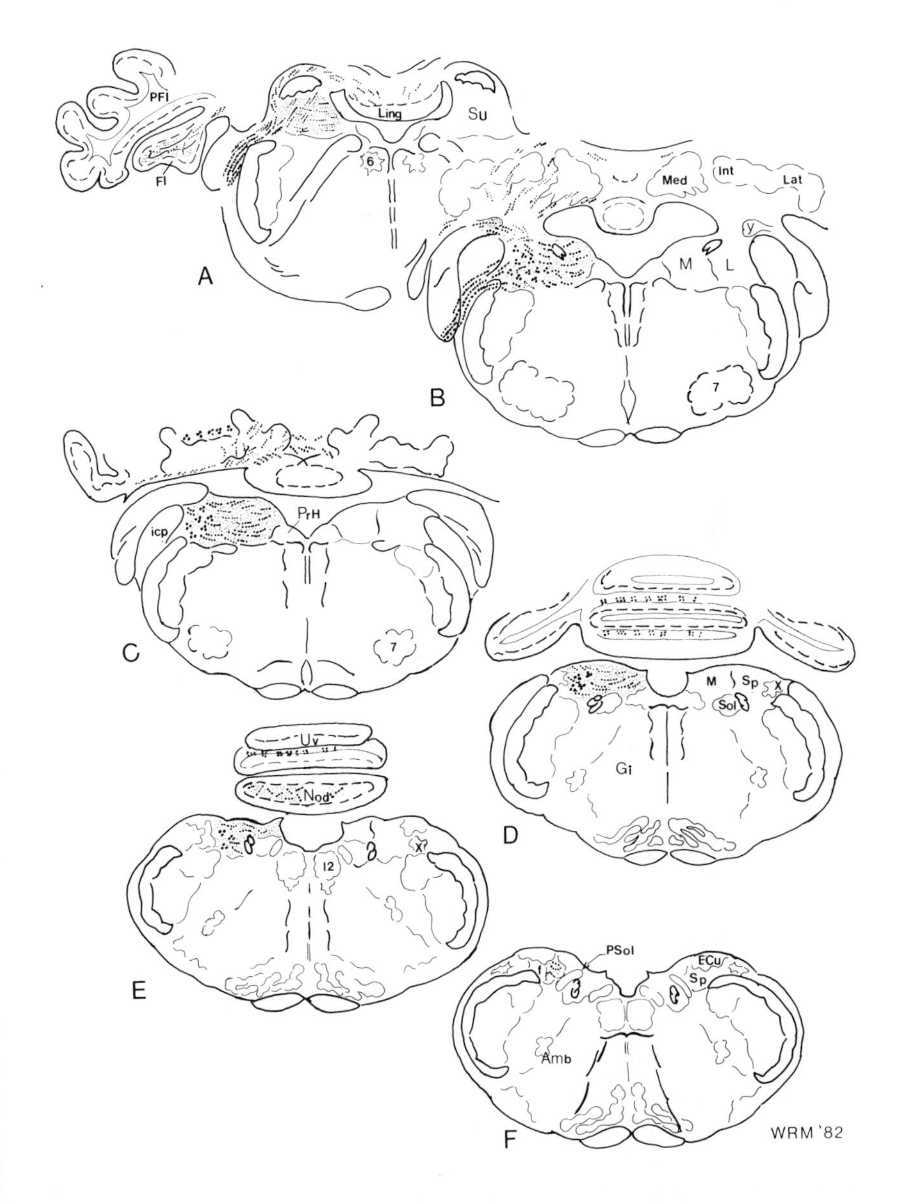

Fig. 2: Charting of the course of degenerating primary vestibular nerve fibers (dots) and the distribution of their terminations (stipple) in the vestibular nuclei and cerebellum of the rat after complete destruction of (Scarpa's) vestibular ganglion. 6, abducens nucleus; 7, facial nucleus; 12, hypoglossal nucleus; Amb, ambiguus nucleus; ECu, external cuneate nucleus; Gi, gigantocellular reticular nucleus; icp, inferior cerebellar peduncle; Int, Lat and Med, interposed, lateral and medial cerebellar nuclei; PSol, parasolitary nucleus; Sp, spinal vestibular nucleus; Su, superior vestibular nucleus; X, nucleus X; Y, nucleus Y.

between the ventrolateral part of MVe and the ventral part of LVe which will be discussed in other contexts.

The large multipolar neurons which characterize LVe show a considerable size variation among different species. They are 30 μm–45 μm in kittens (Brodal and Pompeiano, 1957a), 60 μm–70 μm in adult cats (Mugnaini *et al.*, 1967) and up to 60 μm in diameter in rats (Johnson *et al.*, 1976). There are at least as many small and medium sized cells in the LVe as there are giant cells in most species studied and, although actual quantitative and statistical studies still need to be carried out, the rat does not appear to be an exception to this rule.

Ultrastructural studies of the rat LVe provide information concerning cytoarchitectural and synaptologic features of the nucleus. Unlike large neurons found elsewhere in the central nervous system, those in the rat LVe contain small Nissl bodies interconnected in a dense meshwork. These giant cells were found to be in contact with astroglial processes while oligodendroglial cells were seen only rarely as satellites of the giant or "Deiters" cells (Sotelo and Palay, 1968). Synaptologic data pertaining to the shape, size and location of boutons within the nucleus are disucssed by Sotelo and Palay (1970) and Schwarz *et al.* (1977). Johnson (1975a, b) provides other ultrastructural information regarding the effects of deafferentation, ageing (Johnson and Miquel, 1974) and centrifugation (Johnson *et al.*, 1976) on cells in the rat LVe.

Primary vestibular nerve afferent connections in the rat (Fig. 2) distribute chiefly to the rostroventral part of LVe in a manner similar to that established in the cat (Gacek, 1969; Korte, 1979; Walberg *et al.*, 1958). Ascending spinal projections to LVe in the rat appear to be more numerous and somewhat wider in their distribution than those described in the cat (Mehler, 1969; Pompeiano and Brodal, 1957b). Direct long corticofugal cerebellovestibular projections to LVe originate chiefly from the ipsilateral anterior lobe of the vermis and distribute mainly to the dorsocaudal "hindlimb" region of LVe; the rostroventral "forelimb" region of LVe receives mostly indirect contralateral medial cerebellar nucleus projections. These fibers end mostly on the proximal dendrites of the large LVe cells (Brodal, 1974). There are few, if any, commissural connections between the LVe of the two sides of the brainstem.

All sizes of LVe neurons participate in the formation of the descending, ipsilateral, vestibulospinal projection in a somatotopic fashion, with "hindlimb" spinal levels dorsocaudal and "forelimb" ventrorostral in the rat as well as the cat. Vestibulospinal connections have not been studied in the rat, but in the cat they have been shown to terminate on ventral column lamina 7 and 8 cells (Brodal, 1974; Holstege and Kuypers, 1982). Some large LVe cells participate in the ascending vestibular projection to the ocular motor nuclei in the cat and monkey but evidence of such LVe projections in the rat is still lacking (see below).

D. The medial vestibular nucleus

The oral pole of the medial vestibular nucleus (Schwalbe, MVe; *Atlas* Figs 37 to 40) slightly overlaps the caudal parts of the locus coeruleus, abuts upon the ventral parabrachial nucleus and overlies the supratrigeminal nucleus. The dorsomedial border of MVe lies beneath the ependyma of the fourth ventricle throughout its rostral to caudal extent. Its ventral border overlies the paramedian components of the medullary reticular formation and it is slightly indented by the facial genu in its middle part.

Cytoarchitecturally the MVe can be roughly divided into three unequal parts: (i) a small oral polar region; (ii) a genual level middle part; and (iii) a longer caudal portion coextensive in length with the spinal vestibular nucleus (SpVe). Small cells predominate in the polar region. In the middle part, there is an increase in the number of medium sized cells and a few relatively large cells appear in the ventrolateral part of MVe coextensive with the so-called ventral, lateral vestibular nucleus. The caudal part of MVe is characterized by a dense homogeneous dispersion of small and medium sized cells.

The heterogeneously celled ventrolateral region of the middle part of MVe (Fig. 1B and C) is worth attention. Most authors now consider this region as part of the MVe nucleus (Gacek, 1971, 1977). In their atlas of the brain stem of the rabbit, Meessen and Olszewski (1949) labeled the caudal part of the region as the β part of LVe and they included most of the cells in question at rostral levels to lie within the borders of their α LVe subdivision. Following these authors, Mehler (1969) traced a unique, medially coursing cervical spinovestibular projection to this β region. However, recent HRP studies of ocular motor cell afferents in the rat (Mehler and Kohn, in preparation) indicate that cells in this region correspond best to the ventrolateral MVe cells contributing to the vestibulo-ocular pathway, as described by Gacek (1977, 1979), and others, in the cat. This will be discussed again.

Neurons in the MVe may be influenced by spinal, labyrinthine, cerebellar and upper accessory oculomotor nuclei. This information is then relayed from the MVe to the oculomotor complex via the medial longitudinal fasciculus (mlf) and to cervical spinal cord motor neurons via the medial vestibulospinal tract. Early recognition of this basic connectivity suggested that the MVe functions primarily to integrate movements of the eyes and head.

E. The spinal vestibular nucleus

The spinal vestibular nucleus (SpVe) is characterized in transverse sections by the presence of many longitudinally oriented myelinated fiber bundles originating chiefly from the descending root of the vestibular nerve and from the contralateral medial cerebellar nucleus. The SpVe extends to the dorsal surface of the medulla for the greater part of its rostrocaudal extent except at its oral pole where it lies ventral to part of the dorsal cochlear nucleus and the caudal portion of the LVe. At its caudal extreme in most mammalian brains the SpVe is capped dorsally by subgroup Z and ventrolaterally by cell group F. The caudal pole blends almost imperceptibly with the oral part of the cuneate nucleus. Laterally, it is bounded from rostral to caudal by the restiform body, clusters of group X and the oral extensions of the external cuneate nucleus.

The medial boundary of the SpVe is established throughout by the MVe. The exact border between these two nuclei is sometimes difficult to ascertain in Nissl preparations. This is probably due to the fact that most cells of the SpVe are similar in size and shape to those of the MVe. The presence of large fiber bundles in SpVe and their absence in MVe are less equivocal landmarks. Larger cells are found concentrated in the rostral half of the SpVe in the cat (Hauglie-Hanssen, 1968) and in *Galago senegalensis* (Rubertone and Haines, 1982), and this also appears to be the case in the rat.

Experimental studies suggested that the SpVe is most specifically related to the cerebellum (Brodal, 1974; Dow, 1936). Horseradish peroxidase studies in the rat and other animals have verified that the SpVe projects profusely to the cerebellum (Mehler, 1977), but, as we will see in later discussions of VNC connections, the SpVe also sends axons to the spinal cord and contributes slightly to ascending VNC fibers systems projecting to the ocular motor nuclei and the thalamus. The SpVe receives a moderate input from the contralateral medial cerebellar nucleus (Rubertone and Haroian, 1982) and is reciprocally related to cerebellar lobules 9 and 10 via corticovestibular afferents and vestibulocerebellar efferents projections in the rat (Rubertone and Mehler, 1981). Carpenter (1960a) first described commissural connections between the SpVe nuclei in the cat in Nauta method studies; these connections have been confirmed by studies using HRP.

F. Associated small cell groups: X, Y, Z and F

Cell group X forms a continuous chain of uniformly small stellate cells which form irregular aggregates lying primarily ventrolateral to the spinal vestibular nucleus (SpVe), partially intercalated in the medial edge of the inferior cerebellar peduncle. It is coextensive in length with the SpVe nucleus, extending from the level of entry of the eighth nerve to the level of the obex where it partially overlaps with rostral clusters of the external cuneate nucleus. Cells of the external cuneate nucleus, which are generally larger than X cells, receive heavy cervical dorsal root fibers while group X cells receive finer, secondary ascending lateral funicular fiber connections originating from levels of the spinal cord as far caudal as the third lumbar segment in the cat (Pompeiano and Brodal,

1957b). The latter observation, still to be verified in the rat, raised a question: are these ascending spinal connections with X collaterals of the dorsal spinocerebellar tract, or do they originate independently from spinal cells outside Clarke's nucleus (Snyder *et al.*, 1978)?

Lesions of the medial cerebellar nucleus also produce dense fiber terminations on group X cells. These connections originate chiefly on the contralateral side in the rat and in other species studied (Mehler, 1966).

Efferent projections of group X cells to the cerebellum were first suggested by Dow's (1936) Marchi method investigations of cerebellar afferent connections in the rat and cat. Brodal and Torvik (1957) concluded that their Gudden's method studies (that is, lesion induced retrograde cell changes in kittens) supported Dow's contention that group X cells and cells dispersed in the caudal and lateral parts of SpVe nucleus projected preferentially to vestibulocerebellar (VC) regions. Non-VC vermian HRP injections, however, indicate that X cells in the rat, as well as in the cat and monkey, also project throughout the vermis (Mehler, 1977). Gould (1980), using HRP, found evidence of X cell projections in the paramedian lobules in the cat.

Reports that some group X cell axons ascend in the mlf (Brodal and Pompeiano, 1957a, b) have not been substantiated to our knowledge by retrograde tracer studies. No labeled X cells appear after ocular motor nuclei HRP injections in either rat or monkey, nor after large thalamic injections involving the ventral tier nuclei in monkeys. In the rat, large thalamic HRP injections consistently label a group of cells in the lateral part of SpVe that might be mistaken for X cells (Kevetter *et al.*, 1982).

Another cell group that can be mistaken for an X group is the paratrigeminal nucleus (Chan-Palay, 1978). This variably shaped cluster is intercalated between the spinal trigeminal tract and the inferior cerebellar peduncle just caudal to the dorsal cochlear nucleus. After HRP injections into the cerebellum, these cells appear to label like X cells, but they do not receive spinal or cerebellofugal fiber connections as X cells do and they have been selectively implicated in 2-deoxyglucose studies of hibernation in the ground squirrel (Kilduff *et al.*, 1983).

1. Cell group Y and the infracerebellar nucleus

Group Y (Brodal and Pompeiano, 1957a, b) is a small, but easily discerned, triangular nuclear entity that caps the dorsomedial edge of the inferior cerebellar peduncle just caudal to the entry of the vestibular nerve (Figs 1D and 2B). Fig. 1D shows the close relationship of Y to the ventral parvocellular part of the lateral cerebellar nucleus and the even closer proximity of Y to the infracerebellar nucleus (Inf; arrow Fig. 1D; *Atlas* Fig. 37). Gacek (1977, 1979) utilized the neutral term infracerebellar nucleus to attempt to differentially designate a population of cells lying immediately rostral and dorsal to the vestibular group Y that consistently labeled after injections of HRP into the ocular motor complex (OMC), and which do *not* receive primary vestibular fibers. Similar specificity of labeling appears in the Inf of the rat after HRP injections restricted to the OMC. Group Y projects to the flocculus in the rat (Blanks *et al.*, 1983). In both cat and rat, Y receives primary vestibular connections, but data on primary fiber terminations in Inf are equivocal.

After Nissl stain or HRP injections, two types of small cells can be observed in group Y: round cells predominate in the medial main triangular body part, and spindle-shaped cells predominate in the "apical" part that extends laterally over the inferior cerebellar peduncle (icp) at caudal LVe levels. Both types of cells can be found in the opposite order in the roughly reverse triangular area formed by the caudal part of Inf: round cells are lateral, and spindle cells are medial. They appear, in turn, after Y, but they are subtly disassociated from the icp by fibers of the floccular peduncle. They appear to be in tandem, as are the two parts of the interstitial nucleus of the eighth nerve. Inf is, in essence, the rostral bed nucleus of the floccular peduncle and Y the caudal bed nucleus.

Measurements of Y and Inf in horizontal plane sections in the rat reveal that the caudal to rostral limits of each cell group is less than 200 μm, and that Inf is only half the size of Y in coronal sections where Y measures less than 250 μm dorsoventrally. The proportionate size of group Y compared to other vestibular nuclear complex components remains constant in the rat, cat and monkey. The "infracerebellar" nuclei, however, increase in size dramatically from rat to cat to monkey. This increase is reflected chiefly in the length of Inf in the cat and monkey. The increase in size of Inf may reflect the increase in size of the flocculus from a single folium in the rat to three folia in the cat (Sato *et al.*, 1982), to nine folia in the monkey (Carpenter *et al.*, 1972). The rostral pole of Inf blends imperceptively with the dorsolateral part of the superior vestibular nucleus (SuVe) in the three species.

A monograph could be written about the controversies surrounding the specific cells from which ocular motor complex (OMC) afferents originate that ascend with cerebellofugal fibers in the brachium conjunctivum, rather than via the mlf. However, only a vignette can be presented. Haines (1977a) examines the literature supporting a proposed ventral, parvocellular (PC) lateral cerebellar nucleus (Lat) origin for these OMC afferents based, in the main, on the fact that floccular efferents which relay visual impulses converged on PCLat. Carpenter and Strominger (1964) and Chan-Palay (1977), using Nauta and autoradiographic techniques respectively, suggested that the OMC afferents in question originated in the ventral region of the Lat nucleus (not group Y), but the exact cells of origin are still in contention. We found that only a small, ventral Lat lesion in the rat that also included Inf, but not group Y cells, induced ascending fiber projections to the contralateral OMC (experiments RCA 50–55, Mehler and Faull, unpublished). Graybiel and Hartwieg (1974), however, discovered that HRP injections into the OMC in the cat labeled cells ventral to Lat which at the time were best identified as group Y. In both experiments they describe they show HRP labeled cells divorced from the icp in what we would now term Inf, according to Gacek (1977), to avoid further confusion with vestibular group Y. In their original description of Y, Brodal and Pompeiano (1957a) noted that: "From the dorsal aspect of this group (that is, Y), scattered cells form strands extending to the ventralmost part of the dentate nucleus". This is a good description of Inf in Nissl preparations.

In terms of the present terminology, Brodal and Hoivik (1964) ascribed primary vestibulocerebellar (VC) fiber connections with the following: Y, Inf, ventral Lat, the flocculus (Fl) and ventral paraflocculus (PFl). Gacek (1969) did not describe VC fibers, and neither Korte and Magnaini (1979) nor the present authors found any Inf, ventral Lat or PFl connections. Could cortical damage explain the differences? There is a consensus that ventral PFl projects to ventral Lat (Dietrichs, 1981). Let us also consider the critical but meager data on Fl projections. Angaut and Brodal's (1967) cat 12, with a lesion involving only the distal parts of PFl and Fl, exhibits ventral Lat plus Inf (labeled "Y" in their Fig. 1:55), but no Y *per se* connections. In our rat experiment RCA-30, an almost identical distal PFl–Fl lesion produces the same intracerebellar projection pattern: connections with the ventral part of Lat and Inf but no Y *per se* terminations. Haines' (1977b) Fl lesion suggests Inf connections (but is labeled Y) and projections from his nodular-uvular lesion show both Inf and Y terminations. Our experiment RCA-34 with a nodular-uvular lesion confirms both connections, especially projections to Y. Gould's (1979) cat CHRCB/20 with a rostral floccular HRP injection demonstrates Inf cell labeling, in the main. Her CHRCB/18 experiment with a uvular-nodular injection demonstrates chiefly Y cell labeling.

2. *Cell group Z*

Brodal and Pompeiano (1957a, b) designated a small, somewhat heterogeneous cell group generally intercalated between the rostral pole of the gracile nucleus and the caudal pole of the MVe and dorsal to the caudal end of SpVe as group Z. Since this cell group did not receive primary vestibular nerve connections and apparently did not undergo retrograde cell degeneration after cerebellar ablations as adjacent SpVe cells did, Brodal and Pompeiano (1957a, b) concluded it probably did not belong to the vestibular nuclear complex. Nevertheless, subsequent studies of the vestibular nuclear complex and reviews usually include Z as an accessory vestibular cell group along with X and Y because it receives spinovestibular-like projections ascending in the dorsolateral fasciculus.

Landgren and Silfvenius (1971) demonstrated that Z cells relayed hindlimb Ia muscle afferent impulses to the cerebral cortex in the cat. The thalamic relay cells in the cat lie in the border zone between the ventral posterior lateral and ventral lateral thalamic nuclei (Grant *et al.*, 1973; Berkley, 1983).

Rubertone and Haines (1982) point out that group Z is more readily delineated in the monkey brain that it is in the cat. Mehler (1969) did not differentiate a group Z in the rat, but Zemlan *et al.* (1978), in a more detailed analysis of ascending anterolateral fiber projections to the dorsal funicular and caudal vestibular regions in the rat, labeled a dorsally situated zone lying between the oral poles of the dorsal funicular nuclei and the caudal pole of SpVe as nuclei Z. Although large injections of HRP into the ventral tier nuclei in rats or monkeys label cells in both vestibular and dorsal funicular nuclei, a Z like cell group currently cannot be identified with certainty. In order to identify Z with certainty additional studies along these lines are needed, particularly studies employing HRP injections into the VPL-VL border zone in the rat.

3. *Cell group F*

Brodal and Pompeiano (1957a) followed Meessen and Olszewski's (1949) parcelation of an "F" subdivision of cells lying ventral or ventrolateral to the caudal pole of the spinal vestibular (SpVe) nucleus, medial to external cuneate and X nuclei, and dorsolateral to the solitary nucleus. Cell group F in the rabbit, cat and opossum (Henkel and Martin, 1977a) is characterized and differentiated from surrounding nuclei by being a rather densely packed, variably shaped, cell group of medium to large sized cells in Nissl preparations. An F like group is difficult to localize in the vestibular nuclear complex of the rat brain by cytoarchitectural means alone. However, based upon connectivity features of group F described in the cat, there is reasonable evidence supporting its existence in the rat. In the rat, as has been reported in the opossum (Henkel and Martin, 1977b) and cat (Walberg *et al.*, 1958), F does not receive primary vestibular fibers, but it does receive fastigiobulbar connections via the uncinate tract (hook bundle) and long corticofugal projections from the flocculus and nodulus (Brodal, 1974). Efferent connections with the cerebellum, suggested by ablation studies, have been confirmed by investigations using HRP in both species (Mehler, 1977).

We have found some F like cell labeling after spinal HRP injections in the rat and these are similar to the vestibulo-spinal connections described by Petersen and Coulter (1977) in the cat. These authors found that the group F contained almost one third of the labeled cells found in the vestibular nuclear complex after the injection of HRP into the lumbar spinal cord. They suggest that group F might be the origin of a third vestibulospinal pathway that may be functionally distinct from the classical lateral and medial vestibulospinal pathways that originate, respectively, primarily in the lateral and medial vestibular nuclei.

G. The prepositus hypoglossal nucleus

The prepositus hypoglossal nucleus (PrH; *Atlas* Figs 36 to 39) is not usually considered with the accessory vestibular nuclei. However, based upon its intimate relationship with the medial vestibular nucleus (MVe) and its now established connections with the vestibular nuclear complex, we have included a brief sketch of PrH. The paired nuclei form bilateral, narrow, triangular shaped cell columns in the floor of the fourth ventrical dorsal to the medial longitudinal fasciculus extending from the genu of the facial nerve to the oral pole of the hypoglossal nuclei. Their apices abut the median raphe and their bases coextend with the medial part of the caudal two thirds of MVe. Small cells predominate in the oral part of the nucleus and medium sized cells appear in its caudal part.

Initially PrH was classified as a perihypoglossal nucleus together with Roller's nucleus (*Atlas* Fig. 41) and the intercalated nucleus (*Atlas* Fig. 40) on the assumption that the function of the three nuclei affected motor control of the tongue (Brodal, 1952). New data suggest that PrH is an important noncerebellar relay for visual input to the vestibular nuclei (Cazin *et al.*, 1982).

Afferent connections with PrH originating from the fastigial nucleus (Walberg *et al.*, 1961) and the flocculonodular lobe (Angaut and Brodal, 1967) were first suggested by fiber degeneration studies in the cat. Cerebellar lesion experiments, particularly with a rostral, medial (that is, fastigial) cerebellar nuclear lesion (RCA 26 experiments), confirm this connection in the rat. Sato *et al.* (1982) found a few labeled Purkinje cells in the caudal half of the flocculus following small HRP injections into PrH in agreement with Angaut and Brodal (1967). Another HRP study of PrH in the cat, however, negates cerebellar afferent connections, describing primarily paramedian pontine reticular, superior collicular, nucleus of the optic tract (OT) and the olivary pretectal nucleus (OPT) cell connections (Magnin *et al.*, 1983). A previous autoradiographic study of pretectal projections in the rat, where the injection sites included OT and OPT (*Atlas* Fig. 24) which receive primary optic fiber connections, showed direct terminal projections to PrH (Cazin *et al.*, 1982).

Ascending spinal fiber projections to PrH can be demonstrated by complete hemisections at C1 in the rat (experiments by Mehler). These fibers terminate mainly on cells in the caudal, larger celled part of PrH and some endings extend into the smaller celled rostral part. Hazlett *et al.* (1972) depict such connections in the opossum after C2 hemisections as part of the medial spinovestibular system. These fibers probably originate from the central cervical nucleus (CeC) along with established projections to the MVe (Mehler, 1969; Rubertone and Mehler, 1980). The failure of Pompeiano and Brodal (1957b; that is, their Cat 42), Mehler (1969) and Zemlan *et al.* (1978) to recognize the spinal–PrH connection appears to be the result of incomplete cervical spinal lesions that failed to affect the central cervical nucleus or interrupt its medially ascending projections. Unfortunately neither of the studies of HRP injected into the PrH of the cat, cited above (Sato *et al.*, 1982; Magnin *et al.*, 1983), include observations on cell labeling in the spinal cord, and data from our available rat HRP experiments involving PrH are equivocal because of Mve involvement.

Efferent projections to the cerebellum were first suggested by retrograde cell degeneration studies in the cat (Brodal, 1954) and confirmed by HRP retrograde cell labeling in the rat, cat and monkey (Faull, 1977). PrH projections to the anterior lobe of the vermis, lobules 5–7 and the flocculonodulus (10) have been described (Gould, 1980; Rubertone and Haines, 1981). Studies using injections of HRP have also indicated PrH connections with the three ocular muscle nuclei in the cat (Graybiel and Hartwieg, 1974; Gacek, 1977). To date, we have confirmed only the oculomotor nucleus connections in the rat using HRP. Extensive bilateral PrH connections with the vestibular nuclear complex have been demonstrated by means of HRP injection, in both the cat (Pompeiano *et al.*, 1978) and rat (Rubertone and Mehler, 1980). Blanks *et al.* (1983) have shown extensive PrH and pontine suprageniculate nuclei projections to the rat's flocculus. Brodal and Brodal (1983) review these connections.

Prepositus hypoglossal nucleus efferent connections with the oculomotor nuclei and widespread vestibular and cerebellar connections, coupled with the fact that PrH receives direct secondary retinal input via pretectal nuclei, established PrH as both a direct pathway to the oculomotor nuclei similar to a vestibular nucleus, and also an indirect relay to these motor nuclei via PrH connections with the vestibular nuclei and the cerebellum.

III. AFFERENT CONNECTIONS OF THE VESTIBULAR NUCLEI AND ASSOCIATED CELL GROUPS

A. Primary vestibular nerve fiber connections with the brain stem and cerebellum

Analyses of complete vestibular (Scarpa) ganglion lesions in rats have revealed massive fiber degeneration which envelops the interstitial nucleus of the eighth nerve and enters the ventral part of the lateral vestibular nucleus (LVe) where the root fibers bifurcate into descending and ascending limbs made up of many secondary fascicles (Fig. 2). The descending limb traverses the length and breadth of the spinal vestibular nucleus (SpVe) nucleus and issues large numbers of small degenerating collateral bundles medially which terminate throughout medial vestibular nucleus (MVe) right up to the ependymal lining of the fourth ventricle (Fig. 3E). Fascicles of the ascending limb traverse the superior vestibular nucleus (SuVe), which exhibits terminals throughout, and these ascending fibers also issue other, longer collaterals that terminate in the oral third of MVe. Multiple small fascicles splay out in the dorsal part of SuVe, penetrate or swing around the egressing normal brachium conjunctivum and then, in the main, deflect caudally either through or ventral to the interposed and medial cerebellar

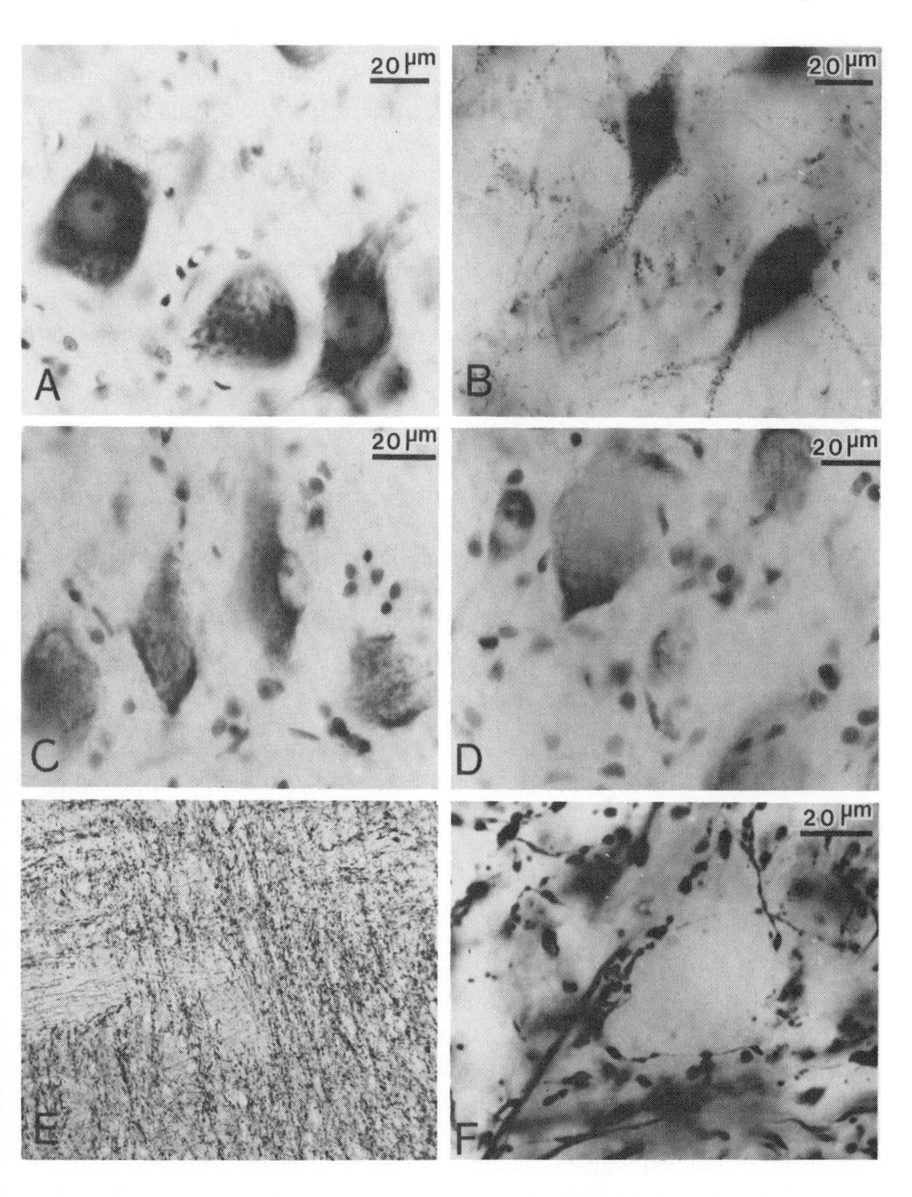

Fig. 3: A: Rat lateral vestibular (LVe) cells. Normal cresylecht violet stain. B: Horseradish peroxidase labeled LVe cells in a rat 44 hours after spinal cord injection (TMB chromagen). C–D: Retrograde cell changes in rat LVe cells five days after cervical spinal cord section (C, Excentric nucleus and peripheral clumping of chromidial substance; D, full blown chromotolysis or tigrolysis). E: Primary vestibular nerve fiber degeneration descending in bundles in the spinal vestibular (SpVe) nucleus (dorsal half of figure) and issuing degenerating collaterals into the medial vestibular nucleus (bottom half of figure). Horizontal series RV-19. Fink and Heimer method (× 180). F: Axosomatic terminal fiber degeneration in ventral part of LVe. RV-11. Nauta method (× 360).

nuclei. Terminal connections with cells of the deep cerebellar nuclei appear to be limited to the caudal, smaller celled part of the medial nucleus. A contingent of these vestibulocerebellar fibers could be traced into the lingula (lobule 1) in three out of four of our "complete" ganglionectomies with only a minor projection going to the flocculus in all cases (Fig. 2A). Caudally, the majority of the vestibulocerebellar fibers arch diffusely, then assemble into intermittent concentrations or bands restricted to the white matter of the nodulus and the ventral half of the uvula from which they enter the granular layer and terminate as mossy fiber endings. These terminations are found chiefly on the side ipsilateral to the ganglionectomy, but some fibers also end in the medial parts of the contralateral halves of these lobules and also in the lingula.

1. Connections with the vestibular nuclei

Ramón y Cajal (1911), using mouse Golgi preparations showed that primary vestibular nerve fibers arriving at the brainstem divided into two branches, one ascending to the cerebellum that gave off collaterals to SuVe and one descending in SpVe and issuing collaterals to SpVe, LVe and MVe. Subsequent normal fiber and Marchi method investigations in nonmammalian (Mehler, 1972) and mammalian species (including cats, dogs, rabbits and the American opossum) confirmed most aspects of Cajal's plan, but also generated controversies, especially in relation to the overall distribution of primary fibers in the VNC and projections to the cerebellum (see Dow, 1936; Walberg *et al.*, 1958). Lorente de Nó (1933) often cited mouse Golgi analysis, while revealing unique detailed information on vestibular nuclear complex connections, also failed to show a single confirmatory vestibular fiber entering the cerebellum in hundreds of brains (*op. cit.*, p. 28).

Twenty five years later Walberg *et al.* (1958), who employed the silver methods of Nauta and Glees, dismissed most of the Marchi data contending that it, as well as Cajal's description, had led to a textbook misconception that primary fibers distributed to the entire territories of the four classical vestibular nuclei.

Walberg *et al.* (1958) concluded that subsequent to complete eighth nerve destruction in cats, terminal degeneration in LVe was confined to the rostroventral region (that is, neck and forelimb projecting region), largely in the central part of SuVe, chiefly in the lateral part of MVe and the SpVe but sparsely in its rostroventral part. Subsequent Nauta and Fink and Heimer method studies in the monkey (Stein and Carpenter, 1967), cat (Gacek, 1969; Korte, 1979) and the present findings in the rat all support the absence of primary fiber connections with LVe cells lying dorsocaudally (that is, in the hindlimb portion). Although Gacek (1969) supported the notion of largely central SuVe terminations, the work of Korte (1979) and the material presented here show widely distributed endings throughout SVe and also complete coverage of MVe at all levels right up to the overlying subependymal granular layer. In the rat overall terminals in SpVe are suggested as Korte (1979) found in the cat. However, due to the large amount of degenerating axonal debris in the descending root coursing through SpVe, it is difficult to assess "terminal" densities in SpVe. The probable reasons for the widespread VNC negativity described by Walberg *et al.* (1958), such as longer postoperative survival times and increased sensitivity of later Nauta method variants, are discussed by Korte (1979), who used cats surviving for three to six days, as we did with our rats.

Although Walberg *et al.* (1958) noted the absence of primary eighth nerve connections with accessory VNC nuclei such as X, Z and F (supported by subsequent authors) they completely overlooked group Y's connections, later shown by Brodal and Hoivik (1964), Gacek (1969) and Korte (1979), and in the cases discussed here. Gacek (1969) discovered that Y afferents originate uniquely from the saccular nerve whose cells lie in the caudal part of Scarpa's ganglion which might easily be spared in subtotal lesions. The differential analyses of Stein and Carpenter (1967) and Gacek (1969) of the topologic distribution in the VNC of axons innervating the various vestibular endorgans (that is, the cristae of the three semicircular canals of the maculae of the utricle and sacculus) are good accounts of a topic beyond the scope of the present treatise. Kevetter and Perachio (1983) have initiated HRP transganglionic transport studies of the central distributions of the individual semicircular canal nerves in the gerbil that eventually should permit comparisons of a rodent's VNC topographic afferent patterns with those described in the primate and the carnivore.

2. The vestibulocerebellum

Experiments on rats have demonstrated that the portion of the cerebellum which receives primary vestibular nerve fibers, termed the vestibulocerebellum (VC), is composed of the flocculonodular lobule (10), the ventral half of the uvula (9) and the lingula (1). Dow (1936) was the first author to confirm (Marchi method) the nodular–ventral uvular connection in the rat. Dow's experiments showed that projections to the flocculus in the rat were sparse (as our experiments and most Nauta method studies in other species also have shown); however Dow reluctantly accepted floccular connections, probably because Olof Larsell, who was his doctoral mentor, had comparative and developmental evidence of such connections (Larsell, 1970). Brodal and Hoivik (1964) concluded that the ventral paraflocculus (ventral PFl) should be added to the VC, but neither the experiments of Korte and Mugnaini (1979) on cats nor any of ours on rats suggest primary fiber projections to ventral PFl. An HRP study of ventral afferents, like the analysis by Blanks *et al.* (1983) of floccular afferents, might resolve this old question if Scarpa's ganglion was examined.

Few authors besides Carpenter (1960a), Brodal and Hoivik (1964) and ourselves have included the lingula in the cortical VC. There also are reservations about subcortical primary vestibular fiber connections with the caudal medial cerebellar nucleus which has reciprocal connections with the vermian VC (that is, the nodulus–ventral uvula).

Voogd (1964), and Angaut and Brodal (1967) review the efferent connections of the VC in the cat. Long corticofugal VC vermian and hemispheric floccular (Purkinje) cell axons project differentially to the ipsilateral SuVe, MVe and SpVe nuclei while the caudal, medial cerebellar nucleus associated with the VC projects bilaterally with a contralateral bias chiefly on ventral LVe, SpVe, group F, the parasolitary nucleus, several "paramedial" reticular nuclei and even parts of the inferior olive.

Functionally, the vermian vestibulocerebellum has been implicated as the critical link between the labyrinth and an ill-defined vomiting trigger zone in the reticular formation by the fact that ablation of this part of the vestibuto-cerebellum suppresses motion induced vomiting (Mehler, 1983; Wang and Chinn, 1956). The flocculus, on the other hand, exerts inhibition on oculomotor projecting vestibular nuclear complex cells (Blanks *et al.*, 1983). Voogd (1964) states: "... Larsell's flocculonodular lode (that is, the archicerebellum) shares both afferent and efferent connections with the VNC but at the same time differs because the disposition of their efferent connections is characteristic for lobules belonging to the folial chains of the hemisphere and the caudal vermis respectively".

B. Non-primary afferent connections of the vestibular nuclei

1. Spinovestibular connections

Ascending spinal fiber projections to various regions of the vestibular nuclear complex have been demonstrated in the rat (Mehler, 1969), cat (Pompeiano and Brodal, 1957b) and several primate species (Rubertone and Haines, 1982). These projections were shown in Nauta method studies to ascend chiefly in the lateral quadrant of the spinal cord in association with the dorsal spinocerebellar tract (dsct) that enters the inferior cerebellar peduncle (icp). Degenerating collateral-like processes emerged dorsomedially from parent fibers ascending in the peduncle and terminated on group X cells and parts of SpVe and LVe in all of the species studied. In the rat and opossum, however, Mehler (1969) found that additional degenerating fibers could also be traced via the medial reticular formation to terminations in medial vestibular nuclear complex regions after cervical level hemisections suggesting dual spinovestibular pathways. Hazlett *et al.* (1972) also described this medial spinovestibular system as cervical in origin in the opposum.

Review of the original rat cordotomy series published by Mehler (1969) and two newer Nauta method C1 level hemisection experiments reveals that while some lateral spinovestibular projections to X, SpVe and LVe can be demonstrated in mid-thoracic level hemicordotomies, the spinal projections to MVe originate solely from deeper regions of the cervical spinal cord. More superficial C2 anterolateral funicular cordotomy by Zemlan *et al.* (1978), while interrupting lateral spinovestibular projections and medial spinoreticular connections (including the caudal pole of the prepositus hypoglossal nucleus), failed to show MVe connections. These observations, coupled with evidence to be presented, that the deeply situated central cervical nucleus (CeC; *Atlas* Plate 66) projects to MVe, confirm earlier contentions that there are two separate spinovestibular systems, a lateral and a medial.

The medial spinovestibular system in the rat terminates in MVe especially the ventrolateral region of MVe at caudal levels of the facial genu that contains large numbers of medium sized cells. Mehler (1969) originally, mistakenly, labeled this region as the β part of LVe according to Meessen and Olszewski's (1949) fallacious parceling of the LVe in the rabbit, as has been discussed. The medium sized cells in this ventrolateral part of MVe also label after HRP injections into the ocular motor nuclei in the rat distinct from the smaller celled, short column, central core cell labeling in MVe at corresponding coronal levels described later in the vestibulo-ocular projection discussion.

a. Origins of spinovestibular projections

Vincent and Rubertone 1984 have analysed the origins of spinovestibular projections in the rat using the HRP method. They utilized a horizontal stereotaxic approach to minimize cerebellar and reticular afferent fiber contamination in their differential vestibular nuclear complex injections. They identified six major spinal loci in the rat: (a) contralateral central cervical nucleus (CeC); (b) intermediate spinal gray cells, bilaterally; (c) cells associated with the medial motor cell column; (d) the dorsolateral funicular, predominently ipsilaterally; (e) the reticular portion of the dorsal horn cells; and (f) the dorsal nucleus of Clarke (D) with an ipsilateral bias. These same loci also have been identified as origins of spinocerebellar pathways in the cat and other species (Matsushita *et al.*, 1979; Snyder *et al.*, 1978), raising the possibility again that the spinovestibular connections are mostly collaterals of one or another of the spinocerebellar projections.

Vincent and Rubertone (1984) found evidence of some spinal projections to all four of the major vestibular nuclear complex nuclei. Individual spinal cord cell labeling patterns are difficult to describe because they vary in intensity, sidedness and spinal levels of appearance. In general, injection of HRP into the SuVe resulted in labeled spinal cord neurons contralaterally in the CeC, bilaterally in the intermediate gray and in an area adjacent to the medial motor cell column. Injections of LVe resulted in labeled cells in loci similar to those found subsequent to SpVe injection but a few positive somata also were found bilaterally in the dorsal nucleus (Clarke's). After injection into the MVe there was little spinal labeling. The few labeled spinal neurons observed in these experiments occupied a position in the contralateral rostral CeC. Injections essentially confined to the SpVe resulted in labeled cells in the dorsal horn, intermediate gray and an area adjacent to the medial motor cell column. Labeled neurons also were observed bilaterally at cervical and lumbar levels in the reticular portion of the dorsal horn and dorsolateral funiculus respectively after SpVe injections.

The suggestion that there are spinal connections with SuVe is a new idea. Evidence from experiments using HRP that there are spinal projections to the other three major vestibular nuclei in the rat, however, confirms and extends conclusions based upon earlier fiber degeneration studies. We are surprised, however, that both Vincent and Rubertone (1984) and Carleton and Carpenter (1983) found only limited numbers of contralateral CeC cells following injections into the MVe in the rat and cat, respectively. The central cervical nucleus (CeC), as we have discussed, is the prime candidate for the origin of the medial spinovestibular system. CeC also receives descending projections from the contralateral MVe according to Carleton and Carpenter (1983) The fact that central cervical nucleus cells are not labeled following injections into the SpVe in both the rat and cat also suggests that separate medial and lateral spinovestibular systems probably exist. Since LVe in the rat appears to receive ascending different fibers via both medial and lateral systems it is not surprising that HRP injections into LVe include some CeC positive cells. Since Carleton and Carpenter (1983) identified only spinovestibular connections originating from the central cervical nucleus, further studies of these projections in the cat are needed to determine whether technical caprices failed to demonstrate the multiple origins of spinovestibular connections shown in the rat or whether there are actual interspecies differences. Vincent and Rubertone (1984) propose double labeling experiments in the rat to resolve which spinal cell groups project to both the VNC and the cerebellum.

2. *Cerebellovestibular projections*

Allen's (1924) Marchi method studies of the rodent brain suggested that there were direct projections to the vestibular nuclei from the cerebellar cortex. Goodman *et al.* (1963) confirmed these long corticofugal connections in the rat with the Nauta method which was also utilized in more detailed studies in the cat (Walberg *et al.*, 1961; Voogd, 1964). In 1967 Eager described variable patterns of anterior vermal corticonuclear connections with the medial and interpositus nuclei in cats and rabbits, but he found that the cortical projections to VNC converged chiefly on the dorsal half of LVe. These findings have been corroborated in the cat (Brodal, 1974) and in several primate species (Haines, 1977b). The lobules of the posterior vermis (6–8) also send some projections to

LVe but axons from the latter region seem to have a predilection for the dorsal part of SpVe. Fibers from the vestibulocerebellum (9–10), on the other hand, essentially ignore LVe and distribute widely but differentially throughout SuVe, SpVe, MVe, PrH and group X (Angaut and Brodal, 1967).

We believe that most of these cerebellovestibular fiber projection patterns may also be found in the rat. Further examination of our large anterior and posterior rat vermian lesions suggest more correspondence than differences in findings. A meticulous anterograde method study of these connections in the rat is needed however, to establish which projections are actually homologous.

Experiments using HRP have been performed to determine the precise location of Purkinje cells giving origin to the long corticofugal projections to the VNC in the rat (Rubertone and Mehler, 1981; Voogd, personal communication). After iontophoretic injections in the SVe, MVe or SpVe, HRP positive neurons were found in posterior vermian lobules, primarily lobules 9 and 10 constituting the vestibulocerebellum. These observations in the rat are in accordance with findings on posterior lobe corticovestibular projections described for other species (Brodal, 1974).

The most interesting aspect of the rat corticovestibular projection is revealed following injections of HRP into the LVe with subsequent perusal of the anterior lobe vermis. The authors would like to acknowledge Dr Jan Voogd for his instructive analysis and elegant reconstruction of the resultant corticovestibular projections to LVe. Voogd reports that after injections into the LVe, labeled Purkinje cells are located in two longitudinal bands (Voogd, Volume 2, Chapter 11) in anterior lobe lobules 1–6. He also noted scattered labeled cells in lobules 9 and 10, a few positive somata in lobule 8 and none in lobule 7. These data substantiate previous studies in the cat (Bigare and Voogd, 1977) which also demonstrated that two separate bands of Purkinje cells in the anterior lobe project to the LVe.

Deep cerebellar nuclei projections to the vestibular nuclear complex in the rat originate chiefly from the medial cerebellar nucleus (Med) of one side and distribute bilaterally, but most heavily to the contralateral ventral LVe, SpVe and parasolitary nucleus (PSol). Autoradiographic studies of Med projections in the monkey, not complicated by the fiber of passage problem encountered in fiber degeneration studies and HRP injection into LVe, suggest that there are bilateral Med projections to the ventral LVe and also to SpVe and PSol that distribute almost symmetrically (Batton *et al.*, 1977). These new findings suggest that the often dense terminations observed in the ipsilateral dorsal half of LVe, and in MVe after lesions of the medial cerebellar nucleus, were corticofugal passage fibers as Voogd (1964) pointed out.

There are a number of discrepancies in the literature concerning the exact cells of origin of the cerebellar nucleovestibular projections. These discrepancies have been attributed to possible species differences and idiosyncrasies of fiber degeneration methods. Studies in the rat involving small iontophoretic injections of HRP into the individual vestibular nuclei provide some new insights into this controversy. In initial studies, Rubertone and Haroian (1982) have found that injections into the SpVe alone result in moderate numbers of labeled cells only in the contralateral medial cerebellar nucleus. Both ipsilateral and contralateral medial cerebellar nuclei, however, contain positive somata after injections which involve both SpVe and MVe. After SuVe and LVe injections, scattered reactive cells are again present bilaterally in the medial cerebellar nuclei, but contralateral cell labeling predominates in all cases. Positive somata are present in the interposed nucleus (Int) only when the parvocellular reticular formation in the rat is involved in the injection site.

3. *Other sources of vestibular nuclei afferent connections*

a. Reticulovestibular connections

A consensus is emerging that there are diffuse vestibular afferent connections that originate from the medullary reticular formation (RF), mainly from cells scattered in the dorsal parts of the gigantocellular (Gi) and caudal pontine (PnC) reticular nuclei and probably also from portions of the parvocellular reticular formation.

Data obtained from anterograde fiber degeneration studies of reticular lesions utilizing either the Marchi or Nauta methods were always equivocal because of the ever present problem of fibers of passage. For example, Mehler (1968) found few unequivocal "reticulovestibular" connections in Nauta method studies in rats with small reticular lesions that could not be explained by the possibility of interruption of fibers of the medial spinovestibular system (Mehler, 1968; Hazlett *et al.*, 1972) which ascends through the medial gigantocellular reticular core and paramedian reticular nuclei such as PnC to reach the vestibular nuclear complex. Ladpli and Brodal's (1968) description of a vast commissural system interlacing the two VNC primarily via the dorsal half of the medullary reticular formation added new uncontrollable (in our belief) dimensions to the fiber of passage problem.

Nevertheless, Hoddevik *et al.* (1975) deduced a pattern of moderate reticulovestibular connections with the Nauta method. These authors also tried tritiated amino acid (TAA) injections without much success, but Graybiel's (1977) experiments utilizing larger TAA injections into the reticular formation demonstrated limited but acceptable evidence of diffuse reticulovestibular connections. Subsequent HRP analysis of vestibular afferents in the cat (Corvaja *et al.*, 1979; Pompeiano *et al.*, 1978) and monkey (Carleton and Carpenter, 1983) support the notion of such connections.

Our large wheatgerm agglutinin–horseradish peroxidase (WAG–HRP) injections of the vestibular nuclear complex in cats (Voogd and Mehler, in preparation) appear to confirm certain reticulovestibular projection patterns described by the foregoing authors, but the ever present problem of HRP spread beyond the VNC in some cases suggests extreme caution in the acceptance of certain reticulovestibular projections. On the other hand, in our iontophoretic rat VNC injection series, the extreme sparcity of reticular cell labeling in the purely vestibular injection experiments raises other questions because of the paucity of reticular cell labeling. Further studies obviously are needed.

b. Inferior olive connections with the VNC

The inferior olivary subnuclei β, dorsal cap of Kooy (k) and dorsomedial cell column (dm) have been shown to receive some MVe and considerable numbers of SpVe projections in the cat (Saint-Cyr and Courville, 1979), rat (Swenson and Castro, 1983) and opossum (Martin *et al.*, 1975). No LVe projections to the inferior olive have been described, but, based upon HRP injection into LVe, Ito *et al.* (1982) have proposed that β and k, which receive SpVe and MVe connections, project to LVe. A projection from k to the dorsal part of LVe was previously described by Groenewegen and Voogd (1977) after selected injections of ^{3}H-leucine into the inferior olive of the cat and such a connection has also been confirmed in the Rhesus monkey (Voogd, personal communication).

IV. EFFERENT CONNECTIONS

A. Vestibulospinal pathways

There are two classical vestibulospinal tracts (vsp): (a) a medial vestibulospinal tract believed to originate chiefly from cells in the MVe nucleus, whose axons join the medial longitudinal fasciculus (mlf) and then descend in the medial part of the anterior funiculus from which they issue terminations into spinal lamina 7 throughout cervical levels, and (b) a lateral vestibulospinal tract, thought to originate solely from LVe, which descends in the ventral part of the anterior funiculus to lumbar levels and terminates in spinal ventral column laminae 7 and 8 (Brodal, 1974). Pompeiano and Brodal (1957a) demonstrated somatotopy in the LVe cell projections in the cat — dorsocaudal cells to hindlimb (HL) spinal levels and ventrorostral cells to neck and forelimb (Fl) levels. Recent intracellular recording studies in cats have verified the anatomic somatotopy in LVe projections including the distribution of thoracic level cell projections from a stratum between the HL-FL cells (Akaike, 1983). In the rat, Shamboul (1980) found that the HL region predominated.

The idea that LVe was the sole source of lateral vsp projections was derived from the lack of demonstrable retrograde cell chromotolysis (Fig. 4) in other vestibular nuclei after spinal lesions (Brodal, 1974). The data accumulated from HRP studies suggest otherwise. Petersen and Coulter (1977), for example,

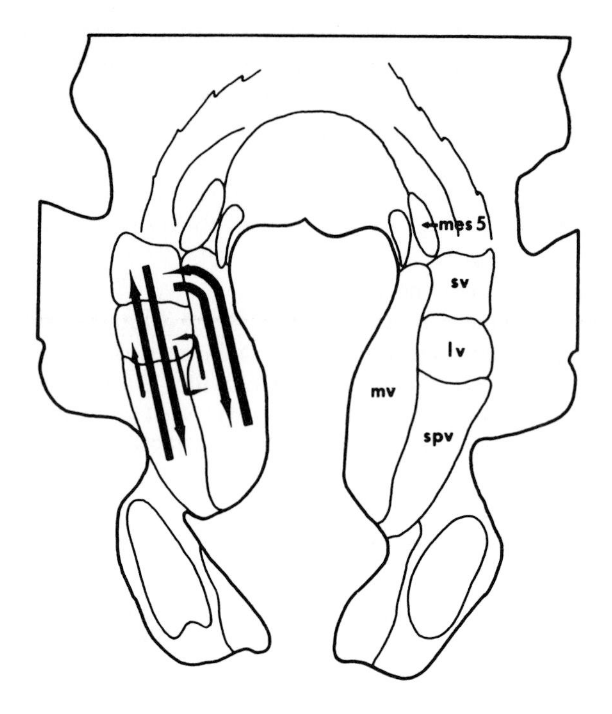

Fig. 4: Summary tracing depicting reciprocal internuclear connections between the vestibular nuclei. Me5, mesencephalic nucleus of the trigeminal; mv, lv, spv, sv, medial, lateral, spinal and superior vestibular nuclei.

showed that lower lumbar HRP injections produced caudal MVe, SpVe and group *f* cell labeling in cats in addition to the dorsocaudal (HL) LVe cell pattern. The spinal funicular trajectory of these caudal VNC cell axons has not been worked out. If they do not join the lateral vestibulospinal tract they may represent a third pathway as Petersen and Coulter (1977) propose. Vestibulospinal pathways in the rat are under investigation (Hagen and Rubertone, in preparation).

B. Vestibulocerebellar projections

Early Marchi method studies of the vestibulocerebellar projection in the rat were performed by Dow (1936). Dow believed that secondary vestibulocerebellar fibers were about three times as numerous as primary eighth nerve fibers. Following a lesion of the juxtarestiform body in the rat, Dow traced what he believed to be secondary fibers to both medial cerebellar nuclei and to the cortex of both flocculi and the nodulus and uvula. Some of these fibers were also thought to end in basilar parts of the anterior lobe. Utilizing a modified Gudden method (cortical ablation in kittens), Brodal and Torvik (1957) concluded that secondary vestibulocerebellar projections originated chiefly from group X cells and cells in the adjoining ventrolateral part of SpVe.

Utilizing the then recently introduced HRP method, injections of lobules 2–9 revealed the widespread origins of the secondary vestibulocerebellar projections in the rat, cat and monkey (Mehler, 1977). Labeled neurons were present throughout the SuVe, MVe and SpVe nuclei, in the interstitial nucleus of the eighth nerve and subgroups X and Y. No labeled somata were found in the LVe of the rat, nor in the LVe of the cats or monkeys studied. Small ipsilateral HRP injections of lobules 9 and 10 in the rat (Rubertone and Mehler, 1981) substantiated the findings of the earlier study and showed conclusively that most of the secondary vestibulocerebellar projections in the rat are probably bilateral (see Gould, 1980, and Kotchabhakdi and Walberg, 1978 for detailed reviewed of the vestibular nuclear complex projection patterns of individual vermian lobules in the cat).

Secondary vestibulocerebellar projections to the deep cerebellar nuclei appear to end chiefly on the medial cerebellar nuclei as Dow (1936) originally proposed. Experiments using injections of HRP in the rat suggest that these projections originate from widespread areas of the vestibular nuclear complex.

C. Commissural connections

Carpenter (1960b) described vestibular "arcuate" fiber projections (that is, commissural) between the SpVe nuclei of the cat that did not enter the contralateral inferior cerebellar peduncle as most crossed reticulocerebellar fibers that course through the vestibular nuclear complex tend to do. The fiber degeneration studies of Ladpli and Brodal (1968) in the cat, however, were the first to show how extensive was the commissural system reciprocally linking most of the VNC nuclei of the two sides of the brainstem. They showed that this intervestibular fiber system stretched across the midline ventral to the medial longitudinal fasciculus (mlf) from the rostral poles of the SuVe nuclei to the caudal poles of the MVe nuclei. Subsequent HRP analyses of vestibular afferent connections have confirmed these interconnections in the cat (Pompeiano *et al.*, 1978), monkey (Carleton and Carpenter, 1983) and rat (Rubertone and Mehler, 1980). The numerically high cell labeling densities found in the major contralateral vestibular nuclear complex nuclei, except for X, F and LVe, attest to the magnitude of the commissural system in the three species. One gains the impression from large VNC injections in both the cat and rat that a very large majority of cells of all sizes in the SuVe and MVe nuclei have commissural connections.

Small iontophoretic HRP injections in the rat, as well as in the cat, indicate that these commissural projections are not merely point to point connections between homologous nuclei on each side of the brain stem. This type of experiment reveals that, while some point to point connections may exist, cells in several different contralateral nuclei contribute commissural connections to an individual ipsilateral nucleus. For example, following injections of the SpVe in the rat, moderate numbers of labeled cells occur in the contralateral SpVe, MVe and SuVe, while only a few reactive small cells are present in the edge of LVe and in subgroup Y. Injections of MVe reveal some contralateral labeled neurons in SpVe, sparse labeling in group Y, moderate numbers of cells in SuVe and an extensive connection with the contralateral MVe. The LVe does not seem to form a commissural connection with its contralateral counterpart or subgroup Y. There is a paucity of reactive cells in SpVe after SuVe injection and the contralateral LVe is completely devoid of reactive cells. These data suggest that there is probably as high a degree of organization in the intervestibular commissural system as is found in the intravestibular intrinsic connectivity of each side (Rubertone *et al.*, 1983).

The commissural connections of individual vestibular nuclei also can be graded on the basis of numbers of contralateral labeled neurons. Following this gradation the MVe establishes strong, commissural connections. SVe and SpVe moderate, and the LVe weak, if any, commissural connections with the contralateral vestibular complex. We would judge group Y as having moderate, and the prepositus hypoglossal nucleus (PrH) as having very strong commissural connections (Rubertone and Mehler, 1980).

D. Vestibulo-ocular projections

Fiber projections ascending from various parts of the vestibular nuclear complex (VNC) to the ocular motor nuclei were studied with anterograde fiber degeneration techniques primarily in nonrodent species (Gacek, 1971; McMasters *et al.*, 1966). Fiber of passage in the VNC and cerebellar contamination plagued initial Marchi and later Nauta method analysis of degeneration produced by lesions of the vestibulo-ocular (V-O) projections. Nevertheless, much valid information about V-O projections was gleaned from such studies. These technical problems, and the relatively small size of the rat's vestibular nuclear complex mitigated against the use of rodents in such experimental studies.

There was little agreement on specific patterns of VNC cells participating in the VO pathways. Viewed in retrospect, however, retrograde cell degeneration patterns induced by lesions of the medial longitudinal fasciculus in the monkey by Carpenter and McMasters (1963) equate remarkably well with the now accepted VO reflex centers in the VNC demonstrated by the recently introduced axon transport methods (Gacek, 1977; Graybiel and Hartwieg, 1974).

Out of curiosity about what the vestibulocerebellum patterns might be in the rat's VNC, and in the hope that we might be able to settle the related group Y *vs.* infracerebellar nucleus *vs.* parvocellular lateral cerebellar (dentate) nucleus controversy (see above), we initiated HRP studies of ocular motor nuclei afferent connections (Mehler and Kohn, in preparation). To date, 24 attempts have been made to iontophorectically inject the ocular motor complex (OMC). In the OMC we include both the rat's third nerve complex and the fourth nerve nuclei. The experiments have produced 14 acceptably restricted OMC injections. Tabulations of the cell labeling found in these experiments have shown consistent, but slightly variable, topological patterns of HRP positive cells chiefly in SuVe and MVe and a few cells on the medial edge of SpVe. Heavily labeled cells appear in the contralateral abducens nucleus and a few just caudal to the abducens nucleus in the dorsal part of the caudal pontine nucleus (PnC) of Paxinos and Watson (1982) and in the prepositus hypoglossal nucleus. A trickle of labeled cells appears in, or on the edge of, the ventral, ophthalmic part of the spinal trigeminal nucleus interpolaris part and some in both the interposed and infracerebellar deep nuclei in the rat as Gacek (1977) found in the cat.

In this series, other experiments with HRP injections, chiefly in the interstitial nucleus of Cajal, show divergent, more diffuse variations on the theme of VNC cell labeling established by injections restricted to the OMC. Injections in the central gray dorsal to OMC show no VNC cell labeling.

Cell labeling in SuVe in some cases, for example, ranges from almost complete dispersion throughout the coronal limits of SuVe to instances where a central core arrangement predominates. Ipsilateral SuVe projections to the ocular motor complex (OMC) predominate as has been suggested by previous anterograde as well as retrograde investigations. We have seen no evidence in the rat so far that LVe contributes to OMC connections. In both the cat HRP studies cited (Gacek, 1977; Graybiel and Hartwieg, 1974), LVe cell labeling, when present, was limited to a small medial marginal part of the ventral half of LVe.

There are few if any MVe nucleus projections to the ocular motor complex (OMC) in the rat and cat that originate from the rostral pole of the MVe. In the rat, as in the cat, MVe–OMC projections form a distinct, chiefly contralateral, column of cells in the MVe extending from levels caudal to the genu of the facial nerve almost to the caudal pole of the MVe. In large OMC injections in the cat, this column of cells often straddles the medial border of the SpVe nucleus which otherwise appears to contribute only scattered cells to the VNC–OMC projections except for a group of cells at the oral pole of SpVe. In our existing rat material, however, the column of labeled OMC afferent cells consistently appears to be located in the center of the middle third of the MVe in coronal sections. Only scattered positive cells appear in the caudal third of the MVe in the rat and cat. Infracerebellar nuclear (Inf) cell labeling shows a contralateral bias and group Y cells *per se* rarely appear labeled.

Cell labeling in the prepositus hypoglossal nucleus (PrH)—treated in this report as an associated vestibular cell group—appears throughout the length of the nucleus blending almost imperceptibly, in some cases, with labeled cells in the intercalated nucleus. We also have seen labeled cells in Roller's nucleus in unpublished monkey OMC injections experiments suggesting that the third component of Brodal's (see Brodal and Brodal, 1983) "perihypoglossal" nuclei also has projections to the OMC yet to be demonstrated in the rat and cat.

E. Vestibulothalamic projections

Hassler (1948) described direct projections from the vestibular nuclear complex (VNC) to the thalamus in the cat based on Marchi method fiber degeneration tracing. Except for Carpenter and Strominger's (1964) positive findings with the Nauta method (based chiefly on mlf lesions) workers studying the problem with comparable silver methods denied the existence of any vestibulothalamic pathway. Convincing evidence that there were vestibular nuclear complex projections to the ventral tier thalamic nuclei was first presented by Lang *et al.* (1979) in the autoradiographic method studies in the monkey. The HRP investigations by Kotchabhakdi *et al.* (1980) of thalamic afferents originating from the VNC in the cat confirmed the ventral tier thalamic nuclei connections shown in the monkey, and demonstrated that there also were other VNC projections to the intralaminar thalamic nuclei and the ventral nucleus of the lateral geniculate. These authors concluded that the thalamic projections to ventral posterolateral-ventrolateral border regions in the ventral tier originated mainly from caudal parts of MVe and SpVe and the intralaminar afferent connections mainly from SuVe and group Y.

The existence of similar vestibular nuclear complex patterns of vestibulothalamic projections in the rat was confirmed by Kevetter *et al.* (1982) and is now being expounded upon (Kevetter, in preparation). Mehler and Voogd's (1984) recent anterograde axon transport study utilizing WGA–HRP injections into the cat VNC have verified the dual principal and intralaminar thalamic nuclei distributions of VNC projections of Kotchabhakdi *et al.* (1980). To date, however, comparable WGA–HRP studies in the rat have failed to reveal equivalent anterograde fiber distributions. This failure probably reflects the lack of involvement of SuVe and the caudal lateral SpVe region which most consistently demonstrate thalamic connections, as we have pointed out.

F. Centrifugal fibers in the vestibular nerve

Vestibular efferent neurons whose axons project in the eighth nerve to the sensory epithelium of the labyrinth have been found to be localized primarily to a region dorsolateral to the facial genu and ventromedial to MVe. A small group also appears medial to the genu ventral to the supragenulate nucleus (Fig.

la). These small cells (15 μm ± 5) are more readily seen in Nissl preparation of primate brains than they are in the rat. They register well with acetylcholinesterase (AchE) staining cells in the same locales (Blanks and Palay, 1978; Gacek and Lyon, 1974; Goldberg and Fernandez, 1980; Volume 1, Chapter 14).

The discovery of centrifugal fibers in the vestibular part of the eighth nerve is intimately related to the scientific unraveling of the auditory system's centrifugal (superior) olivocochlear projections with whose axons vestibular efferent neuron fibers are inextricably mixed in their trajectory out of the brain stem. These cells were first, but erroneously, described in rat AChE studies as possible salivatory preganglionic cells. Their afferent connections are a matter of speculation (White and Warr, 1983).

V. INTRINSIC VESTIBULAR CONNECTIONS

Iontophoretic microinjections of HRP into the individual vestibular nuclei reveal highly organized, rather extensive internuclear connections in the rat (Rubertone *et al.*, 1983). Labeled cells, for example, are present in the ipsilateral subgroup Y and SuVe subsequent to injections of the SpVe. However, there is a conspicuous lack of HRP positive somata in the ipsilateral MVe and LVe following SpVe injections. Injections of MVe result in labeled neurons in LVe, SuVe and Y. Ipsilateral labeled neurons are only present in the SpVe and MVe nuclei subsequent to injection of SuVe. Injections of LVe reveal extensive internuclear connections. In the latter experiments, HRP reactive neurons were present to some degree in all three of the other major ipsilateral VNC nuclei—SuVe, MVe and SpVe. Rubertone *et al.* (1983) concluded that the SuVe appears to have strong reciprocal intrinsic connections with the SpVe and MVe. The LVe is reciprocally related to the MVe but does not reciprocate input from the SpVe. This intrinsic organization between individual vestibular nuclei suggests a high degree of integrative communication within the vestibular nuclear complex of each side of the brain stem. Rubertone *et al.* (1983) summarize a proposed intravestibular circuit suggested in our recent study.

VI. HISTOCHEMICAL OBSERVATIONS

Data are beginning to accumulate on the distribution of neuroactive compounds in the VNC. The medial, superior and lateral vestibular nuclei are AChE positive (*Atlas* Figs). The MVe stains more intensely than the SuVe nucleus and individual giant and smaller cells in LVe stand out. Kimura *et al.* (1981) concluded that only cells in the lateral nucleus in the cat may be choline acetyltransferase (ChAT) positive. They classified the superior and medial nuclei as cholinoceptive (that is, the boutons on the soma and dendritic surfaces not the cytoplasm being ChAT immunoreactive). In the rat, muscarinic cholinergic and histamine receptors have been localized autoradiographically, primarily in the medial vestibular nucleus (Palacios *et al.*, 1981; Wamsley *et al.*, 1981) *not* in the lateral as Peroutka and Snyder (1982) misquoted. Nomura *et al.* (1984) have studied the distributions of VNC structures containing substance P (SP), Leu-enkephalin (EK) and γ-aminobutyric acid (GABA) in the rat. Immunoreactive SP and EK cell bodies were distributed chiefly in the caudal halves of the medial and spinal vestibular nuclei. The distribution of glutamic acid decarboxylase (GAD) positive GABA cells was essentially similar to the distributions of the two neuropeptide cell patterns. Some of all three cell types appeared in the dorsal region of the medial nucleus at levels of the facial nerve genu. Substance P and EK immunoreactive cells appeared sporadically in the perihypoglossal nucleus. No immunoreactive cells were localized in either the superior or lateral vestibular nuclei. Enkephalin cells are depicted in a region that seemed to correspond to nucleus X, but no clues to the histochemistry of the Y group or infracerebellar nuclei appear in their report.

Immunoreactive fibers of all three types appeared in all the major VNC nuclei. Enkephalin fibers, however, were concentrated primarily in the subependymal region of the medial nucleus. GABA fibers terminated heavily in the dorsal half of the lateral, medial, and spinal nuclei, an arrangement reflecting the projection pattern of GABAergic long corticofugal cerebellar fibers issuing from the anterior vermis.

REFERENCES

Akaike, T. (1983). Neuronal organization of the vestibulospinal system in the cat. *Brain Res.* 259, 217–227.

Allen, W. E. (1924). Distribution of the fibers originating from different basal cerebellar nuclei. *J. Comp. Neurol.* 36, 399–439.

Angaut, P. and Brodal, A. (1967). The projection of the "vestibulocerebellum" onto the vestibular nuclei in the cat. *Arch. Ital. Biol.* 105, 441–479.

Bangma, G. C. and Ten Donkelaar, H. J. (1983). Some afferent and efferent connections of the vestibular nuclear complex in the red-eared turtle *Psedemys scripta elegans*. *J. Comp. Neurol.* 220, 453–464.

Batton, R. R., III, Jayaraman, A., Ruggiero, D. and Carpenter, M. B. (1977). Fastigial efferent projections in the monkey: An autoradiographic study. *J. Comp. Neurol.* 174, 281–306.

Berkley, K. J. (1983). Spatial relationships between the terminations of somatic sensory motor pathways in the rostral brainstem of cats and monkeys. II. Cerebellar projections compared with those of the ascending somatic sensory pathways in lateral diencephalon. *J. Comp. Neurol.* 220, 229–251.

Bigare, F. and Voogd, J. (1977). Cerebello-vestibular projections in the cat. *Acta Morphol. Neerl.-Scand.* 15, 323–325.

Blanks, R. H. I. and Palay, S. (1978). The location and form of efferent vestibular neurons in rat. *Anat. Rec.* 190, 34.

Blanks, R. H. I., Precht, W. and Torigoe, Y. (1983). Afferent projections to the cerebellar flocculus in the pigmented rat demonstrated by retrograde transport of horseradish peroxidase. *Exp. Brain Res.* 52, 293–306.

Brodal, A. (1952). Experimental demonstration of cerebellar connections from the perihypoglossal nuclei (nucleus intercalatus, nucleus praepositus hypoglossi and nucleus of Roller) in cat. *J. Anat.* 86, 110–129.

Brodal, A. (1954). Cerebellar afferents from the perihypoglossal nuclei. In J. Jansen and A. Brodal (Eds). *Aspects of cerebellar anatomy*, pp. 158–161. Grunderson, Oslo.
Brodal, A. (1974). Anatomy of the vestibular nuclei and their connections. In H. H. Kornhuber (Ed.). *Handbook of sensory physiology*, Vol. 1, pp. 240–351. Springer-Verlag, Berlin.
Brodal, A. (1983). The perihypoglossal nuclei in the Macaque monkey and the chimpanzee. *J. Comp. Neurol.* 218, 257–269.
Brodal, A. and Brodal, P. (1983). Observations on the projection from the perihypoglossal nuclei onto the cerebellum in the Macaque monkey. *Arch. Ital. Biol.* 121, 151–166.
Brodal, A. and Hoivik, B. (1964). Site and mode of termination of primary vestibulo-cerebellar fibers in the cat. An experimental study with silver impregnation methods. *Arch. Ital. Biol.* 102, 1–21.
Brodal, A. and Pompeiano, O. (1957a). The vestibular nuclei in the cat. *J. Anat.* 91, 438–454.
Brodal, A. and Pompeiano, O. (1957b). The origin of ascending fibers of the medial longitudinal fasciculus from the vestibular nuclei. An experimental study in the cat. *Acta Morphol. Neerl. Scand.* 1, 306–328.
Brodal, A. and Torvik, A. (1957). Uber den Ursprung der sekundaren vestibulo-cerebellaren Fasern bei der Katze: Eine experimentellanatomische Studie. *Arch. Psychiat. Nervenkr.* 195, 550–567.
Brown Gould, B. (1979) The organization of afferents to the cerebellar cortex in the cat: Projections from the deep cerebellar nuclei. *J. Comp. Neurol.* 184, 27–42.
Brown Gould, B. (1980). Organization of afferents from the brain stem nuclei to the cerebellar cortex in the cat. *Adv. Anat. Embryol. Cell. Biol.* 62, 1–90.
Cajal, S. Ramón y (1911). *Histologie du système nerveux de l'homme et des vertébrés.* Paris, Maloine.
Carleton, S. C. and Carpenter, M. B. (1983). Afferent and efferent connections of the medial, inferior and lateral vestibular nuclei in the cat and monkey. *Brain Res.* 278, 29–51.
Carpenter, M. B. (1960a). Experimental anatomical-physiological studies of the vestibular nerve and cerebellar connections. In G. L. Rasmussen and W. Windle (Eds). *Neural mechanisms of the auditory and vestibular systems*, pp. 297–323. C. C. Thomas, Springfield.
Carpenter, M. B. (1960b). Fiber projections from the descending and lateral vestibular nuclei in the cat. *Am. J. Anat.* 107, 1–22.
Carpenter, M. B. and McMasters, R. E. (1963). Disturbances of conjugate horizontal movements in the monkey. II Physiological effects and anatomical degeneration resulting from lesions of the medial longitudinal fasciculus. *Arch. Neurol.* 8, 347–368.
Carpenter, M. B. and Strominger, N. L. (1964). Cerebello-oculomotor fibers in the rhesus monkey. *J. Comp. Neurol.* 123, 211–230.
Carpenter, M. B., Stein, B. M. and Peter, P. (1972). Primary vestibulocerebellar fibers in the monkey: Distribution of fibers arising from distinctive cell groups of the vestibular ganglia. *Am. J. Anat.* 135, 221–249.
Cazin, L., Magnin, M. and Lannou, J. (1982). Non-cerebellar visual afferents to the vestibular nuclei involving the prepositus hypoglossal complex: An autoradiographic study in the rat. *Exp. Brain Res.* 48, 309–313.
Chan-Palay, V. (1978). The paratrigeminal nucleus. I. Neurons and synaptic organization. *J. Neurocytol.* 7, 405–418.
Chan-Palay, V. (1977). *Cerebellar dentate nucleus.* Springer, Berlin.
Corvaja, N., Mergner, T. and Pompeiano, O. (1979). Organization of reticular projections to the vestibular nuclei in the cat. *Prog. Brain Res.* 50, 631–644.
Dietrichs, E. (1981). The cerebellar corticonuclear and nucleocortical projections in the cat as studied with anterograde and retrograde transport of horseradish peroxidase. *Exp. Brain Res.* 44, 235–242.
Dow, R. S. (1936). The fiber connections of the posterior parts of the cerebellum in the rat and cat. *J. Comp. Neurol.* 63, 527–548.
Eager, R. P. (1967). Efferent cortico-nuclear pathways in the cerebellum of the cat. *J. Comp. Neurol.* 120, 81–103.
Faull, R. L. M. (1977). A comparative study of the cells of origin of cerebellar afferents in the rat, cat and monkey studied with the horseradish peroxidase technique: I. The non-vestibular brainstem afferents. *Anat. Rec.* 187, 577.

Gacek, R. R. (1969). The course and central termination of first order neurons supplying vestibular endorgans in the cat. *Acta Otolaryngol. Suppl. (Stockholm)* 254, 1-66.
Gacek, R. R. (1971). Anatomical demonstration of the vestibulo-ocular projections in the cat. *Acta Otolaryngol. Suppl. (Stockholm)* 293, 1-63.
Gacek, R. R. (1977). Location of brain stem neurons projecting to the oculomotor nucleus in the cat. *Exp. Neurol.* 57, 725–749.
Gacek, R. R. (1979). Location of trochlear vestibulo-ocular neurons in the cat. *Exp. Neurol.* 66, 692–706.
Gacek, R. R. and Lyon, M. (1974). The localization of vestibular efferent neurons in the kitten with horseradish peroxidase. *Acta Otolaryngol. (Stockholm)* 77, 92–101.
Goldberg, J. M. and Fernandez, C. (1980). Efferent vestibular system in the squirrel monkey: Anatomical location and influence on afferent activity. *J. Neurophysiol.* 43, 986–1025.
Goodman, D. C., Hallett, R. E. and Welch, R. B. (1963). Patterns of localization in the cerebellar corticonuclear projections of the albino rat. *J. Comp. Neurol.* 121, 51–67.
Grant, G., Boivie, J. and Silfvenius, H. (1973). Course and termination of fibers from the nucleus z of the medulla oblongata: An experimental light microscopic study in the cat. *Brain Res.* 55, 55–70.
Graybiel, A. M. (1977). Organization of oculomotor pathways in the cat and Rhesus monkey. *In* Baker and Berthoz (eds), *Control of gaze by brain stem neurons.* Elsevier, Amsterdam.
Graybiel, A. M. and Hartwieg, E. A. (1974). Some afferent connections of the oculomotor complex in the cat: An experimental study with tracer techniques. *Brain Res.* 81, 543–551.
Groenewegen, H. J. and Voogd, J. (1977). The parasagittal zonation within the olivo-cerebellar projection 1. Climbing fiber distribution in the vermis of cat cerebellum. *J. Comp. Neurol.* 174, 417–488.
Haines, D. E. (1977a). A proposed functional significance of parvicellular regions of the lateral and medial cerebellar nuclei. *BrainBehav. Evol.* 14, 328–340.
Haines, D. E. (1977b). Cerebellar corticonuclear and corticovestibular fibers of the fluccolonodular lobe in a prosimian primate (*Galago senegalensis*). *J. Comp. Neurol.* 174, 607–630.
Hassler, R. (1948). Forels Haubenfaszikel als vestibulare Empfindungsbahn mit Bemerkungen uber einige andere sekundare Bahenen des Vestibularis und Trigeminus. *Arch. Psychiat. Nervenkrank* 180, 23–53.
Hauglie-Hanssen, E. (1968). Intrinsic neuronal organization of the vestibular nuclear complex in the cat: A Golgi study. *Ergeb. Anat. Entwickl.-Gesch.* 40, 1–105.
Hazlett, J. C., Dom, R. and Martin, G. F. (1972). Spino-bulbar, spino-thalamic and medial lemniscal connections in the American opossum. *J. Comp. Neurol.* 146, 95–118.
Henkel, C. K. and Martin, G. F. (1977a). The vestibular complex of the American opossum (*Didelphis virginiana*) I. Conformation cytoarchitecture and primary vestibular input. *J. Comp. Neurol.* 172, 299–320.
Henkel, C. K. and Martin, G. F. (1977b). The vestibular complex of the American opossum (*Didelphis virginiana*) II. Afferent and efferent connections. *J. Comp. Neurol.* 172, 321–348.
Hoddevik, G. H., Brodal, A. and Walberg, F. (1975). The reticulovestibular projection in the cat. An experimental study with silver impregnation methods. *Brain Res.* 94, 383–399.
Holstege, G. and Kuypers, H. G. J. M. (1982). The anatomy of brainstem pathways to the spinal cord in cat. A labeled amino acid tracing study. *Prog. Brain Res.* 57, 145–175.
Ito, J., Sasa, M. Matsuoka, I. and Takaori, S. (1982). Afferent projection from reticular nuclei, inferior olive and cerebellum to lateral vestibular nucleus in cat as demonstrated by horseradish peroxidase. *Brain Res.* 231, 427–437.
Johnson, J. E., Jr (1975a). A fine structural study of degenerative-regenerative pathology in the surgical deafferented lateral vestibular nucleus of the rat. *Acta Neuropathol.* 33, 227–243.
Johnson, J. E., Jr (1975b). The occurrence of dark neurons in the normal deafferented lateral vestibular nucleus in the rat: Observations by light and electron microscopy. *Acta Neuropathol.* 31, 117–127.
Johnson, J. E., Jr and Miquel, J. (1974). Fine structural changes in the lateral vestibular nucleus of aging rats. *Mech. Ageing Develop.* 3, 203–224.
Johnson, J. E., Jr, Mehler, W. R. and Oyama, J. (1976). The effects of centrifugation on the morphology of the lateral vestibular nucleus in the rat: A light and electron microscopic study. *Brain Res.* 106, 205–221.

Kevetter, G. A. and Perachio, A. A. (1983). Termination of vestibular afferents in the vestibular nuclear complex in the Gerbil. *Neurosci. Abst.* 9, 739.
Kevetter, G. A., Mehler, W. R. and Willis, W. D. (1982). Projections to the thalamus from brainstem reticular and vestibular nuclei in the rat. *Neurosci. Abst.* 8, G99.
Kilduff, T. S., Sharp, F. S. and Heller, H. G. (1983). Relative 2-deoxyglucose uptake of the paratrigeminal nucleus increases during hibernation. *Brain Res.* 262, 117–123.
Kimura, H., McGeer, P. H., Peng, J. H. and McGeer, E. G. (1981). The central cholinergic system studied by choline acetyltransferase immunohistochemistry. *J. Comp. Neurol.* 200, 151–201.
Korte, G. (1979). The brainstem projection of the vestibular nerve in the cat. 8J. Comp. Neurol. 184, 279–292.
Korte, G. E. and Mugnaini, E. (1979). The cerebellar projection of the vestibular nerve in the cat. *J. Comp. Neurol.* 184, 265–277.
Kotchabhakdi, N. and Walberg, F. (1978). Primary vestibular afferent projections to the cerebellum as demonstrated by retrograde axonal transport of horseradish peroxidase. *Brain Res.* 142, 142–146.
Kotchabhakdi, N., Rinvik, E., Walberg, F. and Yingchareon, K. (1980). The vestibulothalamic projections in the cat studied by retrograde axonal transport of horseradish peroxidase. *Exp. Brain Res.* 40, 405–418.
Ladpli, R. and Brodal, A. (1968). Experimental studies of commissural and reticular formation projections from the vestibular nuclei in the cat. *Brain Res.* 8, 65–96.
Landgren, A. and Silfvenius, H. (1971). Nucleus z, the medullary relay in the projection path to the cerebral cortex of group I muscle afferents from the cats hind limb. *J. Physiol.* 218, 551–571.
Lang, W., Buttner-Ennever, J. A. and Buttner, U. (1979). Vestibular projections to the monkey thalamus: An autoradiographic study. *Brain Res.* 177, 3–17.
Larsell, O. (1970). *In* J. Jansen (ed.), *The comparative anatomy and histology of the cerebellum from monotremes through apes.* University of Minnesota Press, Minneapolis.
Lorente de Nó, R. (1933). Anatomy of the eighth nerve: The central projection of the nerve endings of the internal ear. *Laryngoscope* 431, 1–38.
McMasters, R. E., Weiss, A. H. and Carpenter, M. B. (1966). Vestibular projections to the nuclei of the extraocular muscles: Degeneration resulting from discrete partial lesions of the vestibular nuclei in the monkey. *Am. J. Anat.* 118, 163–194.
Magnin, M., Courjon, J. H. and Flandrin, J. M. (1983). Possible visual pathways to the cat vestibular nuclei involving the nucleus prepositus hypoglossi. *Exp. Brain Res.* 51, 298–303.
Martin, G. F., Dom, R., King, J. S., Robards, M. and Watson, C. R. R. (1975). The inferior olive nucleus of the opossum (*Didelphis marsupialis virginiana*): Its organization and connections. *J. Comp. Neurol.* 160, 507–533.
Matsushita, M., Hosoya, Y. and Ikeda, M. (1979). Anatomical organization of the spinocerebellar system in the cat as studied by retrograde transport of horseradish peroxidase. *J. Comp. Neurol.* 184, 81–106.
Meessen, H. and Olszewski, J. A. (1949). *A cytoarchitectonic atlas of the rhombencephalon of the rabbit.* S. Karger, Basel.
Mehler, W. R. (1966). *Second symposium on the role of vestibular organs in space exploration.* NASA Special Publication #115, 140–141, Washington, DC.
Mehler, W. R. (1968). Reticulovestibular connections composed with spinovestibular connections in the cat. *Anat. Rec.* 160, 485.
Mehler, W. R. (1969). Some neurological species differences—*a posteriori. Ann. N. Y. Acad. Sci.* 167, 424–468.
Mehler, W. R. (1972). Comparative anatomy of the vestibular nucleus complex in submammalian vertebrates. *Prog. Brain Res.* 37, 55–67.
Mehler, W. R. (1977). A comparative study of the cells of origin of cerebellar afferents in the rat, cat, and monkey studied with the horseradish peroxidase technique: II. The vestibular nuclear complex. *Anat. Rec.* 187, 653.
Mehler, W. R. (1983). Observations on the connectivity of the parvicellular reticular formation with respect to a vomiting center. *Brain Behav. Evol.* 23, 63–80.
Mehler, W. R. and Voogd, J. (1984). Vestibulothalamic connections in the cat: A wheat germ agglutinin-horseradish peroxidase analysis. *Anat. Rec.* 208, 116A.

Mugnaini, E., Walberg, F. and Brodal, A. (1967). Mode of termination of primary vestibular fibers in the lateral vestibular nucleus. An experimental electron microscopical study in the cat. *Exp. Brain Res.* 4, 187–211.

Nomura, I., Senba, E., Kubo, T., Shiraishi, T., Matsunaga, T., Tohyama, H., Shiotani, Y. and Wu, U. Y. (1984). Neuropeptides and γ-aminobutyric acid in the vestibular nuclei of the rat: An immunohistochemical analysis. I. Distribution. *Brain Res.* 311, 109–118.

Palacios, J. M., Wamsley, J. K. and Kuhar, M. J. (1981). The distribution of Histamine H_1 receptors in the rat brain: An autoradiographic study. *Neurosci.* 6, 15–37.

Paxinos, G. and Watson, C. (1982). *The rat brain in stereotaxic coordinates.* Academic Press, Sydney.

Peroutka, S. J. and Snyder, S. H. (1982). Antiemetics: Neurotransmitter receptor binding predicts therapeutic actions. *Lancet*, 658–659.

Petersen, B. W. and Coulter, J. D. (1977). A new long spinal projection from the vestibular nuclei in the cat. *Brain Res.* 122, 351–356.

Pompeiano, O. and Brodal, A. (1957a). Spinovestibular fibers in the cat: An experimental study. *J. Comp. Neurol.* 108, 353–382.

Pompeiano, O. and Brodal, A. (1957b). The origin of vestibulospinal fibers in the cat: An experimental, anatomical study, with comments on the descending medial longitudinal fasciculus. *Arch. Ital. Biol.* 95, 166–195.

Pompeiano, O., Merger, T. and Corvaja, N. (1978). Commissural, perihypoglossal and reticular afferent projections to the vestibular nuclei in the cat. *Arch. Ital. Biol.* 116, 130–172.

Rubertone, J. A. and Haines, D. E. (1981). Secondary vestibulocerebellar projections to flocculonodular lobe in a prosimian primate, *Galago senegalensis. J. Comp. Neurol.* 200, 255–272.

Rubertone, J. A. and Haines, D. E. (1982). The vestibular complex in a prosimian primate (*Galago senegalensis*): Morphological and spinovestibular connections. *Brain Behav. Evol.* 20, 129–155.

Rubertone, J. A. and Haroian, A. J. (1982). Cerebellar nucleovestibular projections (CNVP) in the rat: A horseradish peroxidase (HRP) study. *Anat. Rec.* 202, 162A.

Rubertone, J. A. and Mehler, W. R. (1980). Afferents to the vestibular complex in rat: A horseradish peroxidase study. *Neurosci. Abst.* 6, 225.

Rubertone, J. A. and Mehler, W. R. (1981). Corticovestibular and vestibulocerebellar projections in the rat: A horseradish peroxidase study. *Anat. Rec.* 199, 219A.

Rubertone, J. A., Mehler, W. R. and Cox, G. E. (1983). The intrinsic organization of the vestibular complex: Evidence for internuclear connectivity. *Brain Res.* 263, 137–141.

Saint-Cyr, J. A. nad Courville, J. (1979). Projection from the vestibular nuclei to the inferior olive in the cat: An autoradiographic and horseradish peroxidase study. *Brain Res.* 165, 189–200.

Sato, Y., Kawasaki, T. and Ikarashi, K. (1982). Zonal organization of the floccular Purkinje cells projecting to the vestibular nucleus in cats. *Brain Res.* 235 1–15.

Schwartz, D. W., Schwartz, I. E. and Fredrickson, J. M. (1977). Fine structure of the medial and descending vestibular nuclei in normal rats and after unilateral transection of the vestibular nerve. *Acta Otolaryngol. (Stockholm)* 84, 80–90.

Shamboul, K. M. (1980). Lumbosacral predominance of vestibulospinal fiber projections in the rat. *J. Comp. Neurol.* 192, 519–530.

Snyder, R. L., Faull, R. L. M. and Mehler, W. R. (1978). A comparative study of the neurons of origin of spinocerebellar afferents in rat, cat, and squirrel monkey based on the retrograde transport of horseradish peroxidase. *J. Comp. Neurol.* 181, 833–852.

Sotelo, C. and Palay, S. L. (1968). The fine structure of the lateral vestibular nucleus in the rat. I. Neurons and neuroglial cells. *J. Cell. Biol.* 36, 151–170.

Sotelo, C. and Palay, S. L. (1970). The fine structure of the lateral vestibular nucleus in the rat. II. Synaptic organization. *Brain Res.* 18, 93–115.

Stein, B. M. and Carpenter, M. B. (1967). Central projections of portions of the vestibular ganglia innervating specific parts of the labyrinth in the Rhesus monkey. *Am. J. Anat.* 120, 281–318,

Swenson, R. S. and Castro, A. J. (1983). The afferent connections of the inferior olivary complex in rats: An anterograde study using autoradiographic and axonal degeneration techniques. *Neurosci.* 8, 259–275.

Vincent, S. L. and Rubertone, J. A. (1984). The origin of the spinovestibular pathway: A horseradish peroxidase (HRP) study. *Anat. Rec.* 208, 188A.

Voogd, J. (1964). *The cerebellum of the cat: Structure and fiber connections.* Doctoral thesis. Van Gorcum, Assen.

Walberg, F. (1975). The vestibular nuclei and their connections with the eighth nerve and the cerebellum. In R. Naunton (Ed.). *The vestibular system*, pp. 31–52. Academic Press, New York.

Walberg, F., Bowsher, D. and Brodal, A. (1958). The termination of primary vestibular fibers in the vestibular nuclei of the cat: An experimental study with silver methods. *J. Comp. Neurol.* 110, 391–419.

Walberg, F., Pompeiano, O., Brodal, A. and Jansen, J. (1961). Fastigiovestibular fibers in the cat: An experimental study with silver methods. *J. Comp. Neurol.* 118, 49–75.

Wang, S. C. and Chinn, H. I. (1956). Experimental motion sickness in dogs: Importance of labyrinth and vestibular cerebellum. *Am. J. Physiol.* 185, 617–623.

Wamsley, J. K., Lewis, M. S., Young, W. S., III, and Viuhar, M. J. (1981). Autoradiographic localization of muscarinic cholinergic receptors in rat brainstem. *J. Neurosci.* 1, 176–191.

White, J. S. and Warr, W. B. (1983). The dual origins of the olivocochlear bundle in the albino rat. *J. Comp. Neurol.* 219, 203–214.

Zemlan, F. P., Leonard, C. M., Kow, L.-M. and Pfaff, D. W. (1978). Ascending tracts of the lateral columns of the rat spinal cord: A study using the silver impregnation and horseradish peroxidase techniques. *Exp. Neurol.* 62, 298–334.

10

Precerebellar nuclei and red nucleus

BRIAN A. FLUMERFELT
A.W. HRYCYSHYN

University of Western Ontario
London, Canada

The term precerebellar nuclei refers collectively to those centers that provide projections which terminate within the nuclei and cortex of the cerebellum. Although many such centers occur, it is customary to recognize three or four major sources of cerebellar afferents. The pontine nuclei, situated within the basilar pons, are the largest of the precerebellar nuclei and serve as an important relay for pathways from the cerebral cortex to the cerebellum. The inferior olivary nucleus, located in the ventrocaudal medulla oblongata dorsolateral to the pyramids, is the sole contributor of climbing fibers to the cerebellum. The lateral reticular nucleus, which lies lateral to the caudal two thirds of the inferior olive, is a major relay for spinal pathways and an important source of mossy fibers. The spinal cord also provides mossy fibers to the cerebellum via several projections, including the dorsal and ventral spinocerebellar tracts. A number of other precerebellar centers have also been described, including the vestibular nuclei, the perihypoglossal nuclei, the paramedian reticular nucleus, the cuneate nuclei, the raphe nuclei, the gracile nucleus and the trigeminal nuclei.

In this chapter only the major precerebellar centers will be dealt with: the pontine nuclei; the inferior olivary nucleus; and the lateral reticular nucleus. Information concerning projections to the cerebellum from the other centers is limited in the rat, and their established connections with other areas is dealt with in other chapters.

A second topic to be presented in this chapter is the red nucleus, a prominent center situated in the midbrain tegmentum. The red nucleus is included in this discussion because of the key role that it plays in the interaction between the cerebellum and the precerebellar centers. It not only provides inputs to the inferior olive and the lateral reticular nucleus, but it also receives a major projection from the cerebellar nuclei which it reciprocates via collaterals of the rubrospinal tract.

THE RAT NERVOUS SYSTEM
ISBN 0 12 547632 9

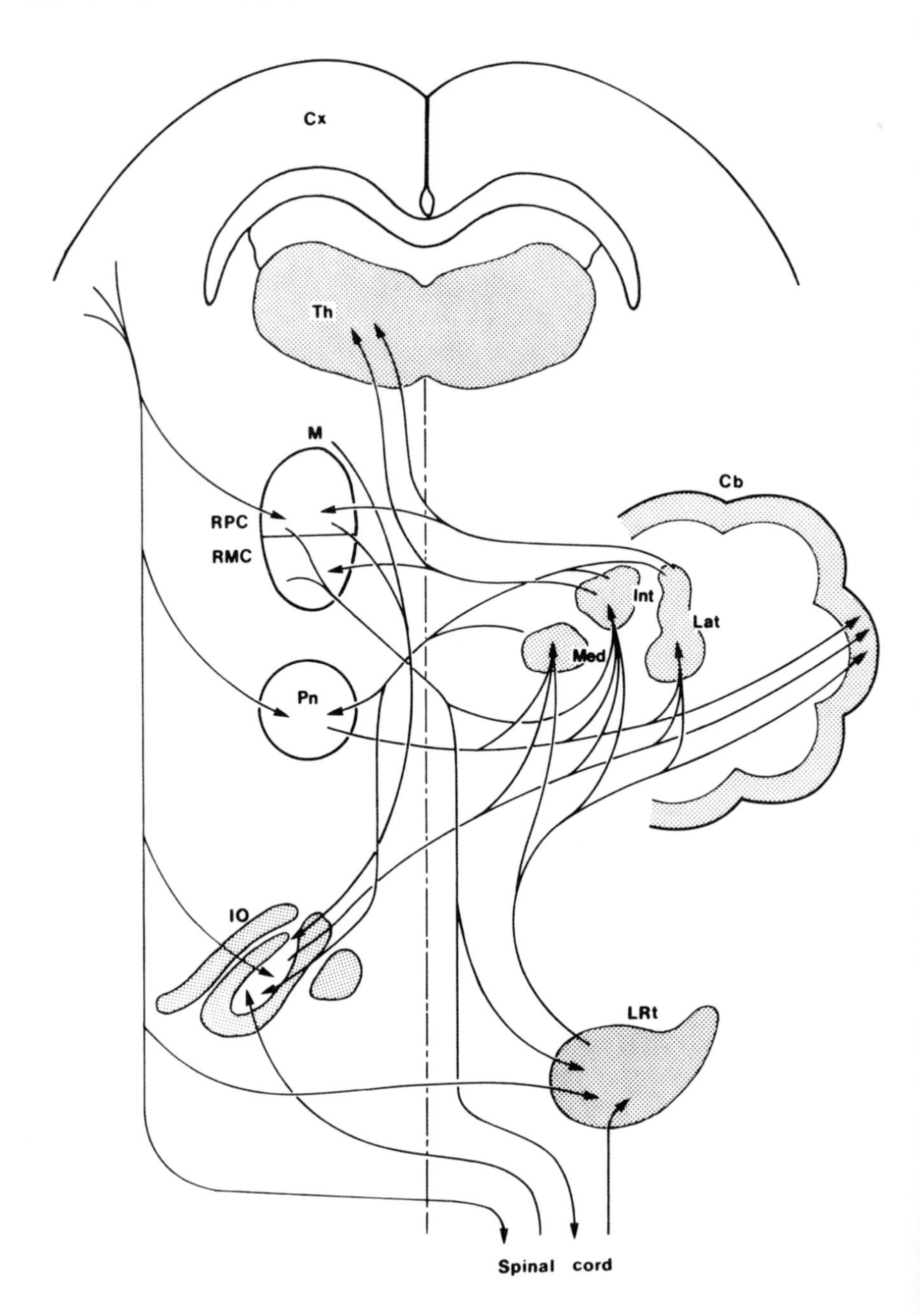

Fig. 1: The main connections of the precerebellar nuclei and red nucleus in the rat. Cb, cerebellum; Cx, cerebral cortex; Int, interpositus nucleus; IO, inferior olive; Lat, lateral nucleus; LRt, lateral reticular nucleus; M, midbrain and diencephalic centers; Med, medial nucleus; Pn, pontine nuclei; RMC, red nucleus, magnocellular division; RPC, red nucleus, parvocellular division; Th, thalamus.

This discussion focuses on the organization and interconnections of these centers in the rat (Fig. 1). When information is sparse, related data on other species will be referred to, but references throughout the text are limited largely to studies on the rat.

I. PONTINE NUCLEI

The pontine nuclei are collectively the most extensive of the precerebellar nuclei, and represent the most important relay for cortical pathways to the cerebellum. They are located in the basilar pons, surrounding the descending fibers of the cerebral peduncle and extending from the trapezoid body to the interpeduncular nucleus. The dorsal boundary is marked by the medial lemniscus, the lateral lemniscus and the tectopontine fibers.

A. Cytoarchitecture

The conformation of the basilar pons in the rat is similar to that in mammals generally, and can be divided into four major subdivisions (Mihailoff *et al.*, 1981b). These include groups of neurons situated around the cerebral peduncle as well as cells immediately adjacent to the peduncular surface or situated within the peduncle itself. These subdivisions are therefore termed the medial, ventral, lateral, and peduncular nuclei, according to their position with respect to the peduncle (Fig. 2).

B. Connections

1. Afferents

The pontine nuclei of the rat receive afferent inputs from a number of sources including various regions of the cerebral cortex, the superior and inferior colliculi, the lateral geniculate body, the deep cerebellar nuclei, the lateral and medial mammillary nuclei, the anterior hypothalamus and the lateral nucleus of the optic tract. In other species, inputs to the pontine nuclei have also been described from the spinal cord and the dorsal column nuclei. The cerebral cortical input to the pontine nuclei is numerically and functionally the most important of the afferent systems.

Corticopontine fibers originate from somatosensory and motor areas, visual areas, auditory areas, granular cingulate cortex and rhinal sulcus. Following injections of horseradish peroxidase (HRP) into the pontine nuclei, the greatest number of labeled cells is observed within the sensorimotor cortex and visual area (Wiesendanger and Wiesendanger, 1982a). Similarly, autoradiography reveals that these regions have the heaviest input, which terminates in a medial to lateral succession from the motor, somatosensory and visual cortices, respectively (Wiesendanger and Wiesendanger, 1982b). Corticopontine neurons

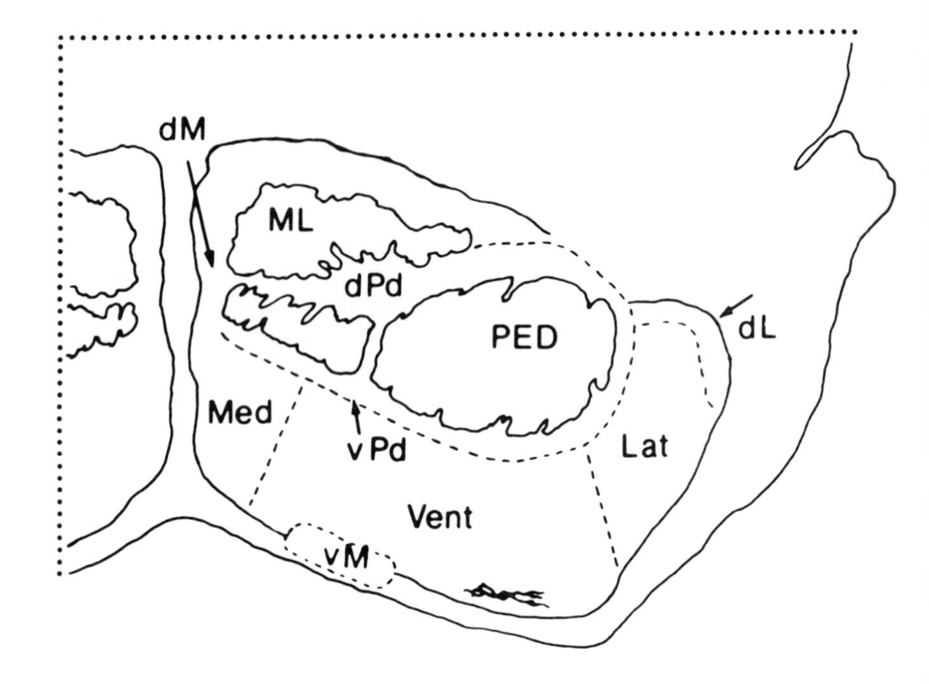

Fig. 2: The subnuclei of the basilar pons in transverse section. dL, dorsolateral area; dM, dorsomedial pontine area; dPd, dorsal peduncular area; Lat, lateral nucleus; Med, medial nucleus; ML, medial lemniscus; PED, cerebral peduncle; Vent, ventral nucleus; vM, ventromedial area; vPd, ventral peduncular area (from Mihailoff *et al.*, 1981b).

in the sensorimotor cortex are situated in layer 5b and include large pyramidal cells and more superficially located small to intermediate sized neurons (Mihailoff *et al.*, 1981c). The large pyramidal cells are believed to give rise to corticospinal neurons, so the pontine inputs from these cells probably arise as collaterals while the smaller cells project primarily to the brain stem (Wise and Jones, 1977). The forelimb sensorimotor cortex projects via the medial third of the cerebral peduncle to terminate ipsilaterally within five longitudinal regions in the rostral two thirds of the pontine nuclei (Mihailoff *et al.*, 1978). The heaviest zones lie just ventral, ventromedial and ventrolateral to the cerebral peduncle while lighter zones of termination are situated between the peduncle and the medial lemniscus, and near the ventral surface in the middle third of the pons. The hindlimb sensorimotor cortex projects via the intermediate and lateral portions of the cerebral peduncle to the same five regions, but the terminal fields are more caudally situated than the forelimb synaptic regions. A small contralateral corticopontine projection which is also topographically organized has been described as well (Mihailoff *et al.*, 1978).

Visual areas also provide a major corticopontine projection in the rat (Wiesendanger and Wiesendanger, 1982a, b). As with sensorimotor neurons, corticopontine cells in visual cortical areas are located in layer 5b (Mihailoff *et al.*, 1981c). Autoradiography has revealed that area 17 projects to three areas in the rostral pons: a medial area, a lateral area below the cerebral peduncle, and a separate ventrolateral region. When area 18 is also injected additional termination sites are labeled dorsolateral and lateral to the cerebral peduncle (Burne *et al.*, 1978). Medial and lateral inputs appear to be unique to the rat. Both anterograde and retrograde methods reveal a sparse cortical projection from the auditory areas, the granular cingulate cortex and the rhinal sulcus (Wiesendanger and Wiesendanger, 1982a, b).

Physiologic data for the rat also reveal strong projections to the pontine nuclei from the sensorimotor, visual and auditory cortices (Potter *et al.*, 1978). The sensorimotor cortex, particularly the face areas, provides the greatest input. Many neurons in the pontine nuclei receive convergent inputs from functionally different cortical areas. Further, physiologic methods do not reveal a point to point projection pattern as seen with anatomic techniques. Considerable cross-talk has been found to occur between the synaptic columns of cortical input demonstrated anatomically. The convergence seen in physiologic studies may be due to overlapping of the zones of corticopontine projections. Alternatively, local convergence may be due to intrinsic neurons. In view of the dendritic pattern seen in Golgi preparations, however, it seems likely that much of the convergence is due to the tendency of the dendritic trees of the pontine neurons to invade neighboring projection columns (Mihailoff *et al.*, 1981b).

The pontine nuclei also receive inputs from other visual centers including the superior colliculus and the lateral geniculate body. The ventral nucleus of the lateral geniculate gives rise to a geniculopontine tract which courses via the basis pedunculi and the lateral lemniscus to terminate in the dorsomedial and lateral areas of the pontine gray (Graybiel, 1974; Ribak and Peters, 1975). The superior colliculus also provides a visual input to the ipsilateral peduncular, dorsolateral and ventrolateral regions in the caudal basilar pons and to the contralateral dorsomedial and medial peduncular areas (Burne *et al.*, 1981). The ipsilateral projection to the lateral basilar pons is topographically organized: the medial superior colliculus to the peduncular region, and the lateral superior colliculus to ventrolateral areas. The pontine projection from the entire tectum also shows a degree of topography: the pretectum to the rostromiddle basilar pons; the superior colliculus to caudal pontine regions; and a sparse projection from the inferior colliculus to even further caudally situated areas.

The pontine nuclei play an important role related to the execution of voluntary movements, as suggested by their position in a major pathway from the sensorimotor cortex to the cerebellum. The involvement of the basilar pontine gray in visual and auditory function has also been well established by both anatomic and physiologic methods. In the rat, a rostrocaudal sequence of inputs from the visual cortex, the lateral geniculate body, the pretectum and the colliculi terminates within the lateral pons (Burne *et al.*, 1981; Eisenman, 1980). The visual areas of the cerebellar cortex, that is, the posterior midvermal lobules and the paraflocculus, receive inputs from the pontine nuclei and the reticulotegmental nucleus. Burne *et al.* (1978) have reported that visual and auditory cortices provide the major source of visual input to the paraflocculus whereas tectal projections terminate primarily in the posterior midvermis. They suggest that the visual projection to the paraflocculus mediates visuomotor integration by initiating or controlling visually guided movement. Physiologic evidence suggests that the tectopontomidvermal system participates in the initiation and execution of saccadic eye movements. On the other hand, the reticulotegmental nucleus contributes to optokinetic responses via its projection to the flocculus.

Most of the cerebellopontine fibers emerge from the cerebellum in the superior cerebellar peduncle, decussate in the caudal midbrain and enter the descending limb. The cerebellopontine axons leave the descending limb along the length of the basilar pons, coursing through and around the medial lemniscus and cerebral peduncle to terminate within the pontine nuclei (Chan-Palay, 1977; Watt and Mihailoff, 1983a). The lateral nucleus gives rise to the largest cerebellar projection, terminating throughout the rostral four fifths of the basilar pons as longitudinally oriented columns. Each of the major subdivisions of the pontine gray thus receive an input via this projection.

The projection from the interpositus complex arises entirely from its anterior division, including the dorsolateral hump region, and terminates in the caudal half of the contralateral basilar pons in the medial, ventral, lateral and peduncular nuclei (Watt and Mihailoff, 1979, 1983a). The medial nucleus terminates primarily in the contralateral dorsomedial basilar pons throughout the rostral four fifths of the pontine gray. A small projection to the contralateral dorsomedial pontine region at midpontine levels has also been described (Watt and Mihailoff, 1983a).

Ipsilateral projections which are more sparse than the contralateral inputs originate from all three deep cerebellar nuclei. In general, the distribution of ipsilateral input mirrors closely the contralateral terminal zones. It has recently been proposed that the deep cerebellar nuclear cells that project to the pontine nuclei also provide collaterals to the cerebellar cortex, thalamus, red nucleus and inferior olive (Watt and Mihailoff, 1983b).

Although the main projections to the pontine nuclei arise from those areas described above, smaller inputs from other sources have also been reported in the rat. The mammillary nuclei provide a projection to the pontine gray (Cruce, 1977), and the nucleus of the optic tract also provides a topographically organized input to the medial third of the ipsilateral, lateral pontine nucleus.

It is apparent from the patterns of termination described above that a good deal of overlap occurs between the afferent systems of the pontine nuclei (Watt and Mihailoff, 1983a). Substantial overlap occurs between those areas that receive lateral cerebellar nucleus inputs and sensorimotor cortex projections, particularly in the medial and ventral pontine regions which are in receipt of motor limb and motor face inputs. In the lateral pontine region the lateral cerebellar nucleus projections overlap with projections from the auditory and visual cortices, and the superior colliculus. The interpositopontine inputs also overlap extensively with projection zones of the sensorimotor cortex. This occurs principally in the medial, ventral and dorsal peduncular pontine areas that receive inputs from hindlimb and trunk regions of sensorimotor cortex. These areas in turn partially overlap with the projection zones of the lateral cerebellar nucleus. The medial nucleus of the cerebellum projects to midpontine and rostral dorsomedial pontine levels which also receive inputs from the medial and lateral mammillary nuclei of the hypothalamus. The medial and lateral cerebellar nucleus inputs also overlap in the dorsomedial region throughout the rostral three fourths of the basilar pons.

2. *Efferents*

The pontine nuclei are a major source of mossy fibers to most regions of the cerebellar cortex and the deep cerebellar nuclei. It is generally accepted that all cells in the pontine nuclei project to the cerebellum via the middle cerebellar peduncle and terminate predominantly on the contralateral side.

The pontine nuclei in the rat provide a major projection to the cerebellar hemispheres and a more restricted one to the vermis. The simple lobule receives an input from the ventral pons at rostral levels with an ipsilateral predominance. Crus 1 receives a projection from the medial, ventral and lateral perimeters of the pontine gray while Crus 2 receives fibers from more central regions, both with a contralateral predominance. Central regions of the basilar pons also project to the paramedian lobule. Projections from the dorsal peduncular region and the lateral and dorsolateral areas of the basilar pons have occasionally been observed as well (Mihailoff *et al.*, 1981a, c). Burne *et al.* (1978) have demonstrated a dual topography within the pontine projection to each lateral cerebellar hemisphere. This projection arises from concentric zones in the central basilar pons so that the center projects to the anterior lateral hemisphere and the cells around the core terminate within the more posterior lobules. The results of Burne *et al.* (1978) also suggest a convergence of pontine inputs within specific lobules.

Projections to the posterior lobe of the vermis arise bilaterally within the pontine nuclei (Azizi *et al.*, 1981). Cells within the rostral half of the basilar pons project to all subdivisions of lobule 6, and limited cell groups in the same area also terminate in lobule 7. Lobule 6a receives its input mainly from rostral and caudal pons while the projection to lobules 6b and 6c arises from medial and ventral parts of the rostral pons and medial, ventral and lateral parts of the midpons. Lobule 7 also receives a projection from the midpons, including medial, ventral, lateral and dorsolateral areas. A dense projection to lobule 8 arises from the caudal pontine gray while a smaller input originates from dorsal and lateral parts of the rostral and midpons, and ventral parts of the rostral pons. Lobule 9a and lobule 9b receive a projection from small collections of cells in the medial and lateral parts of the rostral pons, and peduncular and dorsolateral areas of the caudal and midpons. The projection to lobule 9c arises from dorsolateral and dorsomedial aspects of the caudal basilar pons. In general, a rostrocaudal arrangement occurs such that the rostral pons projects to lobule 6b and lobule 6c, the midpons to lobule 7 and the caudal pons to lobule 8 (Azizi *et al.*, 1981). A topographic pattern in the projection to lobule 9 has also been described whereby lobule 9a receives a projection from a dorsointermediate region and lobules 9b and 9c from more ventral regions (Eisenman and Noback, 1980).

The pontine projection pattern to lobule 8 (pyramis) and its lateral extensions (copula pyramidis) has also been studied in the rat (Eisenman, 1981). Three parasagittal zones occur and receive differing input from the caudal pons. The medial zone receives a projection from medial and ventrolateral regions of the caudal pons, the intermediate zone from the intermediate area, and the lateral zone from medial, ventrolateral and dorsal areas. The paraflocculus receives a projection from a medial and a lateral column of cells oriented rostrocaudally, as well as from scattered cells within the peduncular nucleus. Discrete projections from these areas course separately to the dorsal and the ventral paraflocculus (Eisenman, 1980). The pontine regions projecting to the paraflocculus are mainly in receipt of visual cortical projections and, to a lesser degree, auditory inputs as well (Burne *et al.*, 1978). The area of the basilar pons that provides the largest projection to the posterior vermis is situated dorsolaterally and corresponds to the pontine region that receives a major input from the superior and inferior colliculi. Although the caudal midvermis is commonly held to be part of a midvermal visual–auditory region, it is apparent that this pathway is mediated by a tectopontocerebellar system rather than a visual–auditory corticopontine projection (Azizi *et al.*, 1981). Pontine projections to the anterior lobe have not yet been studied in the rat.

It is generally held that the pontocerebellar projection is very precisely organized and is characterized both by patterns of convergence and divergence. Convergence is readily apparent in the retrograde transport studies described above whereas divergence can be demonstrated by the use of double labeling methods. Using this approach, it has been possible to show that collateral projections from the same pontine neuron often project to two different lobules within the same hemisphere or even to both hemispheres. Intrahemispheric branching occurs most frequently between Crus 1 and the paraflocculus and less frequently between Crus 2 and the simple lobule. Crus 1 and the paramedian lobule, as well as the simple lobule and the paramedian lobule, occasionally receive collateral projections as well. Fewer cells project to both hemispheres but such projections have been observed between the simple lobule bilaterally, Crus 1 bilaterally, and Crus 2 and the opposite paramedian lobule (Mihailoff, 1983).

C. Reticulotegmental nucleus

The reticulotegmental (RtTg) nucleus, also termed the nucleus reticularis tegmenti pontis, lies dorsal to the medial lemniscus rostrally and the trapezoid body caudally. Although it is often considered along with the pontine nuclei, it is a separate center with different connections and correspondingly different functions. In mammals generally, the largest number of afferents to this center arise from the contralateral deep cerebellar nuclei and course in the descending limb of the superior cerebellar peduncle. Most of these fibers originate from the lateral cerebellar nucleus and the anterior interpositus nucleus. The projection from the cerebral cortex originates mainly in the sensorimotor area. Other afferents are also provided by the contralateral superior and lateral vestibular nuclei as well as the mammillary nuclei (Cruce, 1977) and the lateral hypothalamic area (Hosoya and Matsushita, 1981). A number of visual centers, including the superior colliculus (Burne *et al.*, 1981), the nucleus of the optic tract (Terasawa *et al.*, 1979) and the lateral geniculate body (Graybiel, 1974) also provide inputs. In general, projections to the cerebellum from the RtTg are predominantly contralateral and terminate within most parts of the cortex, especially the visual vermal lobules 6 and 7. Recently, double labeling has shown that collateral projections are common in the RtTg cerebellar pathway (Mihailoff, 1983). Although many of the RtTg fibers project onto the vermal visual area, in the cat they arise from cells in areas that do not receive inputs from the pretectum and superior colliculus. On that basis it has been concluded that the RtTg is mainly involved in transmission of nonvisual information. Evidence suggests that the majority of the RtTg cells play a role in locomotion. Nevertheless, a restricted pathway from the superior colliculus to the flocculus via this center is likely since RtTg neurons have been shown physiologically to transfer visual signals to the flocculus. Burne *et al.* (1981) have found that tectal inputs to the cerebellar hemispheres in the rat are relayed primarily through the RtTg rather than the pontine nuclei. Moreover, both of these centers play a role in relaying collicular information to the vermis. The role of the RtTg in visual function thus requires further investigation in the rat.

II. INFERIOR OLIVARY NUCLEUS

The inferior olivary nucleus (IO), is thought to be the sole source of cerebellar climbing fibers (Desclin, 1974). This nucleus is found bilaterally as a folded gray

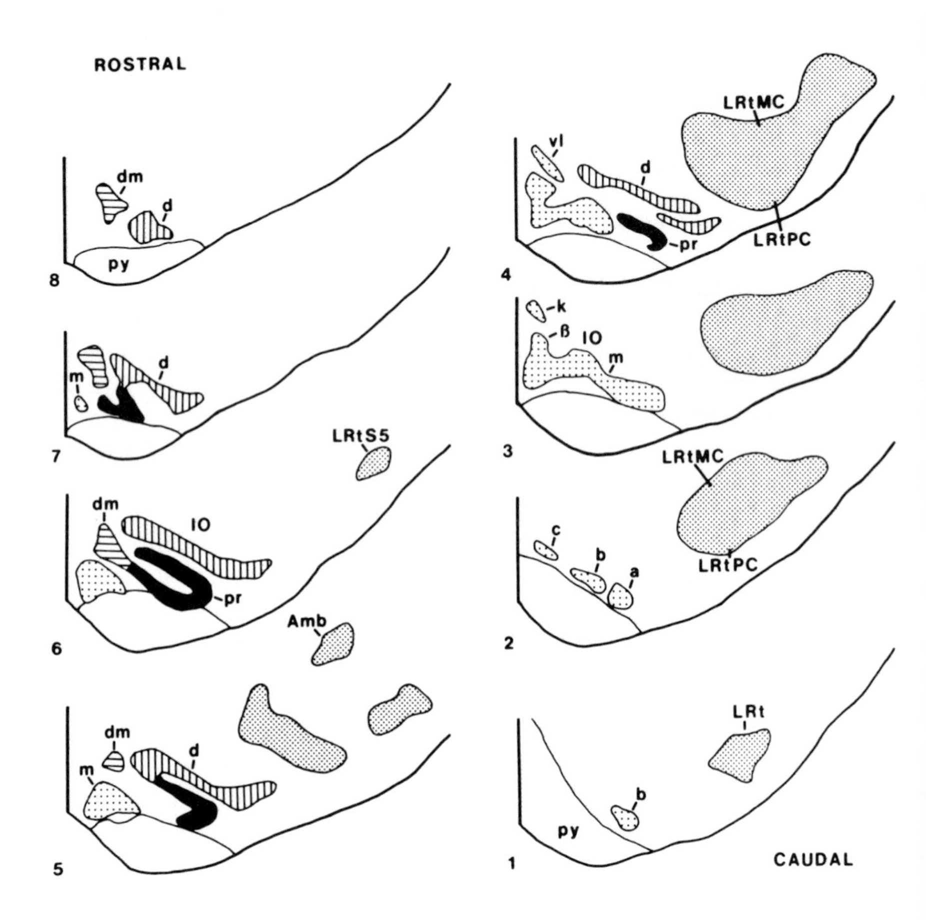

Fig. 3: The inferior olivary nucleus and lateral reticular nucleus at representative levels in transverse section. Amb, ambiguus nucleus; IO, inferior olivary nucleus; a, b, c, subnuclei a, b, c; β, subnucleus β; d, dorsal nucleus; dm, dorsomedial cell column (group); k, dorsal cap (of Kooy); m, medial nucleus; pr, principal nucleus; vl, ventrolateral outgrowth; LRt, lateral reticular nucleus; LRtMC, magnocellular division; LRtPC, parvocellular divison; LRtS5, subtrigeminal division; py, pyramidal tract.

mass of cells lying within the ventromedial medulla oblongata just dorsolateral to the pyramid (Fig. 3). It is situated medial to the lateral reticular nucleus caudally, while its rostral half has the paragigantocellular reticular nucleus as its lateral border. The dorsal boundary of the IO is limited by the reticular nucleus of the ventral medulla, the paramedian reticular nucleus, and the gigantocellular reticular nucleus. The IO extends for a distance of about 2600 μm, beginning about 1500 μm caudal to the obex and running rostrally to end at the level of the caudal pole of the facial nucleus.

A. Cytoarchitecture

The inferior olivary complex is composed of three large subdivisions, the principal, medial accessory and dorsal accessory olive (Kooy, 1917), and four additional smaller subdivisions, the dorsal cap, the ventrolateral outgrowth, the nucleus β and the dorsomedial cell column (Brodal, 1940).

The cytoarchitecture of the IO of the rat consists of all the subdivisions that occur in mammals generally (Gwyn *et al.*, 1977) (Fig. 3). Cell group b of the medial accessory olive forms the caudal pole of the IO (level 1). Further rostrally (level 2), the medial accessory olive is composed of three separate nuclear groups referred to as cell groups a, b and c from lateral to medial (level 2). These three groups eventually merge further rostrally, and a dorsal extension forms from the dorsum of group c. This extension enlarges and the more dorsal cells eventually separate to form the dorsal cap of the medial accessory olive, while the remaining dorsal extension of group c comprises nucleus β (level 3). Further rostrally (level 4), the dorsal cap elongates ventrolaterally and becomes the ventrolateral outgrowth. At this level, the medial accessory nucleus begins to decrease in size until it completely disappears at about the level of the junction of the middle third and the rostral third of the IO (level 7). At the junction between the ventrolateral outgrowth and dorsal cap, the principal and dorsal accessory olive begin (level 4). These two divisions extend throughout the rostral two thirds of the olivary complex. The dorsal accessory olive is situated dorsolateral to the medial accessory olive while the principal olive is located between the two. The dorsal accessory olive eventually increases in size so that its lateral aspect approaches the ventral border of the medulla oblongata and its medial aspect merges with the dorsal lamella of the principal olive (level 5). Within the rostral third of the IO, the dorsal accessory olive begins to decrease in size (level 7) and eventually merges with the dorsomedial cell column to form the rostral pole of the olivary complex (beyond level 8).

The principal olive first appears just rostral to the caudal pole of the dorsal accessory olive as a small compact group of neurons squeezed between the dorsal and medial accessory olive. More rostrally, the principal olive divides into dorsal and ventral lamellae which are joined ventrolaterally. At its rostral limit, the principal olive again appears as a compact group of neurons which eventually disappears, leaving the dorsal accessory olive and the dorsomedial cell column to form the rostral pole of the olivary complex (levels 7, 8).

The dorsomedial cell column first appears at the rostral limit of the ventrolateral outgrowth as a distinct aggregation of neurons just dorsal to the medial accessory olive (level 5). It is located throughout the rostral half of the IO. As the dorsomedial cell column continues rostrally from its caudal pole it enlarges and then fuses with the ventral lamella of the principal olive, where it reaches its greatest mediolateral dimensions. Further rostrally, the principal olive gradually disappears caudal to the rostral pole of the IO.

B. Connections

1. *Afferents*

The afferent connections of the IO have been extensively studied in many mammals and excellent reviews have recently been published (Brown Gould, 1980; Courville *et al.*, 1980). Recent HRP studies of the afferents of the rat IO show that olivary inputs originate from many areas of the central nervous system from the lumbar spinal cord all the way to the cerebral cortex (Brown *et al.*, 1977; Cintas *et al.*, 1980; Swenson and Castro, 1983a).

The spinal projection to the IO is predominantly a contralateral one and originates from the cervical, thoracic and lumbar segments (Swenson and Castro, 1983a). The majority of the spino-olivary cells are located medially at the base of the dorsal and ventral horns, that is, laminae 4, 5 and 7, and their axons probably cross in the anterior commissure before ascending, primarily in the ventral funiculus of the spinal cord, to terminate within the IO (Swenson and Castro, 1983b). The hindlimb input terminates in cell groups a and b and in the caudolateral dorsal accessory olive. The forelimb projection terminates within these same divisions slightly more rostrally and medially than the hindlimb projection. The similarity between spinal cord labeling in the cat and the rat (see Courville *et al.*, 1980; Swenson and Castro, 1983a) suggests that the organization of their spino-olivocerebellar pathways may be similar. It has been shown that spinal inputs may influence olivary neurons both directly and indirectly. Electrophysiologic studies in the cat have determined at least five direct spino-olivary pathways (Oscarsson and Sjölund, 1977). In addition to these direct spino-olivary pathways, indirect spino-olivary pathways have been determined. One such pathway is the dorsal funiculus spino-olivocerebellar pathway (Oscarsson, 1969) which involves a relay at the dorsal column nuclei (gracile and cuneate nuclei). Both anatomic and physiologic studies have shown that both dorsal column nuclei send their axons to the contralateral IO and terminate in the same olivary areas as do the direct spinal inputs (Brown *et al.*, 1977; Oscarsson, 1969; Swenson and Castro, 1983a, b).

The IO receives inputs from other medullary centers such as the lateral reticular nucleus, the reticular formation, the spinal trigeminal nucleus, the vestibular complex, the prepositus hypoglossal nucleus and the raphe obscurus nucleus. Projections from various brain stem reticular nuclei to the IO are thought to be part of a polysynaptic spino-olivary system. A small projection from the lateral reticular nucleus has been found. It is bilateral and terminates within cell groups a and b of the caudolateral medial accessory olive (Brown *et al.*, 1977; Swenson and Castro, 1983a, b). The gigantocellular reticular nucleus also projects to the IO bilaterally, primarily to the ipsilateral caudal medial accessory olive (cell groups b, c and nucleus β) and contralaterally to the rostral dorsal accessory olive. A small projection also terminates within the contralateral caudomedial medial accessory olive (Brown *et al.*, 1977; Swenson and Castro, 1983a, b).

Another pathway which belongs to the somatosensory system of inputs to the IO originates from the spinal trigeminal nucleus. This input is largely contralateral, originating from the interpolar region and terminating within the rostromedial dorsal accessory olive (Brown *et al.*, 1977; Swenson and Castro, 1983a, b). It is thought to be responsible for receiving precise somatotopic information from the face and then relaying this information to the cerebellum (Cook and Wiesendanger, 1976).

The vestibular complex also sends a projection to the IO. This projection originates from the medial and descending (spinal) vestibular nucleus and terminates primarily within the ipsilateral nucleus β and the contralateral dorsomedial cell column. A small portion of this projection terminates within the ipsilateral cell groups b and c of the caudal medial accessory olive. The prepositus hypoglossal nucleus, which is thought to play a role similar to that of the vestibular nuclei, also projects primarily ipsilaterally to the nucleus β and the dorsomedial cell column as well as the dorsal cap and cell group c. A small projection to the IO originates from the raphe obscurus nucleus (Brown *et al.*, 1977; Swenson and Castro, 1983a, b).

Olivary neurons which project to the anterior lobe of the cerebellum are usually activated by the cerebral motor cortex. In the rat, cortico-olivary fibers originate from pyramidal cells which lie in lamina 5 of the frontal and parietal cortex (sensorimotor cortex). This projection is bilateral, with an ipsilateral predominance (Brown *et al.*, 1977; Swenson and Castro, 1983a). The frontal cortex projects to the medial dorsal accessory olive, the dorsomedial cell column and the caudomedial medial accessory olive (Swenson and Castro, 1983b).

The pretectal complex and the superior colliculus, regions which receive a primary optic input, send significant projections to the IO. One such projection originates from the caudal pretectum and from the small cells of the stratum profundum of the superior colliculus. It terminates contralaterally within cell group c and ipsilaterally within the medial dorsal lamella of the principal olive (Swenson and Castro, 1983a, b). A second projection originates from the rostral portions of the pretectum, that is, from the caudolateral part of the anterior pretectal nucleus and the nucleus of the optic tract, the olivary pretectal nucleus, and the posterior pretectal nucleus, and terminates within the ipsilateral dorsal cap (Brown *et al.*, 1977; Swenson and Castro, 1983a, b).

The richest source of afferents to the rat IO originates from the mesencephalon and the caudal diencephalon. Three studies have recently been published in which HRP was injected into the olivary complex, and all studies show labeled cells within the following mesencephalic regions: the interstitial nucleus of Cajal and adjacent mesencephalic reticular formation; the central gray; the ventral tegmental area; the medial terminal nucleus of the accessory optic tract; and the medial and lateral deep mesencephalic nuclei. Labeled cells have also been observed within the following caudal diencephalic regions: the nucleus of Darkschewitsch; the subparafascicular nucleus; the zona incerta; the fields of Forel; and the Edinger-Westphal nucleus (Brown *et al.*, 1977; Cintas *et al.*, 1980; Swenson and Castro, 1983a). All three studies also found that the area within or about the rostromedial red nucleus contained labeled cells (see later). The anterograde autoradiographic tracing method has also been employed to determine the terminal fields of various mesencephalic-caudal diencephalic afferents to the olivary complex (Swenson and Castro, 1983b). It was found that the periaqueductal group, the subparafascicular nucleus and the nucleus of Darkschewitsch send projections to the same olivary areas, that is, the rostromedial accessory olive, both lamellae of the principal olive, the dorsomedial cell column, and weakly to cell groups a and b of the caudal medial accessory olive and the medial portions of cell group c and nucleus β. The medial terminal nucleus of the accessory optic tract projects only to the ventrolateral outgrowth. The projection from the lateral deep mesencephalic nucleus terminates quite heavily within nucleus β and the dorsal accessory olive, and lightly within the dorsomedial cell column, ventrolateral outgrowth and cell group c. The medial deep mesencephalic nucleus projects to different areas, that is, quite substantially to the dorsal lamella of the principal olive and less so to the medial accessory olive, the dorsomedial cell column, the ventral lamella of the principal olive and the ventrolateral outgrowth. The controversial rubro-olivary projection, whether it originates from the rostral red nucleus or the neighboring prerubral field, terminates within the dorsal lamella of the principal olive, the dorsomedial cell column, the ventrolateral outgrowth and the caudolateral medial accessory olive (Anderson *et al.*, 1983; Swenson and Castro, 1983b).

The deep cerebellar nuclei provide a highly organized topographic input to the IO which probably acts as a feedback mechanism into the cerebellar circuits. Studies using either autoradiography or retrograde transport of HRP have demonstrated that all three deep cerebellar nuclei provide inputs to the IO of the rat. Although Brown *et al.* (1977) found that the cerebello-olivary pathway originated only from the lateral and interpositus nuclei, other studies have shown that the medial nucleus in the rat also projects to the IO (Achenbach and Goodman, 1968; Angaut and Cicirata, 1982; Swenson and Castro, 1983a, b). The cerebello-olivary input from the lateral and interpositus nuclei is predominantly contralateral. It originates from small cells within these nuclei as well as some larger cells within the dorsal and caudolateral regions of the lateral nucleus, travels in the brachium conjunctivum to its decussation in the caudal midbrain, and then enters the descending limb of the superior cerebellar peduncle to terminate within the contralateral IO (Chan-Palay, 1977). A small ipsilateral projection has also been described which is similar, but it recrosses in the olivary decussation to the ipsilateral olive (Chan-Palay, 1977). That part of the cerebello-olivary pathway which originates from the medial nucleus travels by way of the uncinate fasciculus to terminate within the contralateral IO (Achenbach and Goodman, 1968).

A clear topographic arrangement of the cerebello-olivary pathway has been demonstrated in many species. In the rat, the topographic organization of this pathway has recently been studied using autoradiographic techniques (Angaut and Cicirata, 1982; Swenson and Castro, 1983b). The rostral half of the interpositus nucleus sends its projection to the dorsal accessory olive, while the caudal half projects to the rostral half of the medial accessory olive and the ventral lamella of the principal olive (Swenson and Castro, 1983b). The anterior interpositus nucleus projects to the dorsal accessory nucleus, while the posterior interpositus nucleus projects to the rostral and central medial accessory olive (Angaut and Cicirata, 1982). A more detailed topographic arrangement has been reported for the lateral nucleus–olivary projection. The rostral lateral nucleus projects to the dorsal lamella while the caudal area projects to the ventral lamella. The lateral nucleus also projects to the ventrolateral outgrowth (Swenson and Castro, 1983b). The ventromedial part of the lateral nucleus projects to the lateral part of the ventral lamella, while the dorsomedial part projects to the medial ventral lamella and the dorsomedial cell column. The ventrolateral lateral nucleus sends its projection to the lateral aspect of the dorsal lamella while the dorsolateral area projects to the medial dorsal lamella of the principal olive. The caudal half of the medial nucleus sends a direct input to the contralateral caudomedial medial accessory nucleus, primarily to its ventrolateral outgrowth and nucleus β, while the rostral half of the medial nucleus does not appear to send any inputs to the IO (Angaut and Cicirata, 1982; Swenson and Castro, 1983b).

2. *Efferents*

Desclin (1974) identified the IO as the sole source of climbing fibers to the rat cerebellum and his view has since been supported by other studies (Anderson and Flumerfelt, 1980). Chan-Palay *et al.* (1977) have shown sagittal banding of labeled terminal fields within the deep cerebellar nuclei and cortex following bilateral labeling of the entire olive by ^{35}S-methionine. Their data suggested that the olivocerebellar climbing fibers terminate within these bands and that the bands which are not labeled contain climbing fiber terminals originating from extraolivary regions. However, Campbell and Armstrong (1983a) have recently shown that all climbing fibers can be labeled with ^{3}H-methionine following injections of the IO.

It is generally accepted that all the neurons within the IO send their axons to the cerebellum. The olivocerebellar pathway is mainly contralateral with a smaller uncrossed ipsilateral component. The contralateral projection crosses in the olivary commissure while the ipsilateral projection ascends uncrossed to reach the cerebellum by way of the lateral part of the inferior cerebellar peduncle. When the olivocerebellar pathway reaches the base of the cerebellum, it breaks up into smaller pathways which surround and penetrate the three deep cerebellar nuclei. Collaterals synapse within the deep cerebellar nuclei, while the main axons of the olivocerebellar projection continue on to terminate within sagittal bands in the molecular layer (Chan-Palay *et al.*, 1977).

The rat olivocerebellar projection, like that of the cat, is topographically organized (see Brodal and Kawamura, 1980; Brown Gould, 1980; Campbell and Armstrong, 1983b). The retrograde tracing studies in the rat by Brown (1980), Eisenman (1981), Hrycyshyn *et al.* (1982b) and most recently by Furber and Watson (1983) have determined the organization of this projection. The rostral anterior lobe vermis (lobules 1 to 3) receives afferents from cells located laterally within the caudal areas of both the medial and dorsal accessory olive, while the caudal anterior lobe vermis (lobules 4 and 5) receives afferents from cells which lie more medially within these two areas (Furber and Watson, 1983; Hrycyshyn *et al.*, 1982b).

The posterior lobe vermis receives a projection from the entire rostrocaudal extent of the medial accessory olive, the dorsal accessory olive, the dorsomedial cell column and neighboring areas of the ventral lamella of the principal olive, as well as the ventrolateral outgrowth (Hrycyshyn *et al.*, 1982b). More specifically, lobule 6 receives its olivocerebellar input from central and medial aspects of the medial accessory olive as well as the medial caudal two thirds of the dorsal accessory olive. The central and medial aspects of the medial accessory olive as well as nucleus β also project to lobule 7 (Furber and Watson, 1983). There also appears to be a small projection to lobules 6 and 7 from the ventral lamella of the principal olive, the dorsomedial cell column and the ventrolateral outgrowth (Hrycyshyn *et al.*, 1982b). Lobule 8 receives its projection mainly from the caudal half of the olivary complex (Hrycyshyn *et al.*, 1982b). Most of the projection arises from cells within two areas of the caudal two thirds of the medial accessory olive. One group lies within the medial aspect and the other within the lateral aspect. Following HRP injections of both lobules 8 and 9, Furber and Watson (1983) found labeled cells within nucleus β, the dorsal cap, and the dorsomedial cell column. It has been reported that lobule 8 (pyramis) of the rat contains three sagittal bands of olivocerebellar terminals on either side, and that these bands receive their projections from discrete areas within the medial accessory olive (Eisenman, 1981).

The lateral part of the simple lobule receives afferents from the medial aspect of the medial accessory olive, discrete cell groups within the ventral lamella of the midportion of the principal olive, the dorsomedial cell column, and the medial aspect of the middle half of the dorsal accessory olive. The medial part of the simple lobule, however, receives afferents from the medial aspect of both the rostral dorsal accessory olive, the medial and central aspects of the rostral principal olive, the dorsomedial cell column, as well as from the rostral medial accessory olive (Hrycyshyn *et al.*, 1982b). Both Crus 1 and 2 (ansiform lobule) receive their inputs primarily from the principal olive (Furber and Watson, 1983; Hrycyshyn *et al.*, 1982b). Olivary projections to the medial part of the paramedian lobule originate from discrete cell groups within the medial aspect of the rostral medial accessory olive and of the middle ventral lamella of the principal olive, as well as the dorsomedial cell column and the most medial part of the more rostral dorsal accessory olive. The lateral part of the paramedian lobule receives afferents from the rostral two thirds of the dorsal accessory olive, the rostral medial accessory olive, the caudal dorsomedial cell column, and the medial and central parts of both lamellae of the principal olive. The copula pyramidis receives its olivary projection mainly from the lateral part of the rostral two thirds of the dorsal accessory olive, the intermediate area of the medial accessory olive and the caudal principal olive. Finally, the paraflocculus receives its input from the rostral medial accessory olive and from the rostral principal olive (Eisenman, 1981; Furber and Watson, 1983; Hrycyshyn *et al.*, 1982b).

A recent autoradiographic study has shown that the olivocerebellar projection in the rat is topographically organized in sagittal strips (Campbell and Armstrong, 1983b). The results of this study are in general similar to those found using the HRP method (Brown, 1980; Eisenman, 1981; Furber and Watson, 1983; Hrycyshyn *et al.*, 1982b). In summary, the caudal medial accessory olive projects to the vermis and flocculus, while the rostral medial accessory olive projects to the hemisphere. The lateral dorsal accessory olive sends a projection to the paravermal cortex of the anterior lobe, while the medial dorsal accessory olive projects to the same area of the posterior lobe. The caudal principal olive projects to the lateral edge of the cerebellar hemisphere, while the rostral principal olive sends its projection to the rest of the hemisphere, that is, the more medial areas, and the dorsal and ventral paraflocculi (Campbell and Armstrong, 1983b).

III. LATERAL RETICULAR NUCLEUS

The lateral reticular nucleus (LRt) is an important precerebellar relay nucleus which sends most of its axons as mossy fibers to the cerebellum. It is comprised of a column of neurons lying ventrolaterally within the medulla between the spinal tract of the trigeminal nerve and the inferior olivary complex. In the rat, the LRt begins at a point about 200 μm below the caudal pole of the inferior olive and extends to the mid-olivary level (Fig. 3) (Kapogianis *et al.*, 1982). Ventrally, it lies next to the surface of the medulla, separated from it by ascending spinal fibers, primarily the ventral spinocerebellar tract which occupies a position ventrolateral to the nucleus. The medial aspect throughout its caudorostral extent is bounded by the inferior olive. The dorsal boundary is surrounded exclusively by the reticular formation at caudal levels, and at more rostral levels where the LRt progressively undergoes separation into a medial and lateral mass, the nucleus is also bounded dorsally by the ambiguus nucleus. The dorsolateral boundary of the caudal half of the LRt lies adjacent to the spinal tract of the trigeminal nerve, while in the more rostral aspect of the nucleus the rubrospinal tract courses lateral to it. At the extreme rostral levels where only the lateral portion of the LRt (subtrigeminal division) remains, the dorsal boundary is formed by the nucleus and spinal tract of the trigeminal nerve.

A. Cytoarchitecture

The comparative study of Walberg (1952) revealed that the cytoarchitecture of the lateral reticular nucleus (LRt) differs substantially in different mammals with regard to its degree of development and organization. Nevertheless, in most animals it consists of three divisions based primarily on the predominance of cells of certain size range within each division. Thus by convention, the LRt of the rat is composed of a dorsomedial magnocellular division, a ventrolateral parvocellular divison, and a rostral subtrigeminal division (Kapogianis *et al.*, 1982).

The caudal pole of the LRt is composed entirely of the parvocellular division. The magnocellular division, which constitutes the greater mass of this nucleus, begins a few hundred micrometers rostral to the caudal pole of the nucleus and rapidly increases in size together with the ventrolaterally placed parvocellular division until the nucleus reaches its maximum size just beyond the halfway mark along its caudorostral extent (Fig. 3; levels 3 and 4). At more rostral levels (level 4), the parvocellular division is reduced to a thin band of cells positioned in the more ventrolateral area, and at the extreme rostral levels (levels 5 and 6), the parvocellular division cannot be easily discerned. However, the magnocellular division begins to subdivide into two parts beyond the level of its maximum size (level 5), one on each side of the ambiguus nucleus. As the nucleus passes still further rostrally, the medial part of the LRt which is positioned ventromedial to the ambiguus nucleus merges and eventually disappears into the reticular formation. The lateral part, that is, that part of the LRt positioned lateral to the ambiguus nucleus, proceeds further rostrally to become the subtrigeminal division, a distinct mass of cells which lies ventral to the nucleus and spinal tract of the trigeminal nerve (level 6).

B. Connections

1. Afferents

Although many studies concerning the afferents to the LRt have been conducted on the cat (Hrycyshyn and Flumerfelt, 1981), only a few studies have been completed on the rat (Flumerfelt and Gwyn, 1974; Flumerfelt *et al.*, 1982; Menetrey *et al.*, 1983). It is generally believed that the LRt receives projections from the spinal cord, red nucleus, cerebellar nuclei, and cerebral cortex. The origin of the spinal projections to the rat LRt has recently been determined in a study using the retrograde HRP technique (Menetrey *et al.*, 1983). Cells from all spinal levels project to the LRt, and particularly large numbers of axons originate from the cervical enlargement and rostral lumbar segments. The more prominent areas include the contralateral ventromedial parts of both intermediate and ventral horns (laminae 7, 8 and 10), the reticular extension of the neck of the dorsal horn bilaterally, and mainly the contralateral superficial layers and nucleus of the dorsolateral funiculus. The spinal projections to the rat LRt travel via the ventrolateral funiculus to terminate within all three divisions of the nucleus (Flumerfelt *et al.*, 1982). The caudal region of the parvocellular division receives the maximum projection, and the magnocellular division, except for the extreme dorsomedial area, also receives a substantial projection. Although the subtrigeminal division receives only a sparse spinal input, it receives a substantial projection from the contralateral red nucleus (Flumerfelt and Gwyn, 1974).

2. Efferents

Although there is no physiologic information on the rat lateral reticular nucleus (LRt), it is thought that information received by the cat LRt is ultimately conveyed to the cerebellum by mossy fibers, where it is employed in the cerebellar control of motor activity. In the rat, reticulocerebellar axons emerge from the LRt and sweep upwards along the lateral surface of the medulla into the ipsilateral inferior cerebellar peduncle. They then pass through the cerebellum between the lateral and interpositus nuclei to travel within the white matter above the deep cerebellar nuclei. Some axons terminate in the ipsilateral lateral and interpositus nuclei while the majority of LRt–cerebellar fibers terminate within the ipsilateral granular cell layer of the cortex in narrow sagittal bands. Still other fibers cross the midline and terminate sparsely in the contralateral lateral and interpositus nuclei and the cortex (Chan-Palay *et al.*, 1977). Most areas of the cortex receive fibers from the LRt and the most abundant projection terminates within the anterior lobe vermis and lobule 8 (pyramis) of the posterior lobe. The LRt projects to areas additional to the “classic” spinal receiving areas and in most cases the projection is topographically arranged (Chan-Palay *et al.*, 1977;

Eisenman, 1982; Hrycyshyn *et al.*, 1982a). Neurons within the more medial LRt send fibers to the caudal lobules of the anterior lobe vermis, and cells more laterally situated send fibers to the rostral lobules of the anterior lobe vermis. That is, lobules 4 and 5 receive inputs primarily from the magnocellular division of the LRt of both the medial and lateral parts, while lobules 2 and 3 receive inputs mainly from cells which lie in the border area between the parvocellular and magnocellular divisions of the medial LRt. The LRt also sends an ample supply of axons to the posterior lobe vermis. The caudal levels of the parvocellular division send a heavy projection to the posterior lobe vermis, especially to lobule 8. Further rostrally, where the LRt divides about the ambiguus nucleus, the ventral areas of the medial and lateral parts both send projections to the posterior lobe. The cerebellar hemispheres also receive a modest input from the dorsal aspect of the LRt. The projection to the simple lobule originates mainly from the caudal two thirds of the magnocellular division, while the projection to the ansiform and paramedian lobules originates mainly from the dorsal aspect of the rostral two thirds of the magnocellular division of the LRt (Hrycyshyn *et al.*, 1982a). The subtrigeminal division of the LRt sends axons which terminate within the caudal anterior lobe vermis, the posterior lobe vermis, the lobulus simplex, and the ansiform and paramedian lobules, as well as the copula pyramidis (Eisenman, 1982; Hrycyshyn *et al.*, 1982a). Thus, there appears to be overlapping of pools of neurons which project to two or three of the main subdivisions of the cerebellum, that is, the anterior and posterior lobe vermis and the hemispheres. Hrycyshyn *et al.* (1982a) suggested that many of the LRt neurons belong to a system whose divergent projections terminate in all three of these regions of the cerebellar cortex. Subsequently, Payne (1983) found collateralization in his investigation of LRt projections to the simple lobule and adjacent Crus I using the fluorochrome double labeling technique. Twelve per cent of the LRt neurons in the rostral half and 5% of neurons in the caudal half project via collaterals to both sides of the cerebellum. Thus it appears, as Cajal suggested in 1911, that each mossy fiber can divide repeatedly into divergent collaterals and that these terminate in different areas within the cerebellar cortex.

A minor pathway originating from the LRt projects to the contralateral inferior olive (Brown *et al.*, 1977). Despite the fact that the LRt is primarily involved in the spinoreticulocerebellar pathway, it also appears to be an intermediate relay to the olive in some of the projections within the spino-olivocerebellar pathway. Other pathways originating from neurons scattered around, but not within, the LRt have recently been reported. These pathways are not related to the role of the LRt as a precerebellar relay nucleus. It has been shown anatomically that they send direct projections to the cord (Ross *et al.*, 1981; Satoh, 1979) or to the hypothalamus (Sakumoto *et al.*, 1978) and belong to either the A1 group of noradrenaline containing cells (Dahlström and Fuxe, 1964) or to the C1 group of adrenalin containing cells (Hökfelt *et al.*, 1974). The neurons that belong to the A1 group lie dorsally and medially to the LRt and are thought to project to the anterior hypothalamus (Sakumoto *et al.*, 1978), and paraventricular hypothalamus (Sawchenko and Swanson, 1982). Some of the noradrenaline containing cells in this vicinity, as well as non-noradrenaline containing cells, have been shown to project to all levels of the spinal cord in the rat (Satoh, 1979). The cells that belong to the C1 group lie in the vicinity of the rostral LRt. These cells project to at least the thoracic spinal cord, especially to the intermediolateral cell column (Ross *et al.*, 1981). Neurons within the LRt have not been shown to contain noradrenaline or adrenaline and are thus not part of the A1 or C1 group. However, there are indications that some LRt cells contain the transmitter proline (Cuénod *et al.*, 1982) and possibly serotonin (Hunt and Lovick, 1982).

IV. RED NUCLEUS

The red nucleus is a major midbrain center which, in mammals generally, receives information from the cerebral cortex and cerebellum, and provides information to the spinal cord and several brain stem structures including the trigeminal nuclei, facial nucleus, inferior olivary nucleus, dorsal column nuclei, vestibular complex, and lateral reticular nucleus. It is a bilateral aggregation of neurons in the midbrain tegmentum extending from the descending oculomotor fibers caudally to the fasciculus retroflexus rostrally. In the rat, it has a round profile in transverse section and appears egg shaped in the parasagittal plane. In the adult, it measures 1200 μm rostrocaudally, 1000 μm mediolaterally and 800 μm dorsoventrally. Fiber tracts which lie in relation to the red nucleus include the medial lemniscus ventrolaterally, tectospinal fibers dorsomedially, and the superior cerebellar peduncle concentrated medially throughout the rostrocaudal extent of the nucleus.

A. Cytoarchitecture

The red nucleus of the rat can be subdivided into a number of neuronal subgroups which are less well defined than in higher order species, but which nevertheless can be distinguished by the predominant cell type therein (Reid *et al.*, 1975a, b). Thus, large neurons predominate in the caudal part of the nucleus while smaller neurons are most numerous in the rostral part. The intermediate area contains a mixture of large and small neurons throughout. On that basis, the caudal half of the nucleus has been designated the magnocellular, large celled portion while the rostral half constitutes the parvocellular, small celled part (Fig. 4). The caudal pole is comprised almostly entirely of large cells clustered among the oculomotor fibers which course ventrally through the

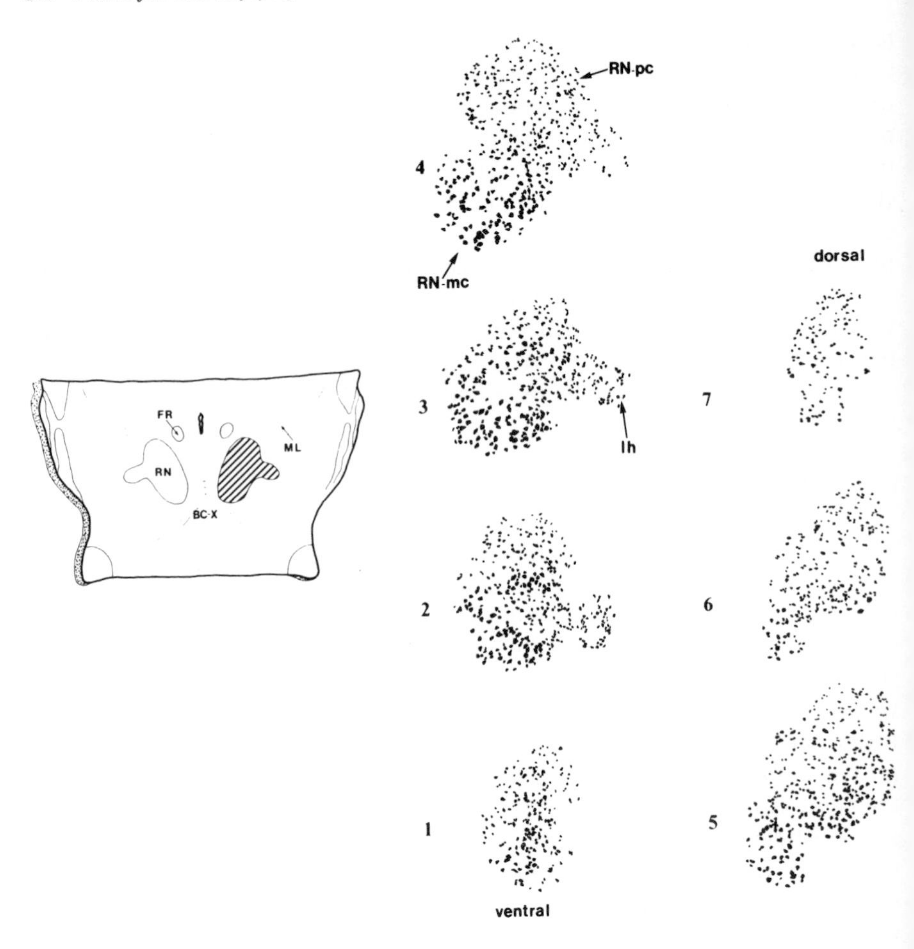

Fig. 4: A horizontal series of cell outline drawings through the red nucleus of the rat (from Reid *et al.*, 1975a). BC-X, decussation of the brachium conjunctivum; FR, fasciculus retroflexus; lh, lateral horn; ML, medial lemniscus; RN, red nucleus; RN-mc, magnocellular part; RN-pc, parvocellular part.

nucleus. Further rostrally, the magnocellular portion divides into dorsomedial and ventrolateral subgroups. Half way up the rostrocaudal extent of the nucleus the cells become concentrated ventrally and project laterally from that region to form the lateral horn. It is noteworthy that a discrete lateral group of small neurons, which has been named the nucleus minimus in the cat and the rabbit, is not present in the rat. The rostral half of the nucleus contains primarily small and medium sized neurons which do not cluster into subgroups in the rat, as seen in primates (Flumerfelt *et al.*, 1973). The rostral pole is therefore difficult to discern with certainty since it blends with the prerubral field. By convention, it is said to lie at the level of the fasciculus retroflexus.

B. Connections

1. Afferents

The major projections to the red nucleus have long been held to arise from the deep cerebellar nuclei and the cerebral cortex, with the cerebellar input being the more extensive of the two. In those species with well developed magnocellular and parvocellular subdivisions, the cerebellar projection is highly organized so that the projection from the interpositus nucleus terminates within the magnocellular portion while the lateral nucleus projects exclusively to the parvocellular portion (Caughell and Flumerfelt, 1977; Flumerfelt *et al.*, 1973). The interpositorubral pathway projects in the form of coarse fibers which terminate on the cell bodies and proximal dendrites of the magnocellular neurons. The pathway from the lateral nucleus terminates entirely as fine fibers on small caliber dendrites in the parvocellular portion. It is now well established that interpositorubral terminals are large and end on cell bodies and proximal dendrites of rubrospinal neurons, while dentatorubral terminals are small and end on intermediate and small dendrites in the parvocellular portion (Dekker, 1981; Flumerfelt, 1978, 1980). While some of the latter may represent distal dendrites of magnocellular neurons, Golgi preparations reveal that few of these dendrites extend into the parvocellular portion. It is therefore likely that most of the dentatorubral terminals end on dendrites of neurons whose cell bodies reside within the parvocellular portion of the nucleus.

The projection from the cerebral cortex to the red nucleus varies with the species. Anatomic studies in the cat have suggested that the motor area projects somatotopically to the caudal two thirds of the ipsilateral red nucleus while the somatosensory area projects to the rostral third. The monkey red nucleus also receives a cortical input throughout its rostrocaudal extent so that motor and premotor areas project bilaterally to the parvocellular portion while the magnocellular part receives an ipsilateral input from the motor cortex. The cortical projection in the rat differs from that in the monkey and the cat as well as most other species for which information is available (Brown, 1974a; Flumerfelt and Gwyn, 1973). It is entirely ipsilateral and projects only to the parvocellular portion. Corticorubral fibers in the rat course ventrocaudally through the thalamus and dorsally from the cerebral peduncle. These fibers terminate most abundantly in the lateral half of the rostral third of the nucleus and do not terminate in any part of its caudal third (Brown, 1974a; Flumerfelt and Gwyn, 1973). Degeneration methods have revealed that the cortical fibers are finer than those from either the lateral or interpositus nuclei and that their terminals are scattered diffusely throughout the neuropil. Electron microscopy has confirmed that the cortical terminals are small and terminate on the distal dendrites and spines of parvocellular neurons (Flumerfelt, 1980).

2. Efferents

The major projection from the red nucleus, which has been recognized since the last century (Held, 1890), is the rubrospinal tract. The rubrospinal fibers emerge from the ventromedial aspect of the nucleus and most cross in the ventral tegmental decussation and take up a position in the ventrolateral brain stem where they descend to various levels of the spinal cord within the lateral funiculus. It is well established that the large cells of the magnocellular red nucleus give rise to rubrospinal fibers. The projection fields of the remaining neurons throughout the nucleus continue to be controversial. For many years it was held that the magnocellular red nucleus gives rise to the entire rubrospinal tract while the parvocellular portion projects to the inferior olivary nucleus. It was then found in the cat that cells of various sizes throughout most of the rostrocaudal extent of the red nucleus provide fibers to this tract. Since that time many studies have provided evidence that at least some parvocellular neurons give rise to rubrospinal fibers (Brown, 1974b; Waldron and Gwyn, 1969). Most recently, retrograde labeling methods have been used to establish the precise origin of the rubrospinal tract in the rat. Shieh *et al.* (1983) have employed HRP to label rubrospinal neurons of various sizes throughout the rostrocaudal extent of the red nucleus following injection at cervical levels. Although this projection is largely contralateral, labeled cells ipsilateral to the injection site were also observed. Huisman *et al.* (1981, 1983) have employed fluorochrome labeling techniques to study the contralateral projection and their findings were similar.

The rubrospinal tract projects as far as the lumbosacral cord in most species, terminating in Rexed's laminae 5 to 7 at various levels. In the rat, it descends in the dorsal part of the lateral funiculus, terminating at the base of the dorsal horn and in intermediate regions of the ventral horn (Brown, 1974b). Through the use of electrophysiologic methods and fluorescent double labeling it has been possible to investigate the tendency for rubrospinal neurons to give rise to collaterals within the cord (Huisman *et al.*, 1981, 1983). In the rat, 20% of rubrocervical neurons provide collaterals to segments caudal to L1. This figure is significantly higher than in the cat and monkey, suggesting that the rat rubrospinal system is less focused than in other species. This is compatible with the role that the rubrospinal tract plays in the control of limb movement, which is relatively less fractionated in the rat.

The red nucleus provides a crossed, rubrobulbar projection, probably via collaterals of the rubrospinal pathway which terminate in a number of brain stem centers including, most notably, the facial nucleus, lateral reticular nucleus, trigeminal nuclei, vestibular complex and dorsal column nuclei. The heaviest input occurs within the lateral part of the facial nucleus, the subtrigeminal and lateral magnocellular part of the lateral reticular nucleus, the nucleus oralis and the main sensory nucleus of the trigeminal complex, the descending vestibular nucleus and the dorsomedial part of the lateral cuneate nucleus (Brown, 1974b; Flumerfelt and Gwyn, 1974; Hinrichsen and Watson, 1983).

The projection from the red nucleus to the inferior olivary nucleus has given rise to considerable controversy in the recent literature. Prior to the advent of the new retrograde tracing methods, this pathway was described in several species. Thus, retrograde and anterograde degeneration following lesions in the central tegmental tract and red nucleus, respectively, suggested an ipsilateral projection from the parvocellular red nucleus to the dorsal lamella of the principal olive. However, several studies in which retrograde tracers have been injected into the inferior olive have suggested that the rubro-olivary pathway is small, and in some species such as the rat it may be entirely absent. Anderson *et al.* (1983) have recently injected lectin conjugated HRP and fluorochromes into the rat inferior olivary nucleus and found no labeled cells in the red nucleus at any survival time. However, these findings are not consistent with the results of Swenson and Castro (1983a, b) who have identified a rubro-olivary tract in the rat using both the autoradiographic and free HRP tracing methods. It has been suggested that the presumptive rubro-olivary projection actually arises from centers adjacent to the rostral red nucleus and that these nuclei were encroached by the lesions or injections employed in the anterograde studies. Thus, following injection of HRP into the inferior olivary nucleus of the rat and the cat, extensive labeling has been observed within the interstitial nucleus of Cajal, the nucleus of Darkschewitsch and the subparafascicular nucleus (Anderson *et al.*, 1983; Cintas *et al.*, 1980; Swenson and Castro, 1983a, b). The latter nucleus has been identified as an important source of olivary afferents in the region of the rostral pole of the red nucleus. It has also been termed the nucleus parafascicularis prerubralis to avoid confusion with the parafascicular and subfascicular nuclei of the thalamus (Carlton *et al.*, 1982). It is likely that lesions of the red nucleus in earlier experimental studies also encroached upon this center or its emerging fibers, thus giving the impression of a substantial rubro-olivary projection in the rat. Nevertheless, the extensive labeling observed within the rostral red nucleus of the cat in several studies provides convincing evidence of the existence of a rubro-olivary pathway in that species. The situation in the rat remains considerably less clear and, until proven otherwise, the very existence of this pathway in this species is open to question.

A reciprocal projection to the deep cerebellar nuclei from the red nucleus in the rat has recently been studied for the first time (Huisman *et al.*, 1983). Double labeling experiments have shown that at least 37% of rubrospinal fibers give rise to rubrocerebellar fibers in the rat and that almost all rubrocerebellar fibers represent collaterals of rubrospinal neurons.

Many studies have shown that the red nucleus is topographically organized in terms of both its afferent and efferent projections. In the monkey and cat, corticorubral fibers terminate in a somatotopic fashion. A somatotopic pattern in the origin of rubrospinal fibers has also been described in the rat (Gwyn, 1971; Shieh *et al.*, 1983). In general, dorsal and dorsomedial regions of the red nucleus project to the cervical cord while ventral and ventrolateral regions project to lumbrosacral segments. The somatotopy of rubral afferents and efferents permits localized motor activity in response to localized peripheral stimuli involving chiefly the flexor muscles of distal segments. It has been shown that the cortical inputs regulate the background excitability of red nucleus cells and this modulates the responsiveness of these cells to other inputs. The motor cortex appears to regulate the effectiveness of other structures on rubrospinal neurons, particularly the cerebellum.

Recent anatomic and physiologic evidence suggests that division of the red nucleus into a dentatorubro-olivary, parvocellular part and an interpositorubrospinal, magnocellular part is probably an oversimplification which merits re-examination. Physiologic data in the cat have revealed that some dentatorubral fibers terminate directly on rubrospinal neurons. A projection from the parvocellular red nucleus to the inferior olive in the rat is now in question and a substantial contribution to the rubrospinal tract from the parvocellular area has been established. As a result, the cortical input is no longer believed to terminate exclusively in the rubro-olivary part of the nucleus and the rat is not unique from other species in that respect. While important new information concerning the anatomy of the red nucleus in the rat has thus provided new insights into its functional arrangement in this species, several details of its organization still remain to be elucidated.

REFERENCES

Achenbach, K. E. and Goodman, D. C. (1968). Cerebellar projection to pons, medulla and spinal cord in the albino rat. *Brain Behav. Evol.* 1, 43–57.

Anderson, W. A. and Flumerfelt, B. A. (1980). A light and electron microscopic study of the effects of 3-acetylpyridine intoxication on the inferior olivary complex and cerebellar cortex. *J. Comp. Neurol.* 190, 157–174.

Anderson, W. A., Rutherford, J. G. and Gwyn, D. C. (1983). Midbrain and diencephalic projections to the inferior olive in rat. *Proc. Can. Fed. Biol. Soc.* 26, 132.

Angaut, P. and Cicirata, F. (1982). Cerebello-olivary projections in the rat. *Brain Behav. Evol.* 21 24–33.

Azizi, S. A., Mihailoff, G. A., Burne, R. A. and Woodward, D. J. (1981). The pontocerebellar system An HRP study in the rat. I. Projections to the posterior vermis. *J. Comp. Neurol.* 197, 543–558

Brodal, A. (1940). Experimentelle Untersuchungen über die olivocerebellare lokalisation. *Z. Ges Neurol. Psychiat.* 169, 1–153.

Brodal, A. and Kawamura, K. (1980). Olivocerebellar projection: A review. *Adv. Anat. Embryol. Cell Biol.* 64, 1–137.

Brown, J. T., Chan-Palay, V. and Palay, S. L. (1977). A study of afferent input to the inferior olivary complex in the rat by retrograde axonal transport of horseradish peroxidase. *J. Comp. Neurol.* 176, 1–22.

Brown, L. T. (1974a). Corticorubral projections in the rat. *J. Comp. Neurol.* 154, 149–167.

Brown, L. T. (1974b). Rubrospinal projections in the rat. *J. Comp. Neurol.* 154, 169–187.

Brown, P. A. (1980). The inferior olivary connections to the cerebellum in the rat studied by retrograde axonal transport of horseradish peroxidase. *Brain Res. Bull.* 5, 267–275.

Brown Gould, B. (1980). Organization of afferents from the brainstem nuclei to the cerebellar cortex in the cat. *Adv. Anat. Embryol. Cell Biol.* 62, 1-90.

Burne, R. A., Mihailoff, G. A. and Woodward, D. J. (1978). Visual corticopontine input to the paraflocculus: A combined autoradiographic and horseradish peroxidase study. *Brain Res.* 143, 139-146.

Burne, R. A., Azizi, S. A., Mihailoff, G. A. and Woodward, D. J. (1981). The tectopontine projection in the rat with comments on visual pathways to the basilar pons. *J. Comp. Neurol.* 202, 287-307.

Cajal, S. Ramón y (1911). *Histologie du système nerveux de l'homme et des vertébrés.* Maloine, Paris.

Campbell, N. C. and Armstrong, D. M. (1983a). The olivocerebellar projection in the rat: An autoradiographic study. *Brain Res.* 275, 215-233.

Campbell, N. C. and Armstrong, D. M. (1983b). Topographical localization in the olivocerebellar projection in the rat: An autoradiographic study. *Brain Res.* 275, 235-249.

Carlton, S. M., Leichnetz, G. R. and Mayer, D. J. (1982). Projections from the nucleus parafascicularis prerubralis to medullary raphe nuclei and inferior olive in the rat: A horseradish peroxidase and autoradiography study. *Neurosci. Lett.* 30, 191-197.

Caughell, K. A. and Flumerfelt, B. A. (1977). The organization of the cerebellorubral projection: An experimental study in the rat. *J. Comp. Neurol.* 176, 295-306.

Chan-Palay, V. (1977). *Cerebellar dentate nucleus: Organization, cytology and transmitters.* Springer, Berlin.

Chan-Palay, V., Palay, S. L., Brown, J. T. and Van Itallie, C. (1977). Sagittal organization of olivocerebellar and reticulocerebellar projections: Autoradiographic studies with [^{35}S]-methionine. *Exp. Brain Res.* 30, 561-576.

Cintas, H. M., Rutherford, J. G. and Gwyn, D. G. (1980). Some midbrain and diencephalic projections to the inferior olive in the rat. *In* J. Courville, C. De Montigny and Y. Lamarre (eds), *The inferior olivary nucleus*, pp. 73-96. Raven Press, New York.

Cook, J. R. and Wiesendanger, M. (1976). Input from trigeminal cutaneous afferents to neurones of the inferior olive in rats. *Exp. Brain Res.* 26, 193-202.

Courville, J., DeMontigny, C. and Lamarre, Y. (eds) (1980). *The inferior olivary nucleus.* Raven Press, New York.

Cruce, J. A. F. (1977). An autoradiographic study of the descending connections to the mammillary nuclei of the rat. *J. Comp. Neurol.* 176, 631-644.

Cuénod, M., Bagnoli, P., Beaudet, A., Rustioni, A., Wiklund, L. and Streit, P. (1982). Transmitter-specific retrograde labeling of neurons. *In* V. Chan-Palay and S. L. Palay (eds), *Cytochemical methods in neuroanatomy*, pp. 17-44. Liss, New York.

Dahlström, A. and Fuxe, K. (1964). Evidence for the existence of monoamine-containing neurons in the central nervous system. I: Demonstration of monoamines in the cell bodies of brain stem neurons. *Acta Physiol. Scand. Suppl.* 232, 5-55.

Dekker, J. J. (1981). Anatomical evidence for direct fiber projections from the cerebellar nucleus interpositus to rubrospinal neurons. A quantitative EM study in the rat combining anterograde and retrograde intra-axonal tracing methods. *Brain Res.* 205, 229-244.

Desclin, J. C. (1974). Histological evidence supporting the inferior olive as a major source of cerebellar climbing fibres in the rat. *Brain Res.* 77, 365-384.

Eisenman, L. M. (1980). Pontocerebellar projections to the paraflocculus in the rat. *Brain Res.* 188, 550-554.

Eisenman, L. M. (1981). Olivocerebellar projections to the pyramis and copula pyramidis in the rat: Differential projections to parasagittal zones. *J. Comp. Neurol.* 199, 65-76.

Eisenman, L. M. (1982). The reticulocerebellar projection to the pyramis and copula pyramidis in the rat: An experimental study using retrograde transport of horseradish peroxidase. *J. Comp. Neurol.* 210, 30-36.

Eisenman, L. M. and Noback, C. R. (1980). The ponto-cerebellar projection in the rat: Differential projections to sublobules of the uvula. *Exp. Brain Res.* 38, 11-17.

Flumerfelt, B. A. (1978). Organization of the mammalian red nucleus and its interconnections with the cerebellum. *Experientia* 34, 1178-1180.

Flumerfelt, B. A. (1980). An ultrastructural investigation of afferent connections of the red nucleus in the rat. *J. Anat.* 131, 621-633.

Flumerfelt, B. A. and Gwyn, D. G. (1973). Synaptology and afferent connections of the red nucleus in the rat. *Anat. Rec.* 175, 321.
Flumerfelt, B. A. and Gwyn, D. C. (1974). The red nucleus of the rat: Its organization and interconnections. *J. Anat.* 118, 374.
Flumerfelt, B. A., Otabe, S. and Courville, J. (1973). Distinct projections to the red nucleus from the dentate and interposed nuclei in the monkey. *Brain Res.* 50, 408–414.
Flumerfelt, B. A., Hrycyshyn, A. W. and Kapogianis, E. M. (1982). Spinal projections to the lateral reticular nucleus in the rat. *Anat. Embryol.* 165, 345–359.
Furber, S. E. and Watson, C. R. R. (1983). Organization of the olivocerebellar projection in the rat. *Brain Behav. Evol.* 22, 132–152.
Graybiel, A. M. (1974). Visuo-cerebellar and cerebello-visual connections involving the ventral lateral geniculate nucleus. *Exp. Brain Res.* 20, 303–306.
Gwyn, D. G. (1971). Acetylcholinesterase activity in the red nucleus of the rat. Effects of rubrospinal tractotomy. *Brain Res.* 35, 447–461.
Gwyn, D. G., Nicholson, G. P. and Flumerfelt, B. A. (1977). The inferior olivary nucleus of the rat: A light and electron microscopic study. *J. Comp. Neurol.* 174, 489–520.
Held, H. (1890). Der Ursprung des tiefen Markes der Vierhügelregion. *Neurol. Zentbl.* 481–483.
Hinrichsen, C. F. L. and Watson, C. D. (1983). Brain stem projections to the facial nucleus of the rat. *Brain Behav. Evol.* 22, 153–163.
Hökfelt, T., Fuxe, K., Goldstein, M. and Johansson, O. (1974). Immunohistochemical evidence for the existence of adrenaline neurons in the rat brain. *Brain Res.* 66, 235–251.
Hosoya, Y. and Matsushita, M. (1981). Brainstem projections from the lateral hypothalamic area in the rat, as studied with autoradiography. *Neurosci. Lett.* 24, 111–116.
Hrycyshyn, A. W. and Flumerfelt, B. A. (1981). A light microscopic investigation of the afferent connections of the lateral reticular nucleus in the cat. *J. Comp. Neurol.* 197, 477–502.
Hrycyshyn, A. W., Flumerfelt, B. A. and Anderson, W. A. (1982a). A horseradish peroxidase study of the projections from the lateral reticular nucleus to the cerebellum in the rat. *Anat. Embryol.* 165, 1–18.
Hrycyshyn, A. W., Flumerfelt, B. A., Anderson, W. A. and Sholdice, J. E. (1982b). A horseradish peroxidase study of the olivocerebellar projection in the rat. *Can. Fed. Biol. Soc.* 25, 324.
Huisman, A. M., Kuypers, H. G. J. M. and Verburgh, C. A. (1981). Quantitative differences in collateralization of the descending spinal pathways from red nucleus and other brain stem cell groups in rat as demonstrated with the multiple fluorescent retrograde tracer technique. *Brain Res.* 209, 271–286.
Huisman, A. M., Kuypers, H. G. J. M., Condé, F. and Keizer, K. (1983). Collaterals of rubrospinal neurons to the cerebellum in rat. A retrograde fluorescent double labelling study. *Brain Res.* 264, 181–196.
Hunt, S. P. and Lovick, T. A. (1982). The distribution of serotonin, metenkephalin and β-lipotropin-like immunoreactivity in neuronal perikarya of the cat brainstem. *Neurosci. Lett.* 30, 139–145.
Kapogianis, E. M., Flumerfelt, B. A. and Hrycyshyn, A. W. (1982). Cytoarchitecture and cytology of the lateral reticular nucleus in the rat. *Anat. Embryol.* 164, 229–242.
Kooy, F. H. (1917). The inferior olive in vertebrates. *Folio Neurobiol.* 10, 205–369.
Menétrey, D., Roudier, F. and Besson, J. M. (1983). Spinal neurons reaching the lateral reticular nucleus, as studied in the rat by retrograde transport of horseradish peroxidase. *J. Comp. Neurol.* 220, 439–452.
Mihailoff, G. A. (1983). Intra- and interhemispheric collateral branching in the rat pontocerebellar system, a fluorescence double-label study. *Neurosci.* 10, 141–160.
Mihailoff, G. A., Burne, R. A. and Woodward, D. J. (1978). Projections of the sensorimotor cortex to the basilar pontine nuclei in the rat: An autoradiographic study. *Brain Res.* 145, 347–354.
Mihailoff, G. A., Burne, R. A., Azizi, S. A., Norell, G. and Woodward, D. J. (1981a). The pontocerebellar system in the rat: An HRP study. II. Hemispheral components. *J. Comp. Neurol.* 197, 559–577.
Mihailoff, G. A., McArdle, C. B. and Adams, C. E. (1981a, b). The cytoarchitecture, cytology, and synaptic organization of the basilar pontine nuclei in the rat. I. Nissl and Golgi studies. *J. Comp. Neurol.* 195, 181–201.

Mihailoff, G. A., Watt, C. B. and Burne, R. A. (1981c). Evidence suggesting that both the corticopontine and cerebellopontine systems are each composed of two separate neuronal populations: An electron microscopic and horseradish peroxidase study in the rat. *J. Comp. Neurol.* 195, 221–242.

Oscarsson, O. (1969). Termination and functional organization of the dorsal spino-olivocerebellar path. *J. Physiol. (London)* 200, 129–149.

Oscarsson, O. and Sjölund, B. (1977). The ventral spino-olivocerebellar system in the cat. I. Identification of five paths and their termination in the cerebellar anterior lobe. *Exp. Brain Res.* 28, 469–486.

Payne, J. N. (1983). Axonal branching in the projections from precerebellar nuclei to the lobulus simplex of the rat's cerebellum investigated by retrograde fluorescent double labeling. *J. Comp. Neurol.* 213, 233–240.

Potter, R. F., Rüegg, D. G. and Wiesendanger, M. (1978). Responses of neurones of the pontine nuclei to stimulation of the sensorimotor, visual and auditory cortex of the rat. *Brain Res. Bull.* 3, 15–19.

Reid, J. M., Gwyn, D. G. and Flumerfelt, B. A. (1975a). A cytoarchitectonic and Golgi study of the red nucleus in the rat. *J. Comp. Neurol.* 162, 337–362.

Reid, J. M., Flumerfelt, B. A. and Gwyn, D. G. (1975b). An ultrastructural study of the red nucleus in the rat. *J. Comp. Neurol.* 162, 363–386.

Ribak, C. E. and Peters, A. (1975). An autoradiographic study of the projections from the lateral geniculate body of the rat. *Brain Res.* 92, 341–368.

Ross, C. A., Armstrong, D. M., Ruggiero, D. A., Pickel, V. M., John, T. H. and Reis, D. J. (1981). Adrenaline neurons in the rostral ventrolateral medulla innervate thoracic spinal cord: A combined immunocytochemical and retrograde transport demonstration. *Neurosci. Lett.* 25, 257–262.

Sakumoto, T., Tohyama, M., Satoh, K., Kimoto, Y., Kinugasa, T., Nanizawa, O., Kurachi, K. and Shimizu, N. (1978). Afferent fibre connections from lower brain stem to hypothalamus studied by the horseradish peroxidase method with special reference to noradrenaline innervation. *Exp. Brain Res.* 31, 81–94.

Satoh, K. (1979). The origin of reticulospinal fibers in the rat: An HRP study. *J. für Hirnforsch.* 20, 313–332.

Sawchenko, P. E. and Swanson, L. W. (1982). The organization of noradrenergic pathways from the brainstem to the paraventricular and supraoptic nuclei in the rat. *Brain Res. Rev.* 4, 275–325.

Shieh, J. Y., Leong, S. K. and Wong, W. C. (1983). Origin of the rubrospinal tract in neonatal, developing, and mature rats. *J. Comp. Neurol.* 214, 79–86.

Swenson, R. S. and Castro, A. J. (1983a). The afferent connections of the inferior olivary complex in rat: A study using the retrograde transport of horseradish peroxidase. *Am. J. Anat.* 166, 329–341.

Swenson, R. S. and Castro, A. J. (1983b). The afferent connections of the inferior olivary complex in rats. An anterograde study using autoradiographic and axonal degeneration techniques. *Neurosci.* 8, 259–275.

Szentágothai, J. and Rajkovits, K. (1959). Uber den ursprung der Kletterfasern des Kleinhirns. *Z. Anat. Ehtwicklungsgesch.* 121, 130–141.

Terasawa, K., Otani, K. and Yamada, J. (1979). Descending pathways of the nucleus of the optic tract in the rat. *Brain Res.* 173, 405–417.

Walberg, F. (1952). The lateral reticular nucleus of the medulla oblongata in mammals: A comparative-anatomical study. *J. Comp. Neurol.* 96, 283–344.

Waldron, H. A. and Gwyn, D. G. (1969). Descending nerve tracts in the spinal cord of the rat. I. Fibres from the midbrain. *J. Comp. Neurol.* 137, 143–154.

Watt, C. B. and Mihailoff, G. A. (1979). The cerebellopontine system in the rat; an autoradiographic and HRP study. *Neurosci. Abst.* 5, 108.

Watt, C. B. and Mihailoff, G. A. (1983a). The cerebellopontine system in the rat. I. Autoradiographic studies. *J. Comp. Neurol.* 215, 312–330.

Watt, C. B. and Mihailoff, G. A. (1983b). The cerebellopontine system in the rat. II. Electron microsopic studies. *J. Comp. Neurol.* 216, 429–437.

Wiesendanger, R. and Wiesendanger, M. (1982a). The corticopontine system in the rat. I. Mapping of corticopontine neurons. *J. Comp. Neurol.* 208, 215–226.

Wiesendanger, R. and Wiesendanger, M. (1982b). The corticopontine system in the rat. II. The projection pattern. *J. Comp. Neurol.* 208, 227–238.

Wise, S. P. and Jones, E. G. (1977). Cells of origin and terminal distribution of descending projections of the rat somatic sensory cortex. *J. Comp. Neurol.* 175, 129–158.

11
Cerebellum

JAN VOOGD
N.M. GERRITTS
E. MARANI

University of Leiden
Leiden, The Netherlands

I. INTRODUCTION

The cerebellum of the rat is used extensively in neurobiologic research, but no systematic description of its morphology is available. This chapter deals with the anatomy of the lobes and lobules, the structure of the deep cerebellar nuclei, and the topography of the white matter of the cerebellum of the rat. The precerebellar nuclei and their cerebellar projections are reviewed in Volume 2, Chapter 10. The efferent connections of the cerebellum, and the relations of the cerebellum with the vestibular nuclear complex and the thalamus, are dealt with in Volume 2, Chapter 9 and Volume 1, Chapter 5. This chapter does not include a review of the histology of the cerebellar cortex of the rodent. For information on this subject the reader is referred to Ramón y Cajal's (1911) original studies and to the monograph of Palay and Chan-Palay (1974).

This description is based on serial, Nissl and Häggqvist stained sections of the rat cerebellum, cut in the conventional planes. For the study of the gross anatomy of the cerebellum and the central cerebellar nuclei, graphic reconstructions were prepared from serial sections according to methods described in Voogd and Feirabend (1981). General reviews of the structure and connections of the cerebellum were published by Brodal (1981), Palay and Chan-Palay (1982) and Ito (1984).

II. THE GROSS ANATOMY OF THE CEREBELLAR COMPLEX

Early references to the gross anatomy of the cerebellum of the rat can be found in the papers of Bradley (1904), Bolk (1906), Ingvar (1919), Riley (1928), Scholten (1946) and Aciron (1951). The development and the adult configuration of the lobes and lobules of the cerebellum of the rat were described by Larsell (1952,

THE RAT NERVOUS SYSTEM
ISBN 0 12 547632 9

1970). Larsell was struck by the close similarity between midsagittal sections of the avian cerebellum (Larsell, 1948) and the cerebellum of the rat. Consequently he subdivided the cerebellum of both species in ten lobules, numbered 1 to 10 from rostral to caudal (Fig. 1e).* The deep primary fissure separates the anterior lobe from the posterior lobe. The primary and preculminate fissures, which subdivide the anterior lobe in the lobules 1 to 3 and 4 to 5, are foliated on their anterior and posterior walls, and reach the lateral margin of the cerebellum. The other interlobular fissures of the anterior lobe do not reach as far laterally (Fig. 1c). Shallow indentations in the surface of lobules 4 and 5 indicate the border between vermis and hemispheres. These indentations are more distinct in lobule 6 (Bolk's [1906] simple lobule), caudal to the primary fissure. A deep paramedian sulcus is present lateral to lobule 7, but absent in lobule 8 (the pyramis). The cortex of the pyramis continues uninterruptedly into the hemisphere as the copula pyramidis. None of the interlobular fissures in the segment of the posterior lobe, located between the primary and prepyramidal fissures, is completely continuous between vermis and hemisphere (Fig. 1b). At the junction of the lobules 6, 7 and the hemisphere, the cerebellar cortex is interrupted and the white matter comes to the surface. Three fissures come together at this point (Figs 1b; 4; 5 and 7): the vermal segment of the posterior superior fissure, located between the lobules 6 and 7; the hemispheral segment of the posterior superior fissure, which separates the simple lobule from the crus 1 of the ansiform lobule; and the intercrural fissure of the ansiform lobule. The ansoparamedian fissure, located between the caudal folium of crus 2 of the ansiform lobule and the rostral folium of the paramedian lobule, ends in the paramedian sulcus lateral to lobule 7.

The relation between vermis and the hemispheres clearly differs for different segments of the cerebellar cortex. Functionally, the mediolateral continuity of the cortex depends on the presence of parallel fibers, that is of a molecular layer (Marani and Voogd, 1979; Voogd, 1975). In the anterior lobe, the simple lobule and, in lobule 8, the cortex of the vermis continues uninterruptedly into the hemispheres. In between lobules 6b and 6c and 7, and the ansiform and paramedian lobules, however, the cortex bridging the paramedian sulcus is greatly constricted or even completely absent. In Bolk's (1906) terms, the folial chains of vermis and hemisphere of this part of the posterior lobe are completely independent of each other.

The cortex of lobules 9 (the uvula) and 10 (the nodule) and the secondary and posterolateral fissures end in a deep paramedian sulcus which separates these lobules from the copula pyramidis (Fig. 1d). Laterally the copula continues into the paraflocculus. The cortex of the paraflocculus constitutes a laterally directed loop, which is continuous with the cortex of the flocculus at the bottom of the hemispheral segment of the posterolateral fissure (Fig. 1f). The cortex of the paraflocculus is interrupted in the center of the loop in the so-called

*The figures in this chapter follow the text, in consecutive order.

intraparafloccular sulcus. These areas, where the central white matter comes to the surface, are found at the caudoventral and rostral aspects of the paraflocculus. For descriptive purposes the dorsal and ventral limbs of the loop are distinguished as the dorsal and ventral paraflocculus, but this distinction is secondary to the essential continuity of the folial chain of the hemisphere. Larsell (1952) has stated that a lateral extension of the secondary fissure separates the dorsal from the ventral paraflocculus. No such continuity exists in the rat. The fissures of this part of the cerebellum develop independently in the cortex of the caudal vermis and in the hemisphere, and end at the white matter in the paramedian and interparafloccular sulci, which separates the caudal vermis from the paraflocculus and the flocculus.

The tenia of the roof of the fourth ventricle is attached to the margin of the nodule, the copula pyramidis and the flocculus. A posterior medullary velum is not present, although the areas devoid of cortex, bordering the tenia in the paramedian sulcus and the ventral aspect of the paraflocculus, could be considered as such. The superior medullary velum is continuous with the cerebellar commissures in the central white matter of the cerebellum. When the cerebellum is dissected from the brain stem, the superior medullary velum is seen to be continuous with the inferior cerebellar peduncle.

The morphology of the cerebellum of the rat conforms to the general mammalian pattern as described by Bolk (1906), Riley (1928), and Voogd (1975). The cortex of vermis and hemispheres is continuous in the anterior lobe and the simple lobule, and in a restricted portion of the posterior lobe between the prepyramidal and secondary fissures. In the intermediate and caudal segments of the posterior lobe, vermis and hemispheres behave as independent folial chains, and mediolateral connections between them are absent or greatly attenuated. In most respects, this description of the cerebellum of the rat closely corresponds to the observations of Larsell (1952) and to the description of the mouse cerebellum by Marani and Voogd (1979), to which the reader is referred for further details. Interruptions of the cerebellar cortex between the ansiform lobule and the vermis, and in the center of the parafloccular loop, were also found to be present in the mouse and in most other mammalian species investigated (Voogd, 1967, 1969). Larsell's account remains indispensable as the basis for the nomenclature of the rat cerebellum. Names should not be taken too literally, however, and there is no reason to attach particular significance to a lobule or to an interlobular fissure because it received a number or a name.

III. THE CENTRAL CEREBELLAR NUCLEI AND THEIR EFFERENT PATHWAYS

A. The subdivision of the central cerebellar nuclei

The central nuclei usually are subdivided according to Weidenreich (1899). His scheme was applied to pinnipedia and catecea by Ogawa (1935) and perfected by

Ohkawa (1957), whose comparative anatomic studies included rodents. These authors divided the central nuclei into two groups of interconnected nuclei. The caudal group consists of the medial cerebellar or fastigial nucleus and the posterior interposed nucleus; the rostral group consists of the anterior interposed and the lateral cerebellar or dentate nucleus. Myelinated fibers occupy the space between the two nuclear groups; the border between the two nuclei within a group often is more difficult to define. Korneliussen (1968) applied this subdivision to the central nuclei of the rat. His description takes account of the presence of certain subnuclei which are peculiar to the rat and which were first described by Goodman *et al.* (1963). His description was adapted in most experimental studies of the connections of the central nuclei. It also served as the starting point for the detailed Golgi and morphometric studies of the dentate nucleus (Chan-Palay, 1977) and the medial cerebellar nucleus (Beitz and Chan-Palay, 1979a, b) of the rat.

The organization of the central nuclear efferents generally supports the distinction of the two groups of central nuclei. Voogd (1964) and Verhaart (1970) described a subdivision of the superior cerebellar peduncle, which contains the ascending fibers of several nuclei, into a smaller medial part and a larger lateral portion, in most mammals studied. The medial third of the superior cerebellar peduncle of the cat contains fibers from the medial cerebellar and posterior interposed nuclei. The lateral two thirds of the peduncle contain efferents from the anterior interposed and lateral cerebellar nuclei. This localization was confirmed by Haroian *et al.* (1981) for the rat (Fig. 3). The superior cerebellar peduncle was included in the graphic reconstruction of the central nuclei of the rat illustrated in Fig. 2. The fibers of the medial part of the superior cerebellar peduncle emerge from the base of the medial cerebellar nucleus and, as a more compact bundle, from the rostral aspect of the posterior interposed nucleus (Fig. 2a). Cell strands separate these fibers from the lateral part of the peduncle; the lateral part can be traced back to the anterior interposed and lateral cerebellar nuclei (Fig. 2b).

B. The medial (fastigial) cerebellar nucleus

The medial cerebellar nucleus is characterized by the prominent dorsolateral protuberance of Goodman *et al.* (1963), a group of large neurons extending far dorsally into the white matter of the posterior lobe (Figs 4, 5 and 7). Korneliussen (1968) subdivided the medial nucleus into middle and caudomedial portions, and the dorsolateral protuberance. The caudomedial subdivision is the most distinct one. Most of its cells are small (Beitz and Chan-Palay, 1979a). The caudomedial subdivision of the medial nucleus is located at the base of the nodule and the uvula. Dorsally, it remains separated from the rest of the nucleus by myelinated fibers; ventrally, where it lines the roof of the fourth ventricle, it merges with the middle portion of the medial nucleus and with the medial, parvocellular part of the posterior interposed nucleus.

The large cells in the rostral and ventromedial portions of the middle subdivision of the medial cerebellar nucleus are indistinguishable from the large cells of the dorsolateral protuberance. Small and medium cells are more numerous in the central and ventrolateral parts of the middle subdivision, where it merges with the superior vestibular nucleus and the posterior interposed nucleus. The middle subdivision is distinguished by its high content of myelinated fibers which belong to two groups. The uncinate tract, which contains a large proportion of coarse fibers, emerges from and traverses the rostral magnocellular region, and smaller, so-called "perforating fibers", traverse its caudal part. The majority of these fibers undoubtedly originate from Purkinje cells of the cerebellar cortex (Voogd, 1964). They emerge from the vermal white matter medial and lateral to the dorsolateral protuberance and traverse the medial nucleus to enter the vestibular nuclei.

The dorsolateral protuberance is usually considered as a separate subdivision of the medial nucleus. According to Korneliussen (1968), lamellae of myelinated fibers separate it from the rest of the medial nucleus and the protuberance is not traversed by bundles of corticofugal or perforating fibers, which characterize the central part of the nucleus. Perforating fibers are also absent among the large cells in the rostral part of the nucleus, which usually are assigned to its middle subdivision. Instead of constituting different subdivisions, the large cells of the rostral pole of the medial nucleus and the dorsolateral protuberance also can be considered as parts of a continuous shell of large neurons surrounding the rostral and dorsolateral aspects of the nucleus. This view is favored by the similarity in morphology and efferent connections (see below) between the two groups. On the other hand, the unique shape of the dorsolateral protuberance and the afferent corticonuclear connections it receives from the hemisphere of the posterior lobe (Armstrong and Schild, 1978a, b; Goodman *et al.*, 1963) set it apart from the rest of the medial nucleus and preclude its identification with any of the subdivisions of the medial cerebellar nucleus of better studied species such as the cat and monkey.

The fibers of the uncinate tract emerge from a groove on the rostrolateral aspect of the medial nucleus. This groove continues as a deep, ventrorostrally directed hilus (Figs 1, 4a, b). The large cells of the rostral pole of the nucleus border the hilus dorsally. The uncinate tract crosses the midline in the caudal part of the cerebellar commissure (Fig. 6), rostral to the gliotic, interfastigial area, in small bundles dorsal and caudal to this area and in the superior medullary velum. Contralaterally, uncinate fibers pass rostral to and through the hilus of the medial nucleus (Fig. 7). The tract arches dorsal to the superior cerebellar peduncle, immediately rostral to the anterior interposed nucleus, to join the inferior cerebellar peduncle in its course lateral to the vestibular nuclei. Some of the fibers of the uncinate tract join the medial part of the superior peduncle as the crossed ascending limb of the uncinate tract (Haroian *et al.*, 1981). Over most of their extent, the efferent fibers of the uncinate tract in the cerebellar commissure, and more laterally in their course over the superior cerebellar peduncle, remain separated from the afferent fibers of the inferior cerebellar peduncle by a layer of thin, olivocerebellar fibers (Fig. 4e).

The uncinate tract carries a large proportion of coarse, myelinated fibers and, therefore, can be traced in normal, Häggqvist stained material. An uncrossed fastigiobulbar tract consisting of large myelinated fibers and descending among the perforating fibers of the juxtarestiform body, as present in the cat (Voogd, 1964), is less distinct in the rat (Fig. 5). It has not been possible to identify these uncrossed fastigiobulbar fibers in experimental studies in the rat. The few experimental studies on the efferents of the medial cerebellar nucleus of the rat (Achenbach and Goodman, 1968; Angaut and Cicirata, 1982; Jansen, 1956; Watt and Mihailoff, 1983) either used silver impregnation methods for degenerated axons after lesions of the medial nucleus, which always interrupt the corticofugal or uncinate tract fibers passing through the nucleus, or did not pay attention to the intracerebellar course of the system.

For a complete account of the distribution of the uncinate and direct fastigiobulbar tracts the reader is referred to the autoradiographic studies of Batton *et al.* (1977) and Carpenter and Batton (1982) in the cat and monkey. According to these authors, both tracts take their origin from the entire fastigial nucleus. Uncinate tract fibers separate from the inferior cerebellar peduncle and arch medially through the rostral and ventral parts of the lateral vestibular nucleus and the lateral pole of the medial vestibular nucleus to enter the medial reticular formation. Other fibers descend in the ventrolateral part of the spinal vestibular nucleus to the level of the parasolitary nucleus. The uncinate and direct fastigiobulbar tracts terminate almost symmetrically in the ventral parts of the lateral vestibular and spinal vestibular nuclei. Projections to the associated cell groups X and F of the vestibular nuclei (Brodal and Pompeiano, 1957) and to the parasolitary nucleus are completely crossed and arise from the caudal two thirds of the medial cerebellar nucleus.

The ipsilateral projections of the medial cerebellar nucleus to the superior vestibular nucleus and to the dorsal parts of the lateral and spinal vestibular nuclei, which have been reported in experiments using axonal degeneration techniques, are probably due to the interruption of corticovestibular fibers. Termination in the perihypoglossal nuclei has not been observed with the autoradiographic tracing method. The fastigioreticular projection is almost completely crossed. The fibers mainly terminate on the gigantocellular nucleus, the dorsal paramedian nucleus and the magnocellular part of the lateral reticular nucleus. Fastigiospinal fibers enter the cord through the bulbar reticular formation. Fastigio-olivary fibers (Angaut and Cicirata, 1982) and fastigiopontine fibers (Watt and Mihailoff, 1983) have been described in the rat and are treated in Volume 2, Chapter 10. Ascending fibers arise from the caudal part of the medial cerebellar nucleus and cross with the uncinate tract. The fibers of the ascending limb of the uncinate tract are incorporated in the medial third of the superior cerebellar peduncle (Haroian *et al.*, 1981; Fig. 3). They do not join the decussation of the superior cerebellar peduncle, but ascend in the dorsal tegmentum, ventrolateral to and within the central gray. They terminate in the deep layers of the superior colliculus. Some of them recross in the posterior commissure or terminate in its nuclei. The diencephalic terminations of the ascending limb of the uncinate tract are summarized in Volume 1, Chapter 5.

Some of the efferent connections of the medial cerebellar nucleus of the rat have been investigated with retrograde transport techniques (Beitz, 1982). Large horseradish peroxidase (HRP) injections in the lateral and superior vestibular nuclei which interrupt the entire projection of the medial nucleus to the brain stem, but spare the superior cerebellar peduncle and the ascending limb of the uncinate tract (Fig. 10d), give an indication of the origin of crossed and uncrossed fastigiobulbar projections (observations by Voogd, Mehler and Rubertone; Fig. 8). The large cells of the rostral pole of the medial nucleus and of the dorsolateral protuberance and the small cells of the caudomedial subdivision are labeled on the contralateral side. Ipsilateral labeling is found in cells of the central and ventrolateral part of the nucleus. The double labeling studies of the central nuclei in the rat by Bentivoglio and Kuypers (1982) included combinations of injections into the thalamus, the superior colliculus, the medial bulbar reticular formation with the inferior olive, and the spinal cord. Diencephalic projections with collaterals to the superior colliculus originate from the central and caudal part of the medial nucleus (Fig. 9a). The crossed projections of the dorsolateral protuberance are directed at the medial bulbar reticular formation; some of the large cells in the ventral part of the medial subdivision also project as far as the spinal cord. Some of the cells located centrally and caudolaterally in the middle subdivision send collaterals to the diencephalon (Fig. 9b, c). Medullary injections that include the inferior olive consistently label small cells in the ventral parts of the central nuclei. These small cells do not show double labeling in any of the combined experiments. The origin of the nucleo-olivary projection from a separate population of central nuclear cells is in accordance with observations in other species (Brown *et al.*, 1977; Haroian, 1982).

C. The posterior interposed (interpositus) cerebellar nucleus

The posterior interposed nucleus is the smallest of the central nuclei of the rat, but it has a very high cell density. It contains rather large cells; small cells are more numerous ventrally and medially. Its borders with the anterior interposed nucleus and lateral cerebellar nucleus are most distinct in horizontal and sagittal sections (Figs 4 and 6). In transverse sections, an area free of cells separates the nucleus from the anterior interposed and lateral nucleus (Fig. 5f). It is easier to define the borders of these nuclei when the origin of the superior cerebellar peduncle is taken into account. Fibers from the posterior interposed nucleus, which constitute the medial part of the peduncle, course through the ventrolateral part of the medial cerebellar nucleus, where they intersect with the perforating fibers and assemble more laterally as a compact bundle which demarcates the border of the posterior and anterior interposed nuclei. The border of the posterior interposed nucleus with the medial cerebellar nucleus is ill-defined.

The efferent connections of the posterior interposed nucleus of the rat have been studied only with respect to their terminations in the thalamus (Haroian *et al.*, 1981). They cross in the dorsal part of the decussation of the superior cerebellar peduncle. In the cat they terminate along the medial margin of the red nucleus, the central gray, the nucleus of Darkschewitsch, and the subparafascicular nucleus (Angaut, 1970; Voogd, 1964; Volume 2, Chapter 10). Their thalamic targets include the ventromedial, ventrolateral and the intralaminar nuclei and are discussed in Volume 1, Chapter 5. The contribution of the posterior interposed nucleus to the crossed descending limb of the superior cerebellar peduncle is small. In the rat, the posterior interposed nucleus does not contribute fibers to the pontine nuclei (Watt and Mihailoff, 1983). Its fibers descend to the level of the inferior olive, where they mainly terminate on the rostral part of the medial accessory olive and the ventral lamella of the principal olive (Swenson and Castro, 1983a, b). In the cat, the rostral part of the medial accessory olive is reciprocally connected with the posterior interposed nucleus (Groenewegen *et al.*, 1979; Tolbert *et al.*, 1976). The recurrent circuit including the posterior interposed nucleus, the central gray of the mesencephalon (the nucleus of Darkschewitsch), and the medial accessory olive, therefore also seems to be present in the rat. The retrograde labeling studies of Bentivoglio and Kuypers (1982) in the rat did not distinguish between the anterior and posterior interposed nuclei. Cells projecting to the spinal cord appeared to be most numerous in the junctional region of the interposed and the medial nuclei (Fig. 9c). In the cat (Matsushita and Hosoya, 1978) the cerebellospinal tract neurons are restricted to the posterior interposed and medial cerebellar nuclei and are absent from the anterior interposed nucleus.

D. The anterior interposed (interpositus) cerebellar nucleus, the dorsomedial crest and the dorsolateral hump

The dorsomedial crest and the dorsolateral hump were described by Goodman *et al.* (1963) as lateral and dorsal protrusions of the undivided interposed nucleus. The cells of the dorsomedial crest and the adjoining medial part of the anterior interposed nucleus are smaller than the cells of the lateral part of this nucleus. A distinct border is present between the small cells of the dorsomedial crest and the larger cells of the posterior interposed nucleus (Fig. 5f). The dorsolateral hump is a ridge of small cells on the rostrolateral and dorsal surface of the anterior interposed nucleus. Korneliussen (1968) included the lateral fourth of the anterior interposed nucleus in the hump. When the hump is defined in this way, it includes the large cells of the caudal pole of the anterior interposed nucleus, which lie intercalated between the posterior interposed and the lateral nucleus. Hump and caudal pole can also be distinguished as separate bulges in the surface relief of the nuclei (Fig. 2b, c). In our opinion, the distinction of the dorsolateral hump as a separate subnucleus, comprising an ill-defined lateral segment of what, in other mammals, would be called the anterior interposed nucleus, is both unnecessary and confusing.

Efferents of the anterior interposed nucleus (Haroian *et al.*, 1981), including those of its dorsolateral hump (Goodman *et al.*, 1963), travel in the middle part of the superior cerebellar peduncle (Fig. 3). The anterior interposed nucleus contributes to the crossed ascending and descending branches of the superior cerebellar peduncle. In other chapters (Volume 1, Chapter 5; Volume 2, Chapter 10) the terminations of the ascending branch in the magnocellular part of the red nucleus and the ventrolateral complex of the thalamus and of the descending branch in the basilar pons, reticulotegmental nucleus and dorsal accessory olive are discussed.

Cells projecting to the ventrolateral complex of the thalamus and the medial bulbar reticular formation, including the inferior olive, are found over the entire anterior interposed nucleus. Cells with collateral projections to the thalamus and superior colliculus are located in the lateral part of the nucleus; cells with double projections to the thalamus and the medial bulbar reticular formation extend more laterally into the dorsolateral hump. Fibers of the anterior interposed nucleus do not descend into the spinal cord. These features are illustrated in Fig. 9 which is taken from the report of Bentivoglio and Kuypers (1982).

Fig. 9 also shows the sparsity of labeled cells in the dorsolateral hump. According to Woodson and Angaut (1984), this region is the main origin of the uncrossed descending branch of the superior cerebellar peduncle. Originally, this system was described by Cajal (1903, 1911) with the Golgi and Marchi methods (Fig. 10a). Mehler (1967, 1969) retraced it with the Nauta method in rats and guinea pigs and conjected that an ipsilateral descending branch of the superior cerebellar peduncle is a characteristic of only the Class Rodentia. The fibers enter the brain stem between the motor and principal sensory nuclei of the trigeminal nerve (Fig. 10b) and descend in the lateral reticular formation to either terminate in the principal and spinal trigeminal nuclei or to enter the spinal cord (Achenbach and Goodman, 1968; Faull, 1978; Woodson and Angaut, 1984). A more extensive uncrossed ascending and descending projection was described for the lateral cerebellar nucleus by Chan-Palay (1977), but the injection sites of her autoradiographic experiments may have included the dorsolateral hump.

E. The lateral (dentate) cerebellar nucleus

The lateral cerebellar nucleus of the rat consists of a dorsolateral magnocellular portion and a ventromedial parvocellular portion (Korneliussen, 1968). Chan-Palay (1977) paid particular attention to the size, shape and orientation of the cells of the lateral nucleus of the rat. She distinguished large cells in the nuclear boundary, with tangentially oriented dendrites, large rounded cells in the rostral and caudal poles and large cells with their long axis oriented at the hilus, in the middle part of the nucleus. Small cells are present in the rostral and caudal poles and in the hilar region; most are short axon cells, a few have long axons which were traced to the hilus in Golgi impregnated material. The cytoarchitecture of the lateral nucleus can be appreciated from the sections depicted in Figs 4, 5 and 7. Small cells are especially numerous in the narrow band, ventral to the hilus and dorsal to the floccular peduncle. Slightly larger, fusiform cells, belonging to the infracerebellar nucleus of Gacek (1977, 1979), are located more ventrally within the floccular peduncle. Group Y cells are located as a compact subnucleus ventromedial to the floccular peduncle, capping the inferior cerebellar peduncle. The cytoarchitecture of this region is discussed in Volume 2, Chapter 9. It is noteworthy that neurons are not only present in the floccular peduncle, but also dispersed in the white matter of the flocculus of the rat.

The efferent connections of the lateral nucleus are contained in the ventral and lateral part of the superior cerebellar peduncle (Haroian *et al.*, 1981; Fig. 3). Some of the afferents of the infracerebellar nucleus and group Y probably take the same route. The group Y neurons also give rise to commissural and cerebellar connections; the infracerebellar nucleus projects to the oculomotor complex (Volume 2, Chapter 9). The scattered neurons in the floccular white matter belong to a cholinergic system terminating in the caudal vermis (Komei *et al.*, 1983). The lateral nucleus contributes to the crossed ascending and descending branches of the superior cerebellar peduncle. The terminations of the crossed ascending fibers in the parvocellular red nucleus and in the thalamus and of the crossed descending fibers in the pontine nuclei, the reticulo-tegmental nucleus and the principal olive are reviewed in Volume 1, Chapter 5 and Volume 2, Chapter 10. Chan-Palay (1977), using autoradiography of ^{35}S-methionine, described a rather extensive ipsilateral projection of the lateral nucleus of the rat to the principal sensory nucleus of the trigeminal nerve, the locus coeruleus, the reticular formation, the central gray, the abducens nerve nucleus, and the pontine and reticulotegmental nuclei. Rostrally, this projection extends into the red nucleus and the oculomotor, Edinger–Westphall, and Darkschewitsch nuclei. Terminations in the ipsilateral thalamus and the ipsilateral inferior olive are the result of recrossing collaterals from the crossed ascending and descending limbs of the superior cerebellar peduncle.

The double labeling study of Bentivoglio and Kuypers (1982) confirms the projection of the lateral, magnocellular part of the lateral nucleus to the thalamus. Collateral projections to the superior colliculus and the spinal cord and the medial bulbar reticular formation arise from different cell groups (Fig. 9). The small cells of the hilar region, which are labeled after medial bulbar injections, mostly project to the inferior olive (Brown *et al.*, 1977). They probably correspond to the small cells with long axons described by Chan-Palay (1977).

IV. THE CORTICONUCLEAR AND CORTICOVESTIBULAR PROJECTIONS

It has been known since the Marchi studies of Klimoff (1899) in the rabbit that the corticonuclear projection is strictly uncrossed and that the cerebellar vermis is connected with the medial cerebellar nucleus and the hemisphere with the interposed and lateral cerebellar nuclei. Corticovestibular fibers originate from the vermis and the flocculus. The main differences between Klimoff's concept and the more recent anterograde degeneration and axonal transport studies in the rat (Armstrong and Schild, 1978a, b; Goodman *et al.*, 1963) concern the existence of projections of the vermis to the interposed nucleus, and of the hemisphere to the dorsolateral protuberance of the medial cerebellar nucleus. Since Klimoff's time, an impressive amount of detail has been assembled, mainly in the cat and rabbit, on the projection of different lobules to different combinations of central cerebellar and vestibular nuclei. Most of the older studies, which were reviewed by Voogd (1964), Larsell and Jansen (1972) and Haines *et al.* (1982), used large lesions and employed arbitrary criteria to define the borders between vermis and hemisphere, and between the different central cerebellar nuclei. Information on the lobular organization of the corticonuclear projection is fairly substantial, but, as pointed out by Armstrong and Schild (1978a), the localization in the corticonuclear projection is sharper in a mediolateral than in a rostrocaudal direction. Attention therefore has to be focused on the existence in the cerebellar cortex of longitudinal zones with a specific projection to the central cerebellar or vestibular nuclei.

A subdivision of the cerebellar cortex into medial (vermal), intermediate and lateral zones projecting to the medial, interposed and lateral cerebellar nuclei, respectively, was introduced by Jansen and Brodal (1940, 1942) and applied to the rat cerebellum by Goodman *et al.* (1963). The position of the borders between these three cortical zones depends on the position of the arbitrary borders between the medial, interposed, and lateral nuclei, and not on specific landmarks in the cortex itself. This also holds for the border between the vermis and the hemisphere in the rat, which is distinct only for the lobules 6b, 6c, 7, 9 and 10. Armstrong and Schild (1978a) used blood vessels overlying the paramedian sulcus as an indication of this border, but, although this method may lead to reproducible results, it remains arbitrary.

Several methods, using intrinsic cortical landmarks, now are available to define longitudinal zones in the cerebellar cortex. The enzyme histochemical studies of Scott (1964, 1965) and Marani (1982) showed the presence of alternating longitudinal zones of high and low 5′-nucleotidase activity in the cerebellar cortex of vermis and hemispheres in mice and rats. This criterion has not been used in cortico-nuclear projection studies, because the fixation procedure for axonal tracer studies is not compatible with 5′-nucleotidase histochemistry. Physiologic and anatomic methods are now available to define longitudinal zones in the cerebellar cortex on the basis of olivary fiber afferents (Armstrong *et al.*, 1974; Oscarsson, 1969; Voogd, 1982). A direct visualization of the longitudinal zonal organization in the corticonuclear projection was given by Voogd (1964, 1969) and Voogd and Bigaré (1980) in the ferret and cat, and by Feirabend *et al.* (1976) and Feirabend (1983) in the chicken. Purkinje cell fibers originating from longitudinal zones in the cerebellar cortex are located in sagittal compartments in the cerebellar white matter. The relatively large Purkinje cell fibers within a compartment are separated from the Purkinje cell fibers in the next compartment by a "raphe", an accumulation of small, presumably afferent, fibers. The borders between the compartments are continuous with the borders between the central nuclei. On the basis of this myeloarchitectonic subdivision of the white matter, a medial A compartment and a lateral B compartment were distinguished in the vermis of the anterior lobe and the simple lobule, which project to the medial cerebellar nucleus and the lateral vestibular nucleus, respectively. The hemisphere was subdivided into three C zones, projecting to the anterior interposed nucleus (C1 and C3) and the posterior interposed nucleus (C2), and two D zones, projecting to the caudal (D1) and rostral (D2) parts of the lateral cerebellar nucleus. The experimental studies of Haines on the corticonuclear projection in primates (for a review, see Haines *et al.*, 1982) and of Voogd and Bigaré (1980) in cats largely confirmed the presence of this longitudinal pattern. Experimental studies of a longitudinal organization in the corticonuclear projection in the rat have not been reported.

A myeloarchitectonic subdivision of the cerebellar white matter in compartments certainly is possible in the rat. Bundles of large Purkinje cell fibers are separated by raphes containing smaller fibers. For the vermis, the localization of the borders between the compartments is indicated in Figs 4 and 5. A and B compartments are present in the vermis of the anterior lobe and the simple lobule. The A compartment is further subdivided into medial and lateral parts. In the dorsal part of the anterior lobe, and in the simple lobule, the area with small fibers between A and B is wide, and can be considered as a separate compartment (Fig. 11a). The identification and the compartmentalization of the Purkinje cell fibers were confirmed using an antibody against cyclic GMP activated proteinkinase (De Camilli *et al.*, 1984) as a specific marker for Purkinje cells and their axons (Fig. 11b and c).

All lobules of the vermis of the rat cerebellum project to all three subdivisions of the medial cerebellar nucleus (Armstrong and Schild, 1978a; Haines and Koletar, 1979), with the exception of the major part of the dorsolateral protuberance. The different lobules share most of their projection fields, but some localization is evident, with rostral lobules projecting more rostrally and caudal lobules more caudally. The localization in a mediolateral direction is more distinct, at least in the posterior lobe. Injections in the lateral part of the vermis of the posterior lobe produce labeling in the medial part of the posterior interposed nucleus. It is evident from a comparison with the retrograde labeling studies of Bentivoglio and Kuypers (1982) that this projection from the posterior lobe vermis terminates on cells of the posterior interposed nucleus, with projections to the spinal cord.

The cortex of the cerebellar hemisphere projects to the lateral, anterior, and posterior interposed nuclei, and the dorsolateral protuberance of the medial cerebellar nucleus (Goodman *et al.*, 1963). Armstrong and Schild (1978b) give further details of the projections of the crura 1 and 2 of the ansiform lobule of the rat. Projections to the lateral nucleus were found to originate more laterally than those to the interposed nucleus with the dorsolateral protuberance. These findings are compatible with the view of Goodman *et al.* (1963) that projections to the dorsolateral protuberance arise from a medial strip of hemispheric cortex. The existence of these projections of the ansiform lobule, as well as the fact that: (i) the paraflocculus is connected with the lateral and posterior interposed nuclei (see also Dow, 1936); and (ii) that medial parts of the hemisphere of the anterior lobe and the paramedian lobule converge on the anterior interposed nucleus, is also compatible with a zonal organization of the corticonuclear projection, as proposed for other species by Voogd (1969), Courville and Faraco-Cantin (1976) and Haines (1977). The published experimental material on the rat, however, does not warrant any definite conclusion on this point.

Corticovestibular projections originate from the vermis and the flocculus. According to Dow (1936), fibers from the flocculus of the rat course through the floccular peduncle and Löwy's (1916) angular bundle to terminate in the lateral and superior vestibular nuclei. Corticovestibular fibers from lobules 9 and 10 course through the middle subdivision of the medial cerebellar nucleus to terminate in all vestibular nuclei. Lobule 8 does not give rise to corticovestibular fibers. Corticovestibular connections have been studied with the Nauta method by Voogd (1964), Angaut and Brodal (1967), and Haines (1977) in other species. Their distribution in the vestibular nuclei was found to be more extensive, and fibers from the flocculus and the caudal vermis were found to terminate in complementary parts of the vestibular nuclei (Voogd, 1964). The caudal part of the flocculus of the cat also projects to the lateral and the posterior interposed nucleus (Bigaré, 1980). In this respect, the presence of three compartments in the white matter of the flocculus of the rat (Fig. 7k) may be of interest. The fibers of the floccular peduncle obviously take their origin from the rostral part of the flocculus. The caudal compartments of the floccular peduncle may contain the fibers terminating in the lateral and posterior interposed nuclei. A zonal origin of the corticovestibular fibers from the caudal vermis of the cat has been advocated by Bigaré (1980) and Balaban (1984). The interrelations of the vestibular nuclei and the cerebellum of the rat are discussed in Volume 2, Chapter 9.

Corticovestibular fibers of the vermis of the anterior lobe and the simple lobule terminate in the lateral vestibular nucleus. In the cat, they originate from the lateral B zone and from the lateral part of the A zone (zone A2; Voogd and Bigaré, 1980). Experiments with injections of HRP in the lateral vestibular nucleus (observations by Voogd, Mehler and Rubertone) demonstrated that the same projections from the A2 and B zones are present in the rat (Fig. 12). Fibers of the B zone course lateral to the medial cerebellar nucleus and the dorsolateral protuberance; fibers from the A2 zone course through the middle subdivision of the medial cerebellar nucleus (Fig. 12d). In the ventral part of the anterior lobe, the A and B zones are contiguous, but in the dorsal part of the anterior lobe and lobule 6 they are separated by a wedge shaped area. In the cat, this area contains the "X" zone (Ekerot and Larson, 1982; Voogd, 1983). In both the rat and the cat, the B zone extends caudally to the border of lobules 6c and 7, where it ends at the area devoid of cortex at the bottom of the intercrural fissure. The localization of the corticovestibular fibers in the cerebellar white matter corresponds to the localization of the A and B compartments as indicated in Figs 4, 5 and 11.

Reciprocal cerebellar nucleocortical connections have been described in several species, but not in the rat (for a review see Tolbert, 1982). The nucleocortical fibers originate as collaterals from the axons of the projection neurons of the central nuclei (McCrea *et al.*, 1978) and terminate as mossy fibers (Tolbert *et al.*, 1980). The topography of the nucleocortical projection roughly parallels the zonal organization of the corticonuclear projection, but is not as sharp. Chan-Palay *et al.* (1979) used retrograde transport of antibodies against glutamic acid decarboxylase to identify the cells of origin of a possible GABAergic nucleocortical projection. Whether these GABAergic cells belong to the same population of cell with extracerebellar axons and collateral projections to the cortex, cannot be decided at present.

V. AFFERENT CEREBELLAR CONNECTIONS

A. The cerebellar peduncles

The origin and the termination of the main afferent systems of the cerebellum are reviewed in Volume 2, Chapters 9 and 10. In this chapter we will address the question of the topographic distribution of some of these tracts. Unfortunately, few anterograde degeneration or transport studies on afferent systems have been reported in the rat. The resolution of most retrograde transport studies in the rat is not sufficient to provide information on these systems below the level of the cerebellar lobules.

Afferent systems enter the cerebellum through the inferior and middle cerebellar peduncles. The ventral spinocerebellar tract, which courses rostral to the entrance of the trigeminal nerve, reaches the cerebellum dorsal to the superior cerebellar peduncle (Figs 4f, 7), where it rejoins the fibers of the inferior cerebellar peduncle.*

The inferior cerebellar peduncle and the ventral spinocerebellar tract enter the cerebellum rostral to the junction of the anterior interposed and lateral cerebellar nuclei. Their fibers sweep medially, rostral and dorsal to the central nuclei (Fig. 7). Some of them enter the cerebellar commissure where they are located rostral and dorsal to the decussation of the uncinate tract (Fig. 6).

Many of the fibers of the inferior cerebellar peduncle and the ventral spinocerebellar tract, which will terminate as mossy fibers in the cerebellar cortex, are large or medium sized. Their caliber distinguishes them from the thin olivocerebellar fibers (Szentagothai, 1942), which occupy the ventromedial part of the inferior cerebellar peduncle before it enters the cerebellum, and from the majority of the fibers from the middle cerebellar peduncle. Large and medium sized fibers, originating from the reticulotegmental and raphe pontis nuclei (Gerrits *et al.*, 1984a), occupy the medial part of the middle peduncle. The majority of the fibers of the middle cerebellar peduncle are very thin and occupy its lateral part (Gerrits *et al.*, 1984a; Voogd, 1964).

The middle cerebellar peduncle enters the cerebellum rostral and lateral to the lateral cerebellar nucleus. Many fibers of the middle cerebellar peduncle terminate at the hemisphere, but some course medially in the medullary gray of the crus 1 of the ansiform lobule, dorsal and caudal to the fibers of the inferior cerebellar peduncle (Fig. 7). The afferent fibers constitute the rostral and dorsal part of the cerebellar commissure, which extends from the superior medullary velum into the white matter of the lobules 6b and c and 7 (Fig. 6). The afferent fibers thus form a continous sheet extending from the entrance of the peduncles to the apex of the crus 1. Laterally, the fibers of the middle cerebellar peduncle extend to the area without cortex in the intraparafloccular fissure (Fig. 4). Dorsally, they touch the area without cortex in the intercrural fissure (Fig. 7h). The mossy fiber systems, contained in the inferior and middle cerebellar

*The terminology for the peduncles of the cerebellum as used in the Atlas of Paxinos and Watson (1982) is also used in this chapter. The term inferior cerebellar peduncle is used as a synonym for the restiform body. This is not correct because it refers both to the restiform body and the juxtarestiform body. The latter is located medial to the restiform body. As "perforating fibers" it intersects the superior cerebellar peduncle in the cerebellar hilus. The term superior cerebellar peduncle, which is used as a synonym for the brachium conjunctivum, should be used in a broad sense to include all fibers leaving or entering the cerebellum through the region bordering the superior medullary velum. It contains the ventral spinocerebellar tract and, in other species, the tectocerebellar tract. The brachium conjunctivum refers to the bundle of central nuclear efferents only.

peduncles, remain separated from the central cerebellar nuclei and their tracts by a layer of olivocerebellar fibers (Figs 4 and 7). Olivocerebellar fibers are located in the ventromedial part of the inferior cerebellar peduncle. At the entrance of the peduncle in the cerebellum, they pass through its mossy fiber components and position themselves immediately rostral to the central cerebellar nuclei and dorsal to the superior cerebellar peduncle and the uncinate tract (Fig. 7h). From this position they enter the white matter of the lobules and the central cerebellar nuclei (Voogd, 1982).

B. Mossy fiber systems

With the exception of the pontocerebellar and lateral reticulocerebellar systems, virtually nothing is known about the exact origin, course and termination of mossy fibers in the rat. Spinocerebellar tracts in the rat were described by Anderson (1943) with the Marchi method and illustrated by Voogd (1967). These tracts are distributed bilaterally to the anterior lobe with a preference for its ventral part, to the bottom of the primary fissure, the pyramis and the dorsal part of the uvula, and to the ventral part of the paramedian lobule and the copula pyramidis. Fibers of the ventral spinocerebellar tract terminate more medially and do not reach lobule 1, the uvula, or the paramedian lobule. From the time of these early studies, several reports of spinocerebellar tracts in other species have appeared. According to the electrophysiologic studies of Oscarsson (1973) four spinocerebellar tracts can be distinguished; the dorsal and ventral spinocerebellar tracts, the rostral spinocerebellar tract, and the cuneocerebellar tract. Of these tracts only the cuneocerebellar tract has been studied in isolation (Gerrits *et al.*, 1984; Grant, 1962). Our knowledge of the distribution of other spinocerebellar tracts rests on degeneration experiments with interruption of the complete spinocerebellar projection at certain levels of the cord. The results of such experiments in other species are very similar to those obtained by Anderson (1943). The only new feature of the spinocerebellar projection which has been added since is its termination in longitudinally arranged aggregates of mossy fiber terminals in the granular layer (Hazlett *et al.*, 1971; van Rossum, 1969; Vielvoye and Voogd, 1977; Voogd *et al.*, 1969; Watson *et al.*, 1976). Recently, a termination in sagittal bands has also been described for the cuneocerebellar tract of the cat with electrophysiologic (Ekerot and Larson, 1980) and anatomic methods (Gerrits *et al.*, 1984b). The same feature is characteristic for the termination of the fibers from the lateral reticular nucleus of the cat (Künzle, 1975; Russchen *et al.*, 1976) and the rat (Chan-Palay *et al.*, 1977). The notion that termination in discrete longitudinal strips is a more general feature of mossy fiber systems, also in the rat, is suggested by Scheibel's (1977) Golgi study of the sagittal organization of mossy fiber terminals in the cerebellum of this species. A more diffuse type of termination was described for the pontocerebellar system, including the projections of the reticulotegmental nucleus (Gerrits and Voogd, 1981; Kawamura and Hashikawa, 1981; Voogd *et al.*, 1969).

The cerebellar distributions of the pontocerebellar, reticulotegmentocerebellar, and lateral reticulocerebellar systems are discussed in Volume 2, Chapter 10. All mossy fiber systems are distributed bilaterally, and the crossing may occur in the spinal cord, in the brain stem, and/or in the cerebellar commissure. The problem of recrossing of pontocerebellar fibers was discussed recently by Rosina and Provini (1984). Pontocerebellar fibers in the cat (Gerrits and Voogd, 1981; Kawamura and Hashikawa, 1981; Voogd, 1964, 1967) are distributed widely in the hemisphere. More medially, their terminal field becomes restricted to the apex of the lobules of the anterior lobe. The termination extends progressively deeper in lobules 5 and 6, to include the entire cortex of lobule 7. Lobule 8 receives pontocerebellar fibers only in its apical portion; lobule 9 receives a heavy projection only to its most dorsal sublobules. Lobule 10 lacks pontocerebellar afferents.

The distribution of spinocerebellar and cuneocerebellar fibers presents a complementary picture (Gerrits and Voogd, 1979). They terminate most heavily in the anterior lobe, the simple lobule, and the pyramis and are absent from lobule 7 and from the lateral part of the hemisphere of the posterior lobe. Their heaviest termination is often found at the base of the lobules, in the bottom of the interlobular fissures. Descriptions of the termination of mossy fiber systems in terms of lobules and fissures tend to ignore some of the most relevant features of these systems in the integrative function of the cerebellum; namely, the mediolateral segregation or overlap of their sagittally oriented terminal fields and the basoapical gradients in their termination.

One feature of the termination of mossy fiber systems which has not been discussed is their termination in the central cerebellar nuclei. This termination has been investigated using injections of retrograde tracers into the central nuclei. The proximity of the afferent tracts to the central nuclei may easily lead to false positive results. The experiments of Eller and Chan-Palay (1976) with injections of HRP into the lateral cerebellar nucleus of the rat showed a multitude of extracerebellar afferents to this nucleus. Of the sources of these afferents, only the reticulotegmental nucleus (Gerrits and Voogd, 1981; Gerrits *et al.*, 1984a), the lateral reticular nucleus (Russchen *et al.*, 1976), and the raphe nucleus (Chan-Palay, 1977) were confirmed with anterograde axonal tracing methods. The pertinent negative results of Eller and Chan-Palay (1976) with respect to the dorsal column nuclei were confirmed in the anterograde transport study of Gerrits *et al.*, (1985) in the cat. Similar data are not available for the other central cerebellar nuclei of the rat. The tentative conclusion is that not all mossy fiber systems distribute collaterals to the central nuclei. The reticular nuclei, including the reticulotegmental and lateral reticular nuclei (but excluding the paramedian reticular nucleus; Eller and Chan-Palay, 1976), and the monoaminergic systems, seem to be the main sources of the extracerebellar central nuclear afferents. The rubrocerebellar pathway (Brodal and Gogstad, 1954), a collateral pathway of the rubrospinal tract, recently was investigated with double labeling techniques and shown to terminate in the anterior interposed nucleus in the rat (Huisman *et al.*, 1983). It is the only pathway known to terminate in the central nuclei and not in the cerebellar cortex.

C. The climbing fiber system: the olivocerebellar projection

The olivocerebellar projection to the cerebellar cortex and the central nuclei is nucleotopically organized. This was already known (Brodal, 1940) before it was realized that the inferior olive is the main source of climbing fibers in the rat (Desclin, 1974). Brodal (1940) investigated the projection of the inferior olive with the retrograde cell degeneration method in young cats and rabbits, and concluded that specific lobules received olivocerebellar fibers from specific subdivisions of the contralateral olivary nucleus. Voogd (1969) and Oscarsson (1969) showed that each subdivision of the inferior olive projects to a particular longitudinal strip of cortex, which can be traced through a number of successive lobules. According to Voogd (1969), the olivocerebellar fibers reach the Purkinje cells of these strips through the same myeloarchitectonic compartments which contain the projection of these Purkinje cells to the central cerebellar nuclei. He concluded that the organizations of the olivocerebellar and the corticonuclear projection are essentially similar. The concept of the longitudinal zonal organization of the olivocerebellar projection was further developed in the anterograde axonal transport studies of Courville *et al.* (1974), Groenewegen and Voogd (1977), Groenewegen *et al.* (1979), and Gerrits and Voogd (1982) in the cat and applied in the retrograde transport studies, summarized in the monograph of Brodal and Kawamura (1980).

Direct evidence for a longitudinal organization of the olivocerebellar projection in the rat is rather meager. Chan-Palay *et al.* (1977) used autoradiography of ^{35}S-methionine to demonstrate the sagittal organization of the olivocerebellar projection in the rat. According to their findings, the projection is bilateral, and banded areas which receive labeled climbing fibers alternate with areas which receive climbing fibers from extraolivary sources. Convincing evidence that, in this case, the injections in the inferior olive must have been incomplete, and that the entire cortex of the cerebellum is provided with climbing fibers from the inferior olive was published by Armstrong *et al.* (1982) and Campbell and Armstrong (1983a). These last authors were unable to confirm the presence of an uncrossed component in the olivocerebellar projection of the rat.

According to Campbell and Armstrong (1983b) the topographic organization of the olivocerebellar projection in the rat and the cat are strikingly similar (Fig.

13). In both species, the caudal half of the medial accessory olive projects to a medial vermal zone (zone A of the cat) and the dorsal accessory olive to a lateral vermal zone (zone B of the cat). The rostral half of the medial and dorsal accessory olives and the principal olive project to discrete zones in the hemisphere. A complete identification of the zones Cl to C3 and Dl to D2 as in the hemisphere of the cerebellum of the cat (Voogd, 1982) is not possible in the rat. Similar conclusions were reached by Sotelo *et al.* (1984) in their autoradiographic studies in neonatal rats.

Detailed nucleotopical maps are present within each olivocerebellar projection zone. For an understanding of these maps, knowledge of the collateralization of the olivocerebellar fibers is indispensable (Rosina and Provini, 1983; Wiklund *et al.*, 1984). Many examples of this intrazonal topography are given in the review on the olivocerebellar projection in Volume 2, Chapter 10. The studies of Eisenman (1981, 1984) and of Furber and Watson (1983) are particularly important in this respect. According to Eisenman (1984), three or four parasagittal zones are present in lobule 9 of the cerebellum of the rat (Fig. 14). The medial zone receives fibers from the group "beta" of the caudal medial accessory olive, the intermediate zone from the dorsomedial cell column and the lateral zone (or zones) from the ventral lamella or the principal olive, and the rostral half of the medial accessory olive. This pattern is in accordance with the situation in the cat (Brodal and Kawamura, 1980; Groenewegen *et al.*, 1979). Some of these zones continue into lobule 8 and the copula pyramidis, and some do not. The medial subzone of lobule 9, receiving fibers from subnucleus "beta", is continuous with a more lateral subzone of lobule 8; the lateral zones, receiving fibers from the medial and principal olives, are continuous with zones in the copula pyramidis. In addition, lobule 8 contains subzones receiving projections from other parts of the caudal medial accessory olive and the dorsal accessory olive. Lobule 9 clearly incorporates longitudinal zones which, in more dorsal lobules, are included in the hemisphere. The border between vermis and hemisphere is usually considered as a boundary between two cerebellar territories with different connections. It appears from these studies, and from the investigations on the corticonuclear projection cited above, that the position of the paramedian sulcus, relative to the sagittal corticonuclear and olivocerebellar projection zones, can shift in different rostrocaudal segments of the cerebellum. Careful comparisons of the olivocerebellar and corticonuclear projections, using the myeloarchitectonic compartments in the white matter as a reference system (Fig. 5h), are necessary to solve the problem of the fundamental longitudinal organization of the cerebellar connections.

VI. CONCLUSIONS

The cerebellum of the rat is often used as a model in neurophysiologic (Bower and Woolston, 1983), neuroembryologic (Wassef and Sotelo, 1984), or immunohistochemical studies (Chan-Palay, 1982; De Camilli *et al.*, 1984) which elaborate on the connectivity of the mammalian cerebellum. Much neuroanatomical research on its connections and its fundamental subdivision is still needed to allow optimal use of the cerebellum of the rat in neurobiologic research. More information is necessary on its more generalized mammalian features and on its anatomic and functional specializations.

ACKNOWLEDGEMENTS

Computer reconstructions for Fig. 2 were made by H. Choufoer and drawn by Mrs Joyce Wetselaar-Whittaker, medical artist. Part of the material reported in this chapter was prepared during the stay of J. Voogd as a research associate of the National Research Council in the neuroanatomical laboratory of Dr W. R. Mehler, NASA-Ames Research Center, Moffett Field, California, U.S.A.

REFERENCES

Achenbach, K. E. and Goodman, D. C. (1968). Cerebellar projection to pons, medulla and spinal cord in the albino rat. *Brain Behav. Evol.* 1, 43–57.

Aciron, E. E. (1951). Fisuracion del cerebelo en la rata blanca. *Arqu. Anat. Antrop.* 27, 215–261.

Anderson, R. F. (1943). Cerebellar distribution of the dorsal and ventral spino-cerebellar tracts in the white rat. *J. Comp. Neurol.* 79, 415–423.

Angaut, P. (1970). The ascending projections of the nucleus interpositus posterior of the cat cerebellum: An experimental anatomical study using silver impregnation methods. *Brain Res.* 377–394.

Angaut, P. and Brodal, A. (1967). The projection of the "vestibulocerebellum" onto the vestibular nuclei in the cat. *Arch. Ital. Biol.* 105, 441–479.

Angaut, P. and Cicirata, F. (1982). Cerebello-olivary projections in the rat. *Brain Behav. Evol.* 21, 24–33.

Armstrong, D. M. and Schild, R. F. (1978a). An investigation of the cerebellar cortico-nuclear projections in the rat using an autoradiographic tracing method. I. Projections from the vermis. *Brain Res.* 141, 1–19.

Armstrong, D. M. and Schild, R. F. (1978b). An investigation of the cerebellar cortico-nuclear projections in the rat using an autoradiographic tracing method. II. Projections from the hemisphere. *Brain Res.* 141, 235–249.

Armstrong, D. M., Harvey, R. J. and Schild, R. F. (1974). Topographical localization in the olivo-cerebellar projection: An electrophysiological study in the cat. *J. Comp. Neurol.* 154, 287–302.

Armstrong, D. M., Campbell, N. C., Edgley, S. A., Schild, R. F. and Trott, Y. R. (1982). Investigations of olivo-cerebellar and spino-olivary pathways. *Exp. Brain Res. Suppl.* 6, 195–232.

Balaban, C. D. (1984). Olivo-vestibular and cerebello-vestibular connections in albino rabbits. *Neurosci.* 12, 129–150.

Batton, R. R., III, Jayaraman, A., Ruggiero, D. and Carpenter, M. B. (1977). Fastigial efferent projections in the monkey: An autoradiographic study. *J. Comp. Neurol.* 174, 281–306.

Beitz, A. J. (1982). Structural organization of the fastigial nucleus. *Exp. Brain Res. Suppl.* 6, 233–246.

Beitz, A. J. and Chan-Palay, V. (1979a). The medial cerebellar nucleus in the rat: Nuclear volume, cell number, density and orientation. *Neurosci.* 4, 31–45.

Beitz, A. J. and Chan-Palay, V. (1979b). A Golgi analysis of neuronal organization in the medial cerebellar nucleus of the rat. *Neurosci.* 4, 47–63.

Bentivolgio, M. and Kuypers, H. G. J. M. (1982). Divergent axon collaterals from rat cerebellar nuclei to diencephalon, mesencephalon, medulla oblongata and cervical cord: A fluorescent double-labeling study. *Exp. Brain Res.* 46, 339–356.

Bigaré, F. (1980). De efferente verbindingen van de cerebellaire schors van de kat. Doctoral thesis, Leiden.

Bolk, L. (1906). *Das Cerebellum der Säugetiere.* Bohn-Fischer, Jena.

Bower, J. M. and Woolston, D. C. (1983). Congruence of spatial organization of tactile projections to granule cell and Purkinje cell layers of cerebellar hemispheres of the albino rat: Vertical organization of cerebellar cortex. *J. Neurophysiol.* 49, 746–766.

Bradley, O. Ch. (1904). The mammalian cerebellum: Its lobes and fissures. *J. Anat. Physiol.* 38, 448–475.

Brodal, A. (1940). Untersuchunger über die Olivo-cerebellaren Lokalisation. *Z. Neurol.* 169, 1–53.

Brodal, A. (1981). *Neurological anatomy in relation to clinical medicine.* Oxford University Press, Oxford.

Brodal, A. and Gogstad, A. C. (1954). Rubro-cerebellar connections. *Anat. Rec.* 118, 455–485.

Brodal, A. and Kawamura, K. (1980). Olivocerebellar projection: A review. *Adv. Anat. Embryol. Cell Biol.* 64, 1–137.

Brodal, A. and Pompeiano, O. (1957). The vestibular nuclei in the cat. *J. Anat.* 91, 438–454.

Brown, J. T., Chan-Palay, V. and Palay, S. L. (1977). A study of afferent input to the inferior olivary complex in the rat by retrograde axonal transport of horseradish peroxidase. *J. Comp. Neurol.* 176, 1–22.

Cajal, S. Ramón y (1903). La dobla via descendente nacida del pedunculo cerebeloso superior. *Trab. Lab. Inst. Biol. Univ. Madrid* 2, 23–29.

Cajal, S. Ramón y (1911). *Histologie du système nerveux de l'homme et des vertébrés.* Maloine, Paris.

Campbell, N. C. and Armstrong, D. M. (1983a). The olivocerebellar projection in the rat: An autoradiographic study. *Brain Res.* 275, 215–233.

Campbell, N. C. and Armstrong, D. M. (1983b). Topographical localization in the olivocerebellar projection in the rat: An autoradiographic study. *Brain Res.* 275, 235–249.

Carpenter, M. B. and Batton, R. R., III (1982). Connections of the fastigial nucleus in the cat and monkey. *Exp. Brain Res. Suppl.* 6, 250–291.

Chan-Palay, V. (1977). *The cerebellar dentate nucleus.* Springer, Berlin.

Chan-Palay, V. (1982). Neurotransmitters and receptors in the cerebellum: Immunocytochemical localization of glutamic acid decarboxylase, GABA-transaminase and cyclic GMP and autoradiography with ^{3}H-muscimol. *Exp. Brain Res. Suppl.* 6, 522–584.

Chan-Palay, V., Palay, S. L., Brown, J. T. and Van Itallie, C. (1977). Sagittal organization of olivocerebellar and reticulocerebellar projections: Autoradiographic studies with ^{35}S-methionine. *Exp. Brain Res.* 30, 561–576.

Chan-Palay, V., Palay, S. L. and Wu, J.-Y. (1979). Gamma-aminobutyric acid pathways in the cerebellum studied by retrograde transport of glutamic acid decarboxylase antibody after in vivo injections. *Anat. Embryol.* 157, 1–14.

Courville, J. and Faraco-Cantin, F. (1976). Cerebellar corticonuclear projection demonstrated by the horseradish peroxidase method. *Neurosci. Abst.* 2, 108.

Courville, J., Faraco-Cantin, F. and Diakew, N. (1974). A functionally important feature of the distribution of the olivo-cerebellar climbing fibers. *Can. J. Physiol. Pharmacol.* 52, 1212–1217.

De Camilli, P., Miller, P. E., Levitt, P., Walter, U. and Greengard, P. (1984). Anatomy of cerebellar Purkinje cells in the rat determined by a specific immunocytochemical marker. *Neurosci.* 11, 761–817.

Desclin, J. C. (1974). Histological evidence supporting the inferior olive as the major source of cerebellar climbing fibers in the rat. *Brain Res.* 77, 365–384.

Dow, R. S. (1936). The fiber connections of the posterior parts of the cerebellum in the cat and rat. *J. Comp. Neurol.* 63, 527–548.

Eisenman, L. M. (1981). Olivocerebellar projections to the pyramis and the copula pyramis in the rat: Differential projections to parasagittal zones. *J. Comp. Neurol.* 199, 65–76.

Eisenman, L. M. (1984). Organization of the olivocerebellar projection to the uvula in the rat. *Brain Behav. Evol.* 24, 1–12.

Ekerot, C.-F. and Larson, B. (1980). Termination in overlapping sagittal zones in cerebellar anterior lobe of mossy and climbing fiber paths activated from dorsal funiculus. *Exp. Brain Res.* 38, 163–172.

Ekerot, C.-F. and Larson, B. (1982). Branching of olivary axons to innervate pairs of sagittal zones in the cerebellar anterior lobe of the cat. *Exp. Brain Res.* 48, 185–198.

Eller, T. and Chan-Palay, V. (1976). Afferents to the cerebellar lateral nucleus: Evidence from retrograde transport of horseradish peroxidase after pressure injections through micropipettes. *J. Comp. Neurol.* 166, 285-302.

Faull, R. L. M. (1978). The cerebellofugal projections in the brachium conjunctivum of the cat. III. The ipsilateral and contralateral descending pathways. *J. Comp. Neurol.* 178, 519-536.

Feirbend, H. K. P. (1983). Anatomy and development of longitudinal patterns in the architecture of the cerebellum of the white leghorn (*Gallus domesticus*). Doctoral thesis, Leiden.

Feirabend, H. K. P., Vielvoye, G. J., Freedman, S. L. and Voogd, J. (1976). Longitudinal organization of afferent and efferent connections of the cerebellar cortex of the chicken. *Exp. Brain Res. Suppl.* 1, 71-78.

Furber, S. E. and Watson, C. R. R. (1983). Organization of the olivocerebellar projection in the rat. *Brain Behav. Evol.* 27, 132-152.

Gacek, R. R. (1977). Localization of brain stem neurons projecting to the oculomotor nucleus in the cat. *Exp. Neurol.* 57, 725-749.

Gacek, R. R. (1979). Localization of trochlear vestibulo-ocular neurons in the cat. *Exp. Neurol.* 66, 692-706.

Gerrits, N. M. and Voogd, J. (1979). Cerebellar mossy fiber systems in the cat. *Acta Morphol. Neerl.-Scand.* 17, 236-237.

Gerrits, N. M. and Voogd, J. (1981). Cerebellar efferents of the nucleus reticularis tegmenti pontis in the cat. *Acta Morphol. Neerl.-Scand.* 19, 57-57.

Gerrits, N. M. and Voogd, J. (1982). The climbing fiber projection to the flocculus and adjacent paraflocculus in the cat. *Neurosci.* 7, 2971-2991.

Gerrits, N. M., Epema, A. H. and Voogd, J. (1984a). The mossy fiber projection of the nucleus reticularis tegmenti pontis to the flocculus and adjacent ventral paraflocculus in the cat. *Neurosci.* 11, 627-644.

Gerrits, N. M., Voogd, J. and Nas, W. S. C. (1984b). Cerebellar and olivary projections of the external and rostral internal cuneate nuclei in the cat. *Exp. Brain Res.* (in press).

Goodman, D. C., Hallett, R. E. and Welch, R. B. (1963). Patterns of localization in the cerebellar corticonuclear projections of the albino rat. *J. Comp. Neurol.* 121, 51-68.

Grant, G. (1962). Projection of the external cuneate nucleus onto the cerebellum in the cat: An experimental study using silver methods. *Exp. Neurol.* 5, 179-195.

Groenwegen, H. J. and Voogd, J. (1977). The parasagittal zonation within the olivo-cerebellar projection. I. Climbing fiber distribution in the vermis of cat cerebellum. *J. Comp. Neurol.* 174, 417-488.

Groenewegen, H. J., Voogd, J. and Freedman, S. L. (1979). The parasagittal zonation within the olivo-cerebellar projection. II. Climbing fiber distribution in the intermediate and hemispheric parts of the cat cerebellum. *J. Comp. Neurol.* 183, 551-602.

Haines, D. E. (1977). Cerebellar corticonuclear and corticovestibular fibers of the flocculonodular lobe in the prosimian primate (*Galago senegalensis*). *J. Comp. Neurol.* 174, 607-630.

Haines, D. E. and Koletar, S. L. (1979). Topography of cerebellar corticonuclear fibers of the albino rat. Vermis of anterior and posterior lobes. *Brain Behav. Evol.* 16, 271-292.

Haines, D. E., Patrick, G. W. and Satrulee, P. (1982). Exp. Brain Res. Suppl. 6, 320-367.

Haroian, A. J. (1982). Cerebello-olivary projections in the rat: An autoradiographic study. *Brain Res.* 235, 125-130.

Haroian, A. J., Massopust, L. C. and Young, P. A. (1981). Cerebellothalamic projections in the rat: An autoradiographic and degeneration study. *J. Comp. Neurol.* 197, 217-236.

Hazlett, J. C., Martin, G. F. and Dom, R. (1971). Spinocerebellar fibers of the opossum *Didelphis marsupialis virginiana*. *Brain Res.* 33, 257-271.

Huisman, A. M., Kuypers, H. J. G. M., Condé, F. and Keizer, K. (1983). Collaterals of rubrospinal neurons to the cerebellum in the rat: A retrograde fluorescent double labeling study. *Brain Res.* 264, 181-196.

Ingvar, S. (1919). Zur Phylo- und Ontogenese des Kleinhirns. *Folia Neurobiol.* 11, 205-495.

Ito, M. (1984). *The cerebellum and neural control.* Raven Press, New York.

Jansen, J. (1956). *On the efferent connections of the cerebellum* (Progress in Neurobiology). Elsevier, Amsterdam.

Jansen, J. and Brodal, A. (1940). Experimental studies on the intrinsic fibers of the cerebellum. II. The cortico-nuclear projection. *J. Comp. Neurol.* 73, 267-321.

Jansen, J. and Brodal, A. (1942). Experimental studies on the intrinsic fibers of the cerebellum. III. The cortico-nuclear projection in the rabbit and the monkey. *Norske Vid. Akad., Avh. I, Math. Nat. Kl.* 3, 1–50.

Kawamura, K. and Hashikawa, T. (1981). Projections from the pontine nuclei proper and reticular tegmental nucleus onto the cerebellar cortex in the cat. *J. Comp. Neurol.* 201, 395–413.

Klimoff, J. (1899). Ueber die Leitungsbahnen des Kleinhirns. *Arch. Anat. Physiol., Anat. Abtheil.* 11–27.

Komei, I., Hajai, T., McGeer, P. L. and McGeer, E. G. (1983). Evidence for an intracerebellar acetylcholinesterase-rich but probably non-cholinergic flocculo-nodula projection in the rat. *Brain Res.* 258, 115–119.

Korneliussen, H. K. (1968). On the morphology and subdivision of the cerebellar nuclei of the rat. *J. für Hirnforsch.* 10, 109–122.

Künzle, H. (1975). Autoradiographic tracing of the cerebellar projections from the lateral reticular nucleus in the cat. *Exp. Brain Res.* 22, 255–266.

Larsell, O. (1948). The development and subdivisions of the cerebellum of birds. *J.Comp. Neurol.* 89, 123–190.

Larsell, O. (1952). The morphogenesis and adult pattern of the lobules and tissues of the cerebellum of the white rat. *J. Comp. Neurol.* 97, 281–356.

Larsell, O. (1970). In J. Jansen (Ed.). *The comparative anatomy and histology of the cerebellum from monotremes through apes.* University of Minnesota Press, Minneapolis.

Larsell, O., and Jansen, J. (1972). *The comparative anatomy and histology of the cerebellum: The human cerebellum, cerebellar connections and the cerebellar cortex.* University of Minnesota Press, Minneapolis.

Löwy, R. (1916). Ueber die Faseranatomie und Physiologie der Formatio vermicularis cerebelli. *Arb. Neurol. Inst. Univ. Wien* 21, 359–382.

McCrea, R. A., Bishop, G. A. and Kitai, S. T. (1978). Morphological and electrophysiological characteristics of projection neurons in the nucleus interpositus of the cat cerebellum. *J.Comp. Neurol.* 181, 397–420.

Marani, E. (1982). Topographical enzyme histochemistry of the mammalian cerebellum. Doctoral thesis, Leiden.

Marani, E. and Voogd, J. (1979). The morphology of the mouse cerebellum. *Acta Morphol. Neerl.-Scand.* 17, 33–52.

Matsushita, M. and Hosoya, Y. (1978). The location of spinal projection neurons in the cerebellar nuclei (cerebello-spinal tract neurons) of the cat: A study with the HRP technique. *Brain Res.* 142, 237–248.

Mehler, W. R. (1967). Double descending pathways originating from the superior cerebellar peduncle: An example of neural species differences. *Anat. Rec.* 157, 374.

Mehler, W. R. (1969). Some neurological species differences—*a posteriori. Ann. N.Y. Acad. Sci.* 167, 424–468.

Ogawa, T. (1935). Beiträge zur vergleichenden Anatomie des Zentralnervensystems der Wassersäugetiere. Ueber die Kleinhirnkerne per Pinnipedien und Cetaceen. *Arb. Anat. Inst. Sendai* 17, 63–136.

Ohkawa, K. (1957). Comparative anatomical studies of cerebellar nuclei in mammals. *Arch. Hist. Jap.* 13, 21–58.

Oscarsson, O. (1969). The sagittal organization of the cerebellar anterior lobe as revealed by the projection patterns of the climbing fiber system. In R. Ilinas (Ed.). *Neurobiology of cerebellar evolution and development*, pp. 525–537. AMA-ERF, Chicago.

Oscarsson, O. (1973). Functional organization of spinocerebellar paths. In: Handbook of sensory physiology. Vol. 2, Somatosensory system, pp. 339–380. Springer, Berlin.

Palay, S. L. and Chan-Palay, V. (1974). *Cerebellar cortex: Cytology and organization.* Springer, Berlin.

Palay, S. L. and Chan-Palay, V. (1982). The cerebellum, new vistas. *Exp. Brain Res. Suppl.* 6,

Paxinos, G. and Watson, C. (1982). *The rat brain in stereotaxic coordinates.* Academic Press, Sydney.

Riley, H. A. (1928). A comparative study of the arbor vitae and the folial pattern of the mammalian cerebellum. *Arch. Neurol. Psychiat.* 20, 895–1034.

Rosina, A. and Provini, L. (1983). Somatotopy of climbing fibers branching to the cerebellar cortex in cat. *Brain Res.* 289, 45–63.

Rosina, A. and Provini, L. (1984). Pontocerebellar system linking the two hemispheres by intracerebellar branching. *Brain Res.* 296, 365–369.
Rossum, J. van (1969). Corticonuclear and corticovestibular projections of the cerebellum. Doctoran thesis, Van Gorcum, Assen.
Russchen, F. T., Groenewegen, H. J. and Voogd, J. (1976). Reticulocerebellar connections in the cat: An autoradiographic study. *Acta Morph. Neerl.-Scand.* 14, 245–246.
Scheibel, A. B. (1977). Sagittal organization of mossy fiber terminal systems in the cerebellum of the rat: A Golgi study. *Exp. Neurol.* 57, 1067–1070.
Scholten, J. M. (1946). De plaats van den paraflocculus in het geheel der cerebellaire correlaties. Doctoral thesis, Noorhollandse Uitg. Mij., Amsterdam.
Scott, Th. G. (1964). A unique pattern of localization within the cerebellum of the mouse. *J. Comp. Neurol.* 122, 1-8.
Scott, Th. G. (1965). The specificity of 5′-nucleotidase in the brain of the mouse. *J. Histochem. Cytochem.* 13, 657–667.
Sotelo, C., Bourrat, F. and Triller, A. (1984). Postnatal development of the inferior olivary complex in the rat. II. Topographic organization of the immature olivocerebellar projection. *J. Comp. Neurol.* 222, 177–199.
Swenson, R. S. and Castro, A. J. (1983a). The afferent connections of the inferior olivary complex in rats: An anterograde study using autoradiographic and axonal degeneration techniques. *Neurosci.* 8, 259–275.
Swenson, R. S. and Castro, A. J. (1983b). The afferent connections of the inferior olivary complex in rats: A study using the retrograde transport of horseradish peroxidase. *Am. J. Anat.* 166, 329–341.
Szentágothai, J. (1942). Die Bedeutung des Faserkalibers und der Markscheidendicke im Zentralnervensystem. *Z. Anat. Entwick. Ges.* 111, 201–223.
Tolbert, D. L. (1982). The cerebellar nucleocortical pathway. *Exp. Brain Res. Suppl.* 6, 296–317.
Tolbert, D. L., Massopust, L. C., Morphy, M. G. and Young, P. A. (1976). The anatomical organization of the cerebello-olivary projection in the cat. *J. Comp. Neurol.* 170, 525–544.
Tolbert, D. L., Kultas-Ilinsky, K., Ilinsky, I. A. and Warton, S. (1980). EM-autoradiography of cerebellar nucleocortical terminals in the cat. *Anat. Embryol.* 215–223.
Verhaart, W. J. C. (1970). *Comparative anatomical aspects of the mammalian brain stem and the cord.* Van Gorcum, Assen.
Vielvoye, G. J. and Voogd, J. (1977). Time dependence of terminal degeneration in spinocerebellar mossy fiber rosettes in the chicken and the application of terminal degeneration in successive degeneration experiments. *J. Comp. Neurol.* 175, 233–242.
Voogd, J. (1964). *The cerebellum of the cat: Structure and fiber connections.* Doctoral thesis, Van Gorcum, Assen.
Voogd, J. (1967). Comparative aspects of the structure and fibre connexions of the mammalian cerebellum. *Prog. Brain Res.* 25, 94–135.
Voogd, J. (1969). The importance of fiber connections in the comparative anatomy of the mammalian cerebellum. In R. Ilinas (Ed.). *Neurobiology of cerebellar evolution and development*, pp. 493–541. AMA-ERF, Chicago.
Voogd, J. (1975). Bolk's subdivision of the mammalian cerebellum. Growth centres and functional zones. *Acta Morphol. Neerl.-Scand.* 13, 35–54.
Voogd, J. (1982). The olivocerebellar projection in the cat. *Exp. Brain Res. Suppl.* 6, 134–161.
Voogd, J. (1983). Anatomical evidence for a cortical "x" zone in the cerebellum of the cat. *Abst. 13th Ann. Meet. Soc. Neurosci. Boston* 1091.
Voogd, J. and Bigaré, F. (1980). Topographical distribution of olivary and cortinonuclear fibers in the cerebellum. A review. In J. Courville, C. de Montigny and Y. Lamarre (Eds). *The inferior olivary nucleus*, pp. 207–234. Raven Press, New York.
Voogd, J. and Feirabend, H. K. P. (1981). Classic methods in neuroanatomy. In R. Lahue (Ed.). *Methods in neurobiology*, pp. 301-364. Plenum Press, New York.
Voogd, J., Broere, G. and Van Rossum, J. (1969). The mediolateral distribution of the spinocerebellar projection in the anterior lobe and the simple lobule in the cat and a comparison with some other afferent fibre systems. *Psychiat. Neurol. Neurochir.* 72, 137–151.

Wassef, M. and Sotelo, C. (1984). Asynchrony in the expression of cylic GMP dependent proteinkinase by clusters of Purkinje cells during the perinatal development of rat cerebellum. *Neurosci.* (in press).

Watson, C. R. R., Broomhead, A. and Holst, M.-C. (1976). Spinocerebellar tracts in the brush-tailed opossum, *Trichosuris vulpecula*. *Brain Behav. Evol.* 13, 142–153.

Watt, C. B. and Mihailoff, G. A. (1983). The cerebellopontine system in the rat. I. Autoradiographic studies. *J. Comp. Neurol.* 215, 312–330.

Weidenreich, F. (1899). Zur Anatomie der zentralen Kleinhirnkerne der Säuger. *Z. Morphol. Anthropol.* 1, 259–312.

Wiklund, L., Toggenburger, G. and Cuénod, M. (1984). Selective retrograde labeling of the rat olivocerebellar climbing fiber system with ^{3}H-D-aspartate. *Neurosci.* (in press).

Woodson, W. and Angaut, P. (1984). The ipsilateral descending limb of the brachium conjunctivum: An autoradiographic and HRP study in rats. *Neurosci. Lett. Suppl.* 18, S58.

Fig. 1: The cerebellum of the rat: graphic reconstructions from serial sections. The anterior lobe has been removed in the right side of the figures to show the simple lobule in the caudal bank of the primary fissure. Interruptions of the cortex are hatched. a: caudal aspect; b: dorsal aspect; c: rostral aspect; d: ventral aspect; e: midsagittal section; f: diagram of the cortical loop of the paraflocculus and the flocculus. Abbreviations for this and subsequent figures: a, subnucleus a of the medial accessory olive; A, myeloarchitectonic compartment or cerebellar zone A; ANT, anterior lobe; ApmF, ansoparamedian fissure; b, subnucleus b of the medial accessory olive; B, myeloarchitectonic compartment or cerebellar zone B; c, subnucleus c of the medial accessory olive; C1, crus 1 of the ansiform lobule; C2, crus 2 of the ansiform lobule; CeC, cerebellar commissure; COP, copula pyramidis; d, dorsal accessory olive; DCo, dorsal cochlear nucleus; dft, direct fastigiobulbar tract; DLH, dorsolateral hump (central cerebellar nuclei); dm, dorsomedial cell column of the inferior olive; DMC, dorsomedial crest (central cerebellar nuclei); FL, flocculus; FLped, floccular peduncle; ic, internal capsule; IcF, intercrural fissure; icp, inferior cerebellar peduncle; Inf, infracerebellar nucleus; Int, interposed cerebellar nucleus; IntA, anterior interposed nucleus; IntP, posterior interposed nucleus; IntPpc, parvocellular part of the posterior interposed nucleus; IPflS, intraparafloccular sulcus; jrb, juxtarestiform body; Lat, lateral cerebellar nucleus; Latpc, parvocellular part of the lateral cerebellar nucleus; LD, laterodorsal nucleus (thalamus); LVe, lateral vestibular nucleus; m, medial accessory olive; mcp, middle cerebellar peduncle; Med, medial cerebellar nucleus; MedCM, caudomedial subdivision of the medial cerebellar nucleus; MedDLP, dorsolateral protuberance of the medial cerebellar nucleus; MedM, middle subdivision of the medial cerebellar nucleus; MedMpc, parvocellular part of the middle subdivision of the medial cerebellar nucleus; ocf, olivocerebellar fibers; PFL, paraflocculus; PlF, posterolateral fissure; PflS, parafloccular sulcus; PM, paramedian lobule; PmS, paramedian sulcus; PnC, nucleus reticularis pontis caudalis; PpF, prepyramidal fissure; pr, principal nucleus of the inferior olive; PreculF, preculminate fissure; PrF, primary fissure; PrV, principal sensory nucleus of the trigeminal nerve; PsF, posterior superior fissure; py, pyramidal tract; SC, superior colliculus; scp, superior cerebellar peduncle; SecF, secondary fissure; Sim, simple lobule; smv, superior medullary velum; unc, uncinate tract; Vco, ventral cochlear nucleus; VL, ventrolateral nucleus (thalamus); VM, ventromedial nucleus (thalamus); vsc, ventral spinocerebellar tract; 'X', myeloarchitectonic compartment or zone 'X'; Y, group Y of the vestibular nuclei; 1 - 10, lobules of the cerebellum; "beta", group "beta" of the medial accessory olive.

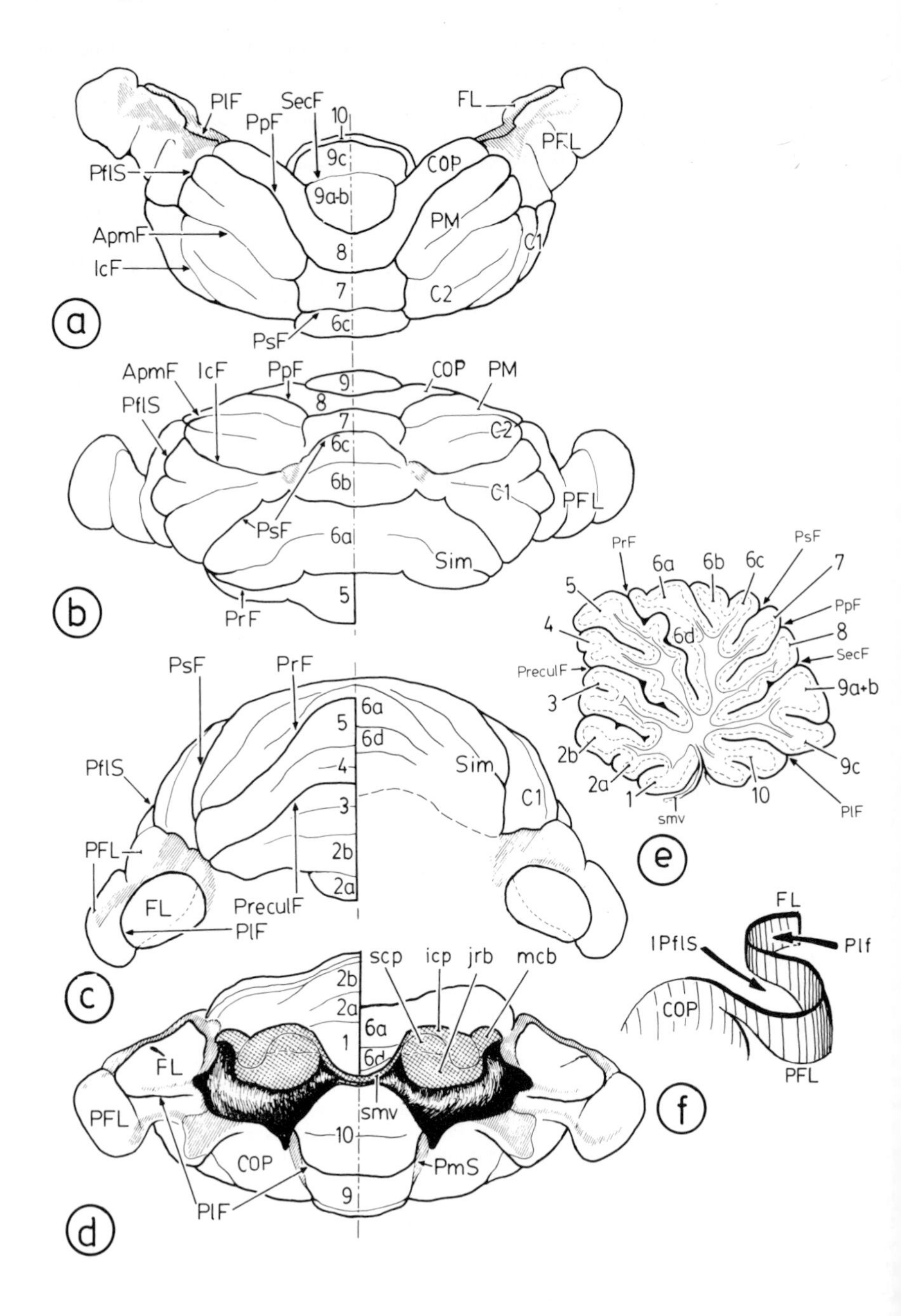

a
PlF
SecF
PpF
10
9c
9a-b
PflS
ApmF
IcF
8
7
6c
PsF
FL
PFL
COP
PM
C1
C2
b
ApmF
IcF
PpF
COP
PM
PflS
9
8
7
6c
6b
6a
5
C2
C1
PFL
Sim
PsF
PrF
c
PsF
PrF
6a
5
6d
4
3
2b
2a
Sim
C1
PflS
PFL
FL
PreculF
PlF
d
scp
icp
jrb
mcb
2b
2a
6a
1
6d
FL
PFL
smv
10
COP
PmS
9
PlF
e
PrF
6a
6b
6c
PsF
7
5
PpF
4
6d
8
SecF
PreculF
3
9a+b
2b
9c
2a
1
10
smv
PlF
f
FL
lPflS
Plf
COP
PFL

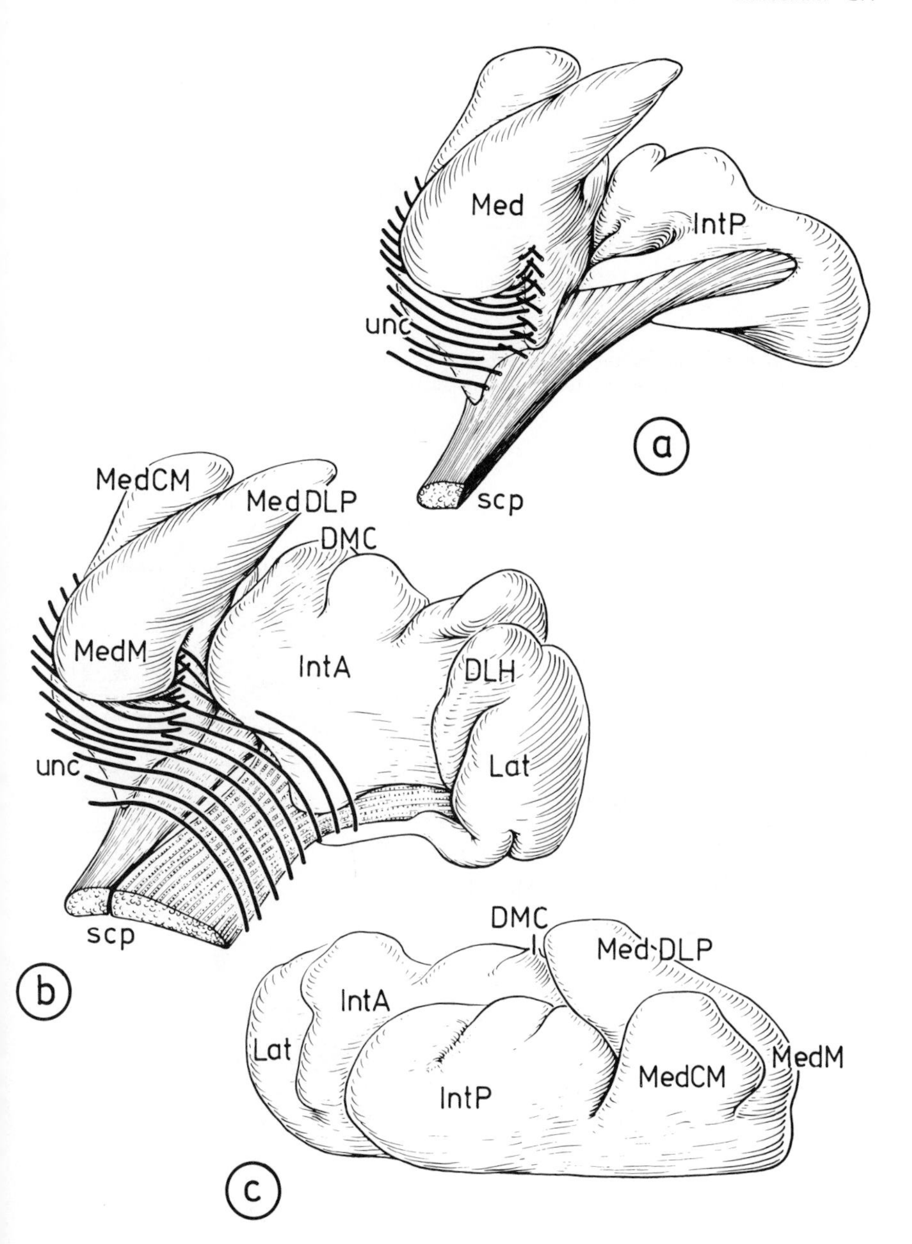

Fig. 2: Rostrolateral (A and B) and caudomedial (C) views of the central cerebellar nuclei of the rat. Drawings from computer reconstructions. Abbreviations as in Fig. 1.

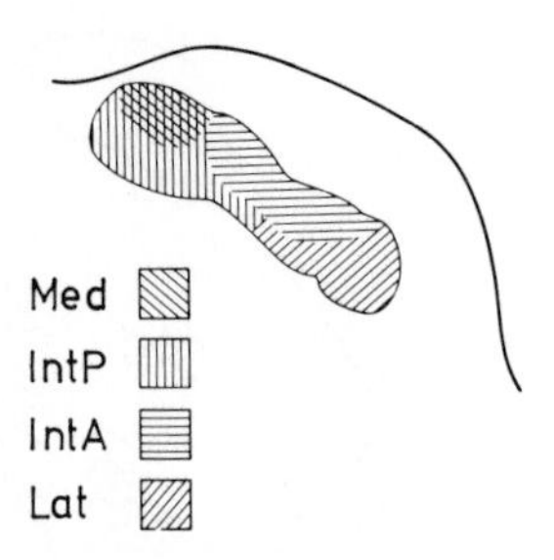

Fig. 3: The fiber composition of the superior cerebellar peduncle in the rat. Relabeled and reproduced from Haroian *et al.* (1981). Abbreviations as in Fig. 1.

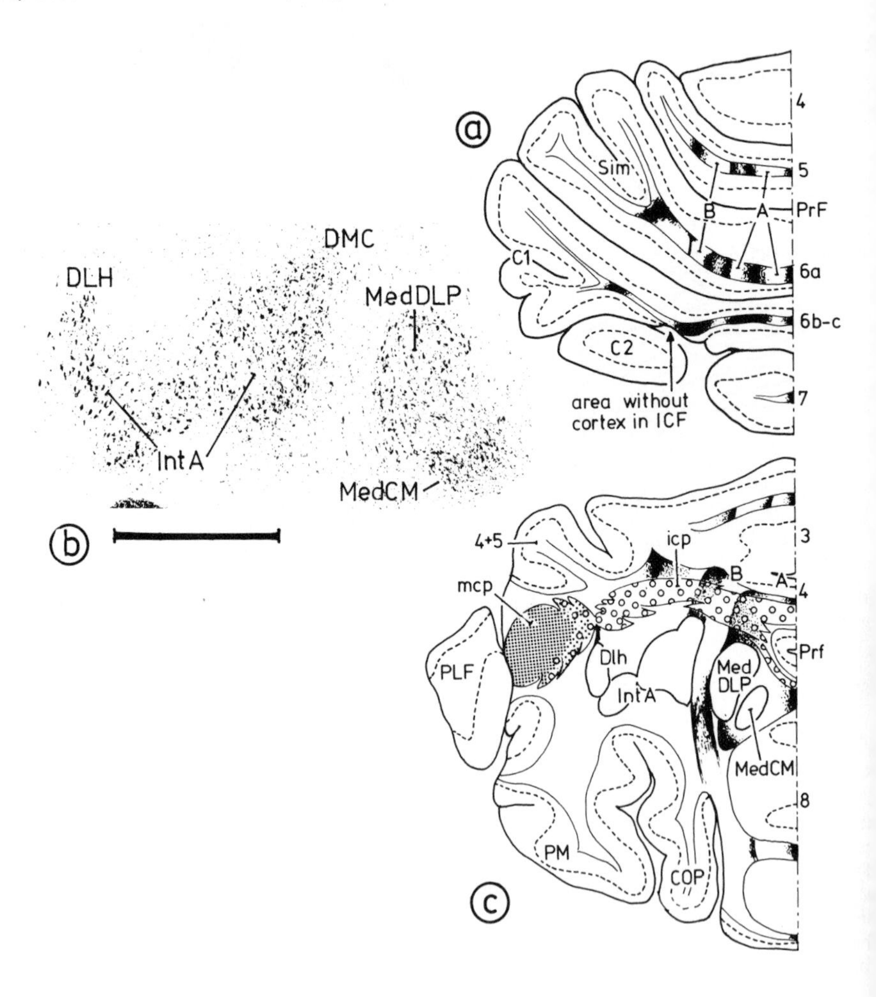

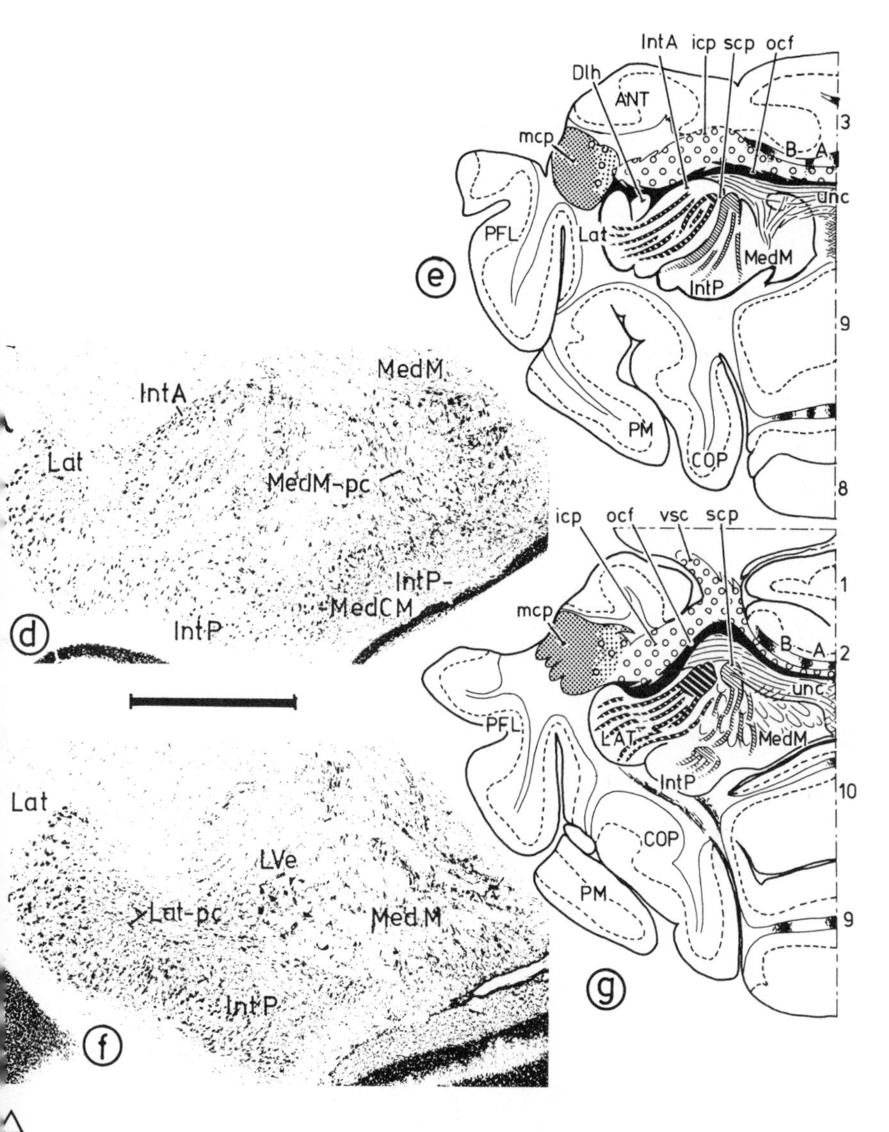

△

ig. 4: A–C: Horizontal sections through the cerebellum of the rat. A, C: drawings of Häggqvist-tained sections. B: Nissl-stained section through the dorsal part of the central nuclei, same level s C. In Figs. 4, 5, 6 and 7 the fibers of the medial and lateral subdivisions of the superior cerebellar eduncle, the inferior cerebellar peduncle and the middle cerebellar peduncle are indicated with ifferent symbols. Olivocerebellar fibers and borders of myeloarchitectonic compartments A and B re indicated in black. Calibration (bar = 1 mm) refers to Nissl stained section. D–G: Horizontal ections through the cerebellum of the rat. D, F: Nissl-stained sections through the ventral part of he central nuclei, same levels as E and G; E, G: drawings of Häggqvist-stained sections. Calibration bar = 1 mm) refers to Nissl-stained sections. Abbreviations as in Fig. 1.

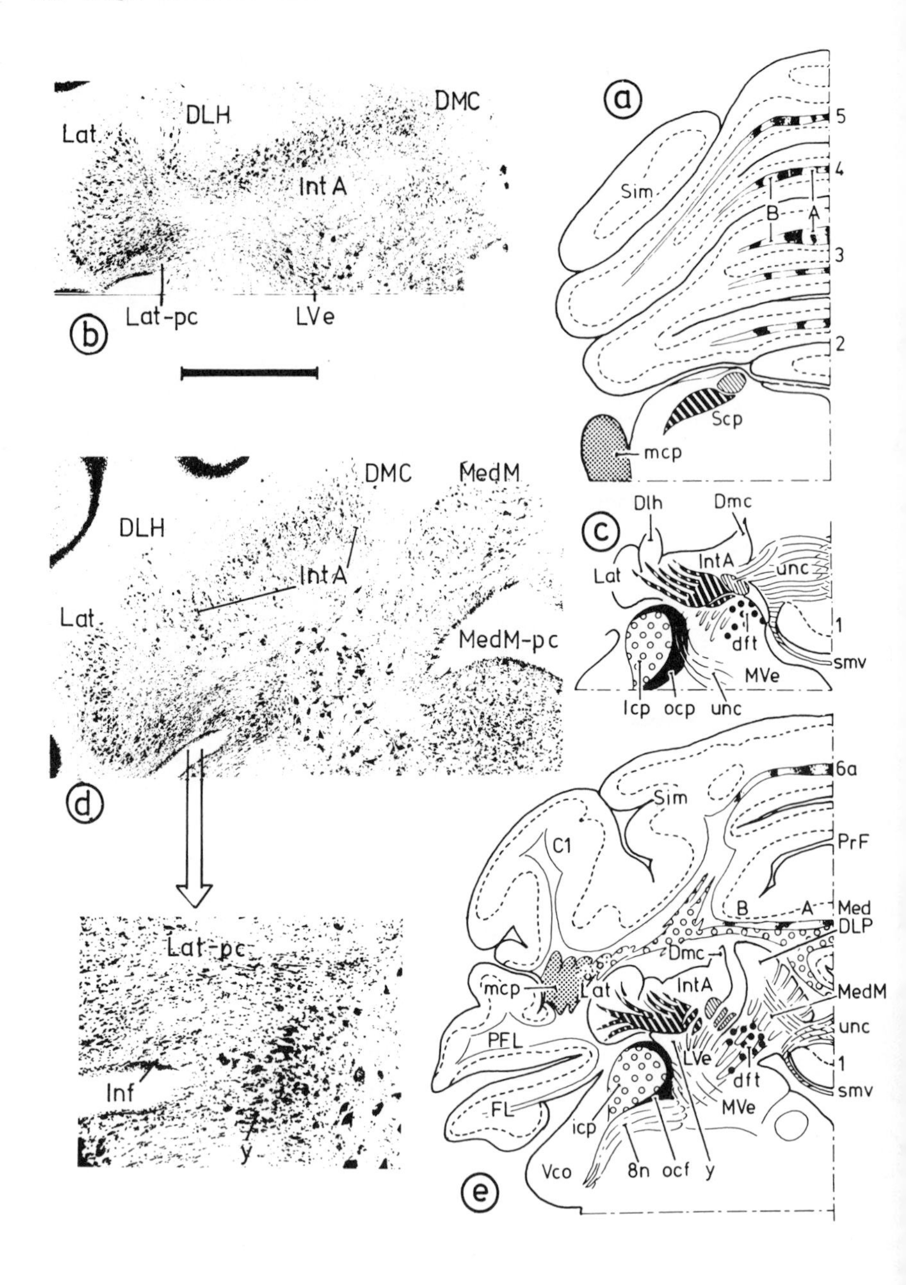

DLH
DMC
Lat
Int A
Lat-pc
LVe
b
a
5
4
3
2
Sim
B
A
Scp
mcp
DMC
MedM
DLH
IntA
Lat
MedM-pc
d
c
Dlh
Dmc
IntA
unc
Lat
1
dft
smv
MVe
Icp
ocp
unc
Lat-pc
Inf
y
6a
Sim
C1
PrF
B
A
Med
DLP
Dmc
mcp
Lat
IntA
MedM
unc
PFL
LVe
1
dft
smv
FL
MVe
icp
Vco
8n
ocf
y
e

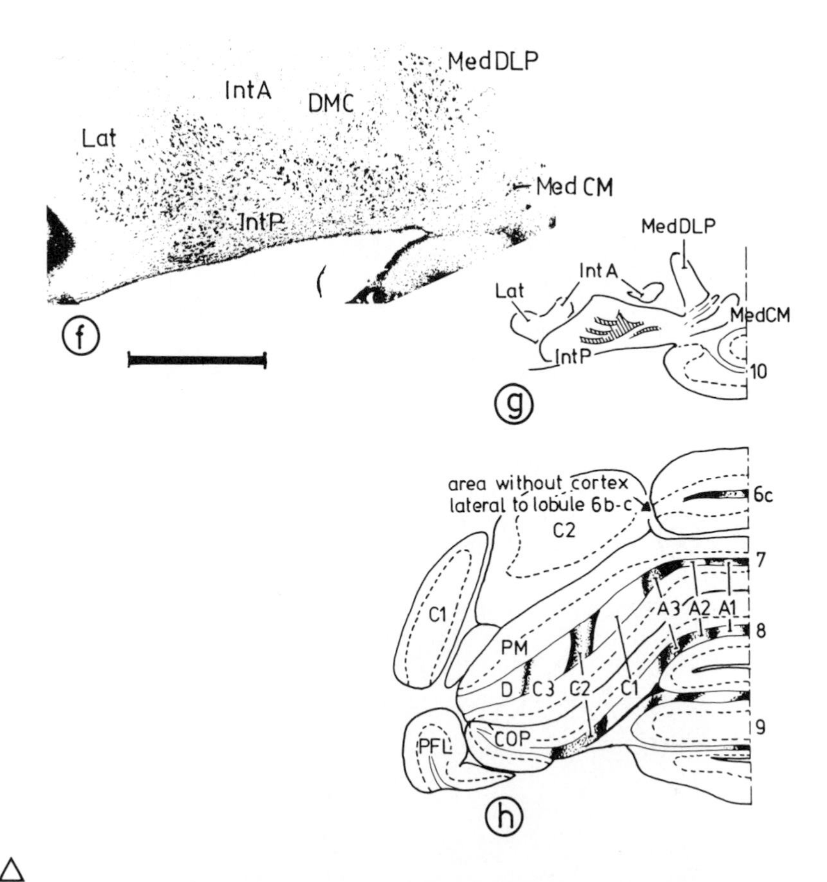

△

Fig. 5: A–E: Transverse sections through the cerebellum of the rat. A, C, E: drawings of Häggqvist-stained sections through the anterior lobe (A. compare Fig. 11A) and the rostral part of the central nuclei (C, E); B, D: Nissl-stained sections, same levels as C and E. Lower left corner: detail of infracerebellar nucleus, group y and parvocellular part of lateral cerebellar nucleus, from section d. Filled circles indicate direct fastigiobulbar tract fibers, for other symbols see Fig. 4. Calibration (bar = 1 mm) refers to Nissl-stained sections. F–H: Transverse sections through the cerebellum of the rat. F: Nissl-stained section through the caudal part of the central nuclei, same level as G; G, H: drawings of Häggqvist-stained sections. For symbols see Fig. 4. Calibration (bar = 1 mm) refers to Nissl-stained section. Abbreviations as in Fig. 1.

Fig. 6: A: midsagittal section through the cerebellum of the rat. For symbols see Fig. 4. B: the cerebellar commissure with contributions from the uncinate tract and the inferior and middle cerebellar peduncles. Häggqvist-stain. Calibration bar = 500 μm. Abbreviations as in Fig. 1. ▷

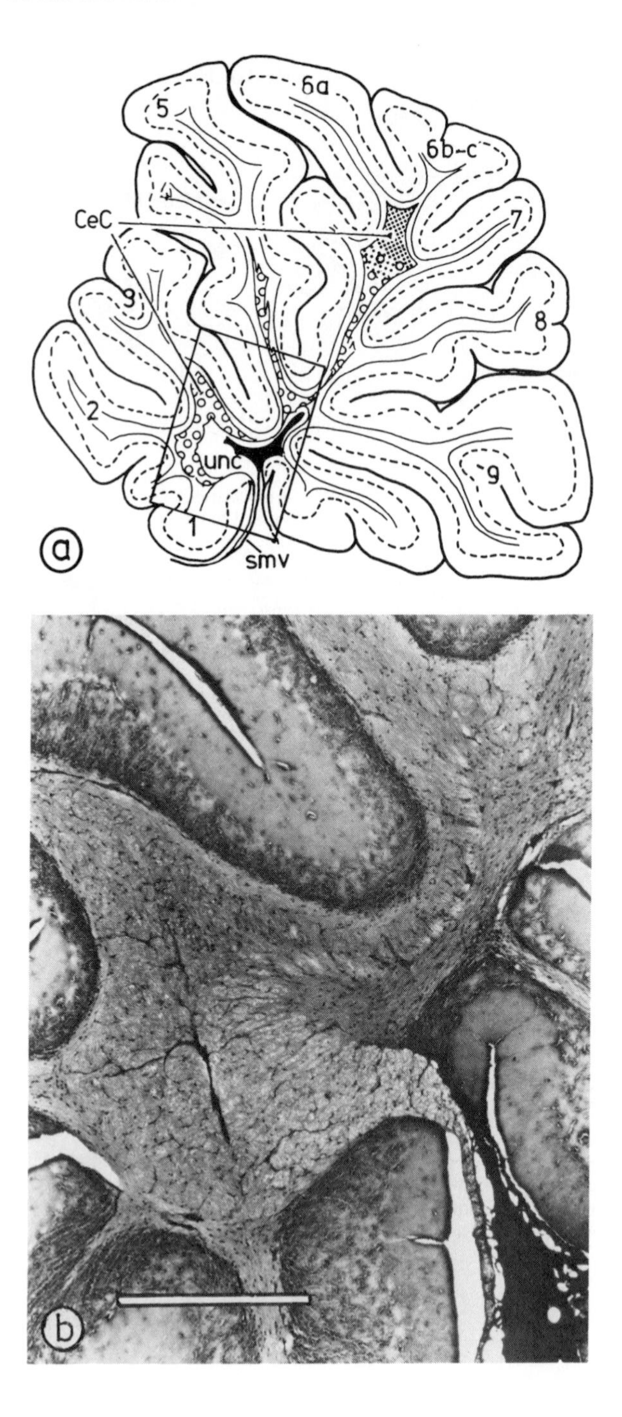

6a
5
6b-c
CeC
7
3
8
2
unc
9
1
smv
a
b

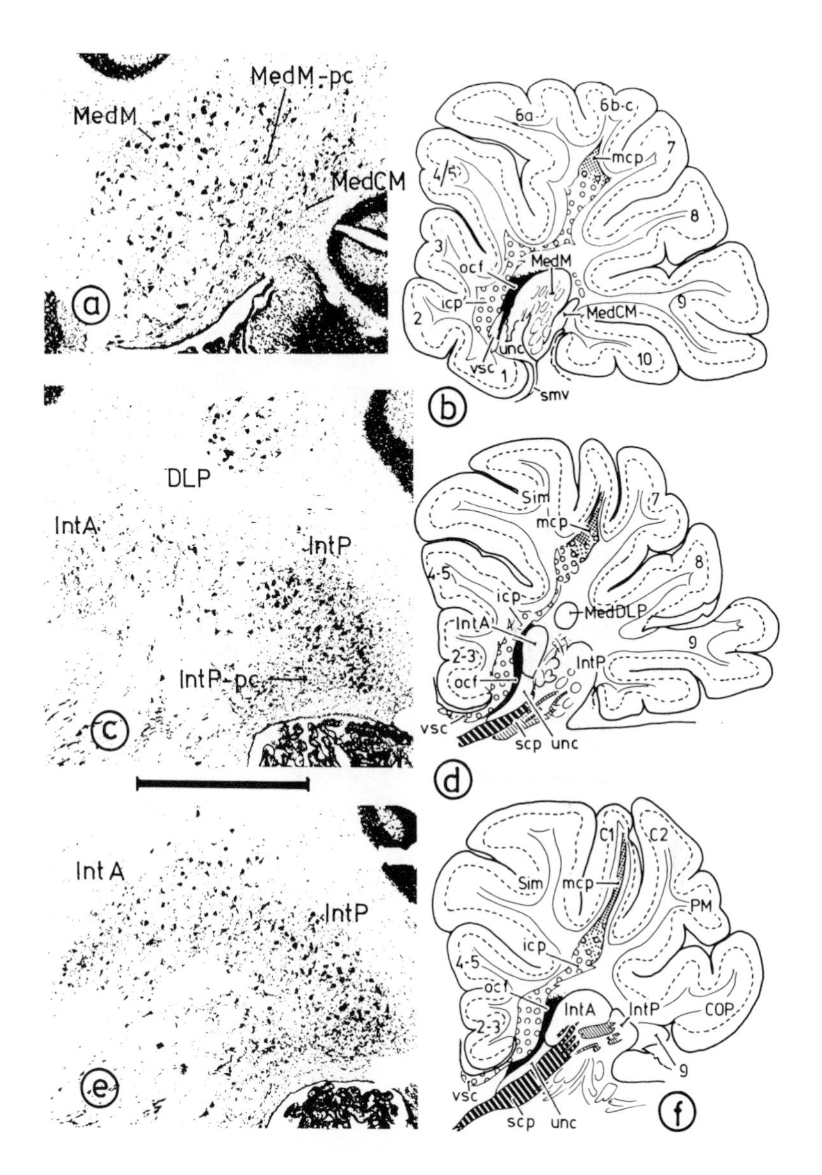

△

Fig. 7: A–F: Sagittal sections through the cerebellum of the rat. A, C, E: Nissl-stained sections ▷ through the medial part of the central nuclei. B, D, F: drawings of parallel Häggqvist-stained sections. For symbols see Fig. 4. Calibration (bar = 1 mm) refers to Nissl-stained sections. G–K: Sagittal sections through the cerebellum of the rat. G, I: Nissl-stained sections through the lateral part of the central nuclei, levels correspond with H and J; H, J, K: drawings of Häggqvist-stained sections. For symbols see Fig. 4. Calibration (bar = 1 mm) refers to Nissl-stained sections. Abbreviations as in Fig. 1.

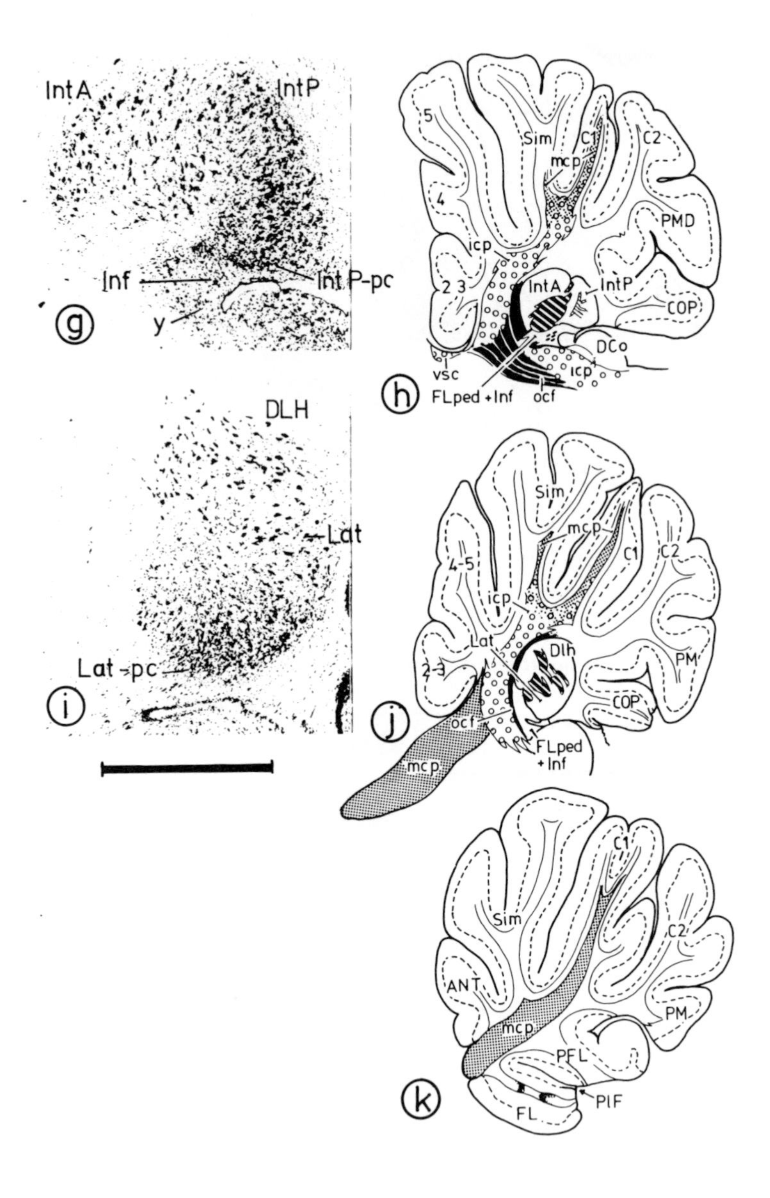

Fig. 8: Retrograde labeling of cells of the ipsilateral and contralateral medial cerebellar nucleus after an injection of horseradish peroxidase in the vestibular nuclei. For injection site see Fig. 12D. Abbreviations as in Fig. 1.

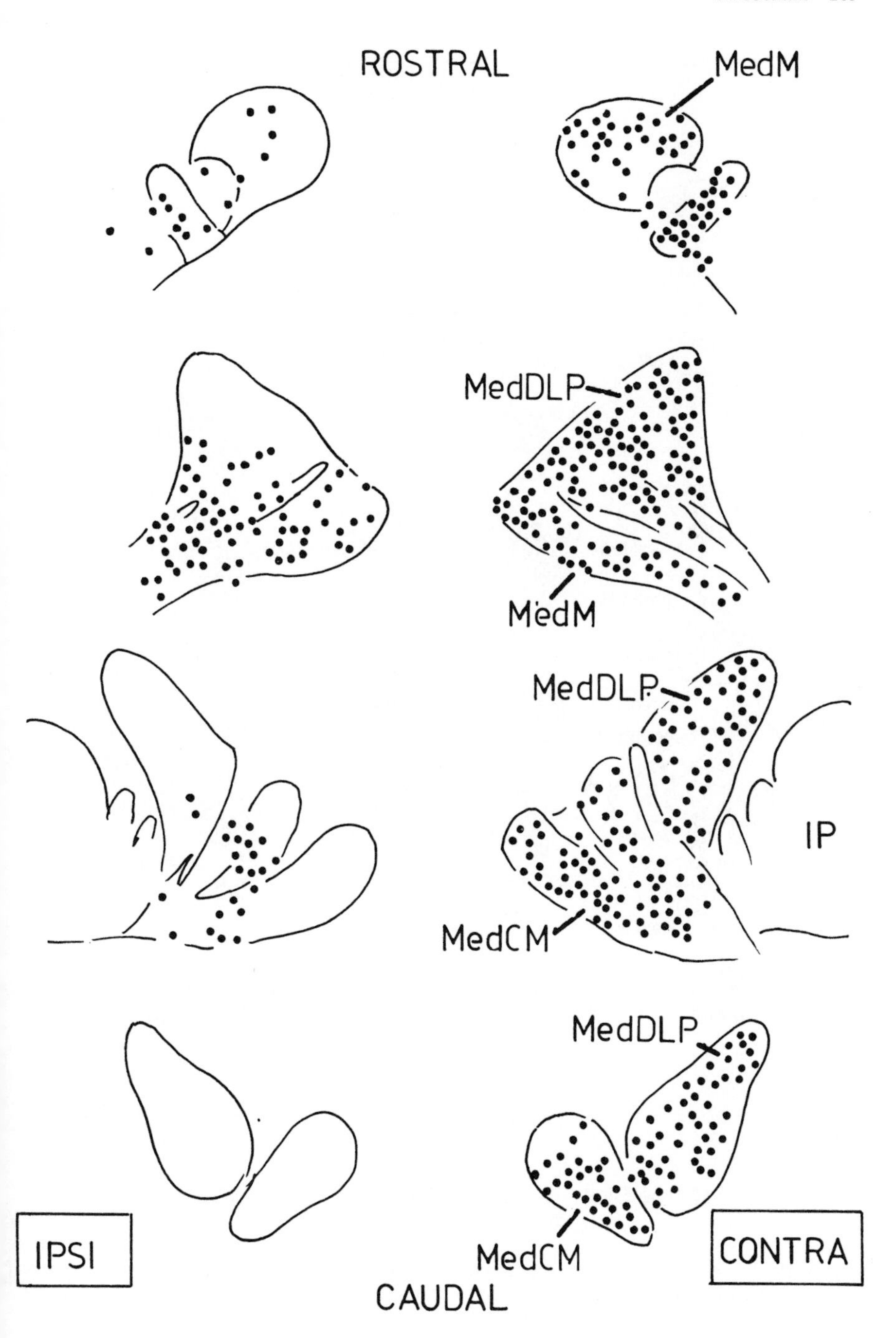
ROSTRAL
MedM
MedDLP
MedM
MedDLP
IP
MedCM
MedDLP
MedCM
IPSI
CONTRA
CAUDAL

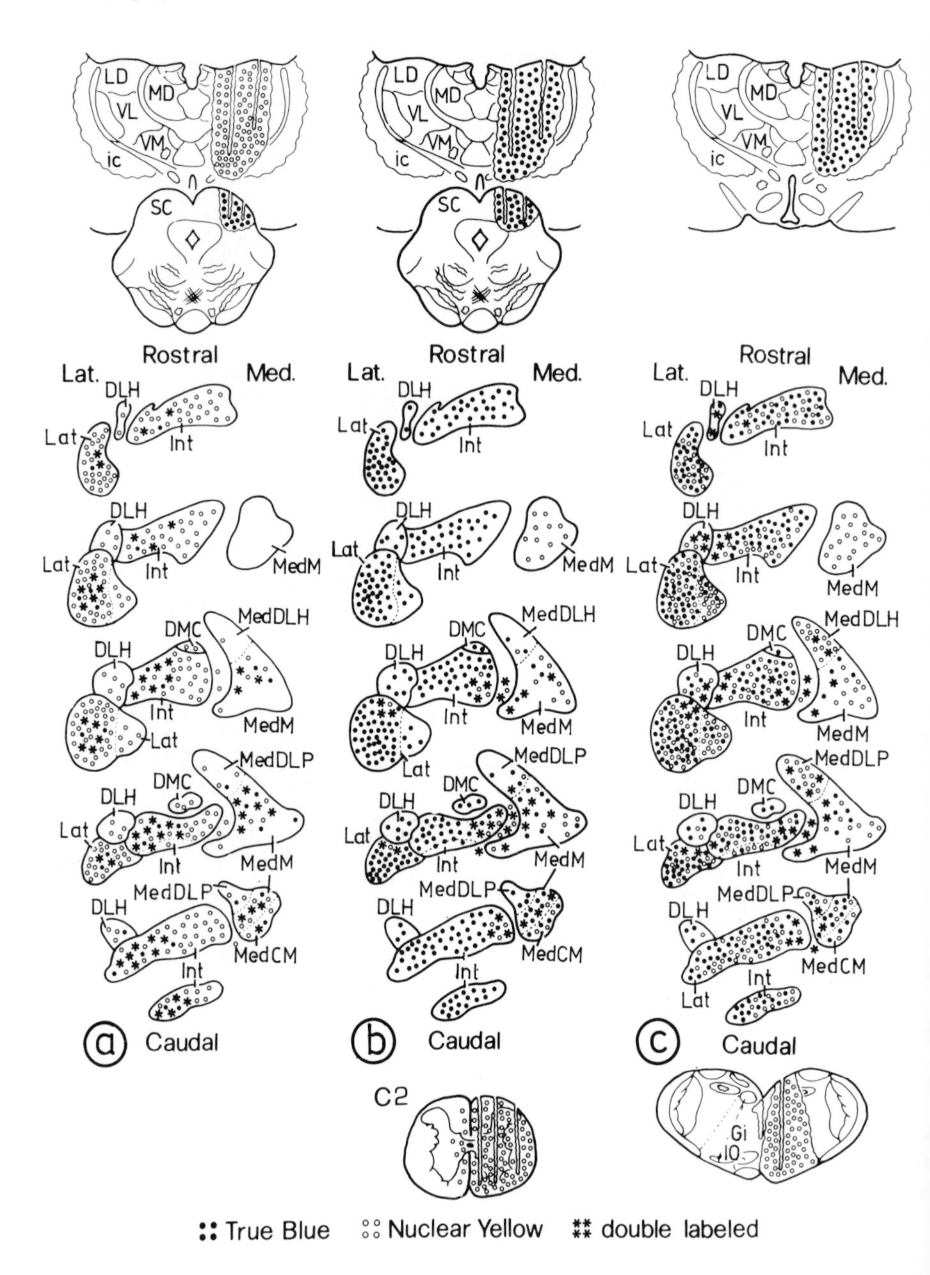

Fig. 9: Double-labeling of cells of the central cerebellar nuclei of the rat after combinations of injections of fluorescent tracers in the thalamus, the superior colliculus, the medial bulbar reticular formation and the spinal cord. Relabeled and reproduced from Bentivoglio and Kuypers (1982). Abbreviations as in Fig. 1.

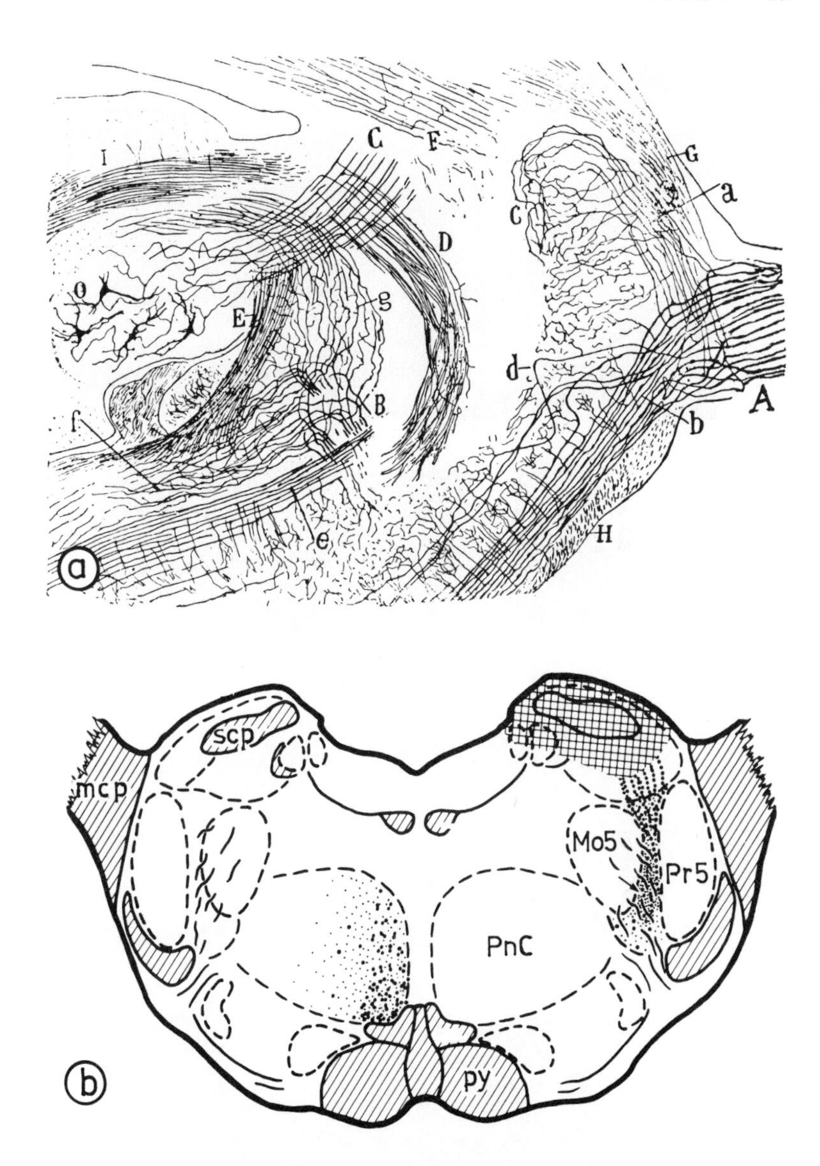

Fig. 10: A: Sagittal section showing the origin of the uncrossed descending branch of the superior cerebellar peduncle, mouse, Golgi method. Reproduced from Cajal (1911). Original labeling: A: Rootfibers of the trigeminal nerve; B: bifurcation of the vestibular nerve; C: superior cerebellar peduncle; D: uncrossed descending branch of the superior cerebellar peduncle; E: inferior cerebellar peduncle; G: middle cerebellar peduncle; H: trapezoid body; O: lateral cerebellar nucleus; a: ascending branch of the trigeminal root; b, d: spinal tract of the trigeminal nerve. B: The localization of the uncrossed (right) and crossed (left) descending branches after a lesion of the superior cerebellar peduncle in the rat. Nauta method. Relabeled and reproduced from Faull (1978). Abbreviations as in Fig. 1.

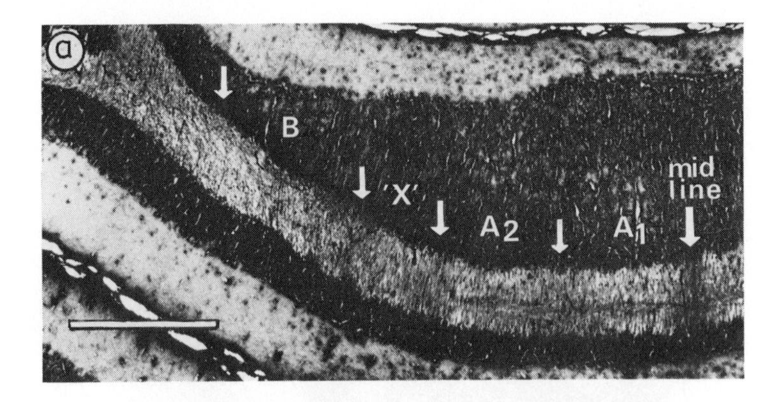

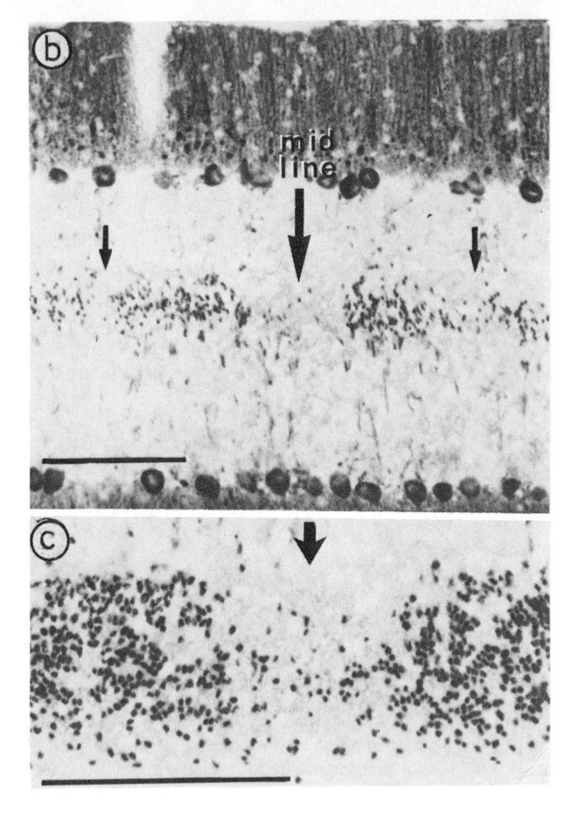

Fig. 11: AP: Myeloarchitectural compartments in the white matter of lobule 6a of the rat; horizontal section, same level as Fig. 4A. Arrows indicate borders ("raphes") between the compartments A1, A2, 'X' and B. Häggqvist-stain. Calibration bar = 500 μm. B: Distribution of Purkinje cells and their axons following immunoperoxidase staining of an antibody against cyclic GMP activated protein kinase. Arrows indicate gaps between the bundles of Purkinje cell axons corresponding to the raphes between the myeloarchitectural compartments; rat anterior lobe. C: Detail of midsagittal raphe. B and C reproduced from De Camilli *et al.* (1984). Calibration bar = 100 μm.

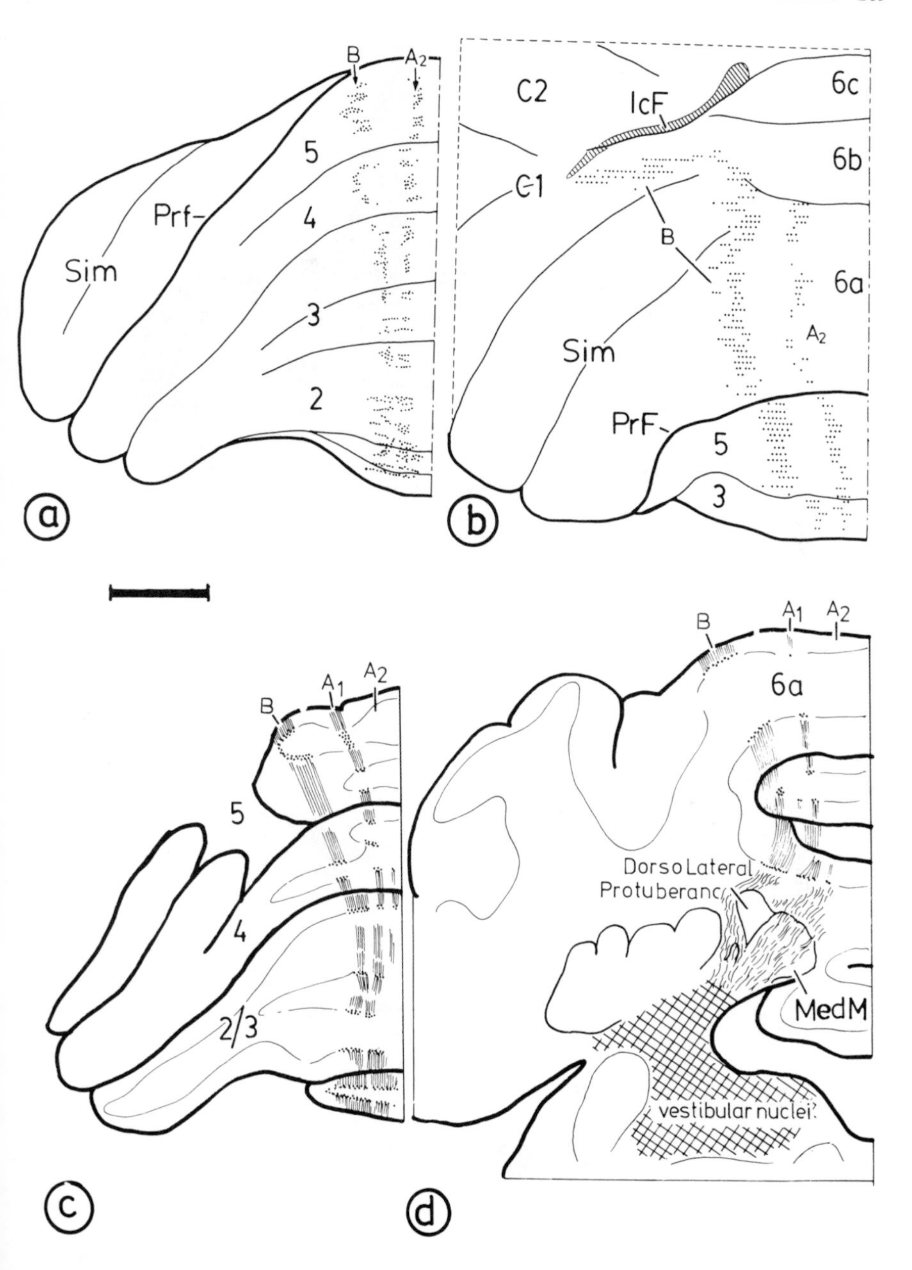

Fig. 12: Localization of labeled Purkinje cells in the anterior lobe and the simple lobule of the rat after an injection of horseradish peroxidase in the vestibular nuclei. A: graphical reconstruction of the rostral aspect of the anterior lobe; B: graphical reconstruction of the dorsal aspect of the cerebellum. The area without cortex in the intercrural fissure is hatched. C: section through the anterior lobe. D: section through the simple lobule and the injection site. Calibration bar = 1 mm. Abbreviations as in Fig. 1.

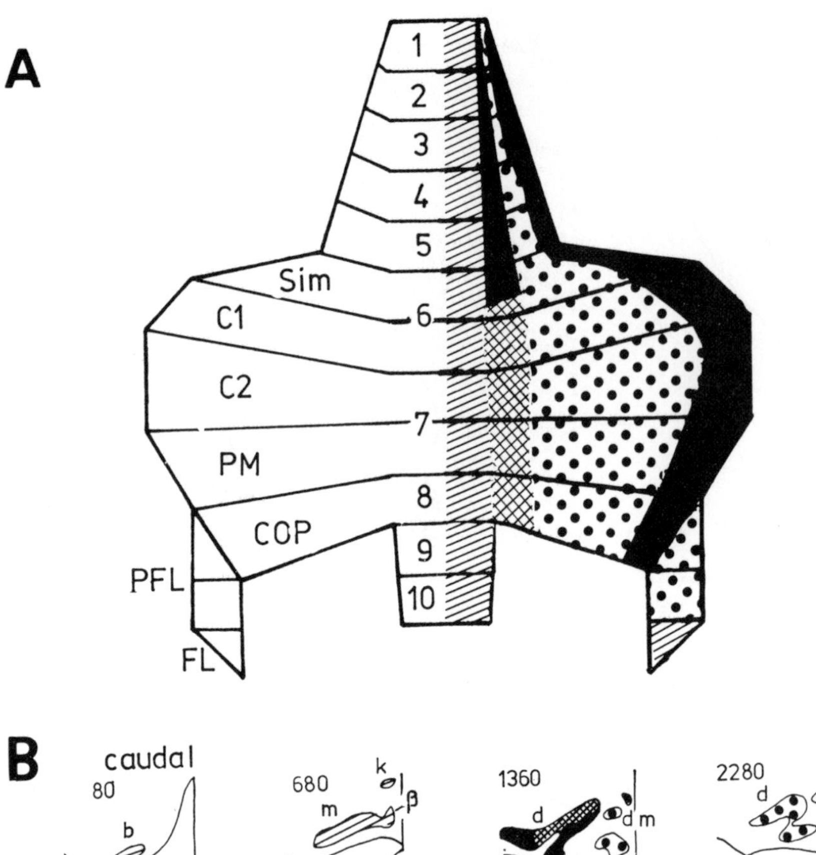
A
1
2
3
4
5
Sim
C1
6
C2
7
PM
8
COP
9
10
PFL
FL

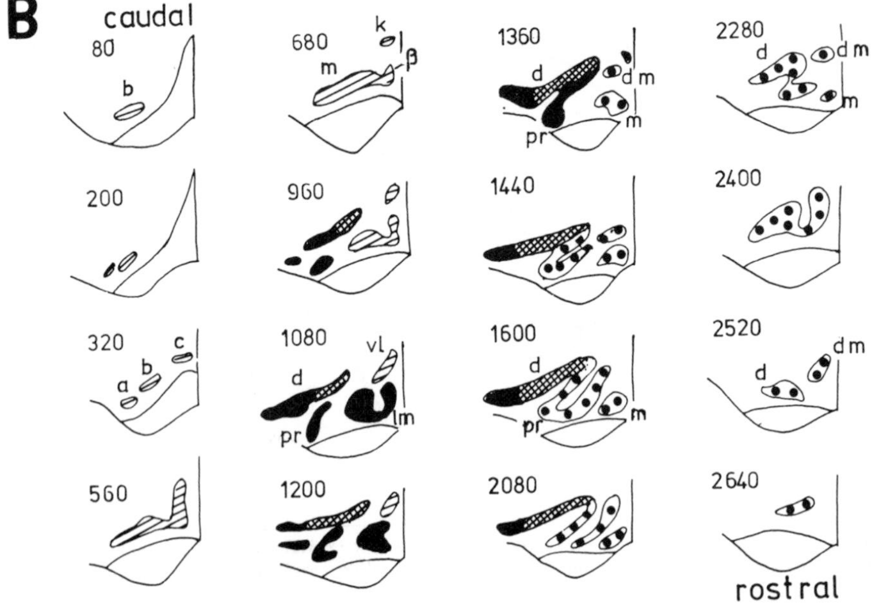
B
caudal
80
b
200
320
a
b
c
560
680
k
m
β
960
1080
d
vl
pr
lm
1200
1360
d
dm
m
pr
1440
1600
d
pr
m
2080
2280
d
dm
m
2400
2520
d
dm
2640
rostral

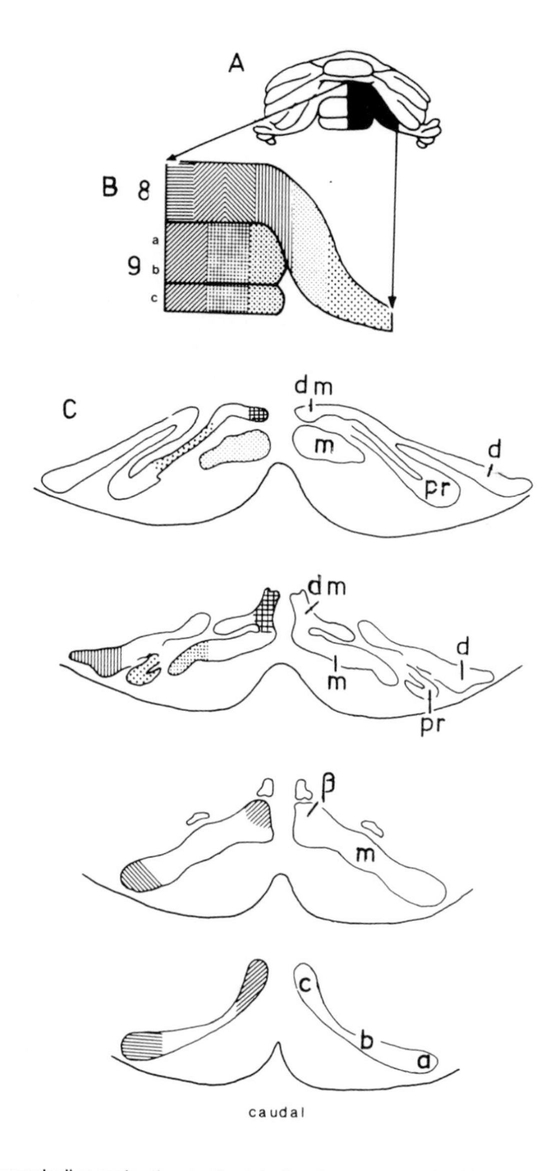

Fig. 14: The olivocerebellar projection to the lobules 8 and 9 in the rat. Relabeled and reproduced from Eisenman (1984). Abbreviations as in Fig. 1.

Fig. 13: The olivocerebellar projection in the rat. A: diagram of the cerebellar cortex, olivocerebellar projection zones are indicated with different symbols. B: diagram of transverse sections through the inferior olive. Relabeled in accordance with Fig. 3 of Volume 2, Chapter 10. Reproduced from Campbell and Armstrong (1983a). Abbreviations as in Fig. 1.

12

Cytoarchitectural organization of the spinal cord

ALAN M. BRICHTA

The Ohio State University
Colombus, Ohio, USA

GUNNAR GRANT

Karolinska Institutet
Stockholm, Sweden

I. INTRODUCTION

Rexed (1952, 1954) divided the cat spinal gray matter into ten layers, most of which extend the entire length of the spinal cord. This division was based on observations made in Nissl stained thick (11 μm) and thin (15 μm–15 μm) sections from spinal cords of young and adult animals. Both transverse and longitudinal sections were used. Being a purely cytoarchitectural scheme without reference to dendritic ramifications or axonal terminations it has still turned out to be of great value in studies of connectivity, both with anatomic and physiologic techniques, and in immunocytochemical and enzyme histochemical studies. His division has also been adopted in most recent textbooks of neuroanatomy. Most, if not all, major features that Rexed described in cats can be seen in the rat spinal cord.

The following scheme for the rat rests heavily on work already completed by others (Fukuyama, 1955; Kuhlenbeck, 1975; McClung and Castro, 1976, 1978; Meikle and Martin, 1980; Molander *et al.*, 1984; Schoenen, 1973; Schrødder, 1977; Steiner and Turner, 1972), but is also the result of our (Brichta) observations on 100 μm sections stained with Luxol Fast blue and cresyl violet (Klüver and Barrera, 1953) (Fig. 1). Rexed (1964) proposed that all the important features of his 10 cell layers (laminae), would be applicable to all higher mammals. For the rat spinal cord this has been confirmed by Steiner and Turner (1972), Schoenen

THE RAT NERVOUS SYSTEM
ISBN 0 12 547632 9

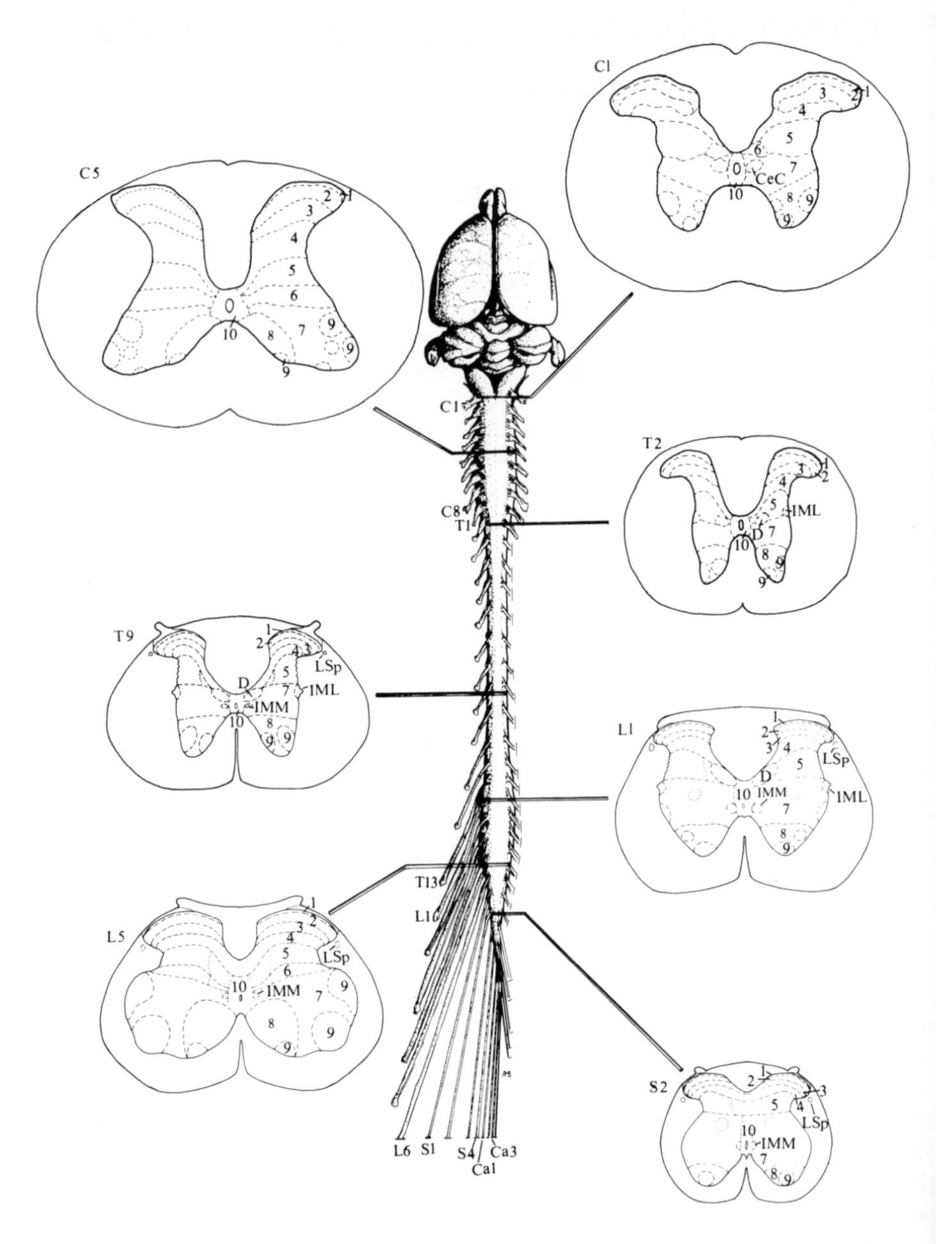

Fig. 1: Diagrams of cross-sections of the spinal cord at cervical (C), thoracic (T), lumbar (L) and coccygeal (Ca) levels. CeC, central cervical nucleus; D, Dorsal nucleus (Clarke); IMM, intermediomedial cell column; IML, intermediolateral cell column; LSp, lateral spinal nucleus; numerals 1–10 show the layers of Rexed.

(1973), and Schrøder (1977), but it was Wall (Wall *et al.*, 1967; Wall, 1968) who first suggested, on a physiologic basis, that Rexed's laminae can be observed in the rat dorsal horn. However, rather than the almost perfectly horizontal placement of most of the laminae in cat lumbar segments, there is a more dorsolateral slant to the laminae in the rat lumbar enlargement (Wall, 1968). Few, if any, exact lines of demarcation can be seen in the spinal cord stained with cresyl violet. Indeed, Rexed made it clear also that he saw the laminar borders as zones of transition. This may be particularly so in the rat (McClung and Castro, 1978); however, if a variety of histologic techniques are used then these zones can be more clearly discriminated (McClung and Castro, 1976, 1978; Molander *et al.*, 1984; Schrøder, 1977). The characteristics of the individual laminae will now be described.

II. LAMINA 1

Lamina 1 is the thinnest layer within the gray matter and is also the most dorsal. It covers the dorsal surface and curves around a variable distance down the lateral surface, appearing largely as a margin between the narrow, white matter zone of Lissauer and the gray matter of the dorsal horn. Several cell types of different sizes and shapes are found in this layer (Lima and Coimbra, 1983). It shows medium to dark staining using Timm's sulfide silver method in the rat (Schrøder, 1977). The neuropil is oriented horizontally. The shift in neuropil orientation offers the best criterion for the transition from this lamina to the subjacent lamina 2 or substantia gelatinosa Rolandi (McClung and Castro, 1976, 1978; Molander *et al.*, 1984).

III. LAMINA 2

Lamina 2 is parallel to lamina 1 and is covered by that layer dorsally and laterally, but not medially. It is substantially wider than lamina 1 and is distinguished easily in cresyl violet sections by its darkly staining, uniformly sized, small cells and tightly packed neuropil (Steiner and Turner, 1972). The lamina is characterized by a relative paucity of myelinated fibers (McClung and Castro, 1978; Molander *et al.*, 1984). Fiber bundles may be seen penetrating from the dorsal columns through the medial part of this layer (Plate 66 and Paxinos and Watson, 1982). As in the cat, lamina 2 can be subdivided into an intensely stained outer zone with densely packed cells and a less compact inner zone (Molander *et al.*, 1984). These zones are now commonly referred to as II_o and II_i or II_a and II_b. The outer zone contains more small myelinated fibers than the inner zone. Except for its most ventral part the latter zone seems to coincide with the fluoride resistant acid phosphatase (FRAP) positive band in the rat (see Nagy and Hunt, 1983).

IV. LAMINA 3

The border between lamina 2 and lamina 3 is again not distinct but the two laminae are cytoarchitecturally different. Lamina 3 is the widest layer so far described and runs parallel to both lamina 1 and lamina 2. Compared to lamina 2, it contains a larger number of myelinated axons which are seen transversely cut in transverse spinal cord sections (McClung and Castro, 1976, 1978; Molander *et al.*, 1984). Layer 3 also contains greater numbers of larger neurons and shows a wider distribution of cell sizes but is organized less compactly than lamina 2.

V. LAMINA 4

Lamina 4 is broader than the preceding laminae. It forms a band beneath laminae 2 and 3 but does not possess their characteristic lateral curvature. The neurons in layer 4 are heterogeneous in size, with scattered larger cells also present (McClung and Castro, 1978). As to the medial part of lamina 4, McClung and Castro (1978) claim that, in the cervical region, the layer ends bordering the dorsal columns, parallel to the previous laminae. Other investigators, however, suggest that lamina 4 differs significantly from the previous laminae in that the medial portion continues ventrally to unite with the contralateral lamina 4 at the region of the dorsal gray commissure (Meikle and Martin, 1980; Molander *et al.*, 1984; Steiner and Turner, 1972). Especially at the mid thoracic, lower lumbar and sacral levels lamina 4 should continue across the midline (Molander *et al.*, 1984; see also Figs 68 and 69 in Paxinos and Watson, 1982). At lower thoracic and upper lumbar levels, however, the column of Clarke (dorsal nucleus, D) seems to extend as far dorsally as to block a continuation of the lamina across the midline (Molander *et al.*, 1984). The exact border between the medial part of lamina 4 and the dorsal funiculus is also an area open to question. It has been proposed that a thin layer of longitudinally orientated lamina 1 type cells separates the two regions (Meikle and Martin, 1980). The existence of a thin layer of separation between lamina 4 and the dorsal funiculus was first indicated, but not commented on, by Steiner and Turner (1972) in their schematic diagrams of the lumbar segments. Schrøder (1977) showed a dark sulfide silver staining medial brim in the thoracic segments, which could be a reflection of such a layer. From our observations this thin band extends throughout the length of the spinal cord but is best seen at lower thoracic and lumbar levels (observations by Brichta).

VI. LAMINA 5

Lamina 5 forms the neck of the dorsal horn. The border between lamina 4 and the reticular structure of especially the lateral part of lamina 5 can often be fairly easily distinguished. The layer extends from the lateral edge of the gray matter towards a region dorsal to the central canal. It has been found to cross the midline in the lower lumbar and sacral regions (Molander *et al.*, 1984; Steiner and Turner, 1972). Neuronal cell sizes are similar to those of layer 4 except that still larger cells are found, especially laterally. The distinctive reticulated appearance of the layer is the result of longitudinal bundles of myelinated axons piercing especially its lateral zone (McClung and Castro, 1976, 1978; Schrøder, 1977). Centrally and medially within the lamina fibers are seen to have a more dorsoventral orientation.

Lamina 5 borders medially with the dorsal nucleus (D; Clarke's column) in the thoracic and upper lumbar segments of the cord. This nucleus lies at the ventromedial base of lamina 5, at the junction with lamina 7 (see Fig. 68 in Paxinos and Watson, 1982). It has its largest extent in the lower thoracic and upper lumbar segments. It does not seem to extend further caudally than the second lumbar segment in the rat (Molander *et al.*, 1984; Waibl, 1973).

VII. LAMINA 6

Rexed claimed that this layer exists in its typical form only in the cervical and lumbar enlargements. In the rat, there have been reports that lamina 6 could not be identified as an individual layer at any level of the spinal cord (Meikle and Martin, 1980; Steiner and Turner, 1972). However, McClung and Castro (1978), in a specific study of the cervical enlargement, described a distinct lamina 6 just ventral to the neck region of the dorsal horn and parallel to lamina 5. This layer was characterized as a band of heavily Nissl stained neurons which were tightly packed medially and more scattered laterally. This topographic organization is similar to that observed by Rexed in the cat (1952, 1954). They also reported in an earlier study (McClung and Castro, 1976) that layer 6 was missing in the thoracic region. In our material, lamina 6 occurs within the cervical and lumbar enlargements, where it appears as a distinct band of darkly Nissl stained cells situated at the ventral border of the neck of the dorsal horn (observations by Brichta). In the lumbar enlargement lamina 6 has been identified in all segments from L3 to S1 (Molander *et al.*, 1984).

VIII. LAMINA 7

The border region between laminae 6 and 7 is demarcated by the transition from an area of cellular diversity to one of cytologic homogeneity. Layer 7 constitutes most of the intermediate zone of the gray matter throughout the entire cord and some of the ventral horn at cervical and lumbar enlargement levels. This lamina is bordering the dorsal horn dorsally, lamina 10 medially and lamina 8 ventrally or ventromedially. Its lateral border is towards the white matter of the lateral funiculus. In the enlargements, lamina 7 adopts a position surrounding both the medially placed lamina 8 and the burgeoning cell columns of lamina 9 of the ventral horn (see Figs 67 and 69 in Paxinos and Watson, 1982). Lamina 7 also borders the dorsal nucleus (D) as commented on in the section on lamina 5, and the intermediolateral nucleus (IML) of the lateral horn, found at the dorsolateral border of lamina 7. This latter nucleus is found not only at thoracic and upper lumbar levels in the rat (Molander *et al.*, 1984; Waibl, 1973; see also Fig. 68 in Paxinos and Watson, 1982). It also appears as the sacral parasympathetic nucleus at L6 and S1 (Hancock and Peveto, 1979a; Nadelhaft and Booth, 1982, 1984; see also Plate 71 in Paxinos and Watson, 1982).

The intermediomedial nucleus (IMM) is another nucleus of lamina 7, located in the medial part of this lamina, adjacent to the central canal (for example, Molander *et al.*, 1984). This nucleus seems to exist along the whole extent of the spinal cord.

In the upper cervical segments the central cervical nucleus (CeC) is found medially in lamina 7, lateral to the central canal (see Fig. 66 in Paxinos and Watson, 1982). This nucleus has been the subject of detailed studies in the cat (Wiksten, 1979a, b, c; Wiksten and Grant, 1983). Recently it has attracted interest in the rat (Ammann *et al.*, 1983; Matsushita and Hosoya, 1979).

IX. LAMINA 8

Lamina 8 can be distinguished from lamina 7 by the darker staining and more variable sizes of its cells. These cells are generally larger than those found in the previous layer. In the thoracic region, layer 8 constitutes most of the ventral horn, excluding clusters of very large neurons seen ventromedially and ventrolaterally in the apex region of the horn (Molander *et al.*, 1984). These clusters of motoneurons are described in the section on lamina 9. In the cervical and lumbar enlargements, the impression is gained that lamina 8 is squeezed towards the medial edge of the ventral horn by lamina 7.

X. LAMINA 9

Lamina 9 is not a true layer, but rather a collection of nuclei containing some of the largest neurons of the rat spinal cord (McClung and Castro, 1978). At thoracic levels it is situated in the most ventral area of the ventral horn. A lateral and a medial nucleus can often be discerned here (Fig. 68 in Paxinos and Watson, 1982; Molander *et al.*, 1984). Within the cervical and lumbar enlargements, additional motor nuclei can be seen laterally. Smaller sized cells are distributed amongst the large neurons, giving the nuclear area a distinct heterogeneous appearance which allows delineation from the surrounding laminae 7 or 8. The larger cells are α-motoneurons, while the smaller cells are presumably γ-motorneurons.

The location of motoneurons innervating different muscles or groups of muscles has been recently studied by several investigators using retrograde tracing methods in the rat (for example, Baulac and Meininger, 1981; Gottschall *et al.*, 1980; Matesz and Székely, 1983; Nicolopoulos-Stournaras and Iles, 1983; Peyronnard and Charron, 1983; Smith and Hollyday, 1983).

XI. LAMINA 10

Lamina 10 surrounds the central canal and comprises most of the dorsal and ventral gray commissures. It seems to contact the ventral white commissure at all levels of the spinal cord, but is mostly separated from the dorsal white commissure and dorsal columns. At thoracic levels and further caudally, the thin lamina 1 type layer described above should contribute to the separation. At lower lumbar and more caudal levels the medial extensions of the deeper laminae of the dorsal horn add to the separation of lamina 10 from the dorsal columns. The lateral bounderies of lamina 10 at different levels of the rat spinal cord have been studied recently by Nahin and Giesler (1982). These authors also described three types of cells in lamina 10: pyramidal; stellate; and fusiform. The pyramidal and stellate cells were found to have uniform distribution, whereas the fusiform neurons were especially frequent in the dorsalmost region of the layer. At L1 and L2, this region seems to correspond to the dorsal commissural nucleus, which has recently been characterized in the rat (Hancock and Peveto, 1979b; Neuhuber, 1982).

In the white matter of the dorsolateral funiculus, extending dorsolaterally towards the surface of the spinal cord, is a group of small cells present at all levels of the rat spinal cord (see Plates 66 to 69 in Paxinos and Watson, 1982). Attention was originally paid to this group by Gwyn and Waldron (1968). These authors suggested that it might be the equivalent of the lateral cervical nucleus of the upper cervical cord of other species, although distributed throughout the whole length of the spinal cord in the rat. However, later studies with retrograde labeling technique, following horseradish peroxidase injections into the thalamus, have demonstrated that a lateral cervical nucleus similar to that of other species exists in the upper cervical cord also in the rat (Giesler *et al.*, 1979). To distinguish between the lateral cervical nucleus and the rest of the rat dorsolateral funiculus nucleus, the term lateral spinal nucleus (LSp) has been proposed for the latter (Giesler, personal communication). This lateral spinal nucleus has recently been shown to have an ascending projection to the midbrain tegmentum (Menétrey *et al.*, 1980, 1982).

REFERENCES

Ammann, B., Gottschall, J. and Zenker, W. (1983). Afferent projections from the rat longus capitis muscle studied by transganglionic transport of HRP. *Anat. Embryol.* 166, 275–289.

Baulac, M. and Meininger, V. (1981). Organisation des motoneurones des muscles pectoraux chez le rat. *Acta Anat.* 109, 209–217.

Fukuyama, U. (1955). On cytoarchitectural lamination of the spinal cord in the albino rat. *Anat. Rec.* 121, 396.

Giesler, G. J., Menétrey, D. and Basbaum, A. I. (1979). Differential origins of spinothalamic tract projections to medial and lateral thalamus in the rat. *J. Comp. Neurol.* 184, 107–126.

Gottschall, J., Neuhuber, W., Müntener, M. and Mysicka, A. (1980). The ansa cervicalis and the infrahyoid muscles of the rat. *Anat. Embryol.* 159, 59–69.

Gwyn, D. G. and Waldron, H. A. (1968). A nucleus in the dorsolateral funiculus of the spinal cord of the rat. *Brain Res.* 10, 342–351.

Hancock, M. B. and Peveto, C. A. (1979a). Preganglionic neurons in the sacral spinal cord of the rat: An HRP study. *Neurosci. Lett.* 11, 1–5.

Hancock, M. B. and Peveto, C. A. (1979b). A preganglionic autonomic nucleus in the dorsal gray commissure of the lumbar spinal cord of the rat. *J. Comp. Neurol.* 183, 65–72.

Klüver, H. and Barrera, E. (1953). A method of the combined staining of cells and fibers in the nervous system. *J. Neuropathol. Exp. Neurol.* 12, 400–403.

Kuhlenbeck, H. (1975). *The central nervous system of vertebrates*, Vol. 4, Spinal Cord and Deuterencephalon. Karger, Basel.

Lima, D. and Coimbra, A. (1983). The neuronal population of the marginal zone (Lamina I) of the rat spinal cord. A study based on reconstructions of serially sectioned cells. *Anat. Embryol.* 167, 273–288.

McClung, J. R. and Castro, A. J. (1976). Neuronal organization in the spinal cord of the rat: An analysis of the nine laminar scheme of Rexed. *Anat. Rec.* 184, 474.

McClung, J. R. and Castro, A. J. (1978). Rexed's laminar scheme as it applies to the rat cervical spinal cord. (19Ex). 58, 145–148.

Matesz, C. and Székely, G. (1983). The motor nuclei of the glossopharyngeal-vagal and the accessorius nerves in the rat. *Acta Biol. Hung.* 34, 215–230.

Matsushita, M. and Hosoya, Y. (1979). Cells of origin of the spinocerebellar tract in the rat, studied with the method of retrograde transport of horseradish peroxidase. *Brain Res.* 173, 185–200.

Meikle, A. D. S. and Martin, A. G. (1980). Rat spinal cord, a cytoarchitectonic study. *Anat. Rec.* 1916, 241A.

Menétrey, D., Chaouch, A. and Besson, J. M. (1980). Location and properties of dorsal horn neurons at origin of spinoreticular tract in lumbar enlargement of the rat. *J. Neurophysiol.* 44, 862–877.

Menétrey, D., Chaouch, A., Binder, D. and Besson, J. M. (1982). The origin of the spinomesencephalic tract in the rat: An anatomical study using the retrograde transport of horseradish peroxidase. *J. Comp. Neurol.* 206, 193–207.

Molander, C., Xu, Q. and Grant, G. (1984). The cytoarchitectonic organization of the spinal cord in the rat. I. The lower thoracic and lumbosacral cord. *J. Comp. Neurol.* 280, 133–141.

Nadelhaft, I. and Booth, A. M. (1982). Preganglionic neurons and visceral afferent fibers in the rat pelvic nerve. *Soc. Neurosci. Abst.* 8, 77.

Nadelhaft, I. and Booth, A. M. (1984). The location and morphology of preganglionic neurons and the distribution of visceral afferents from the rat pelvic nerve: A horseradish peroxidase study. *J. Comp. Neurol.* 226, 238–245.

Nagy, J. I. and Hunt, S. P. (1983). The termination of primary afferents within the rat dorsal horn: Evidence for rearrangement following capsaicin treatment. *J. Comp. Neurol.* 218, 145–158.

Nahin, R. L. and Giesler, G. J. (1982). Anatomical studies of laminar X in the rat. *Soc. Neurosci. Abst.* 8, 93.

Neuhuber, W. (1982). The central projections of visceral primary afferent neurons of the inferior mesenteric plexus and hypogastric nerve and the location of the related sensory and preganglionic sympathetic cell bodies in the rat. *Anat. Embryol.* 164, 413–425.

Nicolopoulos-Stournaras, S. and Iles, J. F. (1983). Motor neuron columns in the lumbar spinal cord of the rat. *J. Comp. Neurol.* 217, 75–85.

Peyronnard, J. M. and Charron, L. (1983). Motoneuronal and motor axonal innervation in the rat hindlimb: A comparative study using horseradish peroxidase. *Exp. Brain Res.* 50, 125–132.

Paxinos, G. and Watson, C. (1982). *The rat brain in stereotaxic coordinates.* Academic Press, Sydney.

Rexed, B. (1952). The cytoarchitectonic organization of the spinal cord in the cat. *J. Comp. Neurol.* 96, 415–496.

Rexed, B. (1954). A cytoarchitectonic atlas of the spinal cord in the cat. *J. Comp. Neurol.* 100, 297–379.

Rexed, B. (1964). Some aspects of the cytoarchitectonics and synaptology of the spinal cord. In J. C. Eccles and J. P. Schadé (Eds). *Organization of the spinal cord*, pp. 58–92 (Progress in Brain Research 11). Elsevier, Amsterdam.

Schoenen, J. (1973). Cytoarchitectonic organization of the spinal cord in different mammals including man. *Acta Neurol. Belg.* 73, 348–358.

Schrøder, H. D. (1977). Sulfide silver architectonics of the rat, cat and guinea pig spinal cord: A light microscopic study with Timm's method for demonstration of heavy metals. *Anat. Embryol.* 150, 251–267.

Smith, C. L. and Hollyday, M. (1983). The development and postnatal organization of motor nuclei in the rat thoracic spinal cord. *J. Comp. Neurol.* 220, 16–28.

Steiner, T. J. and Turner, L. M. (1972). Cytoarchitecture of the rat spinal cord. *J. Physiol. (London)* 222, 123P–125P.

Waibl, H. (1973). Zur Topographie der Medulla spinalis der Albinoratta (*Rattus norvegicus*). *Adv. Anat. Embryol. Cell Biol.* 47.

Wall, P. D. (1968). Organization of cord cells which transmit sensory cutaneous information. In D. R. Kenshalo (Ed.). *The skin senses.* Thomas, Springfield, Ill.

Wall, P. D., Freeman, J. and Major D. (1967). Dorsal horn cells in spinal and in freely moving rats. *Exp. Neurol.* 19, 519–529.

Wiksten, B. (1979a). The central cervical nucleus in the cat. I. A Golgi study. *Exp. Brain Res.* 36, 143–154.

Wiksten, B. (1979b). The central cervical nucleus in the cat. II. The cerebellar connections studied with retrograde transport of horseradish peroxidase. *Exp. Brain Res.* 36, 155–173.

Wiksten, B. (1979c). The central cervical nucleus in the cat. III. The cerebellar connections studied with anterograde transport of H-leucine. *Exp. Brain Res.* 36, 175–189.

Wiksten, B. and Grant, G. (1983). The central cervical nucleus in the cat. IV. Afferent fiber connections. An experimental anatomical study. *Exp. Brain Res.* 51, 405–412.

13

Primary afferent projections to the spinal cord

GUNNAR GRANT

Karolinska Institutet
Stockholm, Sweden

I. INTRODUCTION

The primary afferent fibers projecting to the rat spinal cord enter via the dorsal roots. Ventral roots have been suggested to contribute afferents. It appears from recent studies in the cat that the ventral roots may contain such afferents at sacral levels and that these fibers terminate in the spinal gray matter (Mawe *et al.*, 1984). Afferent unmyelinated fibers have been demonstrated in the L7 ventral root, also in the cat. However, instead of penetrating into the central nervous system these fibers either make U-turns or enter the spinal pia mater (Risling and Hildebrand, 1982; Risling *et al.*, 1984). The situation in the rat is not known.

The dorsal roots in the rat are grouped in pairs of eight cervical, 13 thoracic, six lumbar, four sacral and three caudal (coccygeal; for example, Waibl, 1973). The number of dorsal root ganglion cells in single pairs may vary considerably between the two sides (Ygge *et al.*, 1981). A similar variation can therefore be expected to exist also in the number of dorsal root axons. Their actual number, however, can be assumed to be larger than the number of ganglion cells (see Chung and Coggeshall, 1984; Langford and Coggeshall, 1979). In the monkey, but not in the cat, small calibered dorsal root axons are segregated into a lateral bundle as the dorsal rootlets enter the spinal cord (Snyder, 1977). The organization in the rat is not known, however.

After their entry into the spinal cord, the dorsal root afferents distribute differently depending upon their size. Coarse calibered fibers run medially into the dorsal funiculus, whereas fine calibered fibers approach the dorsal horn via Lissauer's tract. More than two thirds of the axons in Lissauer's tract have been demonstrated at lumbosacral and mid-thoracic levels in the rat to be of primary afferent origin (Chung *et al.*, 1979).

THE RAT NERVOUS SYSTEM
ISBN 0 12 547632 9

Most anatomic studies on primary afferent projections to the spinal cord have been made in the cat; relatively few have been made in the rat. Still it seems safe to conclude that the termination of the primary afferents in the spinal gray matter largely follows two principles of organization also in the rat. First, fine calibered fibers distribute in superficial laminae of the dorsal horn, whereas coarse calibered fibers distribute more ventrally (Light and Perl, 1979a). Second, the primary afferents terminate somatotopically along a mediolateral axis in the dorsal horn (Grant and Ygge, 1981; Grant *et al.*, 1982; Smith, 1983; Ygge and Grant, 1983). The distribution of primary afferents in different laminae and some specific spinal cord nuclei will now be described. Thereafter, their somatotopic arrangement will be considered.*

II. PROJECTION OF PRIMARY AFFERENT FIBERS TO DIFFERENT LAMINAE AND SOME SPINAL CORD NUCLEI

As in the cat and monkey (Beal and Bicknell, 1981; Gobel *et al.*, 1981; LaMotte, 1977; Réthelyi, 1977; Snyder, 1982) both laminae 1 and 2 (substantia gelatinosa) are claimed to receive unmyelinated as well as fine myelinated dorsal root afferents in the rat (Jancsó and Király, 1980; Nagy and Hunt, 1983). Within lamina 2, the finely myelinated afferents, supposedly A-δ fibers with high threshold mechanoreceptors, appear to terminate preferentially in the superficial parts (lamina 2_0; Nagy and Hunt, 1983). This has been reported also in the cat and monkey (Beal and Bicknell, 1981; LaMotte, 1977). Some fine myelinated fibers, presumably A-δ D-hair, may, in addition, reach the deepest part of lamina 2, from the more ventrally located lamina 3. The unmyelinated, C fiber, afferents have been claimed to have their main termination slightly deeper than the first group of fine myelinated fibers, within lamina 2, in the same transverse band where fluoride resistant acid phosphatase (FRAP) positive terminals have been demonstrated (Coimbra *et al.*, 1974; see also Ribeiro-da-Silva and Coimbra, 1982, and Nagy and Hunt, 1983). In the cat the arborizations of these unmyelinated afferents, which enter from the superficial part of the dorsal horn, have been found to be distributed in narrow (150 μm wide) zones within lamina 2 and their terminals in still more narrow (16 μm–28 μm thick), sagittal sheets (Réthelyi, 1977). A zonal organization can be seen for primary afferents in lamina 2 also in the rat (Ygge and Grant, 1983).

The dorsal root afferents to the deep parts of the dorsal horn, which include laminae 3–5 and in the enlargements also lamina 6, are, generally, of coarser caliber than those to the superficial laminae. One exception to this seems to be certain types of afferents, including visceral afferents, terminating in lamina 5, which have been identified in the cat (Cervero and Connell, 1984; Craig and

*The distribution of different kinds of peptides will not be dealt with here. The reader is referred to, for example: Hökfelt *et al.*, 1980; Gibson *et al.*, 1981; Réthelyi and Szentágothai, 1981).

Mense, 1983; Light and Perl, 1979b). The larger diameter primary afferent fibers enter the dorsal horn from the dorsal funiculus. In the cat the course and termination of these fibers has been studied extensively using different methods, including intra-axonal application of horseradish peroxidase (HRP) to physiologically identified units. The findings have been reviewed recently in an extensive monograph on organization in the spinal cord (Brown, 1981; see also Maxwell and Bannatyne, 1983; Ralston *et al.*, 1984; Semba *et al.*, 1983; Willis and Coggeshall, 1978). With regard to larger diameter primary afferent fibers in the rat, only a few studies have been published. Some large primary afferents to the dorsal horn have branches with a recurrent course distributing arborizations in lamina 3 and the most ventral part of lamina 2 (see above; Smith, 1983). These seem to correspond to the recurrent fibers with the "flame-shaped arbors" of Scheibel and Scheibel (1968). Lamina 4 has been found to have a dense plexus of primary afferent fibers at thoracic levels in the rat (Smith, 1983). It is quite clear, however, that lamina 4 receives primary afferents also at other levels of the spinal cord in the rat (Brushart and Mesulam, 1980; Taylor *et al.*, 1982; see also Figs 3 and 4 in Grant *et al.*, 1982). As commented on above, lamina 5 also receives a primary afferent projection in the rat (see also Smith, 1983). The same situation is true for lamina 6 (Craig and Mense, 1983; observations by Robertson and Grant; compare also Fig. 2 in Mysicka and Zenker, 1981, with Fig. 66 in Paxinos and Watson, 1982). At least part of the projection to this lamina seems to be derived from muscle afferents (see Craig and Mense, 1983).

The intermediate gray, lamina 7, has also been shown to receive a primary afferent projection from muscle nerves in the rat (Craig and Mense, 1983; Mesulam and Brushart, 1979; Smith, 1983). This is the case also for the dorsal nucleus (D; Clarke's column; Brushart and Mesulam, 1980; Craig and Mense, 1983; observations by Robertson and Grant) and the central cervical nucleus in the upper cervical segments of the spinal cord (CeC; Ammann *et al.*, 1983; Mysicka and Zenker, 1981). The sacral parasympathetic nucleus, at the dorsolateral border of lamina 7 and located in L6 and S1 in the rat, has recently been shown to have visceral primary afferent connections in this species (Nadelhaft and Booth, 1982, 1984).

Primary afferent projections to the ventral horn from muscle nerves have been studied recently in great detail at the light microscopic level in the rat thoracic cord (Smith, 1983). Such projections have also been demonstrated at lumbar levels (Mesulam and Brushart, 1979) and in the upper cervical cord (Mysicka and Zenker, 1981).

Lamina 10, including the dorsal commissural nucleus, has been shown in several recent studies to receive visceral primary afferents in the rat (Ciriello and Calaresu, 1983; Nadelhaft and Booth, 1982, 1984; Neuhuber, 1982).

It is apparent from the above that the principle of organization of the spinal cord primary afferent fibers, implying that fine calibered fibers would terminate in superficial laminae of the dorsal horn and coarse calibered fibers more ventrally in the spinal gray matter, is not an absolute one. Fine calibered primary afferents are found in lamina 5 and visceral, presumed fine, fibers terminate both in laminae 7 and 10. Indeed, a highly specialized central projection of primary afferent endings related to sensory function and not to fiber diameter has been proposed by Light and Perl (1979b).

III. SOMATOTOPIC ORGANIZATION OF PRIMARY AFFERENT PROJECTIONS

Early physiologic studies in the cat demonstrated a somatotopic organization of cells in the dorsal horn suggesting a similar organization of the termination of primary afferent fibers in the dorsal horn (for reference see Koerber and Brown, 1982; Ygge and Grant, 1983). This suggestion has been confirmed in later physiologic studies (see Brown, 1981, and Koerber and Brown, 1982). There was also anatomic support for a somatotopic arrangement of the terminations of primary afferent fibers in the dorsal horn (see Ygge and Grant, 1983). More recent studies with HRP tracing technique have confirmed such an arrangement and extended our knowledge as to the intricacy of the organization. The thoracic level, where the segmental pattern of the body is best preserved, should be well suited for studies of the general principles of organization of primary afferent projections in the spinal gray matter. Two recent studies in which transganglionic transport of HRP was used have dealt with this level of the spinal cord in the rat (Smith, 1983; Ygge and Grant, 1983). The projections of the dorsal and ventral rami, the two main components of the spinal nerve, were found to be restricted to lateral, and medial and central parts of the dorsal horn, respectively. Very little intermingling of dorsal and ventral ramus afferents was found in laminae 2 to 4, whereas more extensive overlap was observed in lamina 5 (Smith, 1983). No somatotopic arrangement was seen in lamina 1. The projections of the two main branches of the ventral ramus were also found to be somatotopically organized, contained within the projection compartment of their parent nerve (Ygge and Grant, 1983). The analysis of the rostrocaudal extension of the projections showed that the mediolateral compartments extend into neighboring segments; this results in an overlap between compartments from corresponding rami of adjacent spinal nerves (Ygge and Grant, 1983).

A somatotopically organized projection of primary afferent fibers, with a mediolateral arrangement similar to that described for the thoracic spinal nerve, has also been reported in anatomic studies on hindlimb nerves, both in the cat and rat (Grant *et al.*, 1982; Koerber and Brown, 1980, 1982). In the rat, the sciatic nerve, which is composed of fibers from ventral rami has been shown to have its projection restricted to approximately the medial two thirds of the dorsal horn, leaving free the area where the dorsal ramus is presumed to have its projection. Furthermore, the tibial and peroneal nerves, the two main branches of the sciatic nerve, project medially and laterally, respectively, within the projection compartment for the sciatic nerve (Grant *et al.*, 1982). Cutaneous nerve projections from the hindlimb in the cat have also been the subject of detailed studies recently (Koerber and Brown, 1980, 1982).

A somatotopic organization of the projection of primary afferent fibers in the dorsal horn, as a general principle, does not exclude the possibility that some primary afferent fibers also enter areas outside their main domain. Examples of this, although not from the dorsal horn, are projections to the cuneate nucleus from the rat sciatic nerve, demonstrated by transganglionic transport of HRP (Grant *et al.*, 1979).

REFERENCES

Ammann, B., Gottschall, J. and Zenker, W. (1983). Afferent projections from the rat longus capitis muscle studied by transganglionic transport of HRP. *Anat. Embryol.* 166, 275–289.

Beal, J. A. and Bicknell, H. R. (1981). Primary afferent distribution pattern in the marginal zone (lamina I) of adult monkey and cat lumbosacral spinal cord. *J. Comp. Neurol.* 202, 255–263.

Brown, A. G. (1981). *Organization in the spinal cord: The anatomy and physiology of identified neurons.* Springer, Berlin.

Brushart, T. M. and Mesulam, M. -M. (1980). Transganglionic demonstration of central sensory projections from skin and muscle with HRP-lectin conjugates. *Neurosci. Lett.* 17, 1–6.

Cervero, F. and Connell, L. A. (1984). Fine afferent fibers from viscera do not terminate in the substantia gelatinosa of the thoracic spinal cord. *Brain Res.* 294, 370–374.

Chung, K. and Coggeshall, R. E. (1984). The ratio of dorsal root ganglion cells to dorsal root axons in sacral segments of the cat. *J. Comp. Neurol.* 225, 24–30.

Chung, K., Langford, L. A., Applebaum, A. E. and Coggeshall, R. E. (1979). Primary afferent fibers in the tract of Lissauer in the rat. *J. Comp. Neurol.* 184, 587–598.

Ciriello, J. and Calaresu, F. R. (1983). Central projections of afferent renal fibers in the rat: An anterograde transport study of horseradish peroxidase. *J. Auton. Nerv. Sys.* 8, 273–285.

Coimbra, A., Sodré-Borges, B. P. and Magalhães, M. M. (1974). The substantia gelatinosa Rolandi of the rat: Fine structure, cytochemistry (acid phosphatase) and changes after dorsal root section. *J. Neurocytol.* 3, 199–217.

Craig, A. D. and Mense, S. (1983). The distribution of afferent fibers from the gastrocnemius-soleus muscle in the dorsal horn of the cat, as revealed by the transport of horseradish peroxidase. *Neurosci. Lett.* 41, 233–238.

Gibson, S. A., Polak, J. M., Bloom, S. R. and Wall, P. D. (1981). The distribution of nine peptides in rat spinal cord with special emphasis on the substantia gelatinosa and on the area around the central canal (lamina X). *J. Comp. Neurol.* 201, 65–79.

Gobel, S., Falls, W. M. and Humphrey, E. (1981). Morphology and synaptic connections of ultrafine primary axons in lamina I of the spinal dorsal horn: Candidates for the terminal axonal arbors of primary neurons with unmyelinated (C) axons. *J. Neurosci.* 1, 1163–1179.

Grant, G. and Ygge, J. (1981). Somatotopic organization of the thoracic spinal nerve in the dorsal horn demonstrated with transganglionic degeneration. *J. Comp. Neurol.* 202, 357–364.

Grant, G., Arvidsson,J., Robertson, B. and Ygge, J. (1979). Transganglionic transport of horseradish peroxidase in primary sensory neurons. *Neurosci. Lett.* 12, 23–28.

Grant, G., Ygge, J. and Molander, C. (1982). Projection patterns of peripheral sensory nerves in the dorsal horn. *In* A. G. Brown and M. Réthelyi (eds), *Spinal cord sensation: Sensory processing in the dorsal horn*, pp. 33–43. Scottish Academic Press, Edinburgh.

Hökfelt, T., Johansson, O., Ljungdahl, Å., Lundberg, J. M. and Schultzberg, M. (1980). Peptidergic neurons. *Nature*, 284, 515–521.

Janscó, G. and Király, E. (1980). Distribution of chemosensitive primary sensory afferents in the central nervous system of the rat. *J. Comp. Neurol.* 190, 781–792.

Koerber, H. R. and Brown, P. B. (1980). Projections of two hindlimb nerves to cat dorsal horn. *J. Neurophysiol.* 44, 259–269.

Koerber, H. R. and Brown, P. B. (1982). Somatotopic organization of hindlimb cutaneous nerve projections to cat dorsal horn. *J. Neurophysiol.* 48, 481–489.
LaMotte, C. (1977). Distribution of the tract of Lissauer and the dorsal root fibers in the primate spinal cord. *J. Comp. Neurol.* 172, 529–562.
Langford, L. A. and Coggeshall, R. E. (1979). Branching of sensory axons in the dorsal root and evidence for the absence of dorsal root efferent fibers. *J. Comp. Neurol.* 184, 193–204.
Light, A. R. and Perl, E. R. (1979a). Reexamination of the dorsal root projection to the spinal dorsal horn including observations on the differential termination of coarse and fine fibers. *J. Comp. Neurol.* 186, 117–132.
Light, A. R. and Perl, E. R. (1979b). Spinal termination of functionally identified primary afferent neurons with slowly conducting myelinated fibers. *J. Comp. Neurol.* 186, 133–150.
Mawe, G. M., Bresnahan, J. C. and Beattie, M. S. (1984). Primary afferent projections from dorsal and ventral roots to autonomic preganglionic neurons in the cat sacral spinal cord: Light and electron microscopic observations. *Brain Res.* 290, 152–157.
Maxwell, D. J. and Bannatyne, B. A. (1983). Ultrastructure of muscle spindle afferent terminations in lamina VI of the cat spinal cord. *Brain Res.* 288, 297–301.
Mesulam, M. -M. and Brushart, T. M. (1979). Transganglionic and anterograde transport of horseradish peroxidase across dorsal root ganglia: A tetramethylbenzidine method for tracing central sensory connections of muscles and peripheral nerves. *Neurosci.* 4, 1107–1117.
Mysicka, A. and Zenker, W. (1981). Central projections of muscle afferents from the sterno-mastoid nerve in the rat. *Brain Res.* 211, 257–265.
Nadelhaft, I. and Booth, A. M. (1982). Preganglionic neurons and visceral afferent fibers in the rat pelvic nerve. *Soc. Neurosci.Abst.* 8, 77.
Nadelhaft, I. and Booth, A. M. (1984). The location and morphology of preganglionic neurons and the distribution of visceral afferents from the rat pelvic nerve: A horseradish peroxidase study. *J. Comp. Neurol.* 226, 238–245.
Nagy, J. I. and Hunt, S. P. (1983). The termination of primary afferents within the rat dorsal horn: Evidence for rearrangement following capsaicin treatment. *J. Comp. Neurol.* 218, 145–158.
Neuhuber, W. (1982). The central projections of visceral primary afferent neurons of the inferior mesenteric plexus and hypogastric nerve and the location of the related sensory and preganglionic sympathetic cell bodies in the rat. *Anat. Embryol.* 164, 413–425.
Paxinos, G. and Watson, C. (1982). *The rat brain in stereotaxic coordinates.* Academic Press, Sydney.
Ralston, H. J., III, Light, A. R., Ralston, D. D. and Perl, E. R. (1984). Morphology and synaptic relationships of physiologically identified low-threshold dorsal root axons stained with intra-axonal horseradish peroxidase in the cat and monkey. *J. Neurophysiol.* 51, 777–792.
Réthelyi, M. (1977). Preterminal and terminal axon arborizations in the substantia gelatinosa of cat's spinal cord. *J. Comp. Neurol.* 172, 511–528.
Réthelyi, M. and Szentagothai, J. (1981). Peptidergic neurons in the spinal cord. In J. Salanki and T. M. Turpaev (Eds). *Neurotransmitters: Comparative aspects*, pp. 123–148. Akademiai Kiado, Budapest.
Ribeiro-da-Silva, A. and Coimbra, A. (1982). Two types of synaptic glomeruli and their distribution in laminae I–III of the rat spinal cord. *J. Comp. Neurol.* 209, 176–186.
Risling, M. and Hildebrand, C. (1982). Occurrence of unmyelinated axon profiles at distal, middle and proximal levels in the ventral root L7 of cats and kittens. *J. Neurol. Sci.* 56, 219–231.
Risling, M., Dalsgaard, C. -J., Cukierman, A. and Cuello, A. C. (1984). Electron microscopic and immunohistochemical evidence that unmyelinated ventral root axons make U-turns or enter the spinal pia mater. *J. Comp. Neurol.* 225, 53–63.
Scheibel, M. E. and Scheibel, A. B. (1968). Terminal axonal patterns in cat spinal cord. II. The dorsal horn. *Brain Res.* 9, 32–58.
Semba, K., Masarachia, P., Malamed, S., Jacquin, M., Harris, S., Yang, G. and Egger, M. D. (1983). An electron microscopic study of primary afferent terminals from slowly adapting Type I receptors in the cat. *J. Comp. Neurol.* 221, 466–481.
Smith, C. L. (1983). The development and postnatal organization of primary afferent projections to the rat thoracic spinal cord. *J. Comp. Neurol.* 220, 29–43.
Snyder, R. (1977). The organization of the dorsal root entry zone in cats and monkeys. *J. Comp. Neurol.* 174, 47–70.

Snyder, R. (1982). Light and electron microscopic autoradiographic study of the dorsal root projections to the cat dorsal horn. *Neurosci.* 7, 1417–1437.

Taylor, D. C. M., Korf, H. -W. and Pierau, Fr. -K. (1982). Distribution of sensory neurons of the pudendal nerve in the dorsal root ganglia and their projection to the spinal cord. *Cell Tiss. Res.* 226, 555–564.

Waibl, H. (1973). Zur Topographie der Medulla spinalis der Albinoratte (*Rattus norvegicus*). *Adv. Anat. Embryol. Cell Biol.* 47.

Willis, W. D. and Coggeshall, R. E. (1978). *Sensory mechanisms of the spinal cord.* Wiley, Chichester.

Ygge, J. and Grant G. (1983). The organization of the thoracic spinal nerve projection in the rat dorsal horn demonstrated with transganglionic transport of horseradish peroxidase. *J. Comp. Neurol.* 216, 1–9.

Ygge, J., Aldskogius, H. and Grant, G. (1981). Asymmetries and symmetries in the number of thoracic dorsal root ganglion cells. *J. Comp. Neurol.* 202, 365–372.

14

Ascending and descending pathways in the spinal cord

DAVID J. TRACEY

University of New South Wales
Kensington, NSW, Australia

I. INTRODUCTION

The white matter of the spinal cord contains the axons which serve to connect different regions of the spinal gray matter, to carry information from the spinal cord to the brain, and to transmit information from the brain to spinal neurons. The pathways can therefore be divided roughly into three groups: ascending pathways; descending pathways; and local pathways restricted to the spinal cord itself (for reviews, see Kuypers, 1981; Kuypers and Martin, 1982; Willis and Coggeshall, 1978).

Our information on the rat does not cover these pathways evenly, so that much more is known about the spinothalamic pathway and the corticospinal tract than about other pathways. With growing interest in the rat as an experimental animal, we can expect information to accumulate rapidly about lesser known connections such as the propriospinal tract.

II. ASCENDING PATHWAYS

Distinct pathways in the spinal cord generally have their axons located in specific regions of the white matter, such as the dorsal columns, dorsolateral fasciculus, or the ventrolateral quadrant. This provides one way of classifying the pathways; but it is a little artificial in that functionally different pathways may have axons in the same region of the white matter. Therefore, in this account, the pathways will be grouped according to their destinations in the medulla and pons, the midbrain, the thalamus and the cerebellum.

THE RAT NERVOUS SYSTEM
ISBN 0 12 547632 9

A. Pathways from spinal cord to medulla and pons

These include axons projecting from the spinal cord to the dorsal column nuclei, the lateral cervical nucleus, nucleus Z, the vestibular nuclei, and the reticular formation.

1. *Dorsal columns*

The dorsal columns contain two groups of ascending fibers: the ascending collaterals of primary afferents, constituting the direct dorsal column pathway; and the axons of postsynaptic dorsal column neurons, which constitute the second order dorsal column pathway. In addition, there are also descending fibers with cell bodies in the dorsal column nuclei, and, in the rat, the fibers of the corticospinal tract. The majority of axons in both the dorsal columns proper and in the corticospinal tract are unmyelinated (Langford and Coggeshall, 1981).

a. Direct dorsal column pathway

The axons are the ascending collaterals of primary afferents, with cell bodies in the dorsal root ganglia. Some of these collaterals terminate in the dorsal column nuclei; of these, collaterals of primary afferents entering the cord in cervical and upper thoracic spinal nerves terminate in the cuneate nucleus, while those entering in lower thoracic and lumbosacral spinal nerves terminate in the gracile nucleus. However, most of the fibers which enter the dorsal columns do not reach the dorsal column nuclei: for example, many of the fibers in the gracile fasciculus leave the dorsal columns to terminate on neurons in Clarke's column.

The ascending collaterals of primary afferents have a somatotopic organization in the dorsal column, such that fibers from the tail run close to the midline, with fibers from the hindlimb, trunk and forelimb being added to the lateral border of the column at progressively more rostral levels. In addition to this somatotopic pattern, there is an arrangement according to submodality, so that fibers from hair and claw receptors run in the superficial part of the column, while fibers from touch and vibration receptors are deep. Fibers from muscle receptors occur at an intermediate depth.

The terminations of primary afferents have been described in the cuneate nucleus of the rat. As in other mammals, they are not distributed evenly throughout the nucleus, but are concentrated in the ventral part caudally and the dorsal part rostrally. The termination patterns are reflected by differences in cytoarchitecture, so that the nucleus is divided into two regions which differ both anatomically and functionally (Basbaum and Hand, 1973; see Volume 2, Chapter 8).

The terminations of axons of the gracile fasciculus (gr) have also been examined in the rat. Terminal projection fields were located at all levels of the gracile nucleus; terminations were also found in Clarke's column, the commissural nucleus, the external cuneate, and the medial reticular formation of the medulla (Ganchrow and Bernstein, 1981). These results were obtained by examining terminal degeneration following lesions of the gr, so that second order dorsal column fibers would also have been involved.

b. Second order dorsal column neurons

The existence of axons of second order neurons in the dorsal column was first recognized in the cat; recordings from axons in the dorsal columns revealed a population of synaptically activated, nonprimary afferents. The axons of the cells enter the dorsal columns through the dorsal and medial border of the dorsal horn, and their terminations ramify in the rostral and ventral parts of the dorsal column nuclei.

In the rat, injection of horseradish peroxidase (HRP) into the dorsal column nuclei results in retrograde labeling of neurons in the medial parts of the dorsal horn. The cells are relatively large, up to 60 μm in diameter (Giesler *et al.*, 1984; Low and Tracey, observations).

2. *Spinocervical pathway*

The spinocervical pathway is an important pathway for cutaneous information. The pathway consists of spinocervical tract cells in the dorsal horn, which send their axons in the dorsolateral funiculus to the lateral cervical nucleus, just lateral to the dorsal horn in cervical segments C1 and C2. These cells in turn project to the ventrobasal thalamus (Giesler *et al.*, 1979b).

The cells of origin of the spinocervical tract in the rat are located in layers 3 to 5; in the rat, the lateral cervical nucleus also appears to receive a projection from layer 1 and from small cells in the substantia gelatinosa (Giesler *et al.*, 1978).

3. *Nucleus Z pathway*

Axons of the dorsal spinocerebellar tract (dsc) do not project solely to the cerebellum. They also send collateral branches to nucleus Z, a small and rather poorly defined group of cells near the rostral ends of the gracile and cuneate nuclei. Nucleus Z projects in turn to the ventrobasal thalamus and cortex.

Fibers belonging to the dsc in the rat run not only in the dorsolateral fasciculus, but also in the ventrolateral white matter. When these fibers are transected, terminal degeneration is found in nucleus Z (Zemlan *et al.*, 1978). The cells of origin of the dsc are mostly in Clarke's column; injection of retrograde tracers into nucleus Z results in the labeling of about 3% of these dorsal spinocerebellar tract cells (Low and Tracey, 1984).

4. Spinovestibular fibers

Spinal neurons project to all four of the vestibular nuclei, including nucleus X and nucleus Z (Volume 2, Chapter 9). Nucleus X and Z are associated anatomically with the vestibular nuclei, but receive spinal afferents from collaterals of dsc neurons running in the dorsolateral fasciculus (see above). In fact it seems probable that there is a "lateral" spinovestibular system whose projections to the vestibular nuclei, like those to X and Z, are through collaterals of the dorsal spinocerebellar tract.

In addition to these projections there is a "medial" vestibulospinal system which projects to MVe from the central cervical nucleus. For further information the reader should consult Volume 2, Chapter 9.

5. Spinoreticular fibers

Two spinoreticular tracts have been distinguished in the rat: a lateral spinoreticular tract projecting to the lateral reticular nucleus, and a medial spinoreticular tract projecting to the remaining reticular formation of the medulla and pons. Neurons at the origin of the medial spinoreticular tract are located mainly in layers 5, 7 and 8. They send their axons in the ventrolateral quadrant of the spinal cord, and terminate in reticular nuclei throughout the bulbar and pontine reticular formation, including the reticular gigantocellular and paragigantocellular nuclei, raphe magnus and pallidus nuclei, pontine reticular nucleus, and subcoeruleus nucleus. Most of these axons project contralaterally, except for those at cervical levels, which project bilaterally.

The spinoreticular neurons located in the dorsal horn have similar physiologic properties to spinothalamic neurons in the rat: most respond to cutaneous input and have large receptive fields, while a minority are activated only by noxious mechanical stimuli (Menétrey *et al.*, 1980). In fact some spinal neurons send collateral projections to both the thalamus and medullary reticular formation in the rat (Kevetter and Willis, 1983).

B. Pathways from spinal cord to midbrain

Spinal neurons project to three regions of the midbrain. These regions are the superior colliculus, the periaqueductal gray, and the midbrain reticular formation.

1. Spinotectal tract

The spinotectal tract terminates primarily in the superior colliculus, with some terminations in the intercollicular nucleus.

In the rat, the cells of origin of the spinotectal tract have been reported in the upper cervical (Cl to C3) and lumbosacral (L4 to S1) regions of the spinal cord. The neurons are located in the contralateral dorsal horn, in layers 3, 4 and 5, with some found ventrally in layers 7 and 8 (Morrell and Pfaff, 1983). Their axons have been divided into two groups—one in the ventral part of the lateral funiculus which crosses in the spinal cord, and one lying more dorsally in the lateral funiculus which does not cross until it reaches the brain stem. Both sets of fibers end in the caudal part of the superior colliculus, with a somatotopic organization such that axons originating in the cervical cord terminate most rostrally, while axons from the sacrococcygeal cord terminate most caudally (Antonetty and Webster, 1975). Fibers ascending in the dorsal part of the lateral funiculus also terminate in the intercollicular nucleus, whereas the projection to the superior colliculus (and central gray) seems to be concentrated in the ventral part (Zemlan *et al.*, 1978).

2. Spinal cord to central gray (spinoannular tract)

The projection to the central gray from the spinal cord is a relatively minor input in comparison with projections from the forebrain and hypothalamus (Beitz, 1982). The spinal cord projects mainly to the ventrolateral region of the central gray: the cells of origin of the tract are located primarily in layers 1, 3, 5, 7 and 10. They are predominantly contralateral, and there is a strong projection from the lumbosacral region of the cord (Liu, 1983). Their fibers run in the ventral part of the lateral columns (Zemlan *et al.*, 1978).

3. Spinal cord to midbrain reticular formation

The presence of neurons in the spinal cord projecting to the deep mesencephalic nucleus was suggested by retrograde labeling following restricted injections of HRP, but could not be confirmed with anterograde transport of radioactive amino acids (Veazey and Severin, 1982). Larger injections of HRP which extended into the central gray labeled neurons in layers 1, 5 and 10 at all levels of the spinal cord. In addition, neurons in the nucleus of the dorsolateral funiculus and in the lateral cervical nucleus have been shown to project to the midbrain reticular formation (Menétrey *et al.*, 1982).

C. Spinothalamic tract

The importance of the spinothalamic tract (spth) is that it appears to be the main pathway for information from receptors signaling pain and temperature. In the rat, the cells of origin are generally located in the dorsal horn, and their axons cross within a few segments to ascend directly to the thalamus, where they terminate in four distinct regions: the ventrobasal complex; the intralaminar nuclei; the posterior group (Lund and Webster, 1967); and the gelatinosus (submedius) nucleus (Craig and Burton, 1981).

In the rat, spinothalamic cells have laminar locations in the spinal cord which depend on their terminations in the thalamus. Thus neurons projecting to the ventrobasal complex tend to be located at the base of the dorsal horn and in the marginal zone, while neurons projecting to the more medial intralaminar nuclei are found more ventrally, extending into layer 7 (Giesler *et al.*, 1979a; Land, 1977). This difference in laminar location seems to be correlated with a difference in afferent input, so that those neurons projecting to the medial thalamus, said to belong to the "medial" spinothalamic tract, are activated by deep pressure and joint movement, while those projecting in the "lateral" spinothalamic tract are activated by cutaneous stimuli, both noxious and innocuous. In fact the "medial" spth neurons send their axons in the ventral funiculus of the cord, while "lateral" spth axons ascend in the ventrolateral funiculus (Giesler *et al.*, 1981).

Some spinothalamic neurons project not only to the thalamus, but also send collaterals to the medullary reticular formation (Kevetter and Willis, 1983); other spinothalamic neurons send collaterals to the cerebellum (Yezierski and Bowker, 1981).

D. Spinocerebellar pathways

The pathways from the spinal cord to the cerebellum include five direct tracts terminating as mossy fibers. Two of these, the dorsal and ventral spinocerebellar tracts, transmit information from receptors in the hindlimb, while two homologous pathways, the cuneocerebellar and rostral spinocerebellar tracts, transmit information from receptors in the forelimb (Oscarsson, 1973). In addition, a recently described pathway carries information from receptors in the neck and labyrinths. This tract originates in the central cervical nucleus (Wiksten, 1979).

Information from the spinal cord may also reach the cerebellum indirectly. These indirect pathways involve mossy fibers from the lateral reticular nucleus (see above), or climbing fibers originating in the inferior olivary nucleus.

1. *Dorsal spinocerebellar tract*

The neurons which give rise to the mammalian dorsal spinocerebellar tract (dsc) are located primarily in the dorsal nucleus, or Clarke's column. Their axons leave the dorsal nucleus laterally and ascend in the dorsolateral fasciculus, entering the cerebellum via the inferior cerebellar peduncle.

In the rat, several studies have used retrograde transport of HRP to demonstrate spinal neurons which project to the cerebellum (Matsushita and Hosoya, 1979; Snyder *et al.*, 1978; Zemlan *et al.*, 1978). Such studies show that in the rat, neurons of the dorsal nucleus and its caudal continuation into sacral and caudal segments (Stilling's nucleus) project to the cerebellum (Snyder *et al.*, 1978). Injections of HRP into the inferior cerebellar peduncle also label neurons in the dorsal nucleus, confirming that they are the cells of origin of the dsc in the rat as in other mammals (Low and Tracey, 1984). Axons of neurons in the rat dorsal nucleus are not confined to the dorsolateral fasciculus, but course through the entire lateral column (Zemlan *et al.*, 1978).

2. *Ventral spinocerebellar tract*

Neurons which give rise to the ventral spinocerebellar tract (vsc) generally include a group of cells located in layers 5 to 7 of the spinal cord, and the spinal border cells. Their axons cross to the contralateral side of the cord, ascend in the ventrolateral column, and enter the cerebellum via the superior cerebellar peduncle. These results have been partially confirmed in the rat, except that spinal border cells do not appear to contribute to the ventral spinocerebellar tract (Zemlan *et al.*, 1978).

3. *Spino-olivary fibers*

In mammals, neurons in the spinal cord project directly to the contralateral inferior olivary nucleus via the ventral funiculus. The pathways are known as the ventral funiculus–spino-olivary cerebellar pathways (Armstrong and Schild, 1980). In addition, there are indirect pathways to the inferior olive via nuclei such as the dorsal column nuclei and reticular nuclei.

In the rat, direct spino-olivary fibers have been shown to terminate in the medial accessory olive and dorsal accessory olive, referred to by Mehler (1969) as paleo-olivary structures. Spino-olivary neurons in the rat are located in the medial aspect of the nucleus proprius at upper cervical levels of the cord and probably at lower levels as well (Brown *et al.*, 1977). Indirect pathways from the spinal cord to the inferior olive are also discussed in this study, including pathways via the dorsal column nuclei as well as via the lateral reticular and gigantocellular reticular nuclei.

III. DESCENDING PATHWAYS

A. Pathways to the spinal cord from medulla and pons

These include pathways from the dorsal column nuclei and spinal trigeminal nucleus, the medullary and pontine reticular formation, the vestibular nuclei and several other sites.

1. Dorsal column nuclei and spinal trigeminal nuclei

In the rat, as well as in other mammals, some neurons of the gracile, cuneate and spinal trigeminal nuclei project to the spinal gray matter. In the dorsal column nuclei, these neurons are located in the middle third of the nuclei, just ventral to the cell nest region. Most neurons in this ventral zone do not project to the thalamus or cerebellum, but receive terminations from second order dorsal column fibers and from corticobulbar fibers.

Fibers from the dorsal column nuclei descend mainly in the ipsilateral dorsal columns and terminate in layers 4 and 5 of the cord, at least at cervical levels, and in the lateral cervical nucleus (Burton and Loewy, 1977). It is in these layers that most of the cells of origin of the second order dorsal column fibers are found. As one might expect, fibers from the cuneate nucleus tend to project to cervical levels of the cord, while fibers from the gracile nucleus project to lumbosacral levels.

Neurons in the spinal trigeminal nucleus project to the entire length of the spinal cord in the rat (Ruggiero *et al.*, 1981). One group of trigeminospinal neurons is located in the caudal and interpolar subnuclei and projects ipsilaterally to all levels of the cord. There are also trigeminospinal neurons located in the oral subnucleus; these neurons project contralaterally to thoracic and lumbosacral levels of the cord, and bilaterally to the cervical cord. There is evidence that some of these axons travel in the anterior funiculus.

2. Reticular formation

In the rat there are two separate reticulospinal systems from the region of the medulla. The first system arises from neurons at the ventral margin of the ventral reticular nucleus of the medulla and gigantocellular reticular nucleus (ventral MdV and Gi), and projects in the lateral columns to the ipsilateral cord. Neurons in the MdV project only as far as lower thoracic levels, but the greatest number of reticulospinal neurons are located in the ventral Gi, and project to the whole length of the cord, including lumbosacral levels.

The second system arises from neurons in the dorsal portion of the MdV and Gi, and projects in the ventral columns to the cervical cord, but not to more caudal levels (Satoh, 1979; Zemlan and Pfaff, 1979). Reticulospinal neurons are also located in the caudal and oral pontine reticular nuclei (PnC and PnO) (Satoh, 1979).

Reticulospinal axons terminate in layers 7, 8, and 9 (Motorina, 1977), and there is recent evidence that reticulospinal axons establish synaptic contacts with the lateral group of motoneurons at the L5 and L6 level of the cord (Holstege and Kuypers, 1982). A more complete account of the reticulospinal projection is contained in Volume 2, Chapter 2.

3. *Vestibular nuclei*

Fibers descending from the vestibular nuclei to the spinal cord constitute the lateral vestibulospinal tract (from LVe) and the medial vestibular tract (from MVe). The lateral vestibulospinal tract projects in the ipsilateral ventrolateral funiculus to all levels of the cord; its fibers terminate mainly in layer 8, but also in parts of layer 7 and 9. There is electrophysiologic evidence that some of these axons make direct or monosynaptic contact with extensor motoneurons, at least in the lumbosacral cord.

Fibers from the MVe descend towards the cord in the medial longitudinal fasciculus, and then in the sulcomarginal fasciculus of the cord, close to the ventral midline. They project only as far as midthoracic levels, and terminate in the same layers as the fibers of the lateral vestibulospinal tract. They probably influence motoneurons of the neck rather than of the limbs. For a more complete account of the vestibulospinal projection refer to Volume 2, Chapter 9.

4. *Raphe nuclei and serotonergic projections to the spinal cord*

In the rat, neurons in the raphe pallidus (RPa) and raphe obscurus (ROb) nuclei project to the spinal cord in the ventral funiculus, while neurons in the raphe magnus nucleus (RMg) descend in the dorsolateral funiculus (Basbaum and Fields, 1979; Leichnetz *et al.*, 1978). The fibers from RMg are strongly implicated in the production of analgesia. In fact many of the neurons which project from the RMg to the spinal cord send collateral branches to the caudal subnucleus of the nucleus of the spinal tract of the trigeminal Sp5, which is known to receive nociceptive afferents (Lovick and Robinson, 1983).

Neurons of the raphe nuclei, such as nucleus raphe magnus (B3), nucleus raphe pallidus (B1) and nucleus raphe obscurus (B2), possess a high proportion of neurons containing serotonin or 5-hydroxytryptamine (as well as other neurotransmitters). In the rat spinal cord, serotonergic terminals are concentrated in layers 1 and 2a of the dorsal horn, layers 8 and 9 of the ventral horn, and in the intermediolateral cell column of the thoracic cord (Bowker *et al.*, 1982; Steinbusch, 1981). Additional evidence suggests that serotonergic terminals in the dorsal horn are derived from the RMg and the lateral reticular formation, while those in the ventral horn are derived from these nuclei and, in addition, from the RPa and ROb. Serotonergic projections to the intermediolateral cell column are derived from all three medullary raphe nuclei (RMg, ROb, RPa) as well as from neurons located ventrolateral to the inferior olivary nucleus (Bowker *et al.*, 1982). For a more complete account of the raphe–spinal projection refer to Volume 2, Chapter 3.

5. Locus coeruleus and noradrenergic projections to spinal cord

Neurons in the locus coeruleus and in the subcoeruleus nucleus and ventral parabrachial nucleus provide the major noradrenergic projections to the spinal cord. These projections have been implicated in the control of autonomic functions, modulating the perception of pain, and modifying motor behavior such as locomotion. While other noradrenergic neurons are also found in the caudal medulla, they do not appear to project to the spinal cord (Westlund *et al.*, 1982).

In the rat it has been shown that neurons in the ventral part of the locus coeruleus and in the subcoeruleus nucleus project bilaterally to the spinal cord, predominantly to the ventral horn; coerulospinal axons presumably terminate on motoneurons (Commissiong, 1981; Commissiong *et al.*, 1978). Some of these coerulospinal neurons are noradrenergic (Satoh, 1979). Noradrenergic terminals are found in the region of preganglionic sympathetic neurons in the intermediolateral cell column of the thoracic spinal cord (Glazer and Ross, 1980), but these terminals do not appear to be derived from the locus coeruleus (Commissiong, 1981). The noradrenergic innervation of preganglionic sympathetic neurons is more likely derived from the subcoeruleus and ventral parabrachial nucleus, while the locus coeruleus (the A6 cell group) projects not only to motoneurons but also to parasympathetic preganglionic neurons of the sacral cord (Westlund *et al.*, 1982).

B. Pathways from the midbrain to the spinal cord

1. Rubrospinal tract

The rubrospinal tract is comparable in some ways with the corticospinal tract; in carnivores and primates the axons of both tracts exert an excitatory effect on flexor motoneurons, descend in the dorsolateral fasciculus, and terminate in the same region of the spinal cord.

In the rat, the axons of the tract leave the ventromedial aspect of the nucleus, course through the ventral tegmental decussation, and eventually descend in the dorsolateral fasciculus of the spinal cord. The tract is primarily contralateral, and terminates in the ventrolateral aspect of the dorsal horn and intermediate region of the ventral horn (layers 5 and 6), in contrast to the corticospinal tract, which terminates in the dorsomedial aspect of the dorsal horn (Brown, 1974; Waldron and Gwyn, 1969). As in other animals, the rubrospinal tract arises mainly from the caudal or magnocellular part of the red nucleus. There is also evidence of a somatotopic arrangement, such that rubrospinal neurons projecting to the cervical cord are found in dorsomedial regions of the red nucleus, while those projecting to the lumbosacral cord are located ventrolaterally in the nucleus (Murray and Gurule, 1979; Shieh *et al.*, 1983).

Some rubrospinal neurons send collateral branches to separate regions of the spinal cord, and also to the interposed nucleus of the cerebellum (Huisman *et al.*, 1981; 1983).

2. *Mesencephalic reticular formation to spinal cord*

In the rat, retrograde labeling techniques have shown that neurons in the mesencephalic reticular formation project to the lumbosacral region of the spinal cord, and probably to the rest of the cord as well. The majority of these neurons are located in the cuneiform nucleus (Satoh, 1979).

Other neurons in the mesencephalic reticular formation have been shown to project from the medial part of the deep mesencephalic nucleus, at least as far as the cervical spinal cord via the ventral funiculus (Veazey and Severin, 1980). Fibers from the deep mesencephalic nucleus were traced as far as layer 8 of the upper thoracic segments of the cord by Waldron and Gwyn (1969). However, it is worth noting that the cuneiform nucleus and subcuneiform nucleus are essentially dorsal and ventral parts of the deep mesencephalic nucleus (see Veazey and Severin, 1980), so that the apparently distinct populations of reticulospinal neurons reported in the two studies above may well be one and the same.

3. *Tectospinal fibers*

In mammals, neurons in the superior colliculus project to the cervical spinal cord. These tectospinal neurons are located in the intermediate and deep gray of the contralateral superior colliculus, and project in the tectospinal tract located just ventral to the medial longitudinal fasciculus.

In the rat, the terminations of tectospinal fibers are restricted to the upper two or three cervical segments, where they terminate in layer 8, in the medial part of the ventral horn (Waldron and Gwyn, 1969).

C. Hypothalamus to spinal cord

There are direct projections from the hypothalamus to the spinal cord. These originate from wide areas of the hypothalamus, and project ipsilaterally to the intermediolateral cell column (Hosoya and Matsushita, 1979; Saper *et al.*, 1976).

D. Corticospinal fibers

The corticospinal tract shows considerable differences between species. In some mammals, such as marsupials, it projects only as far as cervicothoracic segments; in the rat it projects the whole length of the spinal cord. In mammals such as carnivores and primates, the corticospinal tract descends in the lateral funiculus; in the rat it descends in the dorsal columns (Armand, 1982; Kuypers, 1981).

In the rat, the corticospinal tract originates from the largest pyramidal neurons of layer 5B in the primary somatosensory cortex, SI—that is, Parl, FL and HL—and in addition from the agranular motor cortex adjacent to SI—that is, Fr2 and Fr3 (Wise and Jones, 1977). The majority of corticospinal fibers decussate in the caudal medulla and descend in the ventralmost part of the dorsal columns (Brown, 1971). An additional group of fibers is uncrossed and descends in the ventral funiculus (Vahlsing and Feringa, 1980). The axons of the corticospinal tract terminate in the dorsomedial part of the dorsal horn at all levels of the spinal cord (Brown, 1971). Their terminations also extend into laminae 7 and 8, particularly in the cervical and lumbar enlargements of the cord (Donatelle, 1977). There is even physiologic evidence of monosynaptic connections of some corticospinal axons with motoneurons in the cervical region (Elger *et al.*, 1977).

REFERENCES

Antonetty, C. M. and Webster, K. E. (1975). The organisation of the spinotectal projection: An experimental study in the rat. *J. Comp. Neurol.* 163, 449–465.

Armand, J. (1982). The origin, course and terminations of corticospinal fibers in various mammals. *Prog. Brain Res.* 57, 329–360.

Armstrong, D. M. and Schild, R. F. (1980). Location in the spinal cord of neurons projecting directly to the inferior olive in the cat. *In* J. Courville *et al.* (eds), *The inferior olivary nucleus: Anatomy and physiology.* Raven, New York.

Basbaum, A. I. and Fields, H. L. (1979). The origin of descending pathways in the dorsolateral funiculus of the spinal cord of the cat and rat: Further studies on the anatomy of pain modulation. *J. Comp. Neurol.* 187, 513–532.

Basbaum, A. I. and Hand, P. J. (1973). Projections of cervicothoracic dorsal roots to the cuneate nucleus of the rat, with observations on cellular "bricks". *J. Comp. Neurol.* 148, 347–360.

Beitz, A. J. (1982). The organization of afferent projections to the midbrain periaqueductal gray of the rat. *Neurosci.* 7, 133–159.

Bowker, R. M., Westlund, K. N., Sullivan, M. C. and Coulter, J. D. (1982). Organization of descending serotonergic projections to the spinal cord. *Prog. Brain Res.* 57, 239–265.

Brown, J. T., Chan-Palay, V. and Palay, S. L. (1977). A study of afferent input to the inferior olivary complex in the rat by retrograde axonal transport of horseradish peroxidase. *J. Comp. Neurol.* 176, 1–22.

Brown, L. T. (1971). Projections and termination of the corticospinal tract in rodents. *Exp. Brain Res.* 13, 432–450.

Brown, L. T. (1974). Rubrospinal projections in the rat. *J. Comp. Neurol.* 154, 169–188.

Burton, H. and Loewy, A. D. (1977). Projections to the spinal cord from medullary somatosensory relay nuclei. *J. Comp. Neurol.* 173, 773–792.

Commissiong, J. W. (1981). Evidence that the noradrenergic coerulospinal projection decussates at the spinal level. *Brain Res.* 212, 145–151.

Commissiong, J. W., Hellstrom, S. O. and Neff, N. H. (1978). A new projection from locus coeruleus to the spinal ventral columns: Histochemical and biochemical evidence. *Brain Res.* 148, 207–213.

Craig, A. D. and Burton, H. (1981). Spinal and medullary lamina I projection to nucleus submedius in medial thalamus: A possible pain center. *J. Neurophysiol.* 45, 443–466.

Donatelle, J. M. (1977). Growth of the corticospinal tract and development of placing reactions in the postnatal rat. *J. Comp. Neurol.* 175, 207–232.

Elger, C. E., Speckmann, E. J., Caspers, H. and Janzen, R. W. C. (1977). Corticospinal connections in the rat. I. Monosynaptic and polysynaptic responses of cervical motoneurons to epicortical stimulation. *Exp. Brain Res.* 28, 385–404.

Ganchrow, D. and Bernstein, J. J. (1981). Projections of caudal fasciculus gracilis to nucleus gracilis and other medullary structures, and Clarke's nucleus in the rat. *Brain Res.* 205, 383–390.

Giesler, G. J., Cannon, J. T., Urca, G. and Liebeskind, J. C. (1978). Long ascending projections from substantia gelatinosa Rolandi and the subjacent dorsal horn in the rat. *Science* 202, 984–986.

Giesler, G. J., Menétrey, D. and Basbaum, A. I. (1979a). Differential origins of spinothalamic tract projections to medial and lateral thalamus in the rat. *J. Comp. Neurol.* 184, 107–126.

Giesler, G. J., Urca, G., Cannon, T. and Liebeskind, J. C. (1979b). Response properties of neurons of the lateral cervical nucleus in the rat. *J. Comp. Neurol.* 186, 65–78.

Giesler, G. J., Spiel, H. R. and Willis, W. D. (1981). Organization of spinothalamic tract axons within the rat spinal cord. *J. Comp. Neurol.* 195, 243–252.

Giesler, G. J., Nahin, M. L. and Madsen, A. M. (1984). Postsynaptic dorsal column pathway of the rat. I. Anatomical studies. *J. Neurophysiol.* 51, 260–275.

Glazer, E. J. and Ross, L. L. (1980). Localization of noradrenergic terminals in sympathetic preganglionic nuclei of the rat: Demonstration by immunocytochemical localization of dopamine-beta-hydroxylase. *Brain Res.* 185, 39–49.

Holstege, J. C. and Kuypers, H. G. J. M. (1982). Brain stem projections to spinal motoneuronal cell groups in rat studied by means of electron microscopy autoradiography. *Prog. Brain Res.* 57, 177–183.

Hosoya, Y. and Matsushita, M. (1979). Identification and distribution of the spinal and hypophyseal projection neurons in the paraventricular nucleus of the rat. A light and electron microscopic study with the horseradish peroxidase method. *Exp. Brain Res.* 35, 315–331.

Huisman, A. M., Kuypers, H. G. J. M. and Verburgh, C. A. (1981). Quantitative differences in collateralization of the descending spinal pathways from red nucleus and other brain stem cell groups in rat as demonstrated with the multiple fluorescent retrograde tracer technique. *Brain Res.* 209, 271–286.

Huisman, A. M., Kuypers, H. G. J. M., Conde, F. and Keizer, K. (1983). Collaterals of rubrospinal neurons to the cerebellum in rat: A retrograde fluorescent double labeling study. *Brain Res.* 264, 181–196.

Kevetter, G. A. and Willis, W. D. (1983). Collaterals of spinothalamic cells in the rat. *J. Comp. Neurol.* 215, 453–464.

Kuypers, H. G. J. M. (1981). Anatomy of the descending pathways. In V. B. Brooks (Ed.). *Handbook of physiology. Section 1: The nervous system*, Vol. 2: Motor control, part 1, pp. 597–666. American Physiological Society, Bethesda.

Kuypers, H. G. J. M. and Martin, G. F. (Eds). (1982). Anatomy of descending pathways to the spinal cord. *Prog. Brain Res.* 57.

Land, L. J. (1977). The spinothalamic tract in the rat: Independent projections to the ventrobasal complex and the posterior thalamic region. *Anat. Rec.* 187, 634.

Langford, L. A. and Coggeshall, R. E. (1981). Unmyelinated axons in the posterior funiculi. *Science* 211, 176–177.

Leichnetz, G. R., Watkins, L., Griffin, G., Murfin, R. and Mayer, D. J. (1978). The projection from nucleus raphe magnus and other brainstem nuclei to the spinal cord in the rat: A study using the HRP blue-reaction. *Neurosci. Lett.* 8, 119–124.

Liu, R. P. C. (1983). Laminar origins of spinal projection neurons to the periaqueductal gray of the rat. *Brain Res.* 264, 118–122.

Lovick, T. A. and Robinson, J. P. (1983). Ascending and descending projections from nucleus raphe magnus in the rat. *J. Physiol. (London)* 338, 13P.

Low, J. S. T. and Tracey, D. J. (1984). Spinal afferents to nucleus z are collaterals of dorsal spinocerebellar tract neurones. *Proc. Aust. Physiol. Pharmacol. Soc.* 15, 78P.

Lund, R. D. and Webster, K. E. (1967). Thalamic afferents from the spinal cord and trigeminal nuclei: An experimental anatomical study in the rat. *J. Comp. Neurol.* 130, 313–328.

Matsushita, M. and Hosoya, Y. (1979). Cells of origin of the spinocerebellar tract in the rat, studied with the method of retrograde transport of horseradish peroxidase. *Brain Res.* 173, 185–200.

Mehler, W. R. (1969). Some neurological species differences—*a posteriori*. *Ann. N.Y. Acad. Sci.* 167, 424–468.
Menétrey, D., Chaouch, A. and Besson, J.M. (1980). Location and properties of dorsal horn neurons at origin of spinoreticular tract in lumbar enlargement of the rat. *J. Neurophysiol.* 44, 862–877.
Menétrey, D., Chaouch, A., Binder, D. and Besson, J. M. (1982). The origin of the spinomesencephalic tract in the rat: An anatomical study using the retrograde transport of horseradish peroxidase. *J. Comp. Neurol.* 206, 193–207.
Morrell, J. I. and Pfaff, D. W. (1983). Retrograde HRP identification of neurons in the rhombencephalon and spinal cord of the rat that project to the dorsal mesencephalon. *Am. J. Anat.* 167, 229–240.
Motorina, M. V. (1977). Distribution of reticulospinal fibers and their terminations in lumbar segments of the rat spinal cord. *J. Evol. Biochem. Physiol.* 12, 520–527.
Murray, H. M. and Gurule, M. E. (1979). Origin of the rubrospinal tract of the rat. *Neurosci. Lett.* 14, 19–23.
Oscarsson, O. (1973). Functional organization of spinocerebellar paths. *In* A. Iggo (ed.), *Handbook of sensory physiology*, Vol. 2, pp. 339–380. Springer, New York.
Ruggiero, D. A., Ross, C. A. and Reis, D. J. (1981). Projections from the spinal trigeminal nucleus to the entire length of the spinal cord in the rat. *Brain Res.* 225, 225–233.
Saper, C. B., Loewy, A. D., Swanson, L. W. and Cowan, W. M. (1976). Direct hypothalamo-autonomic connections. *Brain Res.* 117, 305–312.
Satoh, K. (1979). The origin of reticulospinal fibers in the rat: A HRP study. *J. für Hirnforsch.* 20, 313–332.
Shieh, J. Y., Leong, S. K. and Wong, W. C. (1983). Origin of the rubrospinal tract in neonatal, developing, and mature rats. *J. Comp. Neurol.* 214, 79–86.
Snyder, R. L., Faull, R. L. M. and Mehler, W. R. (1978). A comparative study of the neurons of origin of the spinocerebellar afferents in the rat, cat and squirrel monkey based on the retrograde transport of horseradish peroxidase. *J. Comp. Neurol.* 181, 833–852.
Steinbusch, H. W. M. (1981). Distribution of serotinin-immunoreactivity in the central nervous system of the rat—cell bodies and terminals. *Neurosci.* 6, 557–618.
Vahlsing, H. L. and Feringa, E. R. (1980). A ventral uncrossed corticospinal tract in the rat. *Exp. Neurol.* 70, 282–287.
Veazey, R. B. and Severin, C. M. (1980). Efferent projections of the deep mesencephalic nucleus (pars medialis) in the rat. *J. Comp. Neurol.* 190, 245–258.
Veazey, R. B. and Severin, C. M. (1982). Afferent projections to the deep mesencephalic nucleus in the rat. *J. Comp. Neurol.* 204, 134–150.
Waldron, H. A. and Gwyn, D. G. (1969). Descending nerve tracts in the spinal cord of the rat. I. Fibers from the midbrain. *J. Comp. Neurol.* 137, 143–154.
Westlund, K. N., Bowker, R. M., Ziegler, M. G. and Coulter, J. D. (1982). Descending noradrenergic projections and their spinal terminations. *Prog. Brain Res.* 57, 219–238.
Wiksten, B. (1979). The central cervical nucleus in the cat. II. The cerebellar connections studied with retrograde transport of horseradish peroxidase. *Exp. Brain Res.* 36, 155–173.
Willis, W. D. and Coggeshall, R. E. (1978). *Sensory mechanisms of the spinal cord.* Wiley, New York.
Wise, S. P. and Jones, E. G. (1977). Cells of origin and terminal distribution of descending projections of the rat somatic sensory cortex. *J. Comp. Neurol.* 175, 129–158.
Yezierski, R. P. and Bowker, R. M. (1981). A retrograde double label tracing technique using horseradish peroxidase and the fluorescent dye 4′,6-diamidino-2-phenylindole 2HCl (DAPI). *J. Neurosci. Meth.* 4, 53–62.
Zemlan, F. P. and Pfaff, D. W. (1979). Topographical organization in medullary reticulospinal systems as demonstrated by the horseradish peroxidase technique. *Brain Res.* 174, 161–166.
Zemlan, F. P., Leonard, C. M., Kow, L. -M. and Pfaff, D. W. (1978). Ascending tracts of the lateral columns of the rat spinal cord: A study using the silver impregnation and horseradish peroxidase techniques. *Exp. Neurol.* 62, 298–334.

15

Autonomic nervous system

GIORGIO GABELLA

University College London
London, UK

I. DISTRIBUTION OF AUTONOMIC GANGLIA

A. General organization

The peripheral autonomic nervous system of the rat consists of a vast array of nerves and ganglia, connected to the spinal cord and the brain stem on the one side and to the viscera on the other. The latter include the organs of the abdominal and thoracic cavities, the blood vessels, and various organs of the head and tissues of the skin. The autonomic nervous system has a common construction in all mammals, the rat being a species which has been extensively studied in this respect.

Many autonomic ganglia can be recognized with the naked eye; they are situated along nerve trunks and appear as fusiform swellings or as protrusions. Certain ganglia are connected in regular sequences or ganglionated chains (the paravertebral sympathetic chain), while others are linked together by a web of nerve trunks and constitute a plexus (for example, the abdominal plexus or pelvic plexus). There are also innumerable ganglia that are microscopic and are buried within a nerve or inside the wall of viscera.

The autonomic nervous system is organized in several arrays of ganglia that can be schematically subdivided, as shown in Fig. 1, into paravertebral ganglia, prevertebral ganglia, paravisceral ganglia and intramural ganglia. A further group of ganglia is located in the head and is related mainly to the salivary glands and the eye. The paravertebral ganglia form two chains (sympathetic chains) on either side of the vertebral column and are directly connected to the thoracolumbar spinal cord by the rami communicantes. The prevertebral ganglia are assembled into a large plexus (the abdominal plexus, which includes the celiac ganglion and the mesenteric ganglia) located close to the abdominal aorta. The paravisceral ganglia lie in the proximity of the viscera; the main groups are

THE RAT NERVOUS SYSTEM
ISBN 0 12 547632 9

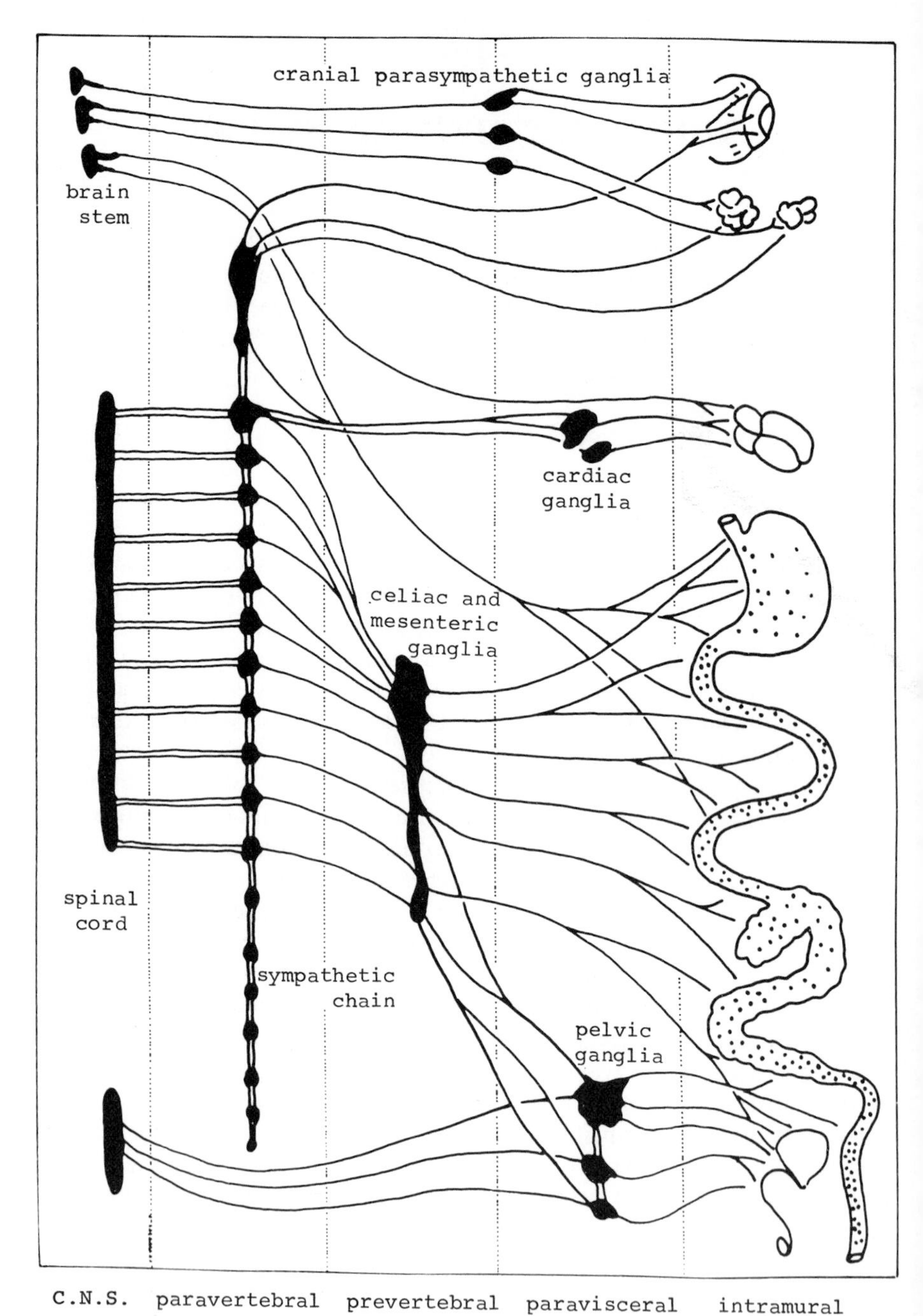

Fig. 1: Representation of the main groups of autonomic ganglia.

in the cardiac plexus and in the pelvic plexus, and there are also small ganglia in the plexus close to the trachea and bronchi. Finally, the intramural ganglia, which can be recognized only microscopically, are located within the wall of the gastrointestinal tract and of the biliary pathways.

The paravertebral and prevertebral ganglia are part of the sympathetic system, or the sympathetic outflow, which emerges from the thoracic and lumbar levels of the spinal cord. The autonomic ganglia of the head are part of the parasympathetic system, or of its cranial portion, which includes also the vagus nerve. The sacral parasympathetic outflow emerges from the sacral segments of the spinal cord and is known as the pelvic nerves. The cardiac and pelvic paravisceral ganglia are mainly situated along parasympathetic outflows; however, they also receive, especially the pelvic ganglia, an input from the sympathetic outflow. The intramural ganglia of the gut cannot be regarded as either sympathetic or parasympathetic; although connected with both these outflows, they are in a class apart, which is referred to as the enteric nervous system (Langley, 1921). The viscera have a plentiful afferent innervation, and if these fibers are regarded as part of the autonomic nervous system, as it seems they should, they cannot be included in either the sympathetic or the parasympathetic system.

B. Sympathetic chains

The sympathetic chain is a bilaterally symmetric structure extending from the base of the skull to the sacrum. In the neck the cervical sympathetic chain lies dorsal to the vagus nerve and the common carotid artery, and ventral to the transverse processes and the prevertebral muscles. It has two prominent ganglia, the superior cervical ganglion and the inferior cervical ganglion (stellate ganglion); sometimes a middle ganglion is present (Baljet and Drukker, 1979; Hedger and Webber, 1976) (Fig. 2). The superior cervical ganglion is spindle shaped, some 5 mm in length (often with a constriction in the middle), and lies dorsal to the bifurcation of the carotid artery. Among its (postganglionic) nerves, a carotid branch leaves the cranial pole of the ganglion and follows the internal carotid artery, while other smaller branches form a plexus around the external carotid artery. Other constant branches (rami) can be traced to the carotid body, to the cranial nerves 9–12 and to the cervical spinal nerves 1–4 (Hedger and Webber, 1976). The stellate ganglion consists of the inferior cervical ganglion and the first two or three thoracic ganglia fused together, and is located at the level of the first two thoracic vertebrae, on the right side, being medial to the innominate artery (Hedger and Webber, 1976) (Fig. 2). Branches (rami) from the stellate ganglion join the lowermost cranial and the uppermost thoracic spinal nerves. Other branches connect with a plexus around the vertebral artery and with a plexus on the ventral aspect of the arch of the aorta (Hedger and Webber, 1976).

In the thorax, the sympathetic chains lie ventral to the head of the ribs and

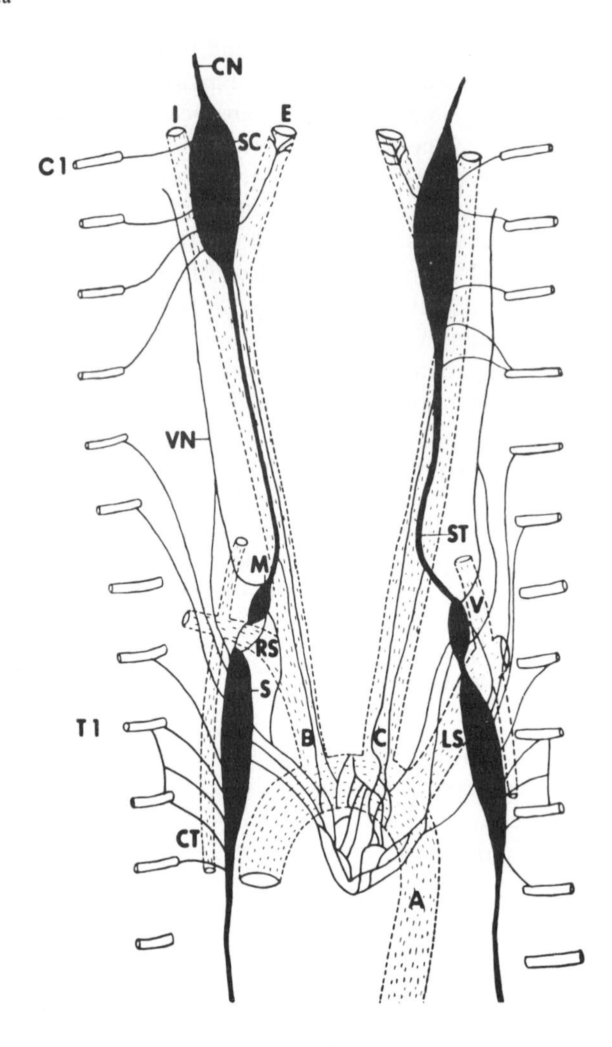

Fig. 2: The cervical sympathetic trunk and its branches in the rat. A, aorta; B, brachiocephalic trunk; C, left common carotid artery; CN, carotid nerve; CT, costocervical trunk; C1, first cervical spinal nerve; I, internal carotid artery; LS, left subclavian artery; S, stellate ganglion; SC, superior cervical ganglion; ST,sympathetic trunk; V, vertebral artery; VN, vertebral nerve; E, external carotid artery; M, middle (intermediate) cervical ganglion; RS, right subclavian artery; T1, first thoracic spinal nerve. It should be noted that cervical sympathetic trunk and superior cervical ganglion are located dorsal to the carotid artery (reproduced with permission from Hedger and Webber, 1975).

dorsal to the parietal pleura (Fig. 3). Each chain is made up of 10 ganglia: the first three are usually fused with one another and with the inferior cervical ganglion. The lowermost ganglion lies opposite the 10th intercostal space. Small horizontal nerve trunks connect the two chains across the midline (Baljet and Drukker, 1979). In addition to the branches to the spinal nerves (rami communi-

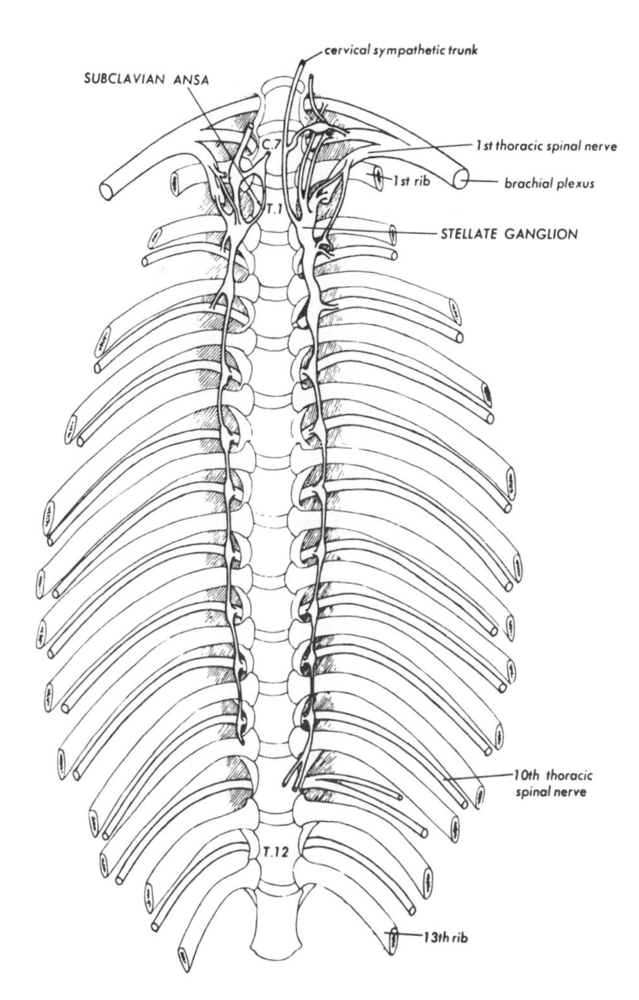

Fig. 3: Lower cervical and thoracic sympathetic trunks of the rat (reproduced with permission from De Lemos and Pick, 1966).

cantes) and to small branches to the blood vessels, the thoracic sympathetic chains issue the splanchnic nerves (see below).

In the abdomen the sympathetic chain is retroperitoneal and is embedded in the psoas muscle. There are five or six pairs of ganglia with rami communicantes to the spinal nerves, branches to blood vessels and small branches to the abdominal plexus (lumbar splanchnic nerves).

C. Rami communicantes

Rami communicantes are short nerve trunks connecting the ganglia of the sympathetic chain to the spinal nerves. The rami are particularly short in the rat (De Lemos and Pick, 1966), and even when they are separated into two or more bundles they cannot be recognized as white and gray rami. Preganglionic and postganglionic fibers are therefore mixed within each ramus, and, since they have no distinctive structural features, their identification can be made only after experiments of selective nerve sections.

D. Splanchnic nerves

Descending branches from the thoracic sympathetic ganglia give rise to the major splanchnic nerve (and occasionally to a minor splanchnic nerve). Its apparent origin is usually from the ninth and the tenth ganglia (Baljet and Drukker, 1979). These nerves (right and left) enter the abdominal cavity, by piercing through the diaphragmatic crus, and end in the abdominal plexus. There is often a para-aortic nerve; this originates in the lowermost thoracic ganglia, enters the abdominal cavity traveling along the aorta, and terminates in the abdominal plexus. The lumbar splanchnic nerves are variable in number (usually four or five), size and origin, and extend from the lumbar sympathetic ganglia to the prevertebral plexus (Baljet and Drukker, 1979).

E. Prevertebral ganglia

The prevertebral ganglia constitute the abdominal plexus, a large assembly of ganglia and nerve trunks lying close to the abdominal aorta and its main branches (Fig. 4). Blood vessels are the chief guide to the identification of prevertebral ganglia. Two sets of vessels originate from the abdominal aorta (Greene, 1935): parietal arteries (inferior phrenic arteries, lumbar arteries, ileolumbar arteries, middle caudal artery and terminal trunk, and common iliac artery) and visceral arteries (celiac artery, superior mesenteric artery, inferior mesenteric artery, renal arteries and ovarian/testicular arteries). The inferior mesenteric artery is often a branch of the right common iliac artery (Baljet and Drukker, 1979).

Two major components can be distinguished in the abdominal plexus, the celiac plexus and the inferior mesenteric plexus, an intermesenteric plexus being interposed between them. The celiac plexus is situated around and between the celiac artery and the superior mesenteric artery and extends dorsally between the adrenal glands and between the cranial half of the kidneys. The left celiac ganglion is crescent shaped and lies on the lateral side of the celiac and superior mesenteric artery; the right celiac ganglion is triangular, smaller than the left one, and lies on the opposite side of the same arteries, dorsal to the inferior vena cava. Both celiac ganglia extend caudally, without distinct boundaries, into the superior mesenteric ganglia. A dorsal extension of the celiac ganglion forms the

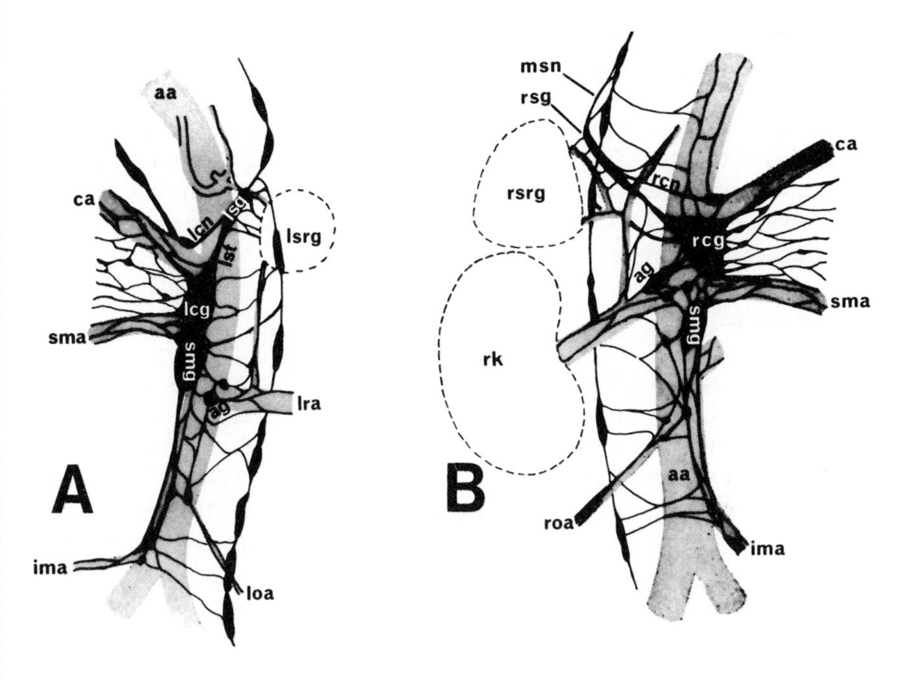

Fig. 4: A left lateral representation (A) and a right lateral representation (B) of the celiac–superior mesenteric ganglion complex of the rat, as derived from serial sections. aa, abdominal aorta; ag, aorticorenal ganglion; ca, celiac artery; cg, celiac ganglion; cn, celiac nerve; ima, inferior mesenteric artery; ipa, inferior phrenic artery; ipv, inferior phrenic vein; isv, inferior suprarenal vein; k, kidney; msn, major splanchnic nerve; oa, ovarian artery; ra, renal artery; sg, suprarenal ganglion; sma, superior mesenteric artery; smg, superior mesenteric ganglion; srg, suprarenal gland; st, splanchnic trunk (reproduced with permission from Hamer and Santer, 1981).

aorticorenal ganglion (Hamer and Santer, 1981) or ganglia; these too are more developed on the left than on the right side (Baljet and Drukker, 1979). Along the terminal part of the major splanchnic nerve, shortly before it joins the celiac ganglion, there is a small ganglion called the suprarenal ganglion (Baljet and Drukker, 1979). Innumerable nerve trunks contribute to, and issue from, the celiac plexus. In addition to the thoracic and abdominal splanchnic nerves, there are nerve trunks (which include smaller or microscopic ganglia) within the plexus itself, including many nerves lying across the midline both ventral and dorsal to the aorta. Nerve trunks emerging from the celiac plexus and directed to abdominal organs reach the suprarenal arteries, the celiac artery, the superior mesenteric artery, the renal arteries and the inferior phrenic arteries. Caudally, the celiac plexus continues into the intermesenteric plexus, an array of nerve trunks and very small ganglia lying on the ventral and lateral aspects of the aorta. The inferior mesenteric plexuses of the two sides are extensively interconnected and lie around the initial segment of the inferior mesenteric artery.

The inferior mesenteric ganglion, which is also called hypogastric ganglion (Langworthy, 1965), is a spindle shaped expansion along the main nerve trunk of the plexus. Caudally, the continuation of the inferior mesenteric ganglion is the hypogastric nerve, a bilaterally symmetric nerve which reaches the pelvic plexus. Nerve branches from the intermesenteric plexus can be followed to the kidneys, the ovaries and the uterus or the testis. The main branches from the inferior mesenteric plexus, apart from the hypogastric nerve, are directed to the periphery along the inferior mesenteric artery.

F. Pelvic plexus

The pelvic plexus is a large and elaborate crossroad of nerves and ganglia supplying the rectum, the lower urinary tract, and the genital tract. The anatomy of this plexus, which is somewhat less complex in the rat than in other species, has been thoroughly investigated by Langworthy (1965), who used microdissection after vital staining with methylene blue (Fig. 5). Other valuable data are found in Puriton *et al.* (1973) and in Hulsebosch and Coggeshall (1982).

In the male rat, the main component of the plexus is a single, large, bilaterally symmetric ganglion, the right and left pelvic ganglion, sometimes referred to as the hypogastric ganglion (for example, by Bentley, 1972; Sjöstrand, 1965). It is relatively flat, measuring about 2 mm × 4 mm, and it lies on the side of the prostate, closely apposed to its fascia, ventral to the rectum and caudal to the ureter and vas deferens. It shows lobulations which protrude in the direction of the surrounding organs. The ganglion is accompanied by a few, small, accessory ganglia mainly related to the seminal vesicles and the vas deferens.

The main incoming nerve trunks to the ganglion are the hypogastric and the pelvic nerves. The hypogastric nerve, carrying the bulk of the sympathetic input, originates as the caudal continuation of the inferior mesenteric ganglion and reaches the cranial pole of the pelvic ganglion. It is retroperitoneal, medial to the external iliac artery, and passes behind the ureter, where it branches into the main and accessory hypogastric nerves. The pelvic nerve, carrying the parasympathetic input to the pelvic ganglion, originates from the last lumbar (L6) and first sacral (S1) spinal nerves (Puriton *et al.*, 1973). It consists of five to seven fascicles (Hulsebosch and Coggeshall, 1982) which accompany the urogenital artery (Puriton *et al.*, 1973) or pelvic artery (Hulsebosch and Coggeshall, 1982), traveling mainly ventrally and reaching the dorsolateral aspect of the ganglion.

Numerous small efferent nerve trunks arise from the ganglion and reach the rectum, the ureter, the vas deferens, the seminal vesicle, the prostate, the bladder and the urethra. The largest trunk is the one that supplies the urethra and then proceeds to innervate the penis. Eight or more small nerves are directed to the bladder; they divide into two groups passing in front and behind the ureter and reaching the ventral and the dorsal surface of the bladder. Minute ganglia can

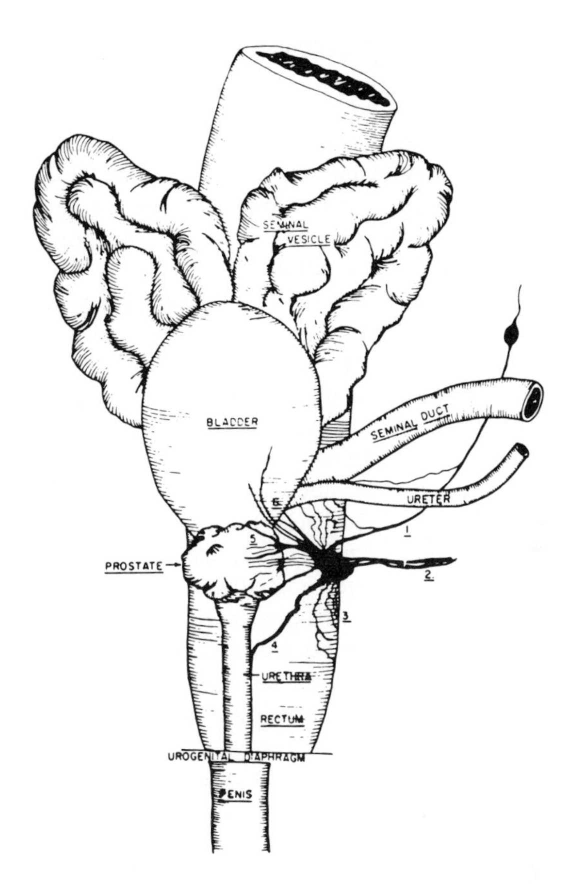

Fig. 5: Schematic representation of the pelvic ganglion in the male rat. 1, hypogastric nerve; 2, pelvic nerves; 3, branches to the rectum; 4, large trunk innervating urethra and penis; 5–7, branches to prostrate, bladder and ureter (reproduced from Langworthy, 1965).

be seen along these nerves and those supplying the vas deferens and seminal vesicle.

In the female rat, the pelvic ganglion, which is also called the paracervical or Fankenhauser ganglion (see Kanerva, 1972; Marshall, 1970), is smaller and more difficult to expose than in the male (Puriton *et al.*, 1973). Its largest branch, arising from the caudal portion of the ganglion, innervates the clitoris after giving branches to the urethra, vagina and rectum. Other fine nerves issuing from the ganglion run to the bladder, the cervix of the uterus, and the upper portion of the vagina. A few small accessory ganglia can be found, usually on the ventrolateral wall of the vagina near the bladder neck (Puriton *et al.*, 1973). Extensive decussation of fibers on the midline occurs over the ventral surface of the cervix.

G. Parasympathetic nerves and ganglia of the head

These ganglia are small or microscopic and are associated with branches of certain cranial nerves.

The ciliary ganglion lies lateral to the optic nerve and is attached to the initial part of the branch of the oculomotor nerve to the medial rectus muscle. Malmfors and Nilsson (1966) have described the exact position of the ganglion and the surgical approach for excision.

The otic ganglion is associated with the glossopharyngeal nerve; its postganglionic nerves reach the parotid gland and the oral mucosa.

The pterygopalatine ganglion is associated with the facial nerve; its postganglionic nerves reach the lacrimal gland and the nasal mucosa.

The submandibular ganglion is a collection of minute ganglia located around the excretory ducts of the submandibular and the sublingual glands, in the connective tissue between these ducts and the lingual nerve, and occasionally within the submandibular gland itself (Snell, 1958).

The vagus nerve emerges from the cranial cavity through the jugular foramen. A prominent swelling of the nerve immediately after its emergence is known as the nodose ganglion; from this two branches originate, the cranial or pharyngeal one forming a plexus with branches from the glossopharyngeal nerve, and the caudal one giving rise to the superior laryngeal nerve (Greene, 1935). From the nodose ganglion the main trunk of the vagus nerve runs caudally into the neck. The main cervical branches are the cardiac branches and the recurrent nerve, which end as the inferior laryngeal nerve; the thoracic vagus issues esophageal and pulmonary branches. The nerve reaches the abdominal cavity and ends spreading over the anterior (left vagus) and the posterior (right vagus) surface of the stomach.

II. STRUCTURE OF AUTONOMIC GANGLIA AND NERVES

A. Preganglionic neurons

The sympathetic preganglionic neurons are located in the spinal cord between the cervical 8 and lumbar 2 levels (C8 and L2). The distribution and the morphology of preganglionic neurons has been studied to great advantage with the horseradish peroxidase (HRP) technique (retrograde filling). This has produced results that are more reliable than those obtained by degeneration methods and silver impregnation methods. The majority of the preganglionic neurons are in a column called the intermediolateral nucleus. A similar column in two or three sacral levels of the spinal cord contains the preganglionic neurons of the sacral parasympathetic outflow. The preganglionic neurons innervating the superior cervical ganglion of the rat are located in segments C8–T5 (90% of them in thoracic 1–3 (T1–T3). Three quarters of the total of 1600 neurons retrogradely filled from the cervical sympathetic trunk of one side of the animal are in the intermediolateral nucleus, 23% in the lateral funiculus and the remainder in the central autonomic area and the intercalated region (Rando *et al.*, 1981). Injected neurons are exclusively homolateral to the injected cervical trunk (Rando *et al.*, 1981). In contrast, Navaratnam and Lewis (1970) found chromatolytic neurons on both sides of the spinal cord after unilateral section of the pelvic nerves.

An additional column of preganglionic neurons, termed the dorsal commissural nucleus, projecting into the hypogastric nerve has been identified in the spinal cord levels L1–L2 (Hancock and Peveto, 1979). Schramm and collaborators (1975) reported labeling of preganglionic neurons at all the levels between T1 and L1 after injection of horseradish peroxidase into the adrenal medulla. Approximately 1000 neurons were labeled after injection into one gland (Schramm *et al.*, 1976). They also reported that the neurons have a marked longitudinal orientation, as has been observed in other species, with the dendrites grouped into two bundles directed cranially and caudally. This high polarization of the dendrites arises rather late in development; it is absent in three-week old rats and is still far from complete in seven-week old rats (Schramm *et al.*, 1976).

Preganglionic neurons in the rat spinal cord display an intense acetylcholinesterase activity (Navaratnam and Lewis, 1970). Nerve endings containing noradrenaline or serotonin are present around the cell bodies (Dahlström and Fuxe, 1965).

B. Preganglionic fibers

Sympathetic preganglionic axons originate in the spinal cord from neurons located mainly in the intermediolateral column (thoracic and lumbar levels). The fibers emerge from the spinal cord within the ventral roots (T1–L2) together with the somatic motor fibers. From the ventral nerves the preganglionic fibers pass to the sympathetic chain via very short connections (rami communicantes). The latter also contain postganglionic fibers which travel to the periphery within somatic nerves. Depending on the level of origin, the preganglionic fibers travel some distance up or down the sympathetic chain forming synaptic contacts with ganglion cells in more than one ganglion. In the lumbar segment of the chain the preganglionic fibers are mainly descending. The length of the preganglionic fibers, therefore, can be considerable. In the upper thoracic segment they are mainly ascending and in the cervical sympathetic trunk all the preganglionic fibers are directed cranially. In this trunk, however, there are also caudally directed postganglionic fibers originating in the superior cervical ganglion and cranially directed postganglionic fibers originating in the middle and lower

Table 1: Number of ganglion neurons in the superior cervical ganglion of the rat

Mean number of neurons ± standard deviation (number of cases)	Study
45 000 ± 600	Davies, 1978
35 000 ± 600	Davies, 1978
39 000 ± 500 (4)	Klingman, 1972
37 000–38 000 (2)	Hedger and Webber, 1976
36 000	Ostberg *et al.*, 1976
32 000 (1)	Levi-Montalcini and Booker, 1960
26 000–32 000 (6)	Eränkö and Soinila, 1981
25 000 ± 1940 (3)	Johnson *et al.*, 1980
15 600 ± 6100 (17)	Brooks-Fournier and Coggeshall, 1981

cervical ganglia (Bowers and Zigmond, 1981). Other sympathetic preganglionic fibers, having reached the paravertebral chain, pass into a splanchnic nerve and travel to prevertebral ganglia in the abdominal cavity and, in a smaller number, as far as the pelvic ganglion.

In the rat, unlike other species, such as humans and the cat, the great majority of preganglionic fibers are unmyelinated. For example, less than 1% of the axons in the cervical sympathetic trunk are myelinated (Brooks-Fournier and Coggeshall, 1981; Dyck and Hopkins, 1972; Hedger and Webber, 1976).

The terminal branches of the preganglionic fibers have varicosities and terminal boutons which synapse on ganglion neurons. In the ganglionic relay there is divergence, that is, each preganglionic fiber innervates several ganglion neurons. The numeric ratio between preganglionic and postganglionic neurons as calculated by different investigators, varies from 1:5 to 1:20. The total number of neurons in the spinal cord that are retrogradely filled with horseradish peroxidase from the cervical sympathetic trunk is about 1600 (Rando *et al.*, 1981), a value that should be compared with the number of neurons they innervate, that is the neurons in the superior cervical ganglion (see Table 1). The divergence accounts for the fact that all ganglion neurons are innervated by preganglionic fibers, in spite of the smaller number of these fibers. But, in addition, each ganglion neuron receives synapses from more than one pregang-lionic neuron, that is, there is convergence of synaptic inputs onto each neuron. The latter process is ideally analyzed with electrophysiologic techniques. Typical values obtained in rabbits (Wallis and North, 1978), hamster (Lichtman and Purves, 1980), and guinea pigs (Nja and Purves, 1977) range between 6 and 11 preganglionic fibers per ganglion neuron; moreover each ganglion neuron receives an input from several levels of the spinal cord (Nja and Purves, 1977).

Parasympathetic preganglionic fibers originate in nuclei of the brain stem (nucleus of Edinger–Westphal, salivatory nuclei, vagal dorsal nucleus) and from a short column at levels L6–S1 of the spinal cord.

C. Sympathetic ganglia

The sympathetic ganglia are scattered along the sympathetic chain (paravertebral ganglia) and in the abdominal plexus (prevertebral ganglia). The best known of them is the superior cervical ganglion (Fig. 2). Because of its large size, its accessibility, its vascular supply and the layout of its preganglion and postganglionic nerves, it has been investigated more extensively than any other ganglion, the rat being one of the species of choice. Many of the structural features of the superior cervical ganglion are reproduced in the other sympathetic ganglia although important differences are being found with more detailed studies, especially between prevertebral and paravertebral ganglia.

In addition to ganglion nerve cells (principal ganglion neurons) (Fig. 6A), the sympathetic ganglia contain several other cell types. These include: small granular cells (or small intensely fluorescent cells); cells of glial type (Schwann cells and satellite cells); vascular cells (mainly endothelial cells); mast cells; and fibroblasts (in thin septa of connective tissue and in the capsule). The capsule, which is in continuation with the sheaths of the incoming and outgoing nerves, is relatively thick and offers a strong barrier to the diffusion of substances (Arvidson, 1979). However, substances (for example, horseradish peroxidase) injected systemically rapidly diffuse from blood vessels and in the intercellular space of sympathetic ganglia (Jacobs, 1977). By contrast, there is no extravasation of the injected tracer in sympathetic nerve trunks, and it has been suggested that at this level (but not within ganglia) there is a blood–nerve barrier to proteins (Jacobs, 1977). The majority of intraganglionic blood vessels are capillaries and some of them are fenestrated.

D. Principal ganglion neurons

The population of sympathetic neurons in the rat probably numbers a few hundred thousand, but accurate counts are available only for the superior cervical ganglion (Table 1). The wide range in the values published by different authors is probably accounted for by a certain amount of experimental error (see Hendry, 1976), but also by some variability between individual animals and possibly also between strains of rats. The variability in the number of neurons in the superior cervical ganglion in mammalian species in general is wide, ranging from about 4200 in a bat (Webber and Kallen, 1968) to nearly a million in a human (Ebbeson, 1968). The final number of neurons is established during fetal life, with few mitoses occurring in ganglion neurons of the rat after birth (Eränkö, 1972). From the first week after birth the number of principal cells remains unchanged (Davies, 1978) or shows a slight decrease (Eränkö and Soinila, 1981; Hendry and Campbell, 1976). Brooks-Fournier and Coggeshall (1981) have obtained by far the lowest counts (15 500 on average, in 17 ganglia) and have also reported large differences in neuron number between right and left ganglion (up to 80% more in one ganglion than in the contralateral one).

Principal ganglion neurons of the rat are multipolar neurons measuring up to 50 μm in diameter (mostly 25 μm–40 μm (Tamarind and Quilliam, 1971). In comparison with other species (especially those of larger body size), the dendritic arborization of sympathetic neurons in the rat is not extensive and the proportional volume of neuropil is relatively small (Fig. 6A). There is little or no evidence of substantial structural differences within the population of ganglion neurons. It is possible that the number and pattern of dendrites characterize neuronal ganglionic subpopulations (as suggested by Cajal, 1911); however, it has proven to be difficult to demonstrate by silver methods the dendritic arborizations in sympathetic ganglia, especially in the rat. More modern techniques, such as a combination of electrophysiology and HRP injection (as used in the ganglia of other species) may prove valuable for sorting out neuronal populations.

Ganglion neurons are individually ensheathed by satellite cells. This glial sheath is continuous over the neuronal soma, although in some areas it is reduced to a very thin cytoplasmic process interposed between the neuron on one side and the basal lamina and connective tissue on the other. A glial sheath extends over the dendrites. Here, however, there are areas where the neuronal cell membrane lies directly apposed to the basal lamina and connective tissue (Fig. 6C). In these dendritic regions there are large clusters of vesicles and, occasionally, membrane specializations similar to dense projections. Similar clusters of vesicles can also be found in the more superficial parts of the cytoplasm of the cell body and along the dendrites (Fig. 6B). Each neuron has an axon, which can be clearly recognized in silver impregnated preparations; the axon travels within the ganglion, often along a tortuous path but without dividing or giving off branches. Under the electron microscope, the axon can be recognized only at some distance from the cell body, and its exact point of origin is therefore not usually seen.

Only a limited topographic subdivision of ganglion neurons into groups is apparent in sympathetic ganglia. The question of the localization of these neurons in relation to the organ they innervate has been investigated through the

---

Fig. 6: Superior cervical ganglion of rat (from Gabella, 1976). A. Araldite section stained with toluidine blue showing ganglion neuron profiles (some nucleated) and the surrounding neuropil. The latter includes satellite cells, neurites, blood vessels and connective tissue × 800. B. Electron micrograph of a ganglion fixed by immersion in glutaraldehyde. A preganglionic nerve ending synpases on a cell process (dendrite) containing a mitochondrion, ribosomes, neurofilaments and a large cluster of electron-lucent vesicles. An ill-defined band of electron-dense material lies beneath the post-synaptic membrane × 56 000. C.Electron micrograph of a dendrite with a large number of vesicles lying directly underneath its surface. The cell membrane has some dense projections attached to it and is here devoid of a satellite cell sheath × 38 000. D. Fluorescence micrograph (by Dr Lars Olson) (Falck-Hillarp method for catecholamines). The nerve cell bodies and some of their processes show specific fluorescence of varying intensity × 270.

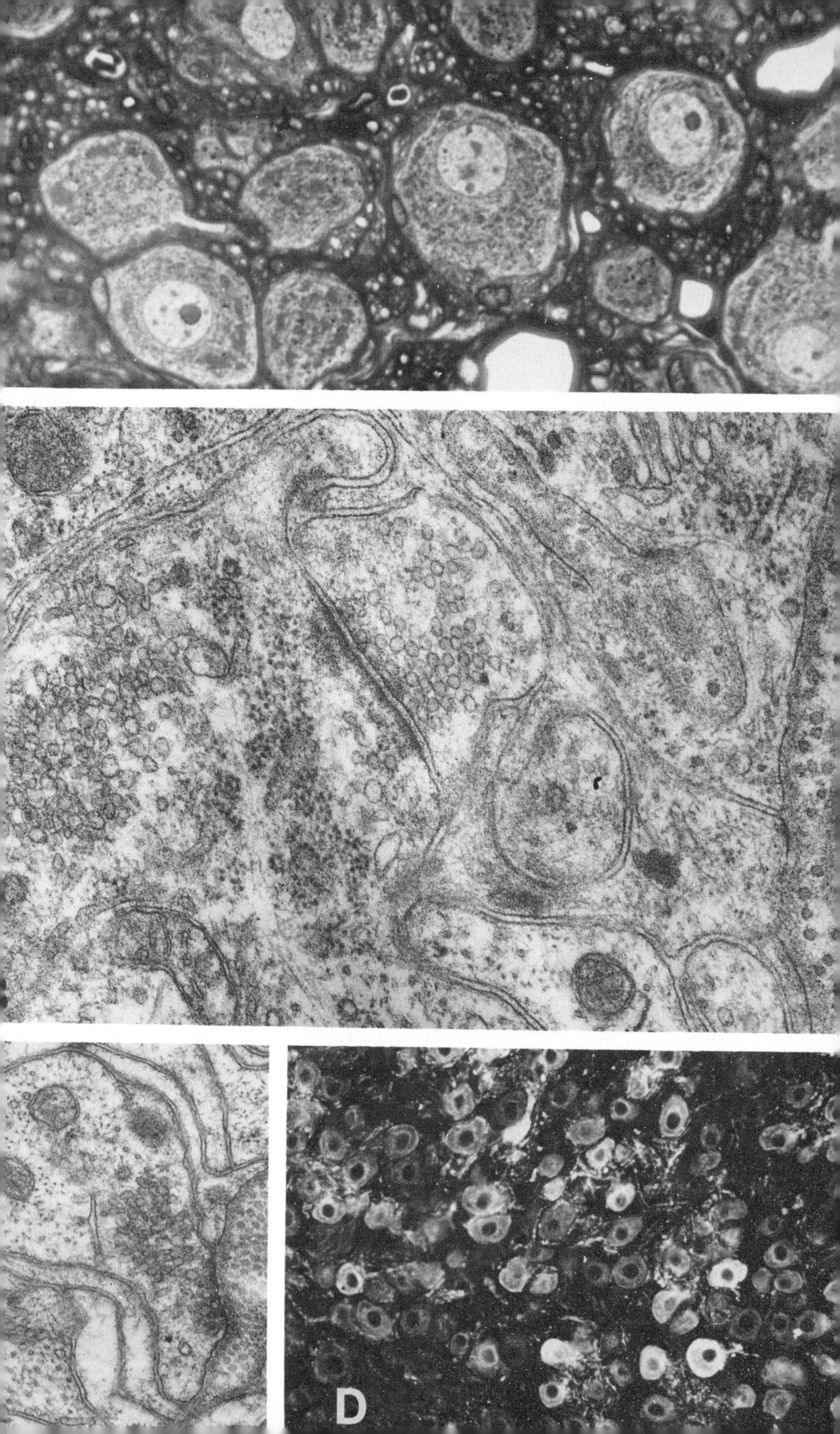
D

retrograde reaction of axotomized neurons (Matthews and Raisman, 1972), the retrograde transport of nerve growth factor (Hendry *et al.*, 1974) and HRP (Bowers and Zigmond, 1979), and the increased utilization of glucose in stimulated neurons (Yarowski *et al.*, 1979). Neurons tend to be located in the part of the ganglion near the site of emergence of their (postganglionic) fibers (Matthews and Raisman, 1972), but this localization is ill defined and in practice neurons projecting to a particular organ may be found in any part of the ganglion (Hendry *et al.*, 1974). However, the neurons whose axons extend in the internal carotid nerve are located mainly in the cranial part of the superior cervical ganglion; those projecting in the external carotid nerve are mainly located in the caudal portion of the ganglion, where there are also neurons sending their axon in the cervical trunk (Bowers and Zigmond, 1979).

E. Small intensely fluorescent cells

Small intensely fluorescent (SIF) cells are identified by their small size and their very intense formaldehyde induced fluorescence (Eränkö and Harkonen, 1965; Norberg *et al.*, 1966). Under the electron microscope their cytoplasm appears rich in large, dense core vesicles (hence the term small granule containing cells; Matthews and Raisman, 1969). These cells are part of the large group of chromaffin and chromaffin like cells (see a review in Taxi, 1979), which includes cells of paraganglia and the adrenal medulla, and cells scattered in many tissues outside the nervous system (for example the enterochromaffin cells).

The number of SIF cells is very variable even in the same ganglion. In the rat superior cervical ganglion up to 1000 (Santer *et al.*, 1975) or 370 (Williams *et al.*, 1977) SIF cells have been counted. Eränkö and Soinila (1981) found an average of about 450 SIF cells per ganglion, and observed a twofold increase in the number of SIF cells during the first three weeks of postnatal life. Therefore, there can be as many as three or four SIF cells for every 100 principal neurons, an unusually high ratio by comparison with other mammalian species. Systemic injection of hydrocortisone in newborn rats induces a dramatic increase in the number of SIF cells (Eränkö and Eränkö, 1972).

The SIF cells of the rat sympathetic ganglia measure 10μm–15 μm in diameter and many of them are grouped into clusters sheathed by satellite glial cells and with extensive membrane to membrane contact between adjacent SIF cells. There are occasional discontinuities in the thin glial sheath and at these points the surface of a SIF cell lies bare and is usually directly opposite to a fenestrated capillary. The cytoplasm contains a large number of dense cored vesicles, rather variable in size (but mostly in the range of 70 nm–120 nm), electron density, and shape; on the basis of these features of the granular vesicles two or three types of SIF cells have been identified (see Taxi, 1979). The SIF cells contain and probably release a biogenic amine; the type of amine varies from species to species, and even between different ganglia of the same species. In the rat superior cervical ganglion the SIF cells have been shown to contain mainly dopamine (Björklund *et al.*, 1970), but more recently it has been reported that in this ganglion there are separate groups of SIF cells, containing dopamine, or noradrenaline or serotonin (5-HT) (Konig, 1979; Verhofstad *et al.*, 1981).

Some cells display two or more processes; a few are tens of micrometers in length and ultrastructurally similar to the cell body. Other cells, especially those in clusters, have no processes. However, the classification of type I and type II SIF cells, on the basis of the occurrence of processes, is not easily applicable to the ganglia of the rat (and other rodents) (Williams *et al.*, 1977).

The SIF cells receive synapses from preganglionic fibers, predominantly on the cell body. A few SIF cells (and notably some in the rat superior cervical ganglion) form specialized contacts with principal ganglion neurons which are described as efferent synapses (Matthews and Raisman, 1969); these are usually somadendritic and large granular vesicles are grouped around dense projections on the presynaptic membrane. Those SIF cells that have afferent and efferent synapses can be regarded as interneurons (Williams, 1967). Many, and possibly the majority of SIF cells, have no efferent synapses, and these, if not all SIF cells, are considered to be endowed with an endocrine role, releasing substances (amines) in the interstitium or in the local circulation of a ganglion. The stimuli can be of central nervous system origin (via preganglionic fibers) or local (via chemoreceptors). A chemoreceptive role is suggested by the structural similarity of certain SIF cells to cells of the carotid body and the aortic glomus (Kondo, 1977). It is noteworthy that, according to Grillo (1978), about 10% of the synapses onto SIF cells in the superior cervical ganglion of the rat are not affected by preganglionic denervation; moreover, some efferent synapses of the SIF cells are not on ganglion neurons, but on large axons originating from the glossopharyngeal nerve and are probably afferent.

F. Nerve endings, synapses and other cell junctions

There are abundant nerve endings synapsing onto principal neurons of sympathetic ganglia (Fig. 6B). The synapses are usually on dendrites or on dendritic spine like processes, and only rarely on the cell soma (axosomatic synapses are common in ganglia of immature rats; Smolen and Raisman, 1980). The ultrastructural features of ganglionic synapses in rat sympathetic ganglia have been thoroughly investigated (see review by Matthews, 1983). Most of the intraganglionic endings are packed with small agranular vesicles (49 nm–60 nm in diameter), in addition to a few mitochondria, endoplasmic reticulum, and microtubules; large granular vesicles, although representing no more than a small proportion of the vesicle population, are usually well in evidence.

In the superior cervical ganglion, virtually all synaptic endings disappear after decentralization of the ganglion, and this confirms that they are of preganglionic origin. The synaptic cleft is rich in acetylcholinesterase (Somogyi and Chubb, 1976), acetylcholine is released (in quantal form) upon stimulation, and decentralization reduces acetylcholine levels by nearly 59% (Klingman and Klingman, 1969). Transmission across the ganglia is achieved by cholinergic synapses operating mainly through nicotinic receptors (see review by Skok, 1983).

There are also synapses from preganglionic fibers onto SIF cells, and occasionally from SIF cells onto a ganglion neuron. Numerous junctions of the adherens type (presumably of mechanical significance) occur between neurons and satellite cells and between neuronal elements. The possibility of dendro-dendritic synapses has been suggested by Kondo *et al.* (1980), who studied neurons of the superior cervical ganglion which had been injected intracellularly with HRP

G. Neurotransmitters and related substances

The great majority of neurons in sympathetic ganglia are adrenergic. Catecholamines are stored in the cell body, dendrites, axon and, in much higher concentration, in the varicosities of the terminal portion of the axon. The catecholamine content of the cell bodies, as detected histochemically by fluorescence microscopy, is variable from neuron to neuron and it tends to decrease with age (Santer, 1979). Ultrastructurally, the biogenic amines are localized in large dense cored vesicles, and in small dense cored vesicles clustered beneath the cell membrane, and in tubules of endoplasmic reticulum (Richards and Tranzer, 1975).

A small proportion of ganglion neurons in paravertebral ganglia (about 4% in the superior cervical ganglion; Yamauchi and Lever, 1971) are intensely positive for acetylcholinesterase and are negative for monoaminoxidase and catecholamines. They supply vasodilator cholinergic fibers to some blood vessels and secretomotor fibers to eccrine sweat glands (Langley, 1922; Wechsler and Fisher, 1968).

Sympathetic neurons (from the superior cervical ganglion) obtained from newborn rats and grown *in vitro* in certain conditions (which include the presence of non-neuronal cells; Patterson and Chung, 1977) undergo a transition from adrenergic to cholinergic (Furshpan *et al.*, 1976, 1982; Johnson *et al.*, 1976). A similar transition seems to occur *in vivo*: the ganglion neurons that innervate sweat glands are adrenergic in very young rats and only adrenergic fibers are found around the developing glands. By the end of the third week of age, however, the same neurons have become cholinergic and only cholinergic fibers are found within the glands (Landis and Keefe, 1983). However, these nerve fibers, which are, both functionally and histochemically, cholinergic, maintain a limited ability to take up and store catecholamines, and some of them remain able to synthesize small amounts of catecholamines (Landis and Keefe, 1983).

Neuropeptides are found within the rat sympathetic ganglia, although their amounts, as assessed by immunofluorescence, are lower than in other species, for example, the guinea pig. A few single fibers immunoreactive for substance P are found in the stellate and the superior cervical ganglion (Hökfelt *et al.*, 1977a) and a small number of fibers weakly immunoreactive for vasoactive intestinal polypeptide (VIP) occur in the superior cervical ganglion (Hökfelt *et al.*, 1977b). There are no positive cell bodies for either peptide. There are also a few fibers immunoreactive for enkephalin and somatostatin (Hökfelt *et al.*, 1977c): after colchicine treatment a positive reaction can be detected in some cell bodies.

I. Prevertebral ganglia

The prevertebral ganglia are in many respects structurally similar to the ganglia of the sympathetic chain. An important difference, established mainly in the guinea pig (Crowcroft and Szurszewski, 1971), is that the ganglion neurons receive inputs not only from the spinal cord via the splanchnic nerves, but also from neurons located in the wall of the gut and from neurons located in adjacent ganglia of the abdominal plexus.

Several neuropeptides are localized in nerve fibers in prevertebral ganglia, including substance P, VIP and enkephalin, although their occurrence is sparser in the rat than in the guinea pig (Schultzberg, 1983). Of particular interest is the localization of substance P fibers in some nerve endings abutting on ganglion neurons in the guinea pig. These are afferent fibers from the dorsal root ganglion cells and they innervate the viscera: in transit through the prevertebral ganglia they issue collateral branches which synapse on ganglion neurons (Matthews and Cuello, 1982). This arrangement allows a reflex involving direct spread of stimuli (for example, nociceptive stimuli) from afferent axons to efferent neurons.

Leranth and Ungvary (1980) have described the presence of several ultrastructural types of axons and have commented on the complexity of the synaptic connections in the rat prevertebral ganglia.

H. Pelvic ganglion

The main component of the ganglion are the principal neurons, measuring 20 μm–40 μm in diameter, sheathed by satellite cells. Both cell types are similar in appearance to those found in the abdominal ganglia (Dail *et al.*, 1975; Kanerva and Teräväinen, 1972). In addition, there are ganglion neurons with large vacuoles (vacuolated neurons). The vacuoles, measuring up to 20 μm in diameter, greatly enlarge the cell and displace the other components of the neuron, which is otherwise similar in structure and synaptic connections to the principal ganglion neurons. The significance of the vacuoles is obscure. The vacuolated neurons are about 0.8% of the neuronal population in the pelvic ganglion of pregnant rats and less than 0.2% in rats which are not pregnant (Lehmann and Stange, 1953). Vacuolated neurons are found also in the pelvic ganglion of male rats (Dail *et al.*, 1975) in the proportion of 0.8%–1.2% (Partanen *et al.*, 1979); they make their appearance around the seventh week of life, but they disappear in castrated animals (Partanen *et al.*, 1979). Small intensely fluorescent cells are also consistently found, usually in large numbers (Kanerva *et al.*, 1972).

Cholinergic and adrenergic neurons are both found in the pelvic ganglion. The adrenergic neurons (identified by formaldehyde induced fluorescence) are about one third of the neuronal population in the pelvic ganglion of the female rat (Kanerva *et al.*, 1972), whereas they constitute the majority of neurons in the male (Dail *et al.*, 1975). Cholinergic neurons (whose identification, based on an intense reaction for acetylcholinesterase, is less certain) are about one fifth of all the neurons. In the female rat cholinergic neurons are among the largest in the ganglion (Kanerva *et al.*, 1972), whereas in the male they are small (15 μm–25 μm) and are mainly found near the entrance of the pelvic nerve into the ganglion (Dail *et al.*, 1975). A large proportion of the small neurons contain VIP; the axons of these neurons (VIPergic fibers) are plentiful in the smooth musculature of the penis and in the helicine arteries (Dail *et al.*, 1983). It has been shown, in some autonomic ganglia of the cat that VIP neurons are often acetylcholinesterase positive (Lundberg, 1981).

Most of the ganglion neurons (including the vacuolated cells) receive synapses, of cholinergic type, from preganglionic fibers: unlike the situation in paravertebral ganglia, the nerve endings mainly abut the soma or somatic spines (Kanerva and Teräväinen, 1972). Some endings are found tunneling deep inside a perikaryon. Other synapsing nerve endings are not readily identified as cholinergic in that they contain a vast number of larger granular vesicles (Kanerva and Teräväinen, 1972).

Some of the cholinergic neurons are surrounded by adrenergic terminals (as seen in fluorescence microscopy), which are regarded as collaterals of adrenergic neurons in the same ganglion (Dail *et al.*, 1975). Adrenergic varicosities abutting onto adrenergic neurons have also been observed by means of fluorescence microscopy (Dail *et al.*, 1975). The origin of these structures remains uncertain: they could be collaterals from other ganglion cells, or processes from SIF cells, or short dendritic processes from the ganglion cell itself. Ultrastructural evidence of adrenergic endings synapsing on pelvic ganglion neurons has been found in the guinea pig (Watanabe, 1971); it is possible that the same situation occurs in the rat.

There is good physiologic evidence of reflex activity mediated through the pelvic ganglion in rats of either sex, without relay to dorsal root ganglia (Puriton *et al.*, 1971). This has led to the suggestion that there are peripheral afferent neurons ("sensory perikarya"), and selective denervation experiments have shown that these neurons are located distal to the pelvic and the hypogastric nerves and are probably situated in the pelvic ganglion itself (Puriton *et al.*, 1971).

The pelvic ganglion has two separate preganglionic inputs: sympathetic cholinergic fibers originating from the lumbar levels of the spinal cord (and reaching the ganglion via rami communicantes, lumbar splanchnic nerves and the hypogastric nerve); and parasympathetic cholinergic fibers originating in the spinal cord at the L6 and S1 levels (and reaching the ganglion via the pelvic nerve). This traditional notion is confirmed by the recent electron microscopy study of Hulsebosch and Coggeshall (1982), which, however, has shown an unexpected complexity in the nerve pathways connected to the pelvic ganglion. Thus, the hypogastric nerve is made up of about 1600 axons, of which 58% are sympathetic postganglionic, 34% are sympathetic preganglionic and 8% are sensory. The pelvic nerve is made up of nearly 5000 axons, of which 34% are sensory and 49% are parasympathetic preganglionic; the remaining 17% are sympathetic postganglionic axons. Sympathetic fibers (preganglionic and postganglionic) are present also in the pudendal nerve, a nerve that is mainly a somatic sensory nerve (Hulsebosch and Coggeshall, 1982). Only 12% of all the preganglionic fibers to the pelvic ganglion are myelinated. A small proportion of the postganglionic fibers are also myelinated.

The cells of origin of the preganglionic fibers, identified with HRP after retrograde transport from the hypogastric nerve, are localized in the spinal segments L1 and L2, but not in the intermediolateral nucleus. Most of the neurons form a column along the midline in the dorsal gray commissure (dorsal commissural nucleus, DNC; Hancock and Peveto, 1979).

J. Cardiac ganglia

The cardiac plexus is formed by a number of minute ganglia located subepicardially and at the base of the aorta and pulmonary artery (King and Coakley, 1958). The neuronal population is uniformly cholinergic neurons. There are also a few SIF cells and some adrenergic fibers in transit, probably originating from the stellate ganglion. In the rat the adrenergic fibers are only in transit, and do not appear to form pericellular synaptic nests around ganglion neurons. The ultrastructure of the rat cardiac ganglia, which is in many respects similar to that of other autonomic ganglia, has been described by Ellison and Hibbs (1976). The incoming synapses are mainly axosomatic and most of the synapsing nerve endings appear to be cholinergic; other endings contain mainly flat and lucent vesicles, whereas dense cored vesicle containing fibers are not seen to make contacts with ganglion neurons in this species.

K. Parasympathetic ganglia of the head

1. Submandibular ganglion

According to Lichtman (1977) there are three to 10 clusters containing from five to about 250 neurons, but isolated neurons are also found. The preganglionic fibers originate in the superior salivatory nucleus and reach the ganglion via the facial nerve, the chorda tympani and the lingual nerve. Three quarters of the ganglion neurons are innervated by a single preganglionic fiber, the remaining neurons being innervated by two or three fibers (Lichtman, 1977). The arrangement whereby most ganglion neurons are driven by a single preganglionic neuron arises during postnatal development.At birth, most ganglion neurons are innervated by four to six preganglionic fibers.During the first six to seven weeks of postnatal life there is a progressive reduction in the number of preganglionic fibers converting on each neuron until the majority of ganglion neurons has a single input. At the same time, however, the total number of synaptic endings increases (Lichtman, 1977). In the adult rat the neurons are usually devoid of large dendrites, but they have numerous minute cytoplasmic projections from the cell body and from the initial portion of the axon. Synaptic boutons are mainly associated with these projections. In preparations stained with zinc–iodide osmium, an average of 44 boutons per neuron was counted (Lichtman, 1977). Two distinct populations of neurons are recognized electrophysiologically (Kawa and Roper, 1984): the neurons innervating the submandibular gland; and those innervating the sublingual gland. About one third of the former (and none of the latter) are electrically coupled (however, gap junctions have not been found; Lichtman, 1977). After decentralization, intrinsic synapses (that is, synapses arising from other ganglion neurons) are found in 72% of the submandibular neurons and in only 12% of the sublingual neurons (Kawa and Roper, 1984). The extent to which the high rate of interneuronal connections among submandibular neurons pre-exists the decentralization remains to be established. In the submandibular ganglion of another species, the mouse, all synapses disappear after decentralization, a clear sign of the absence of interneurons or interneuronal synaptic connections (Yamakado and Yohro, 1977).

L. Intramural ganglia of the gut

Myriads of small intramural ganglia, joined by connecting strands, are gathered into two ganglionated plexuses, the myenteric and the submucosal plexus. The myenteric plexus is intramuscular, being located between the circular and longitudinal muscle layers; it extends without interruptions from the esophagus through the stomach, small, and large, intestine, to the anal canal. In the stomach, myenteric ganglia are larger and more numerous near the lesser curvature. In the small intestine of the rat (unlike other mammalian species) myenteric neurons form cords parallel to the circular musculature, rather than discrete ganglia. By contrast, the myenteric ganglia of the large intestine are larger and with a better defined outline. Myenteric ganglia are found throughout the full length of the anal canal. In the small intestine there are about 9400 myenteric neurons per square centimeter of serosal surface. The submucosal plexus is found in the submucosa of the small and large intestines, usually close to the inner aspect of the circular muscle layer. Its neurons are about half as numerous as in the myenteric plexus and they are also, on average, smaller in size. Submucosal neurons are not found in the stomach.

The intrinsic neurons of the gut form a complex and varied population (Fig. 7A). Several types of neurons have been distinguished on the basis of number of cell processes (as visualized by silver impregnation or by methylene blue staining), affinity for silver salts, and cell size (see a review in Gabella, 1979). Most of the more recent studies have been carried out on the guinea pig; fewer data are available for the rat. For the guinea pig myenteric plexus several investigators have put forward classifications based on ultrastructural features (Cook and Burnstock, 1976), on distribution of neuropeptides (Furness and Costa, 1980) and on electrophysiologic properties (Wood, 1981).

The enteric ganglia have a compact structure with tightly packed cells and cell processes (Fig. 7B). The shape and thickness of myenteric ganglia change greatly with the contraction of the adjacent muscle layers. Collagen fibrils, fibroblasts, interstitial cells and capillaries do not penetrate the ganglia but lie around them without forming a proper capsule. A single basal lamina is spread over the surface of the whole ganglion. Tracers injected intravenously in high concentration diffuse out from periganglиar capillaries and penetrate into the interstices of the enteric ganglia of the rat (Jacobs, 1977).

The cell types found within the ganglia are neurons and glial cells, the latter outnumbering the former by about three to one. The processes of glial cells and the neuronal processes (partly of intrinsic and partly of extrinsic origin) constitute the neuropil. The ganglion neurons, when examined under the electron microscope, have a complex and irregular shape. Characteristically, parts of the neuronal perikarya reach the surface of the ganglion, and their membrane is in direct contact with the basal lamina and the connective tissue surrounding the ganglion. Axosomatic and axodendritic synapses are numerous. The great majority of them survive an extrinsic denervation of the gut and are therefore of intrinsic origin. Tentative classifications of the vesicle containing nerve endings of the enteric ganglia have been proposed for species other than the rat. The distribution of neuropeptides in the enteric ganglia of the rat has been studied in detail by Schultzberg and collaborators (1980). More than 50% of the neurons in the submucosal plexus of the ileum display immunofluorescence for VIP, about 20% for substance P and 19% for somatostatin

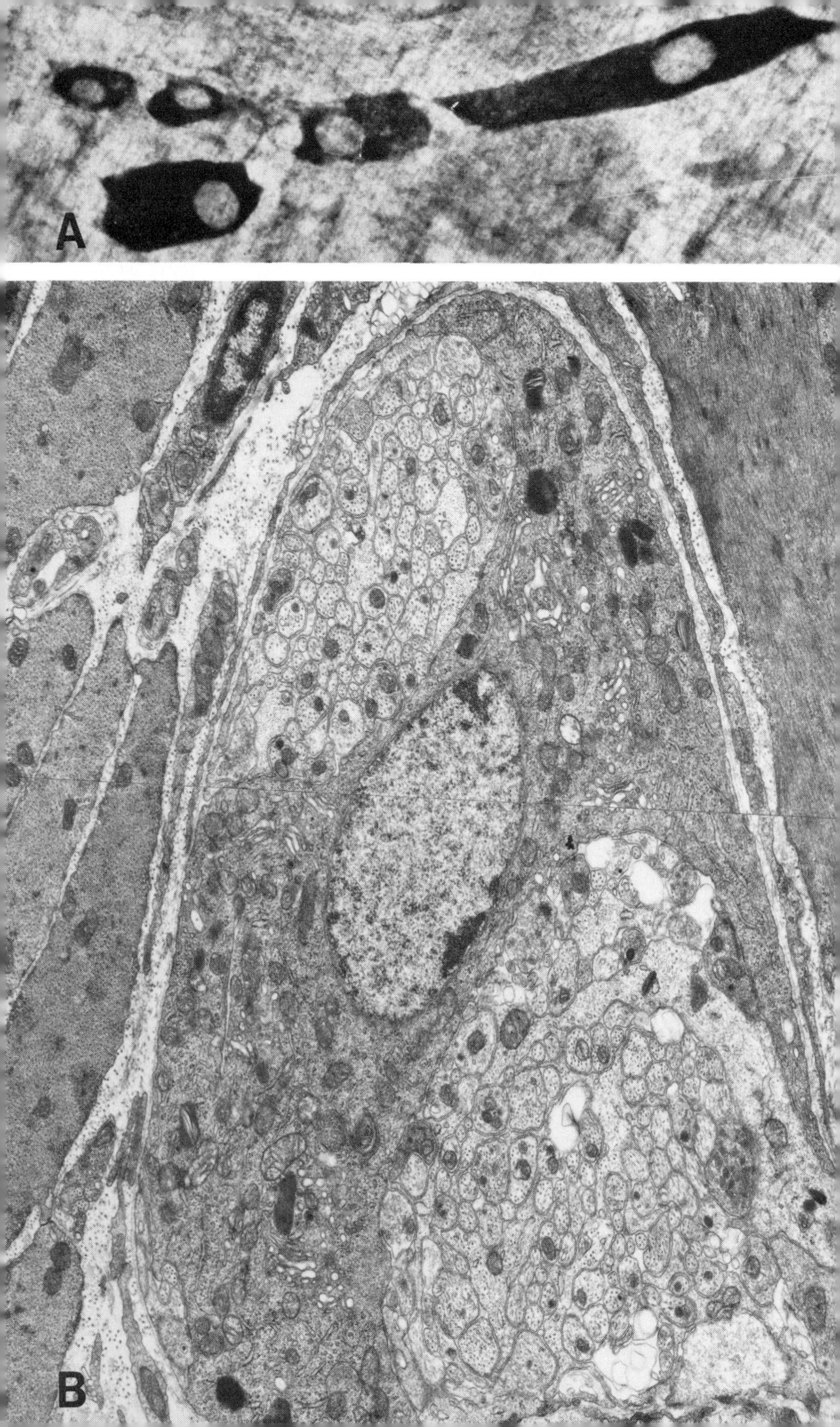
A
B

neurons (Schultzberg *et al.*, 1980). There are no adrenergic ganglion neurons. Adrenergic fibers are seen within the muscle layers, around the intramural blood vessels and within the ganglia (Van Driel and Drukker, 1973).

The enteric glial cells pervade all the spaces within the ganglia, lying over parts of the surface of neurons and between nerve processes. Glial processes reach the surface of the ganglia and are the more conspicuous features of these glial cells. The gliofilaments (intermediate filaments) are inserted in dense plaques anchored to the cell membrane at the surface of the ganglion (Gabella, 1981); they are immunologically identical to the gliofilaments found in astrocytes (Jessen and Mirsky, 1980). There are many specialized contacts between vesicle containing nerve endings and enteric glial cells (Gabella, 1981).

REFERENCES

Arvidson, B. (1979). A study of the peripheral diffusion barrier of a peripheral ganglion. *Acta Neuropathol.* 46, 139–144.

Baljet, B. and Drukker, J. (1979). The extrinsic innervation of the abdominal organs in the female rat. *Acta Anat.* 104, 243-267.

Bentley, G. A. (1972). Pharmacological studies on the hypogastric ganglion of the rat and guinea-pig. *Brit. J. Pharmacol.* 44, 492–509.

Björklund, A., Cegrell, L., Falck, B., Ritzén, M and Rosengren, E. (1970). Dopamine-containing cells in sympathetic ganglia. *Acta Physiol. Scand.* 78, 334–338.

Bowers, C. W. and Zigmond, R. E. (1979). Localization of neurons in the rat superior cervical ganglion that project into different postganglionic trunks. *J. Comp. Neurol.* 185, 381–392.

Bowers, C. W. and Zigmond, R. E. (1981). Sympathetic neurons in lower cervical ganglia send axons through the superior cervical ganglion. *Neurosci.* 6, 1783–1791.

Brooks-Fournier, R. and Coggeshall, R. E. (1981). The ratio of preganglionic axons to postganglionic cells in the sympathetic nervous system of the rat. *J. Comp. Neurol.* 197, 207–216.

Cajal, S. Ramón y (1911). *Histologie du système nerveux de l'homme et des vertébrés* Vol. 2. Paris, Maloine.

Cook, R. D. and Burnstock, G. (1976). The ultrastructure of Auerbach's plexus. I. Neuronal elements. *J. Neurocytol.* 5, 171–194.

Crowcroft, P. J. and Szurszewski, J. H. (1971). A study of the inferior mesenteric and pelvic ganglia of guinea-pigs with intracellular electrodes. *J. Physiol. (London)* 219, 421–441.

Dahlström, A. and Fuxe, K. (1965). Evidence for the existence of monoamine neurons in the central nervous system. II. Experimentally induced changes in the interneuronal amine levels of bulbospinal neuron systems. *Acta Physiol. Scand. Suppl.* 247, 7–34.

Dail, W. G., Jr, Evan, A. P., Jr and Eason, H. R. (1975). The major ganglion in the pelvic plexus of the male rat. *Cell Tiss. Res.* 159, 49–62.

Dail, W. G., Moll, M. A. and Weber, K. (1983). Localization of vasoactive intestinal polypeptide in the penile erectile tissue and in the major pelvic ganglion of the rat. *Neurosci.* 10, 1379–1386.

Davies, D. C. (1978). Neuronal numbers in the superior cervical ganglion of the neonatal rat. *J. Anat.* 127, 43–51.

De Lemos, C. and Pick, J. (1966). The fine structure of thoracic sympathetic neurons in the adult rat. *Z. Zellforsch.* 71, 189–206.

◁ **Fig. 7:** A. Whole mount preparation of the muscle coat of the rat cecum, showing some neurons of the myenteric plexus. The faint staining in the background corresponds to the circular muscle layer. B. Electron micrograph (montage) of a ganglion of the myenteric plexus of the rat small intestine. To the right is the longitudinal musculature, to the left the circular musculature (in transverse section). The ganglion displays a neuronal cell body of complex shape, with its nucleus, and a large number of neuronal and glial processes × 13 000.

Dyck, P. J. and Hopkins, A. P. (1972). Electronmicroscopic observations on degeneration and regeneration of myelinated and unmyelinated fibers. *Brain* 95, 233–234.
Ebberson, S. O. E. (1968). Quantitative studies of superior cervical sympathetic ganglia in a variety of primates including man. I. The ratio of preganglionic fibers to ganglionic neurons. *J. Morphol.* 124, 117–132.
Ellison, J. P. and Hibbs, R. G. (1976). An ultrastructural study of mammalian cardiac ganglia. *J. Mol. Cell. Cardiol.* 8, 89–101.
Eränkö, L. (1972). Ultrastructure of the developing sympathetic nerve cell and the storage of catecholamines. *Brain Res.* 46, 159–175.
Eränkö, L. and Eränkö, O. (1972). Effect of hydrocortisone on histochemically demonstrable catecholamines in the sympathetic ganglia and extra-adrenal chromaffin tissue of the rat. *Acta Physiol. Scand.* 84, 125–133.
Eränkö, O. and Harkonen, M. (1965). Monoamine-containing small cells in the superior cervical ganglion of the rat and on an organ composed of them. *Acta Physiol. Scand.* 63, 511–512.
Eränkö, O. and Soinila, S. (1981). Effect of early postnatal division of the postganglionic nerves on the development of principal ganglion cells and small intensely fluorescent cells in the rat superior cervical ganglion. *J. Neurocytol.* 10, 1–18.
Furness, J. B. and Costa, M. (1980). Types of nerves in the enteric nervous system. *Neurosci.* 5, 1–20.
Furshpan, E. J., MacLeish, P. R., O'Lague, P. H. and Potter, D. D. (1976). Chemical transmission between rat sympathetic neurons and cardiac myocytes development in microcultures: Evidence for cholinergic, adrenergic and dual function neurons. *Proc. Nat. Acad. Sci. USA* 73, 4225–4229.
Furshpan, E. J., Potter, D. D. and Landis, S. C. (1982). On the transmitter repertoire of sympathetic neurons in culture. *Harv. Lect.* 76, 149–191.
Gabella, G. (1976). *Structure of the autonomic nervous system.* Chapman and Hall, London.
Gabella, G. (1979). Innervation of the gastrointestinal tract. *Int. Rev. Cytol.* 59, 129–193.
Gabella, G. (1981). Ultrastructure of the nerve plexuses of the mammalian intestine: The enteric glial cells. *Neurosci.* 6, 425–436.
Greene, E. C. (1935). Anatomy of the rat. *Trans. Am. Phil. Soc.* 1–370.
Grillo, M. A. (1978). Ultrastructural evidence for a sensory innervation of some SIF cells in rat superior cervical ganglia (abstract). *Anat. Rec.* 190, 407.
Hamer, D. W. and Santer, R. M. (1981). Anatomy and blood supply of the coelia-superior mesenteric ganglion complex of the rat. *Anat. Embryol.* 162, 3, 353–362.
Hancock, M. B. and Peveto, C. A. (1979). A preganglionic autonomic nucleus in the dorsal grey commissure of the lumbar spinal cord of the rat. *J. Comp. Neurol.* 183, 65–72.
Hedger, J. H. and Webber, R. H. (1976). Anatomical study of the cervical sympathetic trunk and ganglia in the albino rat (*Mus norvegicus albinus*). *Acta Anat.* 96, 206–217.
Hendry, I. A. (1976). A method to correct adequately for the change in neuronal size when estimating neuronal numbers after nerve growth treatment. *J. Neurocytol.* 5, 337–349.
Hendry, I. A. and Campbell, J. (1976). Morphometric analysis of rat superior cervical ganglion after axotomy and nerve growth factor treatment. *J. Neurocytol.* 5, 351–360.
Hendry, I. A., Stockel, K., Thoenen, H. and Iversen, L. L. (1974). The retrograde axonal transport of nerve growth factor. *Brain Res.* 68, 103–121.
Hökfelt, T., Elfvin, L.- G., Schultzberg, M., Goldstein, M. and Nilsson, G. (1977a). On the occurrence of substance P-containing fibers in sympathetic ganglia: Immunohistochemical evidence. *Brain Res.* 132, 29–41.
Hökfelt, T., Elfvin, L.-G., Schultzberg, M., Fuxe, K., Said, S. I., Mutt, V. and Goldstein, M. (1977b). Immunohistochemical evidence of vasoactive intestinal polypeptide-containing neurons and nerve fibers in sympathetic ganglia. *Neurosci.* 2, 885–896.
Hökfelt, T., Elfvin, L. G., Elde, R., Schultzberg, M., Goldstein, M. and Luft, R. (1977c). Occurrence of somatostatin-like immunoreactivity in some peripheral sympathetic noradrenergic neurons. *Proc. Nat. Acad. Sci. USA* 74, 3587–3591.
Hulsebosch, C. E. and Coggeshall, R. E. (1982). An analysis of axon populations in the nerves to the pelvic viscera in the rat. *J. Comp. Neurol.* 211, 1–10.
Jacobs, J. M. (1977). Penetration of systemically injected horseradish peroxidase into ganglia and nerves of the autonomic nervous system. *J. Neurocytol.* 6, 607–618.
Jessen, K. R. and Mirsky, R. (1980). Glial cells in the enteric nervous system containing glial fibrillary acidic protein. *Nature* 286, 736–737.

Johnson, M., Ross, D., Meyer, M., Ress, R., Bunge, R., Wakshull, E. and Burton, H. (1976). Synaptic vesicle cytochemistry changes when cultured sympathetic neurones develop cholinergic interactions. *Nature*, 262, 308–310.

Johnson, E. M., Jr., Gordin, P. D., Brandeis, L. D. and Pearson, J. (1980). Dorsal root ganglion neurons are destroyed by exposure in utero to maternal antibodies to nerve growth factor. *Science* 210, 916–918.

Kanerva, L. (1972). Ultrastructure of sympathetic ganglion cells and granule-containing cells in the paracervical (Frankenhauser) ganglion of normal and pregnant rats. *Acta Physiol. Scand.* 86, 271–277.

Kanerva, L. and Teräväinen, H. (1972). Electron microscopy of the paracervical ganglion (Frankenhauser) of the adult rat. *Z. Zellforsch.* 129, 161–177.

Kanerva, L., Lietzén, R. and Teräväinen, H. (1972). Catecholamines and cholinesterases in the paracervical (Frankenhauser) ganglion of normal and pregnant rats. *Acta Physiol. Scand.* 86, 271–277.

Kawa, K. and Roper, S. (1984). On the two subdivisions and intrinsic synaptic connexions in the submandibular ganglion of the rat. *J. Physiol. (London)* 346, 301–320.

King, T. S. and Coakley, J. B. (1958). The intrinsic nerve cells of the cardiac atria of mammals and man. *J. Anat.* 92, 353–376.

Klingman, G. I. (1972). In G. Steiner and E. Schonbaum (Eds). *Immunosympathectomy*. Elsevier, Amsterdam.

Klingman, G. I. and Klingman, J. D. (1969). Cholinesterase in rat sympathetic ganglia after immunosympathectomy, decentralization and axotomy. *J. Neurochem.* 16, 261–268.

Kondo, H. (1977). Innervation of SIF cells in the superior cervical and noduse ganglia: An ultrastructural study with serial sections. *Biol. Cell.* 30, 253–264.

Kondo, H., Dun, N. J. and Pappas, G. D. (1980). A light and electron microscopic study of the rat superior cervical ganglion cells by intracellular HRP-labeling. *Brain Res.* 197, 193–199.

Konig, R. (1979). Consecutive demonstration of catecholamines and dopamine-beta-hydroxylase within the same specimen. *Brain Res.* 212, 39–49.

Landis, S. C. and Keefe, D. (1983). Evidence for neurotransmitter plasticity *in vivo*: Development changes after axotomy and nerve growth treatment. *Dev. Biol.* 98, 349–372.

Langley, J. N. (1921). *The autonomic nervous system.* Heffer, Cambridge.

Langley, J. N. (1922). The secretion of sweat. I. Supposed inhibitory nerve fibres on the posterior nerve roots. Secretion after denervation. *J. Physiol. (London)* 56, 110–119.

Langworthy, O. R. (1965). Innervation of the pelvic organs of the rat. *Invest. Urol.* 2, 491–511.

Lehmann, H. J. and Stange, H. H. (1953). Uber das Vorkommen vakuolenhalt iger Ganglienzellen im Ganglion cervicale uteri trachtiger und nichttrachtiger Ratten. *Z. Zellforsch.* 38, 230–236.

Leranth, Cs, and Ungvary, Gy. (1980). Axon types of prevertebral ganglia and the peripheral autonomic reflex arc. *J. Auton. Nerv. Sys.* 1, 265–281.

Levi-Montalcini, R. and Booker, B. (1960). Excessive growth of sympathetic ganglia evoked by a protein isolated from mouse salivary glands. *Proc. Nat. Acad. Sci. USA* 46, 373–384.

Lichtman, J. W. (1977). The reorganization of synaptic connexions in the rat submandibular ganglion during post-natal development. *J. Physiol. (London)* 273, 155–177.

Lichtman, J. W. and Purves, D. (1980). The elimination of redundant preganglionic innervation to hamster sympathetic ganglion cells in early post-natal life. *J. Physiol. (London)* 301, 213–228.

Lundberg, J. M. (1981). Evidence for coexistence of vasoactive intestinal polypeptide (VIP) for acetylcholine in neurons of cat exocrine glands: Morphological, biochemical and functional studies. *Acta Physiol. Scand. Suppl.* 496, 1–57.

Malmfors, T. and Nilsson, O. (1966). Parasympathetic post-ganglionic denervation of the iris and the parotid gland of the rat. *Acta Morphol. Neerl.-Scand.* 6, 81–85.

Marshall, J. M. (1970). Adrenergic innervation of the female reproductive tract: Anatomy, physiology and pharmacology. *Ergebn. Physiol.* 62, 6–67.

Matthews, M. R. (1983). The ultrastructure of junctions in sympathetic ganglia of mammals. In L. G. Elfvin (Ed.), *Autonomic ganglia.* Wiley, New York.

Matthews, M. R. and Cuello, A. C. (1982). Substance P-immunoreactive peripheral branches of sensory neurons innervate guinea pig sympathetic neurons. *Proc. Nat. Acad. Sci. USA* 79, 1668–1672.

Matthews, M. R. and Raisman, G. (1969). The ultrastructure and somatic efferent synapses of small granule-containing cells in the superior cervical ganglion. *J. Anat.* 105, 255–282.
Matthews, M. R. and Raisman, G. (1972). A light and electron microscopy study of the cellular response to axonal injury in the superior cervical ganglion of the rat. *Proc. Roy. Soc. Lond. B* 181, 43–79.
Navaratnam, V. and Lewis, P. R. (1970). Cholinesterase-containing neurons in the spinal cord of the rat. *Brain Res.* 18, 411–425.
Njå, A. and Purves, D. (1977). Specific innervation of guinea-pig superior cervical ganglion by preganglionic fibres arising from different levels of the spinal cord. *J. Physiol. (London)* 264, 565–583.
Norberg, K. A., Ritzén, M. and Ungerstedt, U. (1966). Histochemical studies on a special catecholamine-containing cell type in sympathetic ganglia. *Acta Physiol. Scand.* 67, 260–270.
Ostberg, A. -J. C., Raisman, G., Field, P. M., Iversen, L. L. and Zigmond, R. E. (1976). A quantitative comparison of the formation of synapses in the rat superior cervical sympathetic ganglion by its own and by foreign nerve fibres. *Brain Res.* 107, 445–470.
Partanen, M., Hervonen, A., Vaalasti, A., Kanerva, L. and Hervonen, H. (1979). Vacuolated neurons in the hypogastric ganglion of the rat. *Cell Tiss. Res.* 199, 373–384.
Patterson, P. H. and Chung, L. L. Y. (1977). The induction of acetylcholine synthesis in primary cultures of dissociated sympathetic neurons. I. Effects of conditioned medium. *Dev. Biol.* 56, 263–280.
Puriton, T., Fletcher, T. and Bradley, W. (1971). Sensory perikarya in autonomic ganglia. *Nature New Biol.* 231, 63–64.
Puriton, P. T., Fletcher, T. F. and Bradley, W. E. (1973). Gross and light microscopic features of the pelvic plexus in the rat. *Anat. Rec.* 175, 697–706.
Rando, T. A., Bowers, C. W. and Zigmond, R. E. (1981). Localization of neurons in the rat spinal cord which project to the superior cervical ganglion. *J. Comp. Neurol.* 196, 73–83.
Richards, J. G. and Tranzer, J. P. (1975). Localization of amine storage sites in the adrenergic cell body. *J. Ultrastruct. Res.* 53, 204–216.
Santer, R. M. (1979). Fluorescence histochemical evidence for decreased noradrenaline synthesis in sympathetic neurons of aged rats. *Neurosci. Lett.* 15, 177–180.
Santer, R. M., Lu, K. -S., Lever, J. D. and Presley, R. (1975). A study of the distribution of chromaffin-positive (CH+) and small intensely fluorescent (SIF) cells in sympathetic ganglia of the rat at various ages. *J. Anat.* 119, 589–599.
Schramm, L. P., Adair, J. R., Stribling, J. M. and Gray, L. P. (1975). Preganglionic innervation of the adrenal gland of the rat: A study using horseradish peroxidase. *Exp. Neurol.* 49, 540–553.
Schramm, L. P., Stribling, J. M. and Adair, J. R. (1976). Developmental reorientation of sympathetic preganglionic neurons in the rat. *Brain Res.* 106, 166–171.
Schultzberg, M. (1983). The peripheral nervous system. In L.- G. Elfvin (Ed.). *Autonomic ganglia*, pp. 205–233. Wiley, New York.
Schultzberg, M., Hökfelt, T., Nilsson, G., Terenius, L., Rechfeld, J. R., Brown, M., Elde, R. Goldstein, M. and Said, S. (1980). Distribution of peptide- and catecholamine-containing neurons in the gastrointestinal tract of rat and guinea-pig: Immunohistochemical studies with antisera to substance P, vasoactive intestinal polypeptide, enkephalins, somatostatin, gastrin/cholecystokinin, neurotensin and dopamine-hydroxylase. *Neurosci.* 5, 689–744.
Skok, V. I. (1983). Fast synaptic transmission in autonomic ganglia. In L.- G. Elfvin (Ed.). *Autonomic ganglia*, pp. 265–279. Wiley, New York.
Sjöstrand, N. O. (1965). The adrenergic innervation of the vas deferens and the accessory male genital organs. *Acta Physiol. Scand. Suppl.* 257, 1–82.
Smolen, A. and Raisman, G. (1980). Synapse formation in the rat superior cervical ganglion during normal development and after neonatal deafferentation. *Brain Res.* 181, 315–323.
Snell, R. S. (1958). The histochemical appearances of cholinesterase in the parasympathetic nerve supplying the submandibular and sublingual salivary glands of the rat. *J. Anat.* 92, 534–543.
Somogyi, P. and Chubb, I. W. (1976). The recovery of acetylcholinesterase in the superior cervical ganglion of the rat following its inhibition by diisopropylphosphofluoridate: A biochemical and cytochemical study. *Neurosci.* 1, 413–421.

Tomarind, D. L. and Quilliam, J. P. (1971). Synaptic organization and other ultrastructural features of the superior cervical ganglion of the rat, kitten and rabbit. *Micron* 2, 204–234.

Taxi, J. (1979). The chromaffin and chromaffin-like cells in the autonomic nervous system. *Int. Rev. Cytol.* 57, 283–343.

Van Driel, C. and Drukker, J. (1973). A contribution to the study of the architecture of the autonomic nervous system of the digestive tract of the rat. *J. Neural Trans.* 34, 301–320.

Verhofstad, A. A. J., Steinbusch, H. W. M., Panke, B., Varga, J. and Joosten, H. W. J. (1981). Serotinin-immunoreactive cells in the superior cervical ganglion of the rat: Evidence for the existence of separate serotonin- and catecholamine-containing small ganglionic cells. *Brain Res.* 212, 39–49.

Wallis, D. I. and North, R. A. (1978). Synaptic input to cells of the rabbit superior cervical ganglion. *Pflugers Arch.* 374, 145–152.

Watanabe, H. (1971). Adrenergic nerve elements in the hypogastric ganglion of the guinea-pig. *Am. J. Anat.* 130, 305–330.

Webber, R. and Kallen, R. C. (1968). The sympathetic trunks of a hibernator, the bat *Myotis lucifugus*. *J. Comp. Neurol.* 134, 151–162.

Wechsler, H. L. and Fisher, E. R. (1968). Eccrine glands of the rat. *Arch. Dermatol.* 97, 189–201.

Williams, T. H. (1967). Electron microscopic evidence for an autonomic neuron. *Nature* 214, 309–310.

Williams, T. H., Black, A. C., Jr., Chiba, T. and Jew, J. W. (1977). Species differences in mammalian SIF cells. *Adv. Biochem. Psychopharmacol.* 16, 505–511.

Wood, J. D. (1981). Physiology of the enteric nervous system. In L. R. Johnson (Ed.). *Physiology of the gastrointestinal tract*, pp. 1–37. Raven, New York.

Yamakado, M. and Yohro, T. (1977). Population and structure of nerve cells in mouse submandibular ganglion. *Anat. Embryol.* 150, 301–312.

Yamauchi, A. and Lever, J. D. (1971). Correlations between formol fluorescence and acetylcholinesterase (AChE) staining in the superior cervical ganglion of normal rat, pig and sheep. *J. Anat.* 110, 435–443.

Yarowski, P., Jehle, J., Ingvar, D. H. and Sokoloff, L. (1979). Relationship between functional activity and glucose utilization in the rat superior cervical ganglion *in vivo* (abstract). *Neurosci. Abst.* 5, 421.

Index

R

U

V

Z

5 6 7 8 9 0 1 2 3 4
A B C D E F G H I J